Electric and Hybrid Vehicles

Design Fundamentals

SECOND EDITION

Electric and Hybrid Vehicles

Design Fundamentals

SECOND EDITION

Iqbal Husain

CRC Press is an imprint of the
Taylor & Francis Group, an **informa** business

MATLAB® is a trademark of The MathWorks, Inc. and is used with permission. The MathWorks does not warrant the accuracy of the text or exercises in this book. This book's use or discussion of MATLAB® software or related products does not constitute endorsement or sponsorship by The MathWorks of a particular pedagogical approach or particular use of the MATLAB® software.

CRC Press
Taylor & Francis Group
6000 Broken Sound Parkway NW, Suite 300
Boca Raton, FL 33487-2742

International Standard Book Number: 978-1-4398-1175-7 (Hardback)

Library of Congress Cataloging-in-Publication Data

Husain, Iqbal.
Electric and hybrid vehicles : design fundamentals / Iqbal Husain. -- 2nd ed.
p. cm.
Includes bibliographical references and index.
ISBN 978-1-4398-1175-7 (alk. paper)
1. Electric vehicles. 2. Hybrid electric vehicles. I. Title.

TL220.H87 2010
629.22'93--dc22 2010013378

Visit the Taylor & Francis Web site at
http://www.taylorandfrancis.com

and the CRC Press Web site at
http://www.crcpress.com

Contents

Preface

Environmental concerns and energy challenges have prompted the automotive industry to develop clean, efficient, and sustainable vehicles for urban transportation. Toward the end of the millennium, automotive industries in the United States and the rest of the world became proactive with the production of alternative vehicles. There is significant demand for engineers with experience in designing and engineering alternative vehicles and their components. There is an even greater need today for electric and hybrid vehicle teaching materials that include state-of-the-art technology. This book is intended to be used in a course to educate students on the multidisciplinary aspects of a vehicle system where electrical and mechanical engineers have to work together to complete the system. A technical course addressing the design fundamentals will provide students with the necessary theoretical background and enable them to become automotive engineers capable of meeting today's needs. It has been more than seven years since the publication of the first edition of the book. The technology has advanced significantly since then and, therefore, a revised edition became essential.

The Department of Energy, in collaboration with the automotive industry, has been sponsoring two- to four-year national collegiate-level vehicle design competitions since 1989 to give engineering students an opportunity to participate in hands-on research and development with cutting-edge automotive propulsion systems, fuels, materials, and emission control technologies. The University of Akron participated in one such competition known as *Challenge X: Crossover to Sustainable Mobility*. The four-year student engineering competition with tiered goals for each year called for the teams to reengineer a 2005 Chevrolet Equinox crossover vehicle into an alternative vehicle to reduce energy consumption and decrease the emissions of pollutants while maintaining its performance. The experience gained through the hands-on development of a roadworthy hybrid vehicle and the supervision of a large group of graduate and undergraduate students provided the necessary impetus and served as a guideline for this edition.

This edition maintains the comprehensive systems-level perspective of electric and hybrid vehicles, but covers the hybrid architectures and components of the vehicle in much greater detail. Technical details, mathematical relationships, and design guidelines have been emphasized throughout the book to provide fundamental knowledge to students and engineers. The new additions to this edition include sizing and design guidelines for various hybrid architectures, modeling of energy storage components, control strategies for hybrid vehicles, component cooling systems, emission control systems, and in-vehicle communication methods. The sections on power electronics, electric machines, and motor drives have been reorganized, and sections on DC/DC converters have been added. The sections on mechanical components have also been enhanced with more technical descriptions and example problems. The integration of mechanical and electrical components has been addressed to some extent to emphasize the interdisciplinary nature of automotive engineering. A separate chapter was required to discuss powertrain component-cooling systems given their importance for vehicle safety and operation. The stage has been set in the book for system-level simulations to encourage users to develop models using software such as MATLAB® and Simulink®. A few problems in the book have been resolved using MATLAB.

This book will serve educational aspects of electric and hybrid vehicles, which, in turn, will generate interest to support the development and usage of these vehicles. It will also serve as a reference for a working engineer dealing with the design and improvement of electric and hybrid vehicles. Discussion on most topics has been limited to fundamentals only, considering the wide spectrum of technical aspects related to electric and hybrid vehicle systems. Appropriate references are given to direct the reader toward details on topics for further reading. The intent of the book is not to span the entire spectrum of electric and hybrid electric vehicles, but to prepare students and engineers with the necessary background for starting the design process and evaluating the relevant technologies.

The book, similar to the first edition, starts with a systems-level introduction to electric and hybrid vehicles in Chapter 1, and gives an overview of the components used in these alternative vehicles. The chapter also gives the historical background of electric and hybrid vehicles, well-to-wheel analysis considerations, and comparisons of alternative vehicles with conventional vehicles.

The laws of physics that govern vehicle motion and roadway fundamentals are presented in Chapter 2. The design guidelines for the power and energy requirements based on force–velocity characteristics of ground vehicles are established in this chapter.

The alternative vehicle architectures and powertrain component-sizing requirements are addressed in Chapter 3. Examples of component sizing for electric and hybrid vehicles are given in this chapter.

Chapters 4 and 5 discuss the energy sources and energy storage systems for electric and hybrid vehicles. The electrochemical fundamentals of energy storage devices such as batteries, ultracapacitors, and fuel cells are given in these chapters. Several battery models based on the fundamentals are presented, which will be useful for system simulation and prediction of terminal voltage–current characteristics. The promising new battery technologies for alternative vehicles are presented in Chapter 4. The alternative energy storage systems of fuel cells, ultracapacitors, compressed air, and flywheel are discussed in Chapter 5.

Chapters 6 through 9 cover the basics of electrical components. The various types of electric machines and their basic operating principles are discussed in Chapter 6. The power electronic components required for static power conversion and for electric motor drives are discussed in Chapter 7.

The electric motor drives, both DC and AC types, are presented in Chapter 8. Finally, the controls for AC motor drives are described in Chapter 9.

Chapters 10 through 12 cover the mechanical components. The internal combustion engines suitable for hybrid vehicles are discussed in these chapters. Key objectives in designing hybrid vehicles are high fuel efficiency and reduced emissions. These topics have been addressed in detail in Chapter 10 and emphasized throughout the book. Powertrain components and brakes are presented in Chapter 11, while powertrain component-cooling and vehicle cabin-cooling systems are described in Chapter 12.

Chapter 13 focuses on hybrid control strategies where several example strategies are given with an emphasis on increased fuel economy and reduced emissions.

In-vehicle communication system is an essential element for modern vehicles. Controller area network (CAN) communication network and protocols are described in Chapter 14 after an introduction to open systems interconnection (OSI) seven-layer network model.

The book is intended to be used as a textbook for an undergraduate/graduate-level course on electric and hybrid vehicles. The materials in this book are multidisciplinary enough to teach electrical, mechanical, and chemical engineers all in one course utilizing the systems approach. The 14 chapters can be covered in a three-credit, one semester or in a four-credit, one-quarter course with emphasis on the systems rather than on the components. In this case, Chapters 1 through 5 and Chapter 13 could be covered in depth, while the remaining chapters could be used to introduce the electrical and mechanical components. This type of course will certainly mimic the real situation in many industries where multidisciplinary engineers work together to devise a system and develop a product. Alternatively, the book can be used in a three-credit, one semester or a four-credit, one quarter course for electrical or mechanical engineering students with emphasis on either the electrical or the mechanical components.

The equations and mathematical models presented in the book can be used to develop a system-level modeling and simulation tool for electric and hybrid vehicles on a suitable platform such as MATLAB and Simulink. The book contains several resolved problems and many exercises that are useful to convey the concept to students through numerical examples.

For MATLAB and Simulink product information, please contact

The MathWorks, Inc.
3 Apple Hill Drive
Natick, MA, 01760-2098 USA
Tel: 508-647-7000
Fax: 508-647-7001
E-mail: info@mathworks.com
Web: www.mathworks.com

Acknowledgments

I would like to express my sincere gratitude to all those who devotedly helped me to complete the book. I am thankful to my graduate student, Rajib Mikail, who helped tremendously with many of the figures in the book. I offer my sincere gratitude to the reviewers who provided useful suggestions that helped enhance the quality of the book. I am thankful to Prof. Tom Hartley for reviewing the energy storage chapter (Chapter 4), to Prof. Bob Veillette for reviewing the hybrid vehicle controls chapter (Chapter 13), to Prof. Fabrizio Marignetti for reviewing the chapters on vehicle kinetics and hybrid architectures (Chapters 2 and 3), to Prof. Ikramuddin Ahmed for reviewing the chapters on internal combustion engines and powertrains (Chapters 10 and 11), and to Dr. Nayeem Hasan for reviewing the chapter on power converters (Chapter 7).

I am thankful to the many users of the first edition of the book. The acceptance of the book as a text in several universities, nationally and internationally, provided the incentive to improve and complete this edition. I am especially thankful to Nora Konopka from the Taylor & Francis Group for encouraging me to undertake the writing of this edition. Finally, my sincere apologies and heartfelt gratitude to my wife, Salina, and my children, Inan, Imon, and Rushan, who stood by me patiently and were understanding and supportive of my preoccupation with the project.

Iqbal Husain
Akron, Ohio

Author

Dr. Iqbal Husain is a professor in the Department of Electrical and Computer Engineering at The University of Akron, Ohio, where he is engaged in teaching and research. He received his PhD in electrical engineering from Texas A&M University, College Station, in 1993, where he worked as a lecturer. He also worked as a consulting engineer for Delco Chassis at Dayton, Ohio, before joining The University of Akron in 1994. In 1996 and 1997, Dr. Husain worked as a summer researcher for Wright Patterson AFB Laboratories. He was a visiting professor in the Department of Electrical and Computer Engineering at Oregon State University, Corvallis, in 2002.

Prof. Husain's research interests are in the areas of control and modeling of electrical drives, design of electric machines, development of power conditioning circuits, and electric and hybrid vehicles. He has worked extensively on the development of switched reluctance and permanent magnet motor drives. He has been highly active with research in the areas of automotive and aerospace electronics since joining The University of Akron. He has completed research works and consulted for several automotive and motor drive industries, such as Delphi Chassis, Delphi Saginaw Steering Division, TRW, ITT Automotive, Goodyear Tire and Rubber, Honeybee Robotics, Caterpillar, Continental Group, Avtron Inc., and EBO Group.

Prof. Husain has developed an Electric and Hybrid Vehicle Program at The University of Akron, which includes graduate and undergraduate courses, research on electric drives for electric and hybrid vehicles, and collegiate-level competitions on alternative vehicles. He was the lead faculty advisor for the Department of Energy and General Motors sponsored *Challenge X: Crossover to Sustainable Mobility* competition between 2004 and 2008.

Prof. Husain was the recipient of the National Science Foundation CAREER award in 1997. He also received the 1998 IEEE-IAS Outstanding Young Member award, the 2000 IEEE Third Millennium Medal, the 2004 College of Engineering Outstanding Researcher award, and the 2006 SAE Vincent Bendix Automotive Electronics Engineering award. He is also the recipient of an *IEEE-IAS Magazine* paper award and four IEEE-IAS Committee prize paper awards. He was named an IEEE Fellow in 2009.

1

Introduction to Alternative Vehicles

Environmental as well as economical issues provide a compelling impetus to develop clean, efficient, and sustainable vehicles for urban transportation. Environmental and economical advantages can also be gained by applying the alternative transportation technologies to industrial and commercial off-road vehicles. Passenger vehicles constitute an integral part of our everyday life, yet the exhaust emissions of the conventional internal combustion engine vehicles (ICEVs) are the major source of urban pollution that causes the greenhouse effect, which in turn leads to global warming [1]. The dependence on oil as the sole source of energy for passenger vehicles has economical and political implications, and the crisis will inevitably become acute as the oil reserves of the world diminish. The number of automobiles in the world doubled to about a billion or so in the last 15 years. The increasing number of automobiles being introduced on the road every year is only adding to the increasing pollution levels. There is also an economic factor that is inherent in the poor energy conversion efficiency of the internal combustion (IC) engines. When efficiency is evaluated on the basis of conversion from crude oil to traction effort at the wheels, the number for the alternative electric vehicles is not significantly higher; yet, it does make a difference. The emission due to power generation at localized plants is much easier to regulate than those emanating from IC engine vehicles that are individually maintained and scattered all over the world. People dwelling in cities are not exposed to power plant–related emissions, since these are mostly located outside urban areas. Furthermore, electric power can be generated using renewable sources such as water (hydroelectric), wind, and sun's energy (solar), which would prove to be the most environmentally friendly approach. The electric vehicles (EV) enabled by high-efficiency electric motor and controller, and powered by alternative energy sources, provide the means for a clean, efficient, and environmentally friendly urban transportation system. Electric vehicles have no emission and therefore are capable of curbing the pollution problem in an efficient way. Consequently, electric vehicles are the only zero-emission vehicles (ZEVs) possible.

Electric vehicles made their way into public use as early as in the middle of the nineteenth century, even before the introduction of gasoline-powered vehicles [2]. In 1900, 4200 automobiles were sold, out of which 40% were steam powered, 38% were electric powered, and 22% were gasoline powered.

However, the invention of starter motor, improvements in mass production technology of gas-powered vehicles, and inconvenience in battery charging led to the disappearance of electric vehicles in the early 1900s. However, the environmental issues and the unpleasant dependence on oil led to the resurgence of interest in electric vehicles in the 1960s. The growth in the enabling technologies over the next several decades, and environmental and economic concerns renewed the interest in research and development of electric vehicles.

The interest and research in electric vehicles soared in the 1990s with the major automobile manufacturers embarking on plans for introducing their own electric or hybrid electric vehicles. General Motors introduced the first electric vehicle "Saturn EV1" in 1995, although to a limited market in California and Arizona. The vehicle has since been discontinued, but the event is undoubtedly a milestone in modern vehicle history. The consumer market is not yet ready to accept vehicles that cannot travel long distances on one full charge of the battery-pack. In addition, the battery-packs are also quite expensive and require much longer time for recharging compared to that of filling up a gas tank.

The limited range of the battery-powered electric vehicles led the researchers and auto industry players to search for alternatives. The aggressive efforts by the industry led to the rapid development of hybrid electric vehicles. Toyota led the way among the car manufacturers by introducing the first hybrid vehicle "Toyota Prius" in 1999. The third generation Prius is currently available in the market. Many other car manufacturers have since introduced hybrids for consumers by addressing the concerns about environmental pollution and excessive dependence on fossil fuels. The hybrid electric vehicles (HEVs) or simply hybrid vehicles use both electric machines and an IC engine for delivering the propulsion power; these vehicles have lower emissions compared to a similarly sized conventional IC engine vehicle, resulting in less environmental pollution. The IC engine used in a hybrid vehicle is, of course, downsized compared to an equivalent IC engine vehicle. The IC engine in combination with the electric motor and an energy storage unit provide an extended range for vehicles and bring down pollution. The hybrid vehicle serves as a compromise for the environmental pollution problem and the limited range capability of today's purely electric vehicle. The hybrids are looked upon by many as a shorter term solution until the range limitation and infrastructure problems of zero-emission electric vehicles are solved. Nevertheless, a number of automotive manufacturers are aggressively marketing hybrid vehicles for the general population.

The limited range problem of battery electric vehicles can be overcome by adding a fuel cell engine that will generate electricity for the electric motors as long as fuel is available. The fuel cell is an electrochemical energy converter, but has much higher efficiencies compared to a heat engine such as the IC engine. The fuel for these engines is hydrogen or gases from which hydrogen can be extracted. This concept has led to the development of fuel cell electric vehicles and to the discussions on hydrogen economy.

The trend in using alternative vehicles is increasing today with electric vehicles serving as ZEVs and hybrid vehicles filling in as ultralow emission vehicles. The fuel cells are currently available in very limited quantities, but their mass production is the focus of the auto industry. The core technologies for these alternative vehicles beyond the energy storage system are quite the same. The primary objective of this book is to introduce the system of alternative vehicles and the core enabling technologies embedded in that system.

1.1 Electric Vehicles

An *electric vehicle* is a vehicle that has the following features: (1) the energy source is portable and electrochemical or electromechanical in nature, and (2) traction effort is supplied only by an electric motor. Figure 1.1 shows the block diagram of an electric vehicle system driven by a portable energy source. The electromechanical energy conversion linkage system between the vehicle energy source and the wheels is the powertrain of the vehicle. The powertrain has electrical as well as mechanical components. The components of the electric vehicles will discussed in detail in Section 1.2.

The fuel for electric vehicles is stored in an energy storage device, such as a battery-pack, for energy delivery on demand. The primary source of energy for electricity generation for these vehicles is varied, ranging from fossil fuels to solar energy. The battery electric vehicle requires fuel delivered in the form of electricity to the vehicle through the electric power transmission system. Solar electric vehicles use solar panels and a power converter to charge the batteries on the vehicle. The special feature of these electric

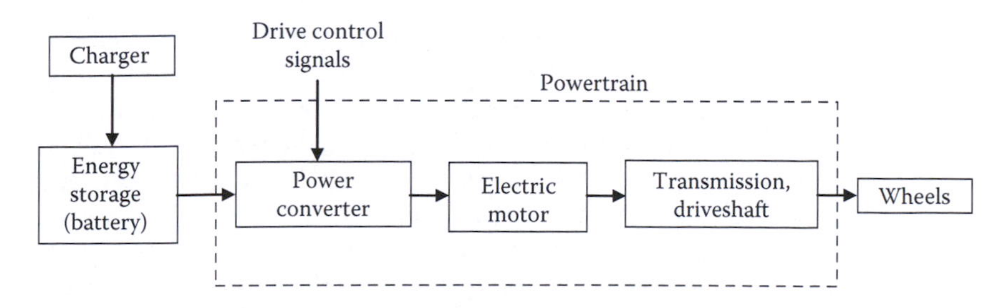

FIGURE 1.1
An electric vehicle system.

vehicles is that these are ZEVs as far as pollution within the vehicle is concerned. The fuel cell electric vehicle can also be ZEV if pure hydrogen is used as fuel on board the vehicle.

1.2 Hybrid Electric Vehicles

The term "hybrid electric vehicle" or "hybrid vehicle" generally refers to vehicles that use an IC engine in conjunction with one or more electric machines for propulsion. A hybrid road vehicle is one in which the propulsion energy during specified operational missions is available from two or more kinds or types of energy stores, sources, or converters, of which at least one store or converter must be on board. The definition of hybrid road vehicle is proposed by the technical committee 69 of Electric Road Vehicles of the International Electrotechnical Commission.

The definition of hybrid vehicles can be extended to vehicles that have engines different from IC engines. The vehicle in which energy from fuel cells and battery-packs are utilized to deliver propulsion power through an electric motor is also a hybrid vehicle. The fuel cell is essentially an engine where stored chemical energy in the fuel is converted into electrical energy directly without the involvement of any combustion process. Triple hybrids are also possible where a fuel cell engine, electric machine, and IC engine are all used to provide propulsion power. The traction electric motors can operate independently or in association with the IC engine to power the wheels of the vehicle depending on the type of vehicle architecture. The blending of the power from two or more devices can be achieved through an electrical or a mechanical device. The power flow control and blending of the mechanical and electrical transmission paths are internal and transparent to the user; the driver in a hybrid vehicle interfaces with the vehicle in the same way as he does with a present-day conventional IC engine vehicle.

The vehicle design complexity increases significantly for hybrid vehicles, since controls and support systems are needed for both the IC engine and the electric machine in addition to the components needed for the controlled blending of power coming from the two sources. The power transmission paths of the hybrid electric vehicle consist of different mechanical and electrical subsystems. The high performance and high efficiency of the electric power transmission path reduces the run time of the IC engine, thereby reducing emissions and increasing fuel efficiency.

The components and the power transmission paths for the electric and hybrid vehicles are discussed in the following section. The hybrid vehicle architectures will be highlighted in Chapter 3 after a discussion on vehicle kinetics in Chapter 2.

1.3 Electric and Hybrid Vehicle Components

The automobile is a complex system made up of numerous hardware components and software algorithms interconnected through mechanical links and electrical communications network. The design of the automobile extends from systems level design to intricate details of a subsystem or component design. The system design fundamentals are in the physics of motion, energy, and power, and in the principles of energy conversion from one form to another, the forms being chemical, electrical, and mechanical. The primary hardware components in the automobile are the energy conversion and power transmission devices; many secondary components are necessary for the functioning of the primary components. This book addresses the system level design fundamentals, the primary hardware components, and the connectivity among these components. The secondary components will be discussed as and when appropriate. A chapter is dedicated to a discussion on the various cooling systems required in a modern automobile.

The primary energy conversion devices in an electric or hybrid vehicle are the *IC engine*, the *electric machine*, and the *energy storage device*. The IC engine is a heat engine that converts chemical energy to mechanical energy. The electric machine can be used either as a *motor* or as a *generator* to convert mechanical power to electrical power or vice versa. In fuel cell electric vehicles, the fuel cell is the engine that converts chemical energy to an electrical form. The *transmission* in the vehicle is a key component for power transfer from the IC engine to the wheels.

With the introduction of electric machines for power and energy transfer in electric and hybrid vehicles, energy storage devices and electrical-to-electrical power/energy conversion devices become essential. A high-energy capacity *battery-pack* is the most common energy storage device in these vehicles. An *ultracapacitor bank* can also be used for energy storage in hybrid vehicles. *Flywheels* have also been used in prototype research hybrid vehicles for energy storage in mechanical form.

The electric machines require an *electric drive* to control the machine and deliver the required power based on requested demands and feedback signals. The electric drives are made of *power electronic devices* and *electronic controllers*. The drives are electrical-to-electrical energy conversion devices that convert steady voltages with fixed-frequency into a variable voltage supply for the electric machine. The drives can also process electrical power in the other direction assisting the electric machine to convert mechanical power into electrical power when the electric machine operates as a generator. The *DC–DC converter* is another electrical power management device used for DC power conversion from high-voltage to low-voltage levels or vice versa. The converter is made of power electronic devices and energy storage inductors; this device can be bidirectional as well. The DC–DC converter is a key component for the fuel cell interface with the electric motor drive.

The energy flow in a vehicle starts from the source of energy and ends at the wheels with the delivery of propulsion power; the path for this power and energy flow is known as the *powertrain* of the vehicle. The energy source within the vehicle could be the diesel or gasoline for the IC engines or the stored energy in batteries for electric motors.

The flow of power and energy in the powertrain is controlled by a set of electronic controllers. In addition to the *electronic controller units* (ECUs) for each of the energy conversion and power transmission devices in the powertrain, there is a master controller for coordinating the system level functions of the vehicle. This controller is termed as the *vehicle supervisory controller* (VSC). The supervisory controller is a key component in hybrid vehicles since it has to coordinate the energy conversion of multiple devices and power transmission through both the electrical and mechanical paths. The supervisory controller is like the brain of the vehicle designed to generate the control commands for the individual powertrain component ECUs. The supervisory controller interacts with the components of the vehicle through a communication network, which is based on a *controller area network* (CAN) protocol. The hybrid vehicle control strategies are discussed in Chapter 13, while the CAN protocols are presented in Chapter 14.

The primary powertrain components in a conventional IC engine vehicle are the engine and the transmission. These components deliver power to the wheels through the driveshaft and other coupling devices. The coupling devices include the differential and final drive in most cases. This mechanical power transmission path (MPTP) is shown in Figure 1.2. The power transmission path in an electric vehicle is mostly electrical except for the coupling devices between the electric propulsion motor and the wheels. This power transfer path will be referred to as the electrical power transmission path (EPTP). This transmission path is shown in Figure 1.3. The coupling device can simply be a gear to match electric machine speeds to vehicle speeds. The

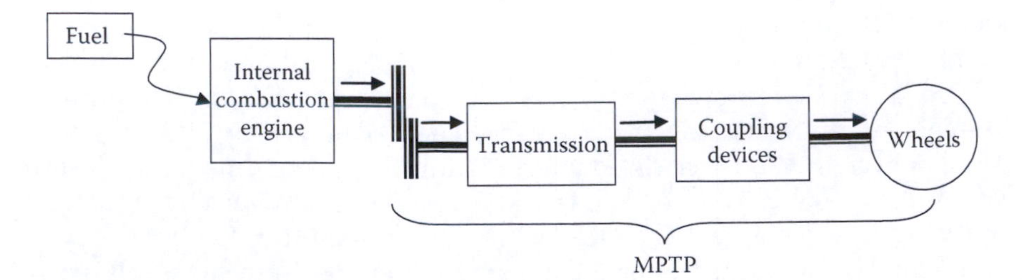

FIGURE 1.2
Power transmission path in a conventional ICE vehicle.

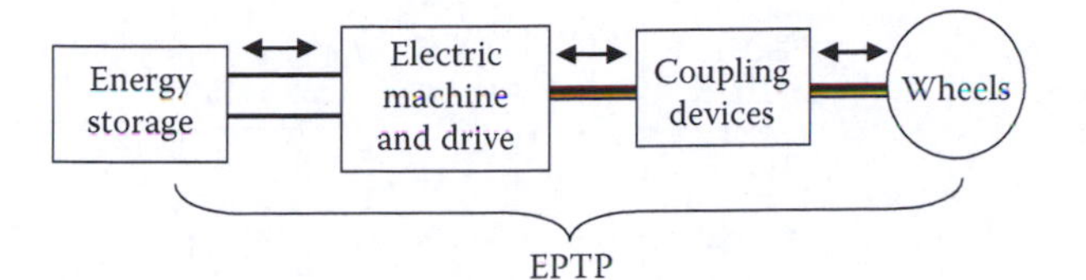

FIGURE 1.3
Power transmission path in an electric vehicle.

coupling can even be direct without any gearing in wheel-mounted motors, which are known as *hub motors*. The striking difference in this path from that in an IC engine vehicle is that power and energy flow can be bidirectional. In electric vehicles, kinetic energy of the vehicle can be processed back to the energy storage device through the electric machine when the vehicle brakes to slow down or stop.

Both the electrical and mechanical transmission paths exist in the powertrain of a hybrid vehicle. The architecture and components of the powertrain of a hybrid electric vehicle is varied depending on the type of the hybrid vehicle. A generic configuration for a charge-sustaining hybrid is shown in Figure 1.4. The *charge-sustaining hybrids* are those that never need to be plugged in for recharging the energy storage system. The only energy source within the vehicle is the stored fuel for the IC engine. All of the propulsion energy gets processed through the engine regardless of whether the power transmission path is electrical or mechanical. The propulsion power comes from one or more electric motors and the IC engine. The propulsion power is transmitted to the wheels through either the MPTP or the EPTP, or the combination of the two. The MPTP is associated with an IC engine and transmission, whereas the EPTP consists of the energy storage system, a generator, a propulsion motor, and transmission. The combination of electric machines and IC engine and the arrangements of the power transmission paths give rise to a variety of architectures for hybrid

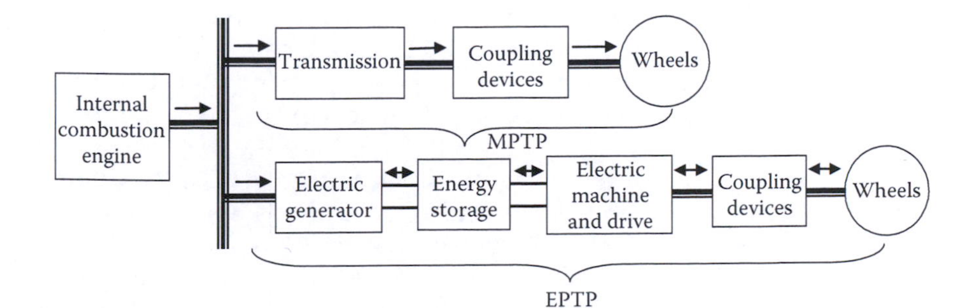

FIGURE 1.4
A hybrid electric vehicle powertrain.

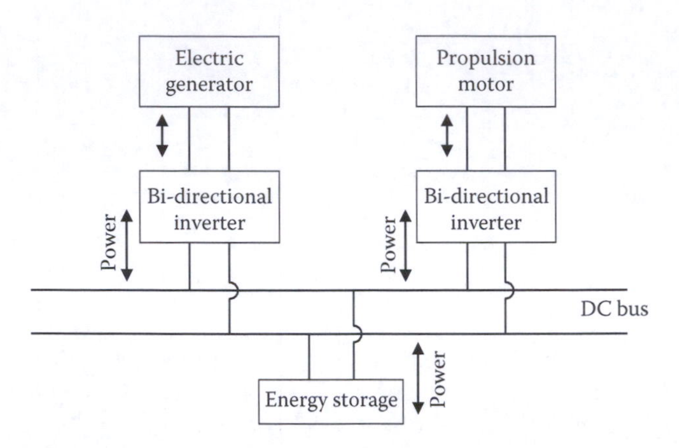

FIGURE 1.5
Electrical components of a hybrid electric vehicle.

electric vehicles. The energy sources and the power conversion devices can be arranged in series, parallel, or a series–parallel combination, giving rise to these various hybrid architectures. The hybrid vehicle architectures are discussed in Chapter 3.

The arrangement of the electrical components of the generic hybrid vehicle is shown in Figure 1.5. The electric generator shown in the hybrid configuration of Figure 1.5 is coupled to the engine and can be operated either as a generator or as a motor. During generation, the power through the generator can be used to charge the energy storage using a bidirectional inverter, or to deliver energy directly to the propulsion motor through the DC bus. The generator can also be operated as a motor during engine starting and torque boosting to meet the driver's peak acceleration demands. The energy storage system will absorb or deliver power depending on the system state of charge and driving conditions. Another bidirectional inverter conditions the power flow for the propulsion motor, which delivers torque to the wheels. The propulsion motor can also capture regenerative energy during vehicle braking. The energy storage devices used in hybrid vehicles are presented in Chapters 4 and 5. The starter/generator and propulsion motor in the EPTP use high-power electric machines. These electric machines need to have motoring and generating capability, high power density, high efficiency, and high starting torque over a wide speed range to meet the performance specifications.

The different types of electric machines are discussed in Chapter 6, while the electric drives and the machine controller methods are presented in Chapters 8 and 9, respectively. The power electronic components for the electric motor drives and the DC–DC converters are presented in Chapter 7. The different types of IC engines and their fundamentals are presented in Chapter 10. The powertrain components and brakes are discussed in Chapter 11.

1.4 Vehicle Mass and Performance

The several vehicle mass definitions used in this book are common to those used in the automotive industry. The *vehicle curb mass* m_v is the total mass of a vehicle with all standard equipment, components, lubricants, and full tank of gas, but without any passenger or cargo. The gross vehicle mass m_{gv} of a vehicle is the curb mass plus the passengers and cargo. The *maximum gross vehicle mass* is the curb mass plus the maximum number of passengers and the maximum mass of the cargo that the vehicle is designed for. Often the term "weight" is generally used for vehicles instead of "mass," although it is the mass of the vehicle that is being referred to. Mass defines the intrinsic physical property of the vehicle whereas weight is the force of the vehicle due to gravitational pull. The scientifically correct term "mass" will be used in this book for the calculations and analysis.

The vehicle curb mass can be distinguished as sprung mass and unsprung mass in relation to the location of the components with respect to the vehicle suspension system. The spring mass is the fraction of the vehicle curb mass that is supported by suspension including suspension members that are in motion. The unsprung mass is the remaining fraction of vehicle curb mass that is carried by the wheels and moving with it.

The front to rear mass distribution is critical for balance and good ride performance of a vehicle. The mass distribution can be defined in terms of axle to axle lengths. Let

l = axle to axle length

a = front axle to vehicle center of gravity, known as front longitudinal length

b = rear axle to vehicle center of gravity, known as rear longitudinal length

The front vehicle mass is

$$m_{vf} = \frac{b}{l} m_v$$

And the rear vehicle mass is

$$m_{vr} = \frac{a}{l} m_v$$

The packaging of components for good balance and ride performance is very important, especially for hybrid vehicles that has more powertrain components than a conventional vehicle. The electric vehicles and hybrids with large energy capacity have significant battery mass requiring efficient packaging within the vehicle. The dynamic effects of braking cause a dynamic mass shift; the front brakes are responsible for about 70% of vehicle braking.

A 60:40 or less ratio of front-to-rear vehicle mass distribution is required considering balance, ride performance, and dynamic braking.

The ratio of sprung to unsprung mass also has to be carefully evaluated during the vehicle packaging and layout design. A 10:1 ratio of sprung to unsprung mass is a desirable target, although a slightly lower ratio can be used for hybrid vehicles that may have more unsprung components. The electric motors mounted on the hub or disk brakes along with the entire caliper assembly add to the unsprung mass of the vehicle.

An equivalent vehicle mass in terms of the curb mass and number of passengers is used in sizing the powertrain components. The equivalent mass to be used in the design calculations is given by

$$m_{eq} = k_m m_v + N_p m_p$$

where

k_m is mass factor related to the translational equivalent of all rotating inertias

N_p is the number of persons in the vehicle

m_p is the average mass of the persons

k_m is a dimensionless mass factor that accounts for the inertia of all the rotating components such as wheels, driveline components, engine, with ancillaries and hybrid electric machines.

The mass factor is given by [3]

$$k_m = 1 + \frac{4J_w}{m_v r_{wh}^2} + \frac{J_{eng}\xi_{eng}^2\xi_{FD}^2}{m_v r_{wh}^2} + \frac{J_{em}\xi_{em}^2\xi_{FD}^2}{m_{cv} r_{wh}^2}$$

where $\xi_{eng/em}$ and ξ_{FD} are the engine/electric machine transmission and final drive gear ratios.

The steering performance of a vehicle is measured in terms of the steering angle, which is the average of the left and right wheel angles [4]. Right to left vehicle mass balance is also needed for good steering performance. Packaging the hybrid components above the vehicle's center of gravity will result in a good steering performance and handling of the vehicle.

1.5 Electric Motor and Engine Ratings

The strengths of electric motors and IC engines are typically described with kilowatt (kW) or horsepower (hp) ratings, although a comparison between electric motors and IC engines in terms of power units only is not fair. The power that an electric motor can continuously deliver without overheating is its rated power, which is typically a derated figure. For short periods of time, the motor can deliver two to three times the rated power. Therefore, a higher power and torque is available from an electric motor for acceleration, and the motor torque can be the maximum under stall conditions, i.e., at zero speed. The motor type determines whether maximum torque is available at zero speed or not. On the contrary, an IC engine is rated at a specific rpm level for maximum torque and maximum power. The maximum torque and power ratings of the IC engine are typically derived under idealized laboratory conditions. In practical situations, it is impossible to achieve the rated power; the maximum power available from an IC engine is always smaller than the rated power.

The torque characteristics of motors are shown in Figure 1.6 along with torque characteristics of IC engines. The characteristics of specific motors and IC engines will differ somewhat from these generalized curves. For electric motors, a high torque is available at starting, which is the rated torque of the motor. The peak or rated power from a motor is obtained at base speed (ω_b) when the motor characteristics enter the constant power region from constant torque region once the voltage limit of the power supply is reached. The motor rated speed (ω_{rated}) is at the end of the constant power region. The IC engine peak power and torque occur at the same speed. At this stage, it will be helpful to review the power and torque relation, which is as follows:

$$\text{Power (W)} = \text{Torque (N m)} \times \text{Speed (rad/s)} \tag{1.1}$$

The power–torque relation in hp and ft·lb is

$$\text{hp} = \frac{\text{Torque (ft}\cdot\text{lb)} \times \text{rpm}}{5252}$$

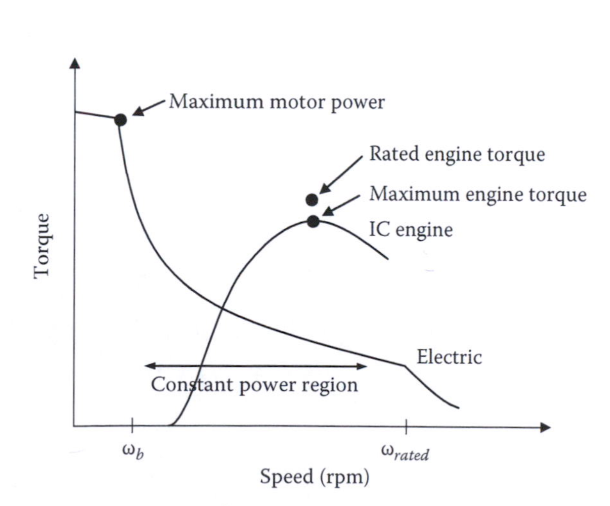

FIGURE 1.6
Electric motor and IC engine torque characteristics.

Figure 1.6 depicts that the IC engine does not produce any torque below a certain speed. A transmission is essential for an IC engine to match the vehicle speed with the narrow high power speed range of the engine. On the other hand, the electric motor produces high torque even at zero speed and typically has constant power characteristics over a wide speed range. Therefore, the electric motor can be attached directly to the drive wheels accelerate the vehicle from zero speed all the way up to the top speed. The motor and, hence, the vehicle speed can be controlled directly through the power electronic converter feeding the current into the motor. There is essentially no requirement for a transmission with an electric motor other than a fixed gear for appropriately sizing the motor.

The important characteristics of an electric or hybrid vehicle motor include flexible drive control, fault tolerance, high efficiency, and low acoustic noise. The motor drive must be capable of handling voltage fluctuations from the source. The requirements of an electric or hybrid vehicle motor, not necessarily in order of importance, are summarized in the following:

- Ruggedness
- High torque to inertia ratio (T_e/J). Large T_e/J results in "good" acceleration capabilities
- Peak torque capability of about 200%–300% of continuous torque rating
- High power to weight ratio (P_e/w)
- High-speed operation, ease of control
- Low acoustic noise, low EMI, low maintenance, and low cost
- Extended constant power region of operation

1.6 Electric and Hybrid Vehicle History

The history of electric vehicles is quite interesting. It includes the emergence of electric vehicles following the discovery of electricity and the means of electromechanical energy conversion, and later being overtaken by gasoline-powered vehicles. People digressed from a potential environmentally friendlier mode of transportation due to the lack of enabling technologies in the early years, but are now revisiting that prospect aided with significant technical developments over the years.

1.6.1 The Early Years

Prior to 1830s, the means of transportation was only through steam power, since the laws of electromagnetic induction, and consequently electric motors and generators, were yet to be discovered. Faraday demonstrated the principle of electric motor as early as in 1820 through a wire rod carrying electric current and a magnet, but it was in 1831 when he discovered the laws of electromagnetic induction that enabled the development and demonstration of electric motors and generators that is essential for electric transportation. The following is a summary of the history of electric vehicles in those early years up to its peak period in the early 1900s:

- Pre-1830—Steam-powered transportation
- 1831—Faraday's law, and shortly thereafter, invention of DC motor
- 1834—Nonrechargeable battery–powered electric car used on a short track
- 1851—Nonrechargeable 19 mi/h electric car
- 1859—Development of lead storage battery
- 1874—Battery-powered carriage
- Early 1870s—Electricity produced by dynamo-generators
- 1885—Gasoline-powered tricycle car
- 1900—4200 automobiles sold:
  40% steam powered
  38% electric powered
  22% gasoline powered

The specifications of some of the early electric vehicles are

- 1897: French Krieger Co. EV—weight, 2230 lb; top speed, 15 mi/h; range, 50 mi/charge
- 1900: French B.G.S. Co. EV—top speed, 40 mi/h; range, 100 mi/charge
- 1915: Woods EV—top speed, 40 mi/h; range, 100 mi/charge
- 1915: Lansden EV—weight, 2460 lb; range, 93 mi/charge; 1 ton payload capacity
- 1912: 34,000 EVs registered; EVs outnumber gas-powered vehicles 2-to-1
- 1920s: EVs disappear and IC engine vehicles become predominant

The factors that led to the disappearance of electric vehicle after its short period of success are

1. Invention of starter motor in 1911 that made gas vehicles easier to start.

2. Improvements in mass production of Henry T (gas-powered car) vehicles, which sold for $260 in 1925 compared to $850 in 1909. EVs were more expensive.
3. Rural areas had very limited access to electricity to charge batteries, whereas gasoline could be sold in those areas.

1.6.2 1960s

Electric vehicles started to resurge in the 1960s primarily due to the environmental hazards caused by the emissions of IC engine vehicles. The major IC engine vehicle manufacturers, General Motors and Ford, became involved in electric vehicle research and development. The General Motors started a $15 million program that culminated in the vehicles called Electrovair and Electrovan. The components and specifications of the two Electrovair vehicles are given below:

Electrovair I (1964) and Electrovair II (1966) by GM

- Systems and characteristics

  Motor: Three-phase induction motor, 115 hp, 13,000 rpm

  Battery: Silver-zinc (Ag-Zn), 512 V, 680 lb

  Motor drive: DC-to-AC inverter using SCR

  Top speed: 80 mi/h

  Range: 40–80 mi

  Acceleration: 0–60 mi/h in 15.6 s

  Vehicle weight: 3400 lb

The Electrovair utilized Chevy Corvair body and chassis. Among the positive features was the acceleration performance that was comparable to IC engine vehicle Corvair. The major disadvantage of the vehicle was the silver-zinc (Ag-Zn) battery-pack that was too expensive and heavy with short cycle life and requiring a long recharge time.

An additional factor in the 1960s that provided the impetus for electric vehicle development include the "The Great Electric Car Race" cross-country competition (3300 mi) between an electric vehicle from Caltech and another one from MIT in August 1968. The race generated great public interest in electric vehicles and provided an extensive road test of the technology. However, the 1960s technology was not mature enough to produce a commercially viable electric vehicle.

1.6.3 1970s

The scenario turned in favor of electric vehicle in the early 1970s as gasoline prices increased dramatically due to energy crisis. The Arab oil embargo of 1973 increased demands for alternate energy sources, which led to an

TABLE 1.1

EV Performance Standardization of 1976

Category	Personal Use	Commercial Use
Acceleration from 0 to 50 km/h	<15 s	<15 s
Gradability at 25 km/h	10%	10%
Gradability at 20 km/h	20%	20%
Forward speed for 5 min	80 km/h	70 km/h
Range: Electric	50 km, C cycle	50 km, B cycle
Hybrid	200 km, C cycle	200 km, B cycle
Nonelectrical energy consumption in hybrid vehicles (consumption of nonelectrical energy must be less than 75% of the total energy consumed)	<1.3 MJ/km	<9.8 MJ/km
Recharge time from 80% discharge	<10 h	<10 h

immense interest in electric vehicles. It became highly desirable to be less dependent on foreign oil as a nation. In 1975, 352 electric vans were delivered to U.S. postal service for testing. In 1976, Congress enacted the Public Law 94-413, the Electric and Hybrid Vehicle Research, Development and Demonstration Act of 1976. This act authorizes a federal program to promote electric and hybrid vehicle technologies and to demonstrate the commercial feasibility of electric vehicles. The Department of Energy (DOE) standardized the electric vehicle performance, which is summarized in Table 1.1.

The case study of a GM electric vehicle of the 1970s is as follows:

- System and characteristics:

  Motor: Separately excited DC, 34 hp, 2400 rpm

  Battery-pack: Ni-Zn, 120 V, 735 lb

  Auxiliary battery: Ni-Zn, 14 V

  Motor drive: Armature DC chopper using SCRs; field DC chopper using BJTs

  Top speed: 60 mi/h

  Range: 60–80 mi

  Acceleration: 0–55 mi/h in 27 s

The vehicle utilized a modified Chevy Chevette chassis and body. This electric vehicle was used mainly as a test bed for Ni-Zn batteries. Over 35,500 mi of on-road testing proved that this electric vehicle is sufficiently road worthy.

1.6.4 1980s and 1990s

The 1980s and the 1990s saw tremendous developments of high power, high frequency semiconductor switches, along with the microprocessor revolution, which led to improved power converter design to drive the electric motors efficiently. The period also contributed to the development of magnetic bearings used in flywheel energy storage systems, although these are not utilized in the mainstream electric vehicles development projects.

In the last two decades, legislative mandates pushed the cause for ZEVs. Legislation passed by the California Air Resources Board in the 1990 stated that by 1998, 2% of vehicles should be ZEVs for each automotive company selling more than 35,000 vehicles. The percentage was to increase to 5% by 2001 and to 10% by 2003. The legislation provided a tremendous impetus to develop electric vehicles by the major automotive manufacturers. The legislation was relaxed somewhat later due to practical limitations and the inability of the manufacturers to meet the 1998 and 2001 requirements. The mandate further relaxed to stand as 4% of all vehicles sold should be ZEV by 2003 and an additional 6% of the sales must be made up of ZEVs and partial ZEVs, which would have required General Motors to sell about 14,000 electric vehicles in California.

Motivated by the pollution concern and potential energy crisis, the government agencies, federal laboratories, and the major automotive manufactures launched a number of initiatives to push for the ZEVs. The partnership for next generation vehicles (PNGV) is such an initiative established in 1993, which is a partnership of federal laboratories and automotive industries to promote and develop electric and hybrid electric vehicles. The most recent initiative by the DOE and the automotive industries is the Freedom CAR initiative.

The trends in electric vehicle developments in recent years can be attributed to the following:

- High level of activity of the major automotive manufacturers
- New independent manufacturers bring vigor
- New prototypes are even better
- High levels of activity overseas
- High levels of hybrid vehicle activity
- A boom in individual or small company IC engine vehicle to EV conversions

The case studies of two GM electric vehicles of the 1990s are given in the following:

1. GM Impact 3 (1993 completed)
    - Based on 1990 Impact displayed at the Los Angeles auto show
    - Two-passenger, two-door coupe, street legal and safe
    - 12 built initially for testing, 50 built by 1995 to be evaluated by 1000 potential customers

- System and characteristics:

  Motor: One, three-phase induction motor, 137 hp, 12,000 rpm

  Battery-pack: Lead-acid (26), 12 V batteries connected in series (312 V), 869 lb

  Motor drive: DC-to-AC inverter using IGBTs

  Top speed: 75 mi/h

  Range: 90 mi on highway

  Acceleration: 0–60 mi in 8.5 s

  Vehicle weight: 2900 lb

The vehicle was used as a testbed for mass production of electric vehicles.

2. Saturn EV1
   - Commercially available electric vehicle made by GM in 1995
   - Leased in California and Arizona for a total cost of about $30,000
   - System and characteristics:

     Motor: one, three-phase induction motor

     Battery-pack: Lead-acid batteries

     Motor drive: DC-to-AC inverter using IGBTs

     Top speed: 75 mi/h

     Range: 90 mi in highway, 70 mi in city

     Acceleration: 0–60 mi in 8.5 s
   - Power consumption:

     30 kW h/100 mi in city, 25 kW h/100 mi in highway

   This vehicle was also used as a testbed for mass production of electric vehicles.

1.6.5 Recent EVs and HEVs

All of the major automotive manufacturers have produced electric vehicles rather recently, some of which were available for sale or lease to the general public. The status of these vehicle programs change rapidly with manufacturers suspending production frequently due to the small existing market demand of such vehicles. Examples of production of electric vehicles, which were available, are GM EV1, Ford Think City, Toyota RAV 4, Nissan Hypermini, Peugeot 106 Electric etc. There are also many prototype and experimental electric vehicles being developed by the major automotive manufacturers. Most of these vehicles used AC induction motors or PM synchronous motors. Also, interestingly, almost all of these vehicles use battery technology other than the lead-acid battery-pack. The list of electric vehicles in production and under development is extensive, and the readers are referred to [3,5,6] for the details of many of these vehicles.

The manufacturers of electric vehicles in the 1990s faced a stiff challenge as their significant research and development efforts on ZEV technologies have been hindered by unsuitable battery technologies. A number of auto industries started developing hybrid vehicles to overcome the battery and range problems of pure electric vehicles. The Japanese auto industries led this trend with Toyota, and Honda entering the market from late 1990s with their hybrid model cars Prius and Insight. The Honda Insight using a mild parallel hybrid system and the Toyota Prius with a series–parallel hybrid architecture typify the two modern schools of thought regarding passenger hybrid vehicles. The Insight utilizes a simple lightweight parallel powertrain featuring a single electric machine and a high-voltage battery-pack. The electric machine is an integrated starter/generator serving as a starter motor, generator, and assist traction motor. The hybrid systems in the Saturn Vue Greenline and Honda Civic Hybrid are similar to that of the Insight. These vehicles have high parts interchangeability with their conventional siblings and deliver the benefits of hybridization without introducing high costs. The hybrid system in the Ford Escape Hybrid and the GM Two-Mode hybrids are similar to that of the Prius in that they have relatively complex and expensive power-split systems. The systems have one IC engine and two electric machines. These systems get the most benefit possible from a hybrid system, and therefore offer the greatest increase in fuel efficiency, but are also more expensive to produce than a mild hybrid. The hybrid vehicles use both an electric motor and an IC engine and thus do not solve the pollution problem, although it does mitigate it. It is perceived by many that the hybrids with their multiple propulsion units and control complexities are not economically viable in the long run, although currently a number of production hybrid vehicle models are available from almost all the major automotive industries around the world. The cost of electric and hybrid vehicles are expected to be high until production volume increases significantly.

While hybrid vehicles have penetrated the market at an extraordinary rate, the battery electric vehicles are also making their way back due to the advances in battery technology. In 2008, Tesla motors delivered the Tesla Roadster, which is the first production vehicle to use Li-ion battery technology and the first battery electric vehicle to travel more than 200 mi on a single charge. The Tesla Roadster has been designed for performance with an acceleration rating of 0–60 mi/h in 3.9 s. The vehicle currently uses a 53 kW h battery-pack made of 6831 Li-ion cells with a nominal voltage of 375 V; this gives the vehicle a nominal range of 244 mi/charge. The vehicle curb mass is 1238 kg with a battery mass of 450 kg. The powertrain consists of a 215 kW, 400 N m peak rated three-phase AC induction motor and a single speed, fixed ratio gearbox. The peak electric motor speed is 14,000 rpm with a base speed of 5,133 rpm. The transmission gear ratio is 8.28:1 in forward motion and 3.12:1 in reverse.

Several major and smaller auto industries are also developing battery electric vehicles for the consumer market. These vehicles are targeted for the urban commuter with ranges around 100 mi or less. Nissan plans to deliver zero-emission battery electric passenger vehicles, called the Leaf, for fleet customers by 2010 and for retail buyers within the next few years. Toyota also plans to produce a 50-mile range electric vehicle in the near future. Myers motors of Ohio is marketing single-passenger electric vehicles called NmG and have plans to produce two passenger vehicles in the future. While battery electric vehicles are slowly entering the market, the auto industry is now more focused on developing production plug-in hybrid vehicles in the near term. General Motors has announced the availability of Chevy Volt, which is a PHEV40 with a range of 40 mi for model year 2010. Volkswagen also plans to introduce the Golf PHEV30 with a range of 30 mi for the same model year. The Ford PHEV30 based on Escape SUV has been in fleet usage for a few years; the production plan is for 2012. Toyota is to deliver 500 Prius PHEVs for fleet customers; the vehicles will be used for market and engineering analysis.

The fuel cell electric vehicles (FCEV) can be a viable alternative to battery electric vehicles to serve as ZEVs without the range problem. DaimlerChrysler built a prototype electric van called NECAR1 in 1994 using a 50 kW Ballard proton exchange membrane (PEM) fuel cell. Toyota built a prototype fuel cell vehicle in 1994 based on RAV4 sports utility vehicle (SUV) using a 20 kW PEM fuel cell. The range of NECAR1 was 81 mi while the range for the Toyota vehicle, called FCHV, was 155 mi. In December 2002, the city of Los Angeles began leasing one of the five Honda FCX vehicles, which is the first fuel cell vehicle authorized for commercial use. The vehicle has an 85 kW PEM fuel cell and a range of 185 mi. Honda has announced their third generation fuel cell vehicle Honda FCX Clarity, which is a fuel cell/battery hybrid using a 100 kW PEM fuel cell and a 288 V Li-ion battery-pack. The vehicle is powered by a 100 kW, 255 N m peak rated PM synchronous motor for propulsion. The vehicle has a 72 mi/kg fuel usage that is equivalent to 74 mi/gge ("gge" stands for gasoline gallon equivalent). Toyota, in collaboration with U.S. Department of Energy, Savannah River National Laboratory, and National Renewable Energy Laboratory, is evaluating the fuel cell vehicle FCHV-adv, which is based on the Toyota Highlander. The vehicle has a range of 431 mi with a fuel economy of 68.3 mi/kg. The major auto industries have production plans of fuel cell electric vehicles within the next decade.

1.7 Well-to-Wheel Analysis

The well-to-wheel (WTW) efficiency is the measure of the overall efficiency of a vehicle starting from the extraction of raw fuel to the wheels including the efficiencies of energy conversion, transport, and delivery at each stage.

The fuel can be extracted from the earth or sea or derived from a renewable source; the transportation can be through the land or sea, or electrically though the transmission lines, and the energy conversion can be through the heat engines, electric machines, or electrochemical devices. The energy transport and conversion path can be divided into two segments of well-to-tank (WTT) and tank-to-wheel (TTW). The fuels for transportation are produced from energy feedstock in the wells through different fuel production pathways. The fuel stored in a vehicle is processed to deliver propulsion power at the wheels. The WTT segment comprises of feedstock-related stages (recovery, processing, transportation, and storage) and fuel-related stages (production, transportation, storage, and distribution). The TTW segment comprises of the energy conversion and delivery stages from the tank to the wheels of a vehicle. This WTW efficiency is consequently the product of the WTT and TTW efficiencies. Figure 1.7 represents the processes involved in evaluating the WTW efficiency. The WTW efficiency is an important factor for evaluating the overall impact, long-term feasibility, and environmental effects of the alternative vehicles such as the electric, hybrid electric, plug-in hybrid, and fuel cell vehicles. The suitable energy paths can be identified and selected using WTW analysis to formulate energy strategies and energy policies. The WTW analysis also provides a level ground for comparing the alternative vehicles with the conventional IC engine vehicles. The fundamental laws of physics and chemistry with a common base for input parameters, numerical procedures, technology standards, and user behavior must be used for these analyses [7].

The TTW efficiency for conventional IC engine vehicles is inherently low (around 25%), because of the low efficiency of the gasoline or diesel engine and the losses in the powertrain components. In the case of hybrid electric vehicles, the engine is downsized and controlled to operate at higher efficiency regions most of the time, which results in lower fuel consumption. The TTW efficiency of hybrid electric vehicles is higher than that of an IC engine vehicle and is estimated to be around 50% [8]. The efficiency of an

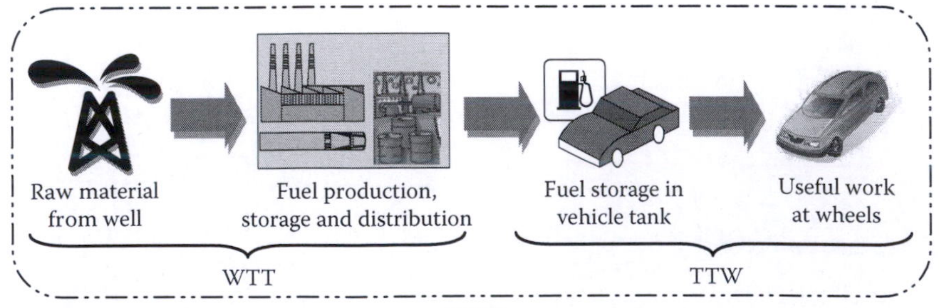

FIGURE 1.7
Processes involved in WTW efficiency calculation.

electric powertrain is around 80%–90%. The TTW efficiency in battery electric vehicles is high due to their large capacity batteries and electric-only mode of operation. However, when WTT efficiencies are included in the comparisons, the overall WTW efficiencies are not significantly different for any type of the vehicles. The WTW efficiency of hybrid and electric vehicles are similar to IC engine vehicles because of the lower WTT efficiencies. The WTT efficiency for the plug-in hybrid vehicle depends on the proportions of the grid energy and the gasoline energy used during a drive cycle.

The WTW efficiencies and emission impacts can be evaluated for a comparison using the GREET model developed at the Argonne National Laboratory [9]. GREET model is essentially a multidimensional spreadsheet model available for public use for analyzing fuel cycles from the source to the wheel of a vehicle. GREET is the acronym for greenhouse gases, regulated emissions, and energy use in transportation. The model has more than 100 fuel production pathways and more than 70 vehicle systems.

Table 1.2 shows the WTW efficiencies and emissions for conventional IC engine, electric, and plug-in hybrid vehicles analyzed for a typical mid-size sedan using the GREET model [8,10]. The CG and RFG represents conventional gasoline and reformulated gasoline vehicles. The WTT efficiency was found to be 66.5% for plug-in hybrid vehicles (grid-connected hybrids) using 33% grid energy and 67% gasoline for operation. On the other hand, the WTT efficiency for conventional and regular hybrid vehicles is 79.5%. The reason for low efficiency of plug-in hybrid vehicle is due to the fact that electricity is mostly generated from conventional energy sources that have very low efficiency.

TABLE 1.2

Energy Consumption, Efficiency, and Emissions for Passenger Cars Using GREET

	SI ICEV (Baseline CG and RFG)	**Plug-In SI HEV (Gasoline and Electricity)**	**Battery EV (Electricity)**
Total energy (W h)	257,551	526,261	1,632,131
WTT efficiency	79.5%	66.5%	38.0%
TTW efficiency	21.9%	23%	48.51%
WTW efficiency	17.41%	15.29%	18.43%
CO_2 (g/million BTU)	17,495	57,024	219,704
CH_4 (g/million BTU)	109.120	145.658	296.031
N_2O (g/million BTU)	1.152	1.535	3.111
VOC: Total (g/million BTU)	27.077	25.630	19.679
CO: Total (g/million BTU)	15.074	23.553	58.448
NO_x: Total (g/million BTU)	50.052	87.100	239.571

The greenhouse gas (CO_2, CH_4, N_2O, CO, VOC, and NO_x) emissions are also found to be higher in plug-in hybrid vehicles than in conventional gasoline vehicles according to the results in Table 1.2. The TTW efficiency for the plug-in hybrid is not significantly different from the baseline vehicle since the IC engine usage is still quite high. Poor efficiencies in both WTW and TTW result in an overall low WTW efficiency for plug-in hybrids. More on efficiency calculation is presented in Section 1.8.

1.8 EV/ICEV Comparison

The relative advantages and disadvantages of alternative vehicles over conventional IC engine vehicles can be better appreciated from a comparison of the two on the basis of efficiency, pollution, cost, and dependence on oil.

1.8.1 Efficiency Comparison

In order to evaluate the efficiencies of different types of vehicles on a level ground, the complete process in both systems starting from crude oil to power available at the wheels must be considered, i.e., the analysis must be carried out based on WTW efficiencies. The electric vehicle process starts not at the vehicles, but at the source of raw power whose conversion efficiency must be considered to calculate the overall efficiency of electric vehicles. The power input P_{IN} to any vehicle ultimately comes from a primary energy source even before it is stored in a vehicle tank. The power extracted from a piece of coal by burning it is an example of primary power obtained from a primary energy source. The power that is available in a vehicle from an energy storage tank or device is applied power obtained from a secondary source of energy. The applied or secondary power is obtained indirectly from raw materials. The electricity generated from crude oil and delivered to an electric car for battery charging is an example of secondary power. The raw or primary power is labeled as $P_{IN\ RAW}$, while the secondary power is designated as $P_{IN\ PROCESS}$.

An example efficiency comparison is presented here based on the complete WTW processes involved in an electric vehicle and an IC engine vehicle. The complete electric vehicle process can be broken down into its constituent stages involving power generation, transmission, and usage, as shown in Figure 1.8. The primary power from the source is fed to the system only at the first stage, although secondary power can be added in each stage. Each stage has its efficiency based on total input to that stage and output delivered to the following stage. The efficiency of each stage must be calculated from input output power considerations, although the efficiency may vary widely depending on the technology being used. Finally, the overall

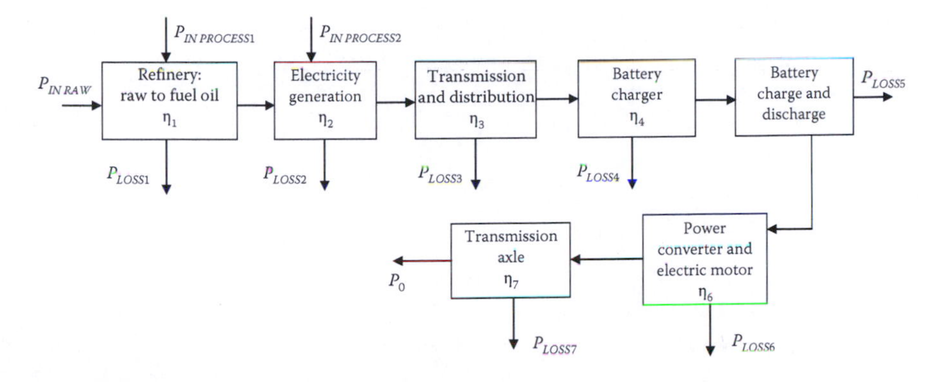

FIGURE 1.8
The complete electric vehicle process from crude oil to power at wheels.

efficiency can be calculated by multiplying the efficiencies of all the individual stages. The overall efficiency of the electric vehicle system shown in Figure 1.8 is

$$\eta_{EV} = \frac{P_0}{P_{IN}} = \frac{P_0}{P_0 + \sum_{i=1}^{7} P_{LOSSi}} = \frac{P_0}{P_6}\frac{P_6}{P_5}\frac{P_5}{P_4}\frac{P_4}{P_3}\frac{P_3}{P_2}\frac{P_2}{P_1}\frac{P_1}{P_{IN}} = \eta_1\eta_2\eta_3\eta_4\eta_5\eta_6\eta_7$$

The IC engine vehicle process details are illustrated in Figure 1.9. The process starts from the conversion of crude oil to fuel oil in the refinery, and then includes the transmission of fuel oil from refinery to gas stations, power conversion in the IC engine of the vehicle, and power transfer from the engine to the wheels through the transmission. The efficiency of the IC engine vehicle process is the product of the efficiencies of the individual stages indicated in Figure 1.9 and is given by

$$\eta_{ICEV} = \eta_1\eta_2\eta_3\eta_4$$

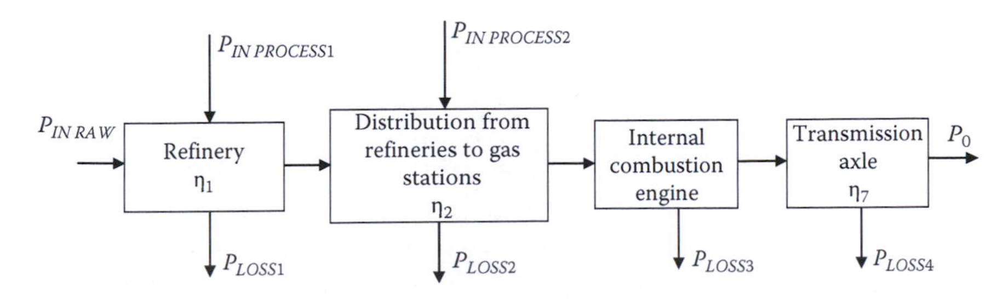

FIGURE 1.9
The complete IC engine vehicle process from crude oil to power at the wheels.

TABLE 1.3

Battery Electric and IC Engine Vehicle WTW Efficiencies

	Battery EV Efficiency (%)		ICEV Efficiency (%)	
	High	High	Low	Low
Refinery (petroleum)	97	90	85	95
Distribution to fuel tank		99	95	
Engine		22	20	
Electricity generation	40			33
Transmission to wall outlet	92			90
Battery charger	90			85
Battery (lead/acid)	75			75
Motor/controller	85			80
FC fuel processor				
PEM fuel cell				
Plug-in grid gasoline ratio				
Transmission/axle	98	98	95	95
WTW	20	19	15	14

A sample comparison of electric vehicle and IC engine vehicle process efficiencies based on the diagrams of Figures 1.8 and 1.9 is given in Table 1.3. Representative numbers have been used for the energy conversion stages to convey a general idea about the efficiencies of the two systems. Table 1.3 shows that the overall efficiency of an electric vehicle is comparable to the overall efficiency of IC engine vehicle.

1.8.2 Pollution Comparison

Transportation accounts for one-third of all energy usage, making it the leading cause of environmental pollution through carbon emissions [11]. The DOE has projected that if 10% of automobiles nationwide were ZEVs, regulated air pollutants would be cut by 1 million tons/year and 60 million tons of greenhouse carbon dioxide gas would be eliminated. With 100% electrification, i.e., every IC engine vehicle replaced by electric vehicle, the following has been claimed:

- Carbon dioxide in air, which is linked to global warming, would be cut by half.
- Nitrogen oxides (a greenhouse gas causing global warming) would be cut slightly depending on government-regulated utility emission standards.
- Sulfur dioxide, which is linked to acid rain, would increase slightly.

- Waste oil dumping would decrease, since electric vehicle do not require crankcase oil.
- Electric vehicles reduce noise pollution since they are quieter than IC engine vehicles.
- Thermal pollution by large power plants would increase with increased electric vehicle usage.

The electric vehicles will considerably reduce the major causes of smog, substantially eliminate ozone depletion, and reduce greenhouse gases. With stricter SO_2 power plant emission standards, electric vehicles would have little impact on SO_2 levels. Pollution reduction is the driving force behind electric vehicle usage. Pollution can be cut drastically when electric vehicles batteries are charged from electricity produced from renewable sources.

1.8.3 Capital and Operating Cost Comparison

The initial electric vehicle capital costs are higher than IC engine vehicle capital cost primarily due to the lack of mass production opportunities. However, electric vehicle capital costs are expected to decrease as volume increases. Capital costs of electric vehicles easily exceed capital costs of IC engine vehicles due to the cost of the battery. The power electronics stages are also expensive, although not at the same level as batteries. The total life cycle cost of an electric vehicle is projected to be less than that of a comparable IC engine vehicle. The electric vehicles are more reliable and will require less maintenance making it a favorable over IC engine vehicle as far as the operating cost is concerned.

1.8.4 U.S. Dependence on Foreign Oil

The importance of search for alternative energy sources cannot be overemphasized, and sooner or later there will be an energy crisis if we, the people of the planet, do not reduce our dependence on oil. Today's industries, particularly the transportation industry, are heavily dependent on oil, the reserve of which will eventually deplete in the not so distant future. Today about 42% of petroleum used for transportation in the United States is imported. An average IC engine vehicle in its lifetime uses 94 barrels of oil based on 28 mi/gal fuel consumption. On the other hand, an average electric vehicle uses 2 barrels of oil in its lifetime based on 4 mi/kW h. The oil is used in the electric vehicle process during electricity generation, although only 4% of electricity generated is from oil. The energy sources for electricity generation are shown in the pie chart of Figure 1.10.

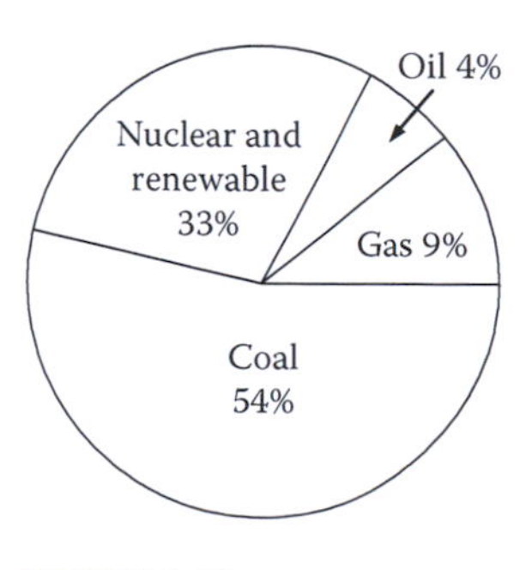

FIGURE 1.10
Electricity generation pie chart.

1.9 Electric Vehicle Market

We typically discuss about electric vehicles for passenger and public transportation, but tend to forget about their use as off-road vehicles in specialty applications where range is not an issue. The electric vehicles have penetrated the off-road vehicles market successfully over the years for cleaner technology as well as for cost advantages. The examples of such applications are airport vehicles for passenger and ground support; recreational vehicles as in golf carts and theme parks, plant operation vehicles like fork lifts, loader trucks; vehicles for disabled persons; utility vehicles for ground transportation in closed but large compounds etc. There are also electric vehicles that run on tracks for material haulage in mines. There is potential for electric vehicle use for construction vehicles. The locomotives that run on tracks with electricity supplied from transmission lines are theoretically no different from other electric vehicles, the major difference being in the way energy is fed for the propulsion motors.

Motivated by the growing concern about global pollution and the success of electric motor driven transportation in various sectors, the interest is ever increasing for road electric vehicles that can deliver the performance of its IC engine vehicle counterparts. The major impediments for mass acceptance of electric vehicles by the general public are the limited electric vehicle range and the lack of electric vehicle infrastructure. The solution for the range problem may come from extensive research and development efforts in batteries, fuel cells, and other alternative energy storage devices. An alternative approach is to create awareness among people on the problems of global warming and the advantages of electric vehicles, while considering the fact that majority of population drive less than 50 mi a day, a requirement that can be easily met by today's technology.

The appropriate infrastructure must also be in place for electric vehicles to become more popular. The issues related to infrastructure are

- Battery charging facilities. Residential and public charging facilities/stations
- Standardization of electric vehicle plugs, cords, outlets, and safety issues
- Sales and distribution
- Service and technical support
- Parts supply

The current initial cost of an electric vehicle is also a big disadvantage for the electric vehicle market. The replacement of the batteries even for hybrid electric vehicles is quite expensive, added to which is the limited life issue of these batteries. The cost of electric vehicles will come down as volume goes up, but in the meantime subsidies and incentives from the government can create a momentum.

The increasing use of electric vehicles will improve the job prospects of electrical, mechanical, and chemical engineers. The new jobs related to electric vehicle will be in the areas of

- *Power electronics and motor drives*: Design and development of the electrical systems of an electric vehicle
- *Power generation*: Additional resources required for increased utility demand due to electric vehicle usage
- *Packaging and cooling*: Design of components and systems for thermal management of electric powertrain components
- *Electric vehicle infrastructure*: Design and development of battery charging stations, hydrogen generation, and storage and distribution systems

References

1. California Air Resources Board Office of Strategic Planning, Air-pollution transportation linkage, 1989.
2. E.H. Wakefield, *History of Electric Automobile*, SAE International, Warrendale, PA, 1994.
3. T.D. Gillespie, *Fundamentals of Vehicle Dynamics*, SAE International, Warrendale, PA, 1992.
4. J.M. Miller, *Propulsions Systems for Hybrid Vehicles*, Institute of Electrical Engineers, London, U.K., 2004.
5. M.H. Westbrook, *The Electric Car*, The Institute of Electrical Engineers, London, U.K. and SAE International, Warrendale, PA, 2001.
6. R. Hodkinson and J. Fenton, *Lightweight Electric/Hybrid Vehicle Design*, SAE International, Warrendale, PA, 2001.
7. U. Bossel, Well-to-wheel studies, heating values, and the energy conservation principle, European Fuel Cell Forum, October 22, 2003, accessible online: www.efcf.com/ reports (E10)
8. F. Kreith, R.E. West, and B.E. Isler, Efficiency of advanced ground transportation technologies, *Transactions of the ASME, Journal of Energy Resources Technology*, 124, 173–179, September 2002.
9. http://www.transportation.anl.gov/modeling_simulation/GREET/index.html (accessed June 24, 2009).
10. S. Chanda, Powertrain sizing and energy usage adaptation strategy for plug-in hybrid electric vehicles, MS thesis, University of Akron, Akron, OH, 2008.
11. The Energy Foundation, 2001 annual report.

Assignment

Search through reference materials and write a short report on the following topics:

1. Commercial and research electric and hybrid vehicle programs around the world over the last 5 years describing the various programs, goals, power range, motor used, type of IC engine, battery source, etc.
2. Case study of a recent electric and hybrid vehicles.
3. State and/or Federal legislations and standardizations.

2

Vehicle Mechanics

The fundamentals of a vehicle's design are embedded in the basic mechanics of physics, particularly in Newton's second law of motion, which relates force and acceleration. Newton's second law states that the acceleration of an object is proportional to the net force exerted on it. The object accelerates when the net force is nonzero, where the term net force refers to the resultant of the forces acting on the object. In the vehicle system, several forces act on it with the resultant or net force dictating the motion according to Newton's second law. A vehicle propels forward with the aid of the force delivered by the propulsion unit overcoming the resisting forces due to gravity, and air and tire resistance. The acceleration and speed of the vehicle depends on the power available from the traction unit and the existing road and aerodynamic conditions. The acceleration also depends on the composite mass of the vehicle including the propulsion unit, all mechanical and electrical components, and the batteries.

A vehicle is designed based on certain given specifications and requirements. Furthermore, the electric and hybrid vehicle system is large and complex involving multidisciplinary knowledge. The key to designing such a large system is to divide and conquer. The system level perspective helps in mastering the design skills for a complex system where the broad requirements are first determined and then system components are designed with more focused guidelines. For example, first the power and energy requirement from the propulsion unit is determined from a given set of vehicle cruising and acceleration specifications. The component level design begins in the second stage where the propulsion unit, the energy source, and other auxiliary units are specified and designed. In this stage, the electrical and mechanical engineers start designing the electric motor for electric vehicles or the combination of electric motor and internal combustion engine for hybrid electric vehicles. The power electronics engineers design the power conversion unit, which links the energy source with the electric motor. The controls engineer works in conjunction with the power electronics engineer to develop the propulsion control system. The chemists and the chemical engineers have the primary responsibility of designing the energy source based on the energy requirement and guidelines of the vehicle manufacturer. Many of the component designs are carried out in an iterative manner, where various designers have to interact with each other to ensure that the design goals are met.

The design is an iterative process, which starts with some known factors and other educated guesses or assumptions to be followed by scientific analysis in order to verify that the requirements are met. In this chapter, we will develop the tools for scientific analysis of vehicle mechanics based on Newton's second law of motion. After defining and describing a roadway, the vehicle kinetics issues will be addressed. The roadway and kinetics will be linked to establish the equation for the force required from the propulsion unit. The force from the propulsion unit, which can be an electric motor or an internal combustion engine, or a combination of the two, is known as the *tractive force*. Once force requirement is defined, one would proceed to calculate the power and energy required for a vehicle under consideration. The emphasis of study in this book is on the broad design goals, such as finding the power and energy requirement, and predicting the range for a given energy source, thereby maintaining a top level perspective. The design details of a subsystem are beyond the scope of this book and the readers are referred to the literature of the respective areas for further details.

2.1 Roadway Fundamentals

A vehicle moves on a level road and also up and down the slope of a roadway. We can simplify our description of the roadway by considering a straight roadway, since the horizontal maneuvering has minimal impact on the force and power requirement of the propulsion unit. Furthermore, we will define a tangential coordinate system that moves along with the vehicle with respect to a fixed two-dimensional system. The roadway description will be utilized to calculate the distance traversed by a vehicle along the roadway.

The fixed coordinate system is attached to the earth such that the force of gravity is perpendicular to the unit vector $\bar{i}_F$. Let us consider a straight roadway, i.e., the steering wheel is locked straight along the x_F-direction. The roadway is then on the $x_F y_F$ plane of the fixed coordinate system (Figure 2.1).

The two-dimensional roadway can be described as $y_F = f(x_F)$. The *roadway position vector* $\bar{r}(x_F)$ between two points a and b along the horizontal direction is

$$\bar{r}(x_F) = x_F \bar{i}_F + f(x_F)\bar{j}_F \quad \text{for } a \le x_F \le b$$

The direction of motion and distance traversed by the vehicle is easier to express in terms of the *tangent vector* of the roadway position vector given as

$$\bar{T}(x_F) = \frac{d\bar{r}}{dx_F} = \bar{i}_F + \frac{df}{dx_F}\bar{j}_F$$

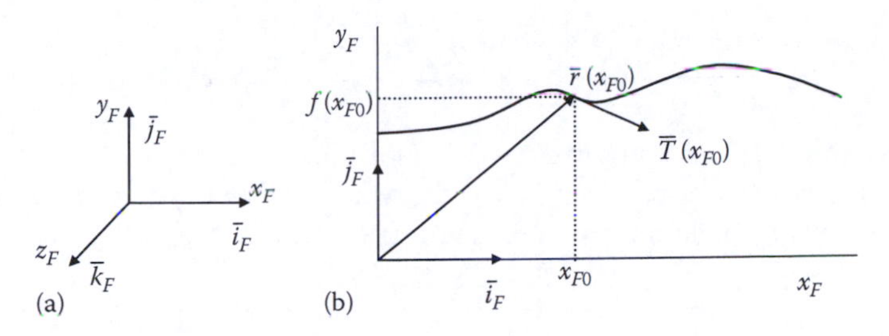

FIGURE 2.1
(a) Fixed coordinate system. (b) Roadway on the fixed coordinate system.

The distance-norm of the tangent vector $\|\bar{T}(x_F)\|$ is

$$\left\|\bar{T}(x_F)\right\| = \sqrt{1+\left[\frac{df}{dx_F}\right]^2}$$

The *tangential roadway length* s is the distance traversed along the roadway. Mathematically, s is the arc length of $y_F = f(x_F)$ over $a \le x_F \le b$. Therefore,

$$s = \int_a^b \left\|\bar{T}(x_F)\right\| dx_F$$

The roadway percent grade can be described as a function of the roadway as

$$\beta(x_F) = \tan^{-1}\left[\frac{df(x_F)}{dx_F}\right]$$

The *average roadway percent grade* is the vertical rise per 100 horizontal distance of roadway with both distances expressed in the same unit (Figure 2.2). The angle β of the roadway associated with the slope or grade is the angle between the tangent vector and the horizontal axis x_F. If Δy is the vertical rise in meters, then

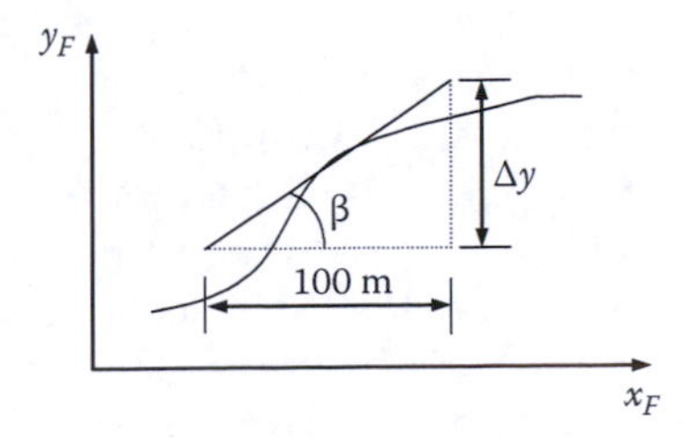

FIGURE 2.2
Grade of the roadway.

$$\% \text{ grade} = \frac{\Delta y}{100\,\text{m}} 100\% = \Delta y\%$$

The tangent of the slope angle is $\tan \beta = \Delta y/100\,\text{m}$.

The percent grade or β is greater than zero when the vehicle is on an upward slope and is less than zero when the vehicle is going downhill.

Exercise 2.1

A straight roadway has a profile in the $x_F y_F$ plane given by $f(x_F) = 3.9\sqrt{x_F}$ for $0 \le x_F \le 2$ mi. x_F and y_F are given in feet.

(a) Plot the roadway, (b) find $\beta(x_F)$, (c) find the percent grade at $x_F = 1$ mi, and (d) find the tangential road length between 0 and 2 mi.

Answer. (b) $\tan^{-1} \frac{1.95}{\sqrt{x_F}}$; (c) 2.68%, and (d) 10,580 ft.

2.2 Laws of Motion

Newton's second law of motion can be expressed in equation form as

$$\sum_i \bar{F}_i = m\bar{a}$$

where

$\sum_i \bar{F}_i$ is the net force

m is the mass

$\bar{a}$ is the acceleration

The law is applied to the vehicle by considering a number of objects located at the several points of contact of the vehicle with the outside world on which the individual forces act. Examples of such points of contact are the front and rear wheels touching the roadway surface, the frontal area that meets the force from the air resistance, etc. We shall simplify the problem by merging all these points of contact into one location at the center of gravity (cg) of the vehicle, which is justified since the extent of the object is immaterial. For all the force calculations to follow, we shall consider the vehicle to be a particle mass located at the cg of the vehicle. The cg can be considered to be within the vehicle, as shown in Figure 2.3.

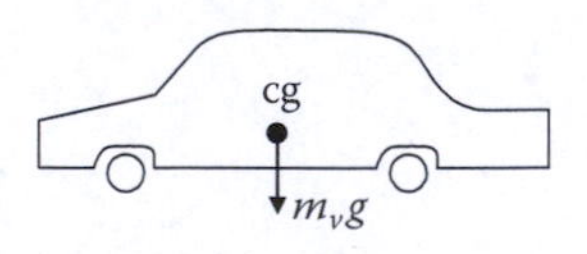

FIGURE 2.3
Center of gravity (cg) of a vehicle.

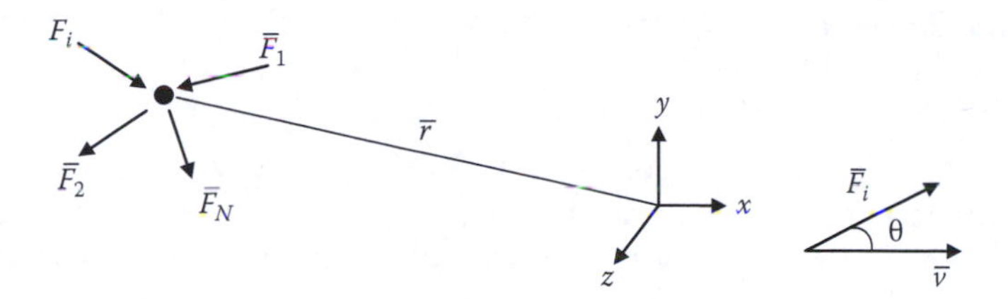

FIGURE 2.4
Forces on a particle.

The particle motion is described by the particle velocity and acceleration characteristics. For the position vector $\overline{r}$ for the particle mass on which several forces are working as shown in Figure 2.4, the velocity v and acceleration a are

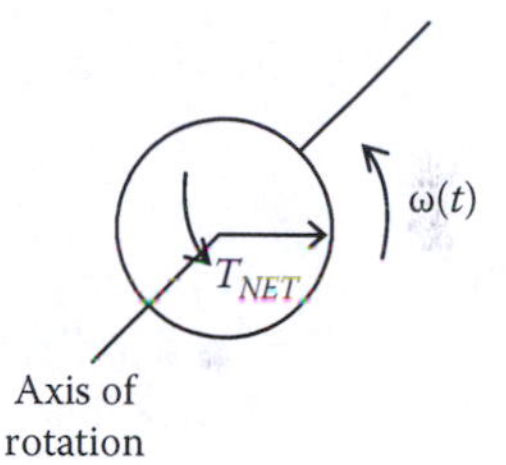

FIGURE 2.5
Rigid body rotation.

$$\overline{v} = \frac{d\overline{r}}{dt} \quad \text{and} \quad \overline{a} = \frac{d\overline{v}}{dt}$$

The power input to the particle for the ith force is

$$P_i = \overline{F}_i \cdot \overline{v} = |\overline{F}_i||\overline{v}|\cos\theta$$

where θ is the angle between F_i and the resultant velocity v.

For a rigid body rotating about a fixed axis, the equivalent terms relating motion and power are torque, angular velocity, and angular acceleration. Let there be a number of independent torques acting on a rigid body causing it to rotate about one of its central principal axes of inertia as shown in Figure 2.5. If J (kg m^2) is the polar moment of inertia of the rigid body, then the rotational form of Newton's second law of motion in the scalar form is

$$\sum_i T_i = T_{NET} = J\alpha$$

where

$$\omega = \text{angular speed (rad/s)}$$

$$\alpha = \frac{d\omega}{dt} = \frac{d^2\theta}{dt^2} = \text{angular acceleration (rad/s}^2\text{)}$$

The power input for the ith torque is $P_i = T_i\omega$.

2.3 Vehicle Kinetics

The tangential direction of forward motion of a vehicle changes with the slope of the roadway. In order to simplify the equations, a tangential coordinate system is defined below so that the forces acting on the vehicle can be defined in through a one-dimensional equation. Let $\bar{u}_T(x_F)$ be the *unit tangent vector* in the fixed coordinate system pointing in the direction of increasing x_F. Therefore,

$$\bar{u}_T(x_F) = \frac{\bar{T}_F(x_F)}{\left\|\bar{T}_F(x_F)\right\|} = \frac{\bar{i}_F + (df/dx_F)\bar{j}_F}{\sqrt{1+[df/dx_F]^2}}$$

The tangential coordinate system shown in Figure 2.6 has the same origin as the fixed coordinate system. The *z*-direction unit vector is the same as in the fixed coordinate system, but the *x*- and *y*-direction vectors are constantly changing with the slope of the roadway.

Newton's second law of motion can now be applied to the cg of a vehicle in the tangential coordinate system as

$$\sum \bar{F}_T = m\bar{a}_T = m\frac{d\bar{v}_T}{dt}$$

where m is the total vehicle mass. In terms of the components of the coordinate system

$$\sum \bar{F}_{xT} = m\frac{d\bar{v}_{xT}}{dt} \quad \text{(component tangent to the road)},$$

$$\sum \bar{F}_{yT} = m\frac{d\bar{v}_{yT}}{dt} \quad \text{(component normal to the road)},$$

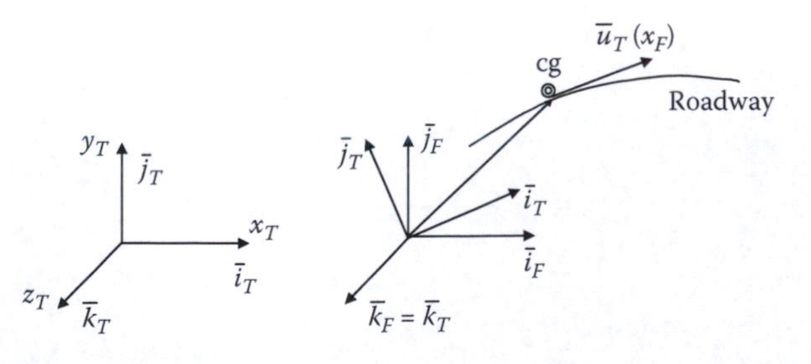

FIGURE 2.6
Tangential coordinate system and the unit tangent vector on a roadway.

$$\sum \bar{F}_{zT} = m\frac{d\bar{v}_{zT}}{dt} = 0 \quad \text{(since motion is assumed confined to the } x\text{–}y \text{ plane).}$$

v_{xT} is the vehicle tangential velocity. The gravitational force in the normal direction is balanced by the road reaction force, and hence, there will be no motion in the y_T normal direction. In other words, the tire always remains in contact with the road. Therefore, the normal velocity v_{yT} is zero. The vehicle motion has been assumed confined to the $x_F y_F$ or $x_T y_T$ plane, and hence, there is neither force nor velocity acting in the z-direction. These justified simplifications allow us to use a one-directional analysis for vehicle propulsion in the x_T-direction. It is shown in the following that the vehicle tractive force and the opposing forces are all in the x_T-direction and hence the vector sign "–" will not be used in the symbols for simplicity.

The propulsion unit of the vehicle exerts the *tractive force* F_{TR} to propel the vehicle forward at a desired velocity. The tractive force must overcome the opposing forces, which are summed together and labeled as the *road load force* F_{RL}. The road load force consists of the gravitational force, rolling resistance of the tires, and the aerodynamic drag force. The road load force is

$$F_{RL} = F_{gxT} + F_{roll} + F_{AD} \tag{2.1}$$

where x_T is the tangential direction along the roadway. The forces acting on the vehicle are shown in Figure 2.7.

The gravitational force depends on the slope of the roadway. The force is positive when climbing a grade and is negative when descending a downgrade roadway. The gravitational force to be overcome by the vehicle moving forward is

$$F_{gxT} = mg\sin\beta \tag{2.2}$$

where

m is the total mass of the vehicle
g is the gravitational acceleration constant
β is the grade angle with respect to the horizon

The rolling resistance is produced by the flattening of the tire at the contact surface with the roadway. In a perfectly round tire, the normal force to the road balances the distributed weight borne by the wheel at the point of contact along the vertical line beneath the axle. When the tire flattens, the instantaneous center of rotation at the wheel

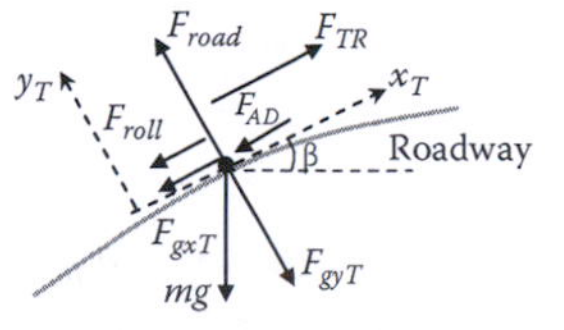

FIGURE 2.7
Forces acting on a vehicle.

moves forward from beneath the axle toward the direction of motion of the vehicle, as shown in Figure 2.8. The weight on the wheel and the road normal force are misaligned due to the flattening of the tire and form a couple that exerts a retarding torque on the wheel. The *rolling resistance force* F_{roll} is the force due to the couple, which opposes the motion of the wheel. The force F_{roll} is tangential to the roadway and always assists in braking or retarding the motion of the vehicle. The tractive force F_{TR} must overcome this force F_{roll} along with the gravitational force and the aerodynamic drag force. The rolling resistance can be minimized by keeping the tires as much inflated as possible. The ratio of the retarding force due to rolling resistance and the vertical load on the wheel is known as the *coefficient of rolling resistance* C_0. The rolling resistance force on a roadway of slope β is

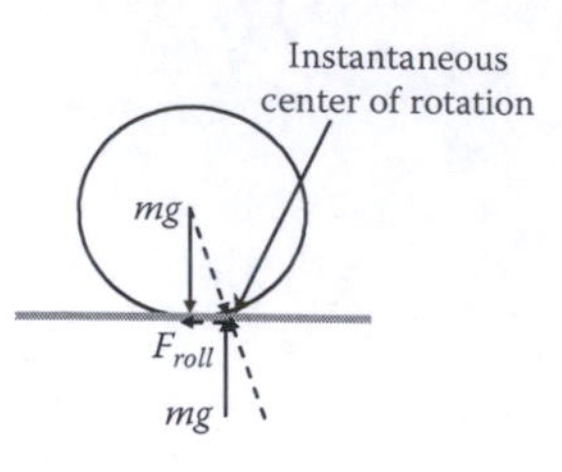

FIGURE 2.8
Rolling resistance force of wheels.

$$F_{roll} = \begin{cases} \text{sgn}[v_{xT}]mg\cos\beta(C_0 + C_1 v_{xT}^2) & \text{if } v_{xT} \neq 0 \\ (F_{TR} - F_{gxT}) & \text{if } v_{xT} = 0 \text{ and } |F_{TR} - F_{gxT}| \leq C_0 mg\cos\beta \\ \text{sgn}[F_{TR} - F_{gxT}](C_0 mg\cos\beta) & \text{if } v_{xT} = 0 \text{ and } |F_{TR} - F_{gxT}| > C_0 mg\cos\beta \end{cases} \tag{2.3}$$

Typically $0.004 < C_0 < .02$ (unitless) and $C_1 \ll C_0$ (s^2/m^2). $C_0 mg$ is the maximum rolling resistance at standstill. The $\text{sgn}[v_{xT}]$ is the signum function given as

$$\text{sgn}[v_{xT}] = \begin{cases} 1 & \text{if } v_{xT} \geq 0 \\ -1 & \text{if } v_{xT} < 0 \end{cases}$$

The aerodynamic drag force is the viscous resistance of the air working against the motion of the vehicle. The force is given by

$$F_{AD} = \text{sgn}[v_{xT}]\{0.5\rho C_D A_F (v_{xT} + v_0)^2\} \tag{2.4}$$

where

ρ is the air density in kg/m^3

C_D is the aerodynamic drag coefficient (dimensionless and typically is $0.2 < C_D < 0.4$)

A_F is the equivalent frontal area of the vehicle

v_0 is the head-wind velocity

2.4 Dynamics of Vehicle Motion

The tractive force is the force supplied by the electric motor in an electric vehicle, and by the combination of electric motor and internal combustion engine in an electric vehicle to overcome the road load. The dynamic equation of motion in the tangential direction is given by

$$k_m m \frac{dv_{xT}}{dt} = F_{TR} - F_{RL} \tag{2.5}$$

where k_m is the rotational inertia coefficient to compensate for the apparent increase in the vehicle's mass due to the onboard rotating mass. Typical value of k_m is between 1.08 and 1.1 and it is dimensionless. Additional explanation on rotational inertia appears in Section 2.7.4. dv_{xT}/dt is the acceleration of the vehicle.

The dynamic equations can be represented in the state space format for simulation of an electric or hybrid electric vehicle system. The motion described by Equation 2.5 is the fundamental equation required for dynamic simulation of the vehicle system. v_{xT} is one of the state variables of the vehicle dynamical system. The second equation needed for modeling and simulation is the velocity equation where either s or x_F can be used as the state variable. The slope of the roadway β will be an input to the simulation model, which may be given in terms of the tangential roadway distance s as $\beta = \beta(s)$ or in terms of the horizontal distance as $\beta = \beta(x_F)$. If β is given in terms of s, then the second state variable equation is

$$\frac{ds}{dt} = v_{xT}. \tag{2.6}$$

If β is given in terms of x_F, then the second state variable equation is

$$\frac{dx_F}{dt} = \frac{v_{xT}}{\sqrt{1 + [df/dx_F]^2}} \tag{2.7}$$

The input–output relational diagram for simulating the vehicle kinetics is shown in Figure 2.9.

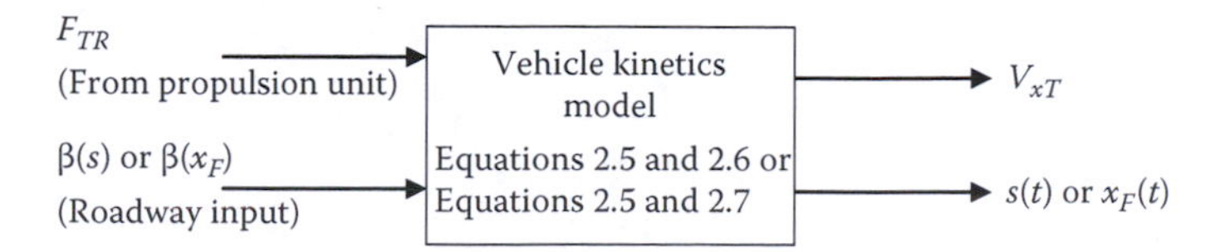

FIGURE 2.9
Modeling of vehicle kinetics and roadway.

2.5 Propulsion Power

The desired power rating of the electric motor or the power required from the combination of electric motor and internal combustion engine (i.e., the propulsion unit) can be calculated from the above equations based on the system constraints of starting acceleration, vehicle rated, and maximum velocity and vehicle gradability. The torque at the wheels of the vehicle can be obtained from the power relation

$$\text{Power} = T_{TR} \cdot \omega_{wh} = F_{TR} \cdot v_{xT} \text{ [W]} \tag{2.8}$$

where
- T_{TR} is the tractive torque in N m
- ω_{wh} is the angular velocity of the wheel in rad/s
- F_{TR} is in N
- v_{xT} is in m/s

The angular velocity and the vehicle speed is related by

$$v_{xT} = \omega_{wh} \cdot r_{wh} \tag{2.9}$$

where r_{wh} is the radius of the wheel in meters. The losses between the propulsion unit and wheels in the transmission and the differential have to be appropriately accounted for while specifying the power requirement of the propulsion unit.

A big advantage of an electrically driven propulsion system is the elimination of multiple gears to match the vehicle speed and the engine speed. The wide speed range operation of electric motors enabled by power electronics control makes it possible to use a single gear-ratio transmission for instantaneous matching of the available motor torque T_{motor} with the desired tractive torque T_{TR}. The gear ratio and the size depend on the maximum motor speed, the maximum vehicle speed, and the wheel radius. A higher motor speed relative to the vehicle speed means a higher gear-ratio, larger size, and higher cost. However, higher motor speed is also desired in order to increase the power density of the motor. Therefore, a compromise is necessary between the maximum motor speed and the gear-ratio to optimize the cost. Planetary gears are typically used for electric vehicles with the gear ratio rarely exceeding 10.

2.5.1 Force–Velocity Characteristics

Having identified the fundamental forces and the associated dynamics for electric and hybrid-electric vehicles, let us now attempt to relate these equations

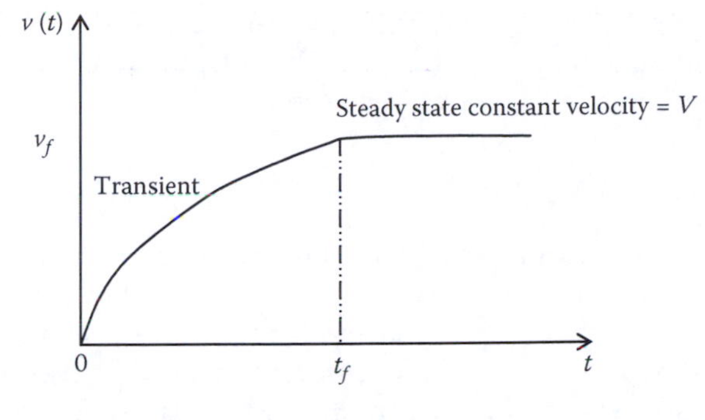

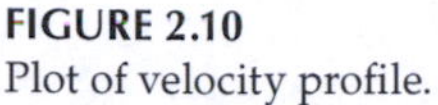
FIGURE 2.10
Plot of velocity profile.

to a vehicle design specification and requirements. For an efficient design of the propulsion unit, the designer must know the force required to accelerate the vehicle to a cruising speed within a certain time and then to propel the vehicle forward at the rated steady-state cruising speed and at the maximum speed on a specified slope. The useful design information is contained in the vehicle speed versus time and the steady-state tractive force versus constant velocity characteristics, illustrated in Figures 2.10 and 2.11. In the sections to follow and in the remainder of the book, we will always assume the velocity to be in the tangential direction and denote it by v instead of v_{xT} for simplicity. The steady-state constant velocity will be denoted by the upper case V.

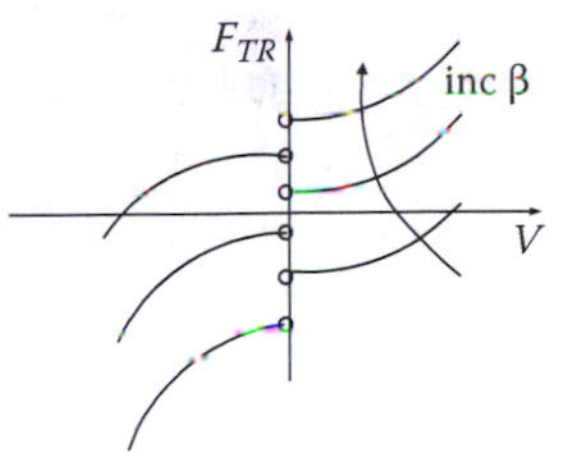

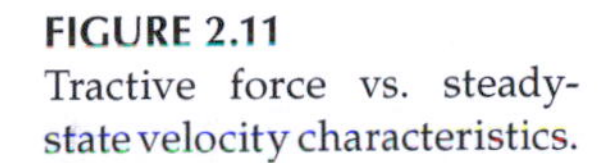
FIGURE 2.11
Tractive force vs. steady-state velocity characteristics.

The tractive force vs. steady-state velocity characteristics can be obtained from the equation of motion Equation 2.5. When steady-state velocity is reached, $dv/dt = 0$; and hence, $\Sigma F = 0$. Therefore, we have

$$F_{TR} - F_{AD} - F_{roll} - F_{gxT} = 0$$

$$\Rightarrow F_{TR} = mg\left[\sin\beta + C_0\,\mathrm{sgn}(V)\right] + \mathrm{sgn}(V)\left[mgC_1 + \frac{\rho}{2}C_D A_F\right]V^2$$

Note that

1. $dF_{TR}/dV = 2V\mathrm{sgn}(V)((\rho C_D A_F/2) + mgC_1) > 0\ \forall V$
2. $\lim_{V \to 0^+} F_{TR} \neq \lim_{V \to 0^-} F_{TR}$

The first equation reveals that the slope of F_{TR} vs. V is always positive, meaning that the force requirement increases more than the velocity increase as the vehicle speed increases, which is due to the aerodynamic drag force opposing the vehicle motion. Also, the discontinuity of the curves at zero velocity is due to the rolling resistance force.

2.5.2 Maximum Gradability

The maximum grade that a vehicle will be able to overcome with the maximum force available from the propulsion unit is an important design criterion as well as performance measure. The vehicle is expected to move forward very slowly when climbing a steep slope, and hence we can make the following assumptions for maximum gradability:

1. The vehicle moves very slowly $\Rightarrow v \cong 0$.
2. F_{AD}, F_{roll} are negligible.
3. The vehicle is not accelerating, i.e., $dv/dt = 0$.
4. F_{TR} is the maximum tractive force delivered by motor (or motors) at near zero speed.

At near stall condition, under the above assumptions,

$$\Sigma F = 0 \Rightarrow F_{TR} - F_{gxT} = 0 \Rightarrow F_{TR} = mg\sin\beta$$

Therefore, $\sin\beta = F_{TR}/mg$. The maximum percent grade is

$$\max \%\ \text{grade} = 100\tan\beta$$

$$\Rightarrow \max \%\ \text{grade} = \frac{100F_{TR}}{\sqrt{(mg)^2 - F_{TR}^2}} \tag{2.10}$$

The force diagram for the maximum gradability condition is shown in Figure 2.12.

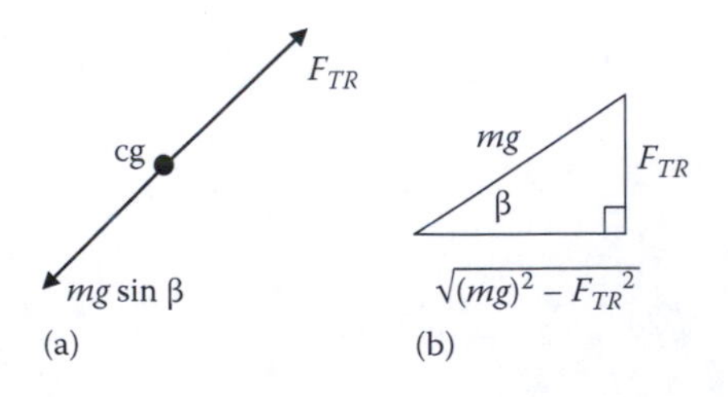

FIGURE 2.12
(a) Force diagram to determine maximum gradability. (b) Tractive force and mg with respect to the grade.

2.6 Velocity and Acceleration

The energy required from the propulsion unit depends on the desired acceleration and the road load force that the vehicle has to overcome. The maximum acceleration is limited by the maximum tractive power available and the roadway condition at the time of vehicle operation. Although the road load force is unknown in a real world roadway, significant insight about the vehicle velocity profile and energy requirement can be obtained through studies of assumed scenarios. The vehicles are typically designed with a certain objective, such as maximum acceleration on a given roadway slope on a typical weather condition. Two simplified scenarios that will set forth the stage for designing electric and hybrid-electric vehicles are discussed in the following.

2.6.1 Constant F_{TR}, Level Road

In the first case, we will assume a level road condition where the propulsion unit for an EV exerts a constant tractive force. The level road condition implies that $\beta(s)=0$. We will assume that the EV is initially at rest, which implies $v(0)=0$. The free body diagram at $t=0$ is shown in Figure 2.13a.

Assume $F_{TR}(0)=F_{TR}>C_0mg$, i.e., the initial tractive force is capable of overcoming the initial rolling resistance. Therefore,

$$\Sigma F(0) = ma(0) = m\frac{dv(0)}{dt}$$

$$\Rightarrow F_{TR} - C_0mg = m\frac{dv(0)}{dt}$$

Since $F_{TR}(0)>C_0mg$, $dv(0)/dt>0$ and the velocity v increases. Once the vehicle starts to move, the forces acting on it changes. At $t>0$,

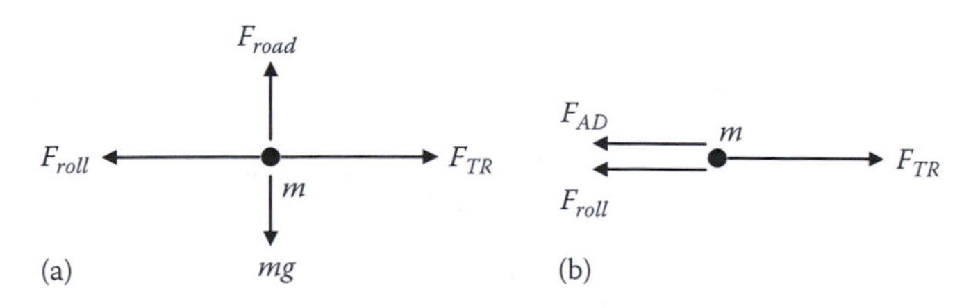

FIGURE 2.13
Forces acting on the vehicle on a level road: (a) free body diagram at $t=0$ and (b) forces on the vehicle at $t>0$.

$$\Sigma F = m\frac{dv}{dt} \Rightarrow F_{TR} - F_{AD} - F_{roll} = m\frac{dv}{dt}$$

$$\Rightarrow F_{TR} - \text{sgn}[v(t)]\frac{\rho}{2}C_D A_F v^2(t) - \text{sgn}[v(t)]mg\left(C_0 + C_1 v^2(t)\right) = m\frac{dv}{dt}$$

Assuming $v(t) > 0$ for $t > 0$ and solving for dv/dt,

$$\frac{dv}{dt} = \left(\frac{F_{TR}}{m} - gC_0\right) - \left[\frac{\rho}{2m}C_D A_F + gC_1\right]v^2$$

Let us define the following constants for a constant F_{TR} acceleration:

$$K_1 = \frac{F_{TR}}{m} - gC_0 > 0$$

$$K_2 = \frac{\rho}{2m}C_D A_F + gC_1 > 0$$

which gives $dv/dt = K_1 - K_2 v^2$.

2.6.1.1 Velocity Profile

The velocity profile for the constant F_{TR}, level road case can be obtained from solving for v from the dv/dt equation above, which gives (Figure 2.14)

$$v(t) = \sqrt{\frac{K_1}{K_2}}\tanh\left(\sqrt{K_1 K_2}\,t\right) \tag{2.11}$$

The terminal velocity can be obtained by taking the limit of the velocity profile as time approaches infinity. The terminal velocity is

$$V_T = \lim_{t\to\infty} v(t) = \sqrt{\frac{K_1}{K_2}} \Rightarrow \sqrt{K_1 K_2} = K_2 V_T$$

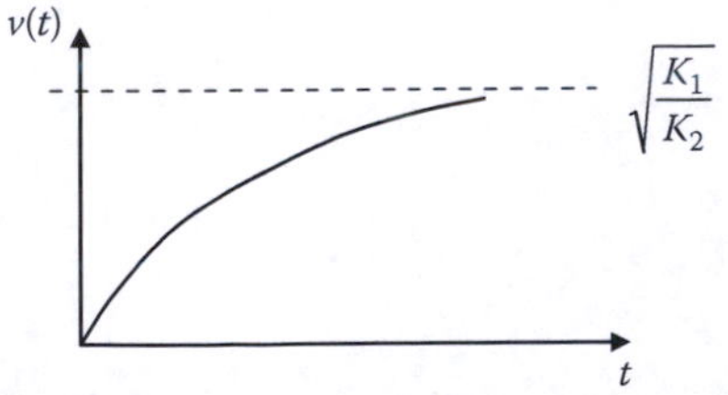

FIGURE 2.14
Velocity profile for a constant F_{TR} on a level road.

2.6.1.2 Distance Traversed

The distance traversed by the vehicle can be obtained from the following relation:

$$\frac{ds(t)}{dt} = v(t) = V_T \tanh(K_2 V_T t)$$

The distance as a function of time is obtained by integrating the above equation:

$$s(t) = \frac{1}{K_2} \ln\left[\cosh(K_2 V_T t)\right] \tag{2.12}$$

The starting acceleration is often specified as 0 to v_f m/s in t_f s, where v_f is the desired velocity at the end of the specified time t_f s. The time to reach the desired velocity and distance traversed during that time is given by

$$t_f = \frac{1}{K_2 V_T} \cosh^{-1}\left[e^{(K_2 s_f)}\right] \tag{2.13}$$

and

$$s_f = \frac{1}{K_2} \ln\left[\cosh\left(K_2 V_T t_f\right)\right] \tag{2.14}$$

respectively.
The desired time can also be expressed as

$$t_f = \frac{1}{\sqrt{K_1 K_2}} \tanh^{-1}\left(\sqrt{\frac{K_2}{K_1}} v_f\right) \tag{2.15}$$

where v_f is the velocity after time t_f.

For example, let t_{V_T} = time to reach 98% of the terminal velocity V_T. Therefore,

$$t_{V_T} = \frac{1}{\sqrt{K_1 K_2}} \tanh^{-1}\left(\frac{0.98 V_T}{V_T}\right)$$

or

$$t_{V_T} = \frac{2.3}{\sqrt{K_1 K_2}} \quad \text{or} \quad \frac{2.3}{K_2 V_T} \tag{2.16}$$

2.6.1.3 *Tractive Power*

The instantaneous tractive power delivered by the propulsion unit is

$$P_{TR}(t) = F_{TR}v(t)$$

Substituting for $v(t)$,

$$P_{TR}(t) = F_{TR}V_T \tanh\left(K_2V_Tt\right)$$

$$\Rightarrow P_{TR}(t) = F_{TR}V_T \tanh\left(\sqrt{K_1K_2}t\right) = P_T \tanh\left(\sqrt{K_1K_2}t\right) \quad (2.17)$$

The terminal power can be expressed as $P_T = F_{TR}V_T$. The tractive power required to reach the desired velocity v_f over the acceleration interval Δt (refer to Figure 2.15) is

$$P_{TRpk} = P_T \tanh\left(\sqrt{K_1K_2}t_f\right) \quad (2.18)$$

The mean tractive power over the acceleration interval Δt is

$$\overline{P_{TR}} = \frac{1}{t_f}\int_0^{t_f} P_{TR}(t)dt$$

$$\Rightarrow \overline{P_{TR}} = \frac{P_T}{t_f}\frac{1}{\sqrt{K_1K_2}}\ln\left[\cosh\left(\sqrt{K_1K_2}t_f\right)\right] \quad (2.19)$$

2.6.1.4 *Energy Required*

The energy requirement for a given acceleration and constant steady state velocity is necessary for the design and selection of the energy source or

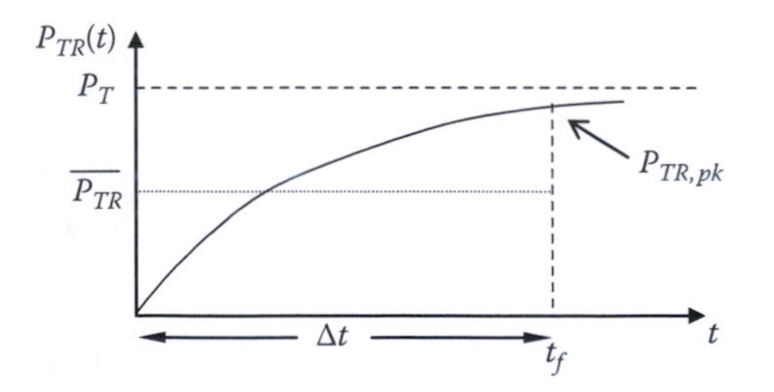

FIGURE 2.15
Acceleration interval $\Delta t = t_f - 0$.

batteries to cover a certain distance. The rate of change of energy is the tractive power given as

$$P_{TR}(t) = \frac{de_{TR}}{dt}$$

where e_{TR} is the instantaneous tractive energy. The energy required or energy change during an interval of the vehicle can be obtained from the integration of the instantaneous power equation as

$$\int_{e_{TR}(0)}^{e_{TR}(t_f)} de_{TR} = \int_{t=0}^{t_f} P_{TR} dt$$

$$\Rightarrow \Delta e_{TR} = t_f \overline{P_{TR}} \tag{2.20}$$

Example 2.1

An electric vehicle has the following parameter values:
$m = 800$ kg, $C_D = 0.2$, $A_F = 2.2$ m^2, $C_0 = 0.008$, $C_1 = 1 \times 0.6 \times 10^{-6}$ s^2/m^2,
Also, take density of air $\rho = 1.18$ kg/m^3, and acceleration due to gravity $g = 9.81$ m/s^2,
The vehicle is on level road. It accelerates from 0 to 65 mi/h in 10 s such that its velocity profile is given by
$v(t) = 0.29055t^2$ for $0 \le t \le 10$ s.

(a) Calculate $F_{TR}(t)$ for $0 \le t \le 10$ s.
(b) Calculate $P_{TR}(t)$ for $0 \le t \le 10$ s.
(c) Calculate the energy loss due to nonconservative forces E_{loss}.
(d) Calculate Δe_{TR}.

Solution

(a) From the force balance equation,

$$F_{TR} - F_{AD} - F_{roll} = m\frac{dv}{dt}$$

$$\Rightarrow F_{TR}(t) = m\frac{dv}{dt} + \frac{\rho}{2}C_D A_F v^2 + mg\left(C_0 + C_1 v^2\right)$$

$$= 464.88t + 0.02192t^4 + 62.78 \text{ N}$$

(b) The instantaneous power is

$$P_{TR}(t) = F_{TR}(t) * v(t)$$

$$= 135.07t^3 + 0.00637t^6 + 18.24t^2 \text{ W}$$

(c) The energy lost due to nonconservative forces

$$E_{loss} = \int_0^{10} v\left(F_{AD} + F_{roll}\right)dt = \int_0^{10} 0.29055t^2(0.0219t^4 + 62.78)dt$$

$$= 15{,}180 \text{ J}$$

(d) The kinetic energy of the vehicle is

$$\Delta \text{KE} = \frac{1}{2}m\left[v(10)^2 - v(0)^2\right] = 337{,}677 \text{ J}$$

Therefore, the change in tractive energy is

$$\Delta e_{TR} = 15{,}180 + 337{,}677$$

$$= 352{,}857 \text{ J}$$

Exercise 2.2

An electric vehicle has the following parameter values $\rho = 1.16\,\text{kg/m}^3$, $m = 692\,\text{kg}$, $C_D = 0.2$, $A_F = 2\,\text{m}^2$, $g = 9.81\,\text{m/s}^2$, $C_0 = 0.009$, and $C_1 = 1.75 \times 10^{-6}\,\text{s}^2/\text{m}^2$. The electric vehicle undergoes constant F_{TR} acceleration on a level road starting from rest at $t = 0$. The maximum continuous F_{TR} that the electric motor is capable of delivering to the wheels is 1548 N.

(a) Find V_T (F_{TR}) and plot it.
(b) If $F_{TR} = 350\,\text{N}$, (i) find V_T, (ii) plot $v(t)$ for $t \geq 0$, (iii) find t_{VT}, (iv) calculate the time required to accelerate from 0 to 60 mi/h, (v) calculate P_{TRpk}, $\overline{P_{TR}}$, Δe_{TR} corresponding to acceleration to $0.98V_T$.

Answer. (a) $V_T\left(F_{TR}\right) = 53.2\sqrt{1.45 \times 10^{-3} F_{TR} - 0.0883}\ \text{m/s}$, (b) (i) 34.4 m/s, (ii) $v(t) = 34.4 \tanh\ (1.22 \times 10^{-2}t)\,[m/s]$, (iii) 189 s, (iv) 85.6 s, (v) $P_{TRpk} = 11.8\,\text{kW}$, $\overline{P_{TR}} = 8.46\,\text{kW}$, $\Delta e_{TR} = 1.61\,\text{MJ}$.

2.6.2 Nonconstant F_{TR}, General Acceleration

In the general case, with a nonconstant F_{TR} and an arbitrary velocity profile as shown in Figure 2.16, the force can be calculated as

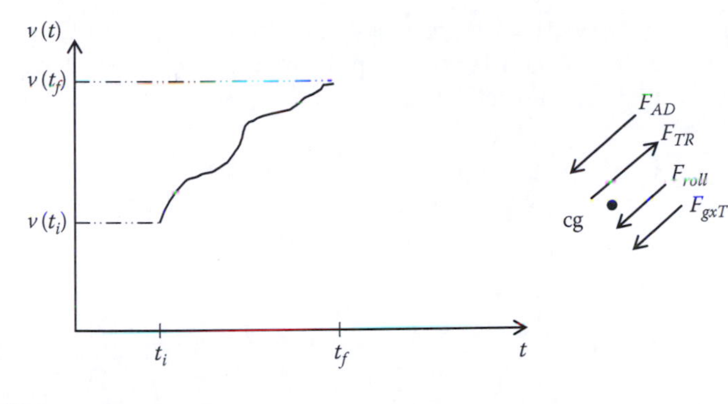

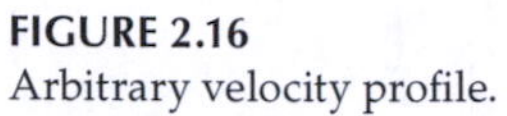

FIGURE 2.16
Arbitrary velocity profile.

$$\Sigma F = m\frac{dv}{dt}$$

$$\Rightarrow F_{TR} - F_{AD} - F_{roll} - F_{gxT} = m\frac{dv}{dt}$$

$$\Rightarrow F_{TR} = m\frac{dv}{dt} + mg\sin\beta + F_{AD} + F_{roll}$$

$$= m\frac{dv}{dt} + mg\sin\beta + \left[mgC_1 + \frac{\rho}{2}A_F C_D\right]v^2 + mgC_0 \tag{2.21}$$

The instantaneous tractive power $P_{TR}(t)$ is

$$P_{TR}(t) = F_{TR}(t)v(t)$$

$$= mv\frac{dv}{dt} + v\left(F_{gxT} + F_{AD} + F_{roll}\right) \tag{2.22}$$

The change in tractive energy Δe_{TR} is

$$\Delta e_{TR} = \int_{t_i}^{t_f} P_{TR}(t)dt$$

$$= m\int_{v(t_i)}^{v(t_f)} v\,dv + \int_{t_i}^{t_f}(v)F_{gxT}dt + \int_{t_i}^{t_f}(v)\left(F_{AD} + F_{roll}\right)dt \tag{2.23}$$

The energy supplied by the propulsion unit is converted into various forms of energy, some of which are stored in the vehicle system, while others are lost due to the nonconstructive forces. It is interesting to note the type of energy associated with each term in Equation 2.23. Let us consider the first term on the right side of Equation 2.23:

$$m\int_{v(t_i)}^{v(t_f)} v\,dv = \frac{1}{2}m\left[v^2(t_f)-v^2(t_i)\right] = \Delta \quad \text{(kinetic energy)}$$

Also,

$$\int_{t_i}^{t_f} (v)F_{gxT}dt = mg\int_{t_i}^{t_f} v\sin\beta dt = mg\int_{s(t_i)}^{s(t_f)} \sin\beta ds = mg\int_{f(t_i)}^{f(t_f)} df$$

$$= mg\left[f(t_f)-f(t_i)\right]$$

$$= \Delta \quad \text{(potential energy)}$$

The above term represents change in vertical displacement multiplied by mg, which is the change in potential energy.

The third and fourth terms on the right side of Equation 2.23 represents the energy required to overcome the nonconstructive forces that include the rolling resistance and the aerodynamic drag force. The energy represented in these terms is essentially the loss term. Therefore,

$$\int_{t_i}^{t_f} (v)\left(F_{AD}+F_{roll}\right)dt = E_{loss}$$

Let $K_3 = mgC_0$, $K_4 = mgC_1 + (\rho/2)C_D A_F$. For, $v(t) > 0$, $t_i \le t \le t_f$,

$$E_{loss} = K_3\int_{t_i}^{t_f} \frac{ds}{dt}dt + K_4\int_{t_i}^{t_f} v^3 dt$$

$$= K_3\Delta s + K_4\int_{t_i}^{t_f} v^3 dt$$

In summary, we can write

$$\Delta e_{TR} = \frac{1}{2}m\left[v^2(t_f)-v^2(t_i)\right] + mg\left[f(t_f)-f(t_i)\right] + \int_{t_i}^{t_f} (v)(F_{AD}+F_{roll})dt$$

or

$$\Delta e_{TR} = \Delta(\text{Kinetic Energy}) + \Delta(\text{Potential Energy}) + E_{loss}$$

Exercise 2.3

The vehicle with parameters given in Exercise 2.2 accelerates from 0 to 60 mi/h in 13.0 s for the two following acceleration types: (i) constant F_{TR} and (ii) uniform acceleration.

(a) Plot on the same graph the velocity profile of each acceleration type.
(b) Calculate and compare the tractive energy required for the two types of acceleration.

$$F_{TR} = \text{const.} = 1548 \text{ N}$$

Answer. (b) $\Delta e_{TR} = 0.2752$ MJ for constant F_{TR} and $\Delta e_{TR} = 0.2744$ MJ for uniform acceleration.

2.7 Tire–Road Force Mechanics

A number of forces including the traction force act at the tire-ground contact plane of a moving vehicle. The traction torque from the propulsion system is converted into a traction force through an interaction of the pneumatic tire and the road surface at the tire–road interface. The pneumatic tire is designed to perform several functions of a ground vehicle including support of the vehicle weight and provide traction for driving and braking. The performance potential of a vehicle depends not only on the characteristics of the propulsion system and transmission, but also on the maximum traction (or braking) force that can be sustained at the tire–road interface. The maximum friction force sustainable is determined by the friction coefficient of the pneumatic tire, which depends on its road adhesion characteristics and the normal load on the drive axle or axles. The smaller of the propulsion system traction (or braking) force and the maximum sustainable friction force determines the overall behavior of the vehicle. The traction properties at the tire–road interface are fundamental in evaluating the dynamic behavior of a vehicle. The study of tire mechanics helps understand the mechanism of force transmission between the axle and the road. The subject is complex and several theories have been proposed to provide the understanding of the process involved [1–3]; a basic overview is presented in this book.

2.7.1 Slip

The friction forces that are fundamental to vehicle traction depend on the difference between the tire rolling speed and its linear speed of travel. The tire rolling speed is related to the wheel angular speed and is given by

$$v_{tire} = \omega_{wh} r_{wh}$$

where

ω_{wh} is the wheel speed

r_{wh} is the driven wheel radius

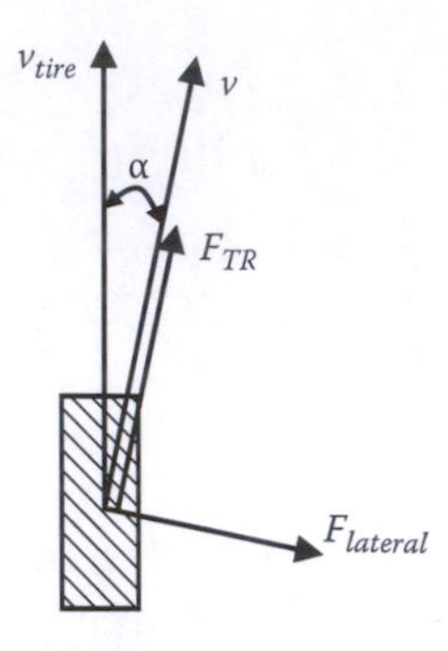

FIGURE 2.17
Speed and forces in the tire–road contact patch area.

The wheel speed of travel is equivalent to the vehicle translatory speed v. Vehicle linear velocity v and the tire speed v_{tire} differ in magnitude and direction as shown in Figure 2.17. The angle between the tire velocity and the vehicle velocity is known as slip angle α, shown in Figure 2.17 with exaggeration. The difference in speed itself is generated due to the tire properties and the interaction of forces at the tire–road interface. The tire forces do not act through a point, but are distributed over the contact patch area of the tire with the road surface. Furthermore, the forces are not uniform along the contact patch in either longitudinal or lateral directions. The vehicle traction force F_{TR} works in the longitudinal direction. Longitudinal and lateral forces will be discussed further later; first, let us discuss the vehicle slip in the longitudinal direction.

The tire treads gradually enter the contact patch of a moving vehicle when traction torque is applied to the driven wheel. The tire treads in front end of the contact patch undergoes high levels of compression compared to the rear end due to the traction torque applied at the axle. The rolling of the wheels and the traction torque also causes a shift in instantaneous center of rotation. The sidewall of the tire simultaneously goes through shear deformation. The distance that the tire travels when subject to propulsion torque will be less than that in a free rolling tire due to the compression. This results in a difference between the wheel angular speed and vehicle speed, which is what is known as *longitudinal slip* or simply *slip*. Slip is defined as the ratio of linear vehicle velocity and the spin velocity of the wheel. Mathematically, slip s of a vehicle is given as

$$s = \left(1 - \frac{v}{\omega_{wh} r_{wh}}\right) \tag{2.24}$$

where v is the vehicle velocity. Since vehicle velocity v is smaller than wheel angular velocity converted to linear velocity $\omega_{wh} r_{wh}$, the slip is a positive number between 0 and 1.0 for propulsion. Slip is often expressed as a percentage.

The vehicle velocity is then given by

$$v = \omega_{wh} r_{wh} (1 - s)$$

During vehicle braking, the tire is subject to similar compression due to the applied braking torque, and a difference in vehicle speed and wheel angular speed occurs enabling the vehicle stop. The slip during braking is given as

$$s = \left(1 - \frac{\omega_{wh} r_{wh}}{v}\right) \tag{2.25}$$

2.7.2 Traction Force at Tire–Road Interface

The rotary motion of the axles is converted to linear motion of the vehicle at the tire–road interface to enable vehicle motion. When a torque is applied at a driven wheel through the axle of a vehicle, friction forces proportional to the traction torque are developed at the tire–road interface. The forces on the tire are not localized at a point as mentioned previously, but are the resultant of the normal and shear stresses exerted by each element of the tire tread distributed in the contact patch. The stress under the tire is distributed non-uniformly in the x and z directions along the two-dimensional contact patch at the tire–road interface; this two-dimensional stress distribution gives rise to both longitudinal and lateral forces. The longitudinal and lateral forces can be obtained by integrating the shear stress in the contact patch over the contact area. The longitudinal force is the traction force responsible for the forward velocity of the vehicle. We will focus our discussion on the longitudinal force; this force has been referred to as traction force F_{TR} earlier in the chapter.

The forces on a wheel and the stress distribution at the tire–road interface are shown in Figure 2.18. The stress distribution and the resultant forces are shown in relation with the contact patch area at the tire-road interface. The traction torque T_{TR} working around the axis of rotation of the wheel results in the wheel angular velocity ω_{wh}; this traction torque is primarily responsible for the distributed pressure or stress at the tire–road interface. The longitudinal stress distribution over the contact patch is shown to have generated the traction force. In addition, there is normal stress due the weight borne at the axle of wheel, which works in the vertical y-direction. The vertical or normal pressure shifts forward due to the deformation of the tire at the tire–road interface. Consequently, the centroid of the normal road reaction force is shifted forward and is not aligned with the spin axis of the wheel. This results in the rolling resistance force F_{roll}. Figure 2.18 also shows the normal stress distribution over the contact patch area of the rolling tire. The normal stress distribution integrated over the contact patch area results in the

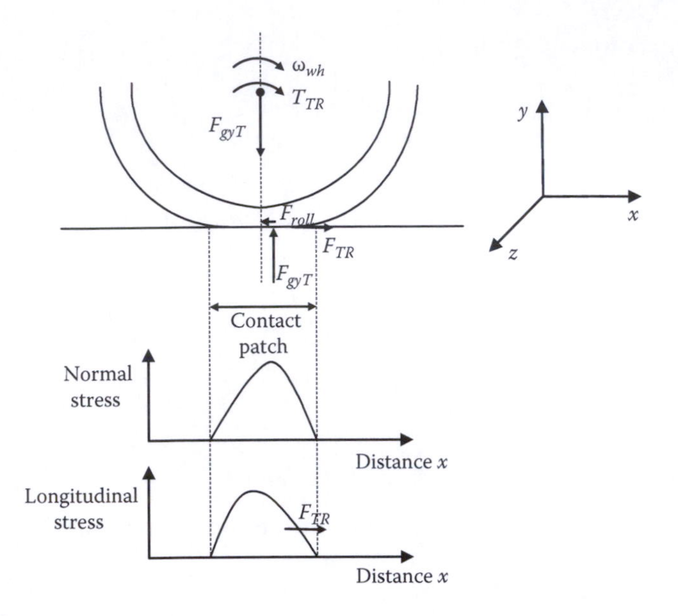

FIGURE 2.18
Forces on the wheel and stress distribution at the tire–road interface.

normal road reaction force F_{gyT}. The normal force F_{gyT} plays a strong role in determining the traction force limit, which is addressed later in Section 2.7.5.

2.7.3 Force Transmission at Tire–Road Interface

The tire–road interface can be thought of a gear mechanism responsible for the generation of the traction force. The conversion of rotary motion of the wheels to linear motion of the vehicle can be compared to the ball–screw arrangement shown in Figure 2.19, which is a gear mechanism used for force transmission along with conversion of rotary motion to linear motion. Let us consider that power is being transmitted from the rotary gear to the linear gear with the latter held stationary. The rotary gear is driven along its spin axis by a rotary motor, which enables the rotary system of motor and gear to

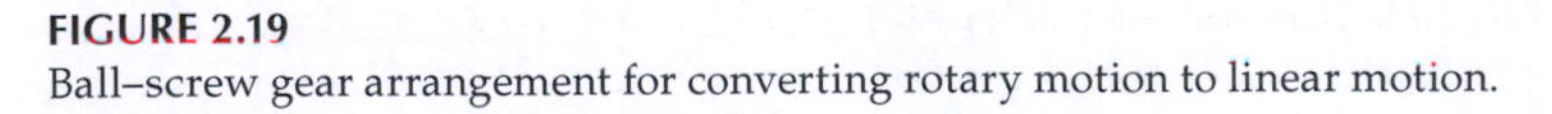

FIGURE 2.19
Ball–screw gear arrangement for converting rotary motion to linear motion.

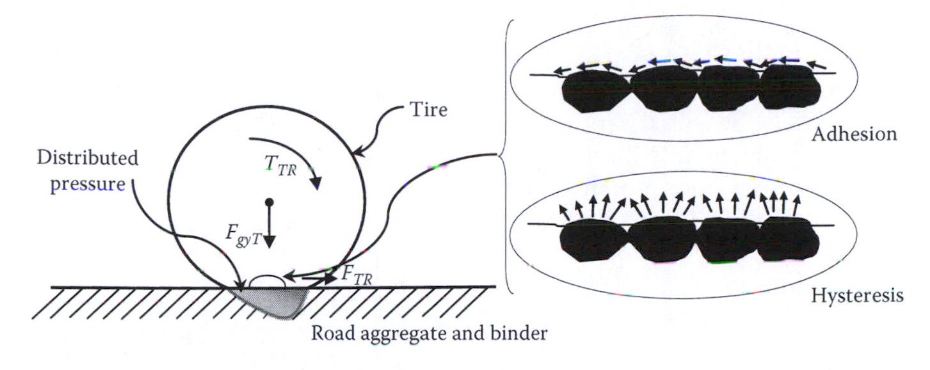

FIGURE 2.20
Forces at tire–road interface on a rolling tire.

move along the linear gear train. In this system, when torque is applied by the rotary motor, friction force is developed in the gear train that helps the rotary system move forward or backward. This type of friction force enables vehicle propulsion along the road surface except that a pneumatic tire and a paved surface is used for developing the friction force has been labeled as traction force F_{TR}.

Two mechanisms contribute to the friction coupling at the tire-road interface in a vehicle; these are *adhesion* and *hysteresis* [3]. Adhesion and hysteresis friction forces generated at the tire–road interface due to the vehicle propulsion torque is shown in Figure 2.20. Adhesion to the road results from the intermolecular bonds between the rubber and road surface aggregates. In the microscopic level, the road surface is not smooth but has the peaks and valley of the surface aggregates as shown in Figure 2.20. The adhesion friction dominates in dry pavements, but the contribution decreases under wet conditions. The hysteresis is the deformation of the rubber when it is sliding over the road aggregates; the friction component comes from the energy loss due to hysteresis. Water does not affect hysteresis friction significantly; tires with improved wet traction are designed with high-hysteresis rubber in the tread.

The friction forces of adhesion and hysteresis depend heavily on the deformation of the tire at the tire–road interface due to the tire properties and the interaction of forces. This tire deformation results in the vehicle longitudinal slip, and the vehicle slip itself is fundamental to vehicle traction. The slip determines the traction force that the tire–road interface can support. Therefore, some slip is essential for vehicle traction, but too much slippage reduces the traction force limit.

2.7.4 Quarter-Car Model

The study of the laws of motion often starts from the consideration of a quarter-car, i.e., one-quarter of the vehicle. The quarter-car model is shown in Figure 2.21. The quarter-car model allows analyzing separately the behavior of driving (powered) wheels and trailing (unpowered) wheels. Moreover, as the chassis and wheel dynamics are treated separately, this model is especially useful to study suspension dynamics if vertical components of forces are taken into account.

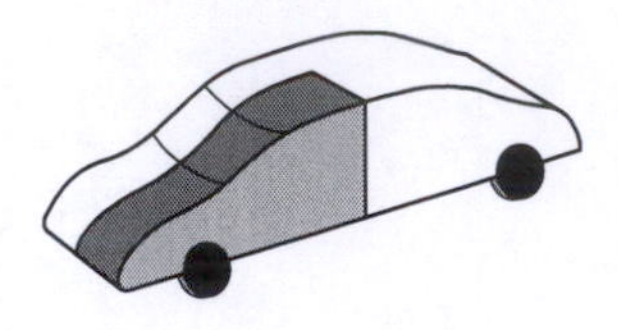

FIGURE 2.21
Quarter car model.

Using the same assumptions as in Section 2.3, the x-component of Newton's second law of motion applied to the chassis is

$$F_{Rw} - F_{AD} - F_{gxT} - F_{Rr} = m\frac{dv}{dt} \tag{2.26}$$

where

F_{Rw} is the reaction force exchanged between the wheel and the chassis
m is the mass of a quarter of the chassis
F_{Rr} is the reaction force between the front and the rear portion of the car, as shown in Figure 2.22

The components on the y-axis are omitted as suspension dynamics will not be analyzed here.

The wheel dynamics can be achieved from the balance of forces and torques as

$$-F_{Rw} - F_{roll} = m_w\frac{dv}{dt} \tag{2.27}$$

$$T_{TR} - F_{gyT}b + F_{roll}r_{wh} = J_w\frac{d\omega_{wh}}{dt} \tag{2.28}$$

where

T_{TR} is the driving or tractive torque
m_w is the mass of the wheel
r_{wh} is the wheel radius

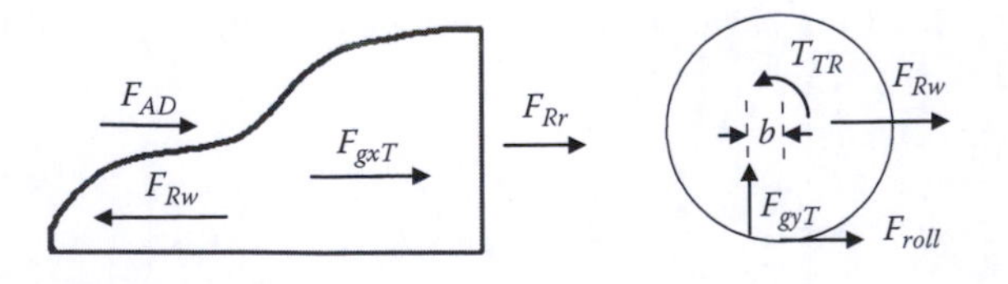

FIGURE 2.22
x-Axis force components of chassis and forces and torques of wheels in the quarter car model.

T_{TR} is zero for the trailing wheel. The tractive force is the force supplied by the electric motor in an electric vehicle and by the combination of electric motor and IC engine in a hybrid vehicle to overcome the road load.

Substituting Equations 2.27 and 2.28 in Equation 2.26, we can write

$$\frac{T_{TR}}{r_{wh}} - F_{gyT}\frac{b}{r_{wh}} - F_{AD} - F_{gxT} - F_{Rr} = (m + m_w)\frac{dv}{dt} + \frac{J_w}{r_{wh}}\frac{d\omega_{wh}}{dt} \tag{2.29}$$

It can be easily concluded that the road load force of Equation 2.29 has the same components as that of Equation 2.1 and that the term T_{TR}/r_{wh} plays the role of the traction force. To accelerate the vehicle, the traction force has to overcome rolling resistance, aerodynamic load, weight component, and the equivalent force of the rear portion of the car. The right-hand side of Equation 2.29 can be written in simplified form as

$$(m + m_w)\frac{dv}{dt} + \frac{J_w}{r_{wh}}\frac{d\omega_{wh}}{dt} = k_m m\frac{dv}{dt} \tag{2.30}$$

The term $k_m m(dv/dt)$ is an equivalent force of inertia accounting for the translating mass of the quarter car plus the inertia of the rotating masses. Finally, the general form of the dynamic equation of motion in the tangential direction assumes the same form as Equation 2.5:

$$k_m m\frac{dv}{dt} = F_{TR} - F_{RL} - F_{Rr} \tag{2.31}$$

where k_m is the rotational inertia coefficient. In Equation 2.30, the tangential speed of the wheel can also differ from the spin velocity of the wheel, according to the value of the slip.

2.7.5 Traction Limit and Control

The maximum rolling resistance force that can be sustained at the tire–road interface depends on the amount of vehicle slip and the vertical load on the wheel. A slip-dependent friction coefficient $\mu(s)$ and the vertical load define the rolling resistance on the wheel as

$$F_{roll}(s) = \mu(s)F_{gyT}$$

where F_{gyT} is the is the normal force on the wheel.

Accordingly, wheel equations can be rewritten as

$$-F_{Rw} - F_{roll}(s) = m_w\frac{dv}{dt}$$

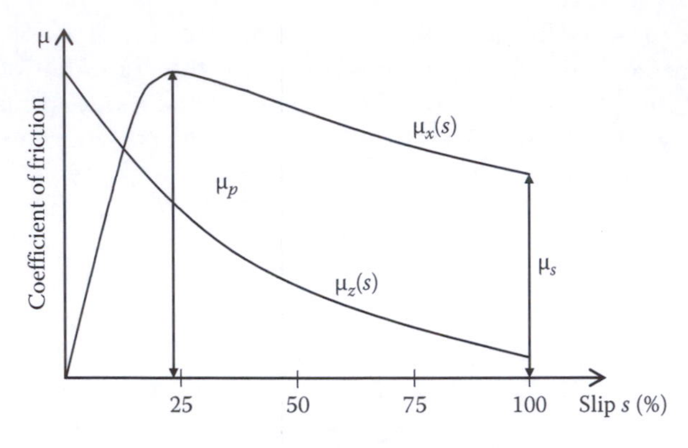

FIGURE 2.23
Longitudinal and lateral traction force coefficient variation.

$$T_{TR} - F_{gyT}b + F_{roll}(s)r_{wh} = J_w \frac{d\omega_{wh}}{dt}$$

Typical friction coefficients for both longitudinal and lateral forces as a function of slip for a fixed slip angle are shown in Figure 2.23. The traction force limit in the contact patch increases almost linearly with slip at the beginning as the majority of tire tread elements are worked most effectively without significant slip. The traction force limit reaches its peak at around 20% of the slip. The relationship between traction effort limit and slip becomes nonlinear beyond the peak; when the slip reaches 100%, the vehicle is sliding only without any linear velocity.

On a dry pavement, the traction force limit peak is about 80%–90% of the vertical load. Peak friction coefficient μ_p defines this limit. The friction coefficient when the slip reaches 100% is known as sliding coefficient μ_s. The friction coefficients vary depending on the road condition, although the shape of the curves remains the same. The nominal peak and sliding friction coefficients for several ground conditions are given in Table 2.1. The peak and

TABLE 2.1
Nominal Longitudinal Peak and Sliding Friction Coefficients

Surface	Peak Coefficient, μ_p	Sliding Coefficient, μ_s
Asphalt and concrete (dry)	0.9	0.75
Concrete (wet)	0.8	0.7
Asphalt (wet)	0.6	0.5
Gravel	0.6	0.55
Snow	0.2	0.15
Ice	0.1	0.07

sliding friction coefficients of an icy road will be much smaller than that of a dry pavement, meaning the wheel will slip much more easily for the same traction torque from the propulsion system. In wheel slippage conditions, the tire–road interface cannot develop the traction force due to the lower friction force limits even though there is traction torque at the wheels. The location of where the peak will appear depends on the slip angle. For smaller slip angles, the peak appears at smaller slip values.

Traction control system (TSC) is incorporated in today's vehicles for controlling both longitudinal and lateral forces. The longitudinal control algorithm limits the traction force to prevent slippage. The lateral control is achieved by keeping the yaw motion to zero. This can be achieved, for example, by controlling the steering angle. The independent slippage prevention on each wheel of a four-wheel-drive vehicle greatly improves the lateral stability. The dynamics of traction systems are highly nonlinear and time-variant; consequently, the design of model-based TSC is also nonlinear and time-varying. The major difficulty for designing such systems is the real-time estimation of friction-coefficient versus slip characteristics for different tires and road surfaces.

2.8 Propulsion System Design

The steady-state maximum velocity, maximum gradability, and the velocity equations can be used in the design stage to specify the power requirement of a particular vehicle.

Let us consider the tractive power requirement for initial acceleration, which plays a significant role in determining the rated power of the propulsion unit. The initial acceleration is specified as 0 to v_f in t_f s. The design problem is to solve for F_{TR} starting with a set of variables including vehicle mass, rolling resistance, aerodynamic drag coefficient, percent grade, wheel radius, etc., some of which are known, while others have to be assumed. The acceleration of the vehicle in terms of these variables is given by

$$a = \frac{dv}{dt} = \frac{F_{TR} - F_{RL}}{m} \tag{2.32}$$

The tractive force output of the electric motor for an electric vehicle or the combination of electric motor and internal combustion engine for a hybrid electric vehicle will be a function of the vehicle velocity. Furthermore, the road load characteristics are also function of velocity, resulting in a transcendental equation to be solved to determine the desired tractive power from the propulsion unit. The other design requirements also play a significant role in determining the tractive power. The problem is best handled by a computer simulation where the various equations of Sections 2.5 and 2.6 can be used iteratively to calculate the tractive force and power requirements from the propulsion unit for the given set of specifications.

The insight gained from the scenarios discussed in this chapter will be used later to specify the power ratings of the electric motor and internal combustion engines for electric and hybrid vehicles in the next chapter.

Problems

2.1 A length of road is straight and it has a profile in the x–y plane described by

$$f(x) = 200\ln\left[7.06\times10^{-4}(x+1416)\right]$$

where $0 \le x \le 3\,\text{mi} = 15{,}840\,\text{ft}$; $f(x)$ and x are in feet.

(a) Plot the road profile in the x–y plane for $0 \le x \le 15{,}840\,\text{ft}$.
(b) Derive an expression for $\beta(x)$. Calculate $\beta(500\,\text{ft})$.
(c) Derive an expression for *percent grade*(x). Calculate *percent grade* (500 ft).
(d) Derive an expression for tangential road length $s(x)$, such that $s(0) = 0$. Calculate $s(500\,\text{ft})$.
(e) Can you find an expression for $x(s)$, i.e., can you express x as a function of s? Show some steps in your attempt.

2.2 A straight Roadway has a profile in the x–y plane given by

$$f(x_f) = 4.1\sqrt{x_f} \quad \text{for } 0 \le x_f \le 2\,\text{mi} = 10{,}560\,\text{ft}.$$

$f(x_f)$ and x_f are in ft.

(a) Derive an expression for $\beta(x_f)$. Calculate $\beta(1\,\text{mi})$.
(b) Calculate the tangential road length, s from 0 to 2 mi.

2.3 An electric vehicle has the following parameter values:

$m = 692\,\text{kg}$, $C_D = 0.2$, $A_F = 2\,\text{m}^2$, $C_0 = 0.009$, $C_1 = 1.75\times10^{-6}\,\text{s}^2/\text{m}^2$.
Also, take $\rho = 1.16\,\text{kg/m}^3$, $g = 9.81\,\text{m/s}^2$

(a) *EV at rest*: The EV is stopped at a stop sign at a point in the road where the grade is +15%. The tractive force of the vehicle is supplied by the vehicle brakes.
  (i) Calculate the tractive force necessary for zero rolling resistance. (The vehicle is at rest.)

(ii) Calculate the minimum tractive force required from the brakes to keep the EV from rolling down the grade.

(b) *EV at constant velocity*: The EV is moving at a constant velocity along a road that has a constant grade of −12%.

(i) Plot, on the same graph, the magnitudes of the tangential gravitational force (F_{gxT}), the aerodynamic drag force (F_{AD}), and the rolling resistance force (F_{roll}) versus velocity for $0 < V \leq 180\,\text{mi/h}$. Over that range of velocity, does F_{gxT} dominate? When does F_{AD} dominate? When does F_{roll} dominate? Label these regions on the graph.

(ii) Derive an expression for the tractive force as a function of velocity. Plot this expression on its own graph. Is the tractive force always in the same direction?

2.4 Showing all steps, derive and plot the velocity profile (i.e., $v(t)$) for constant F_{TR}-constant grade acceleration. (Constant grade means that β is constant but not necessarily zero.) Given

1. That EV starts from rest at $t - 0$
2. The resultant of F_{gxT} and F_{TR} is enough to overcome the rolling resistance to get the EV moving
3. $v(t) \geq 0$ for $t \geq 0$

what effect does gravity have on the velocity profile?

2.5 A vehicle accelerates from 0 to 60 mi/h in 10 s. with the velocity profile given by

$$v(t) = 20\ln(0.282t + 1)\ \text{m/s} \quad \text{for } 0 \leq t \leq 10\ \text{s}$$

The vehicle is on level road. For the problem use the parameters given in Problem 2.3.

(a) Calculate and plot $F_{TR}(t)$ and $P_{TR}(t)$ for $0 \leq t \leq 10$ s. (b) Calculate Δe_{TR}. How much of Δe_{TR} is ΔKE? How much is E_{loss}?

2.6 Using the vehicle parameters given in Problem 2.3, calculate and plot, on the same graph, steady-state F_{TR} vs. V characteristics for $\beta = \pm 4°$ and $-60\,\text{mi/h} \leq V \leq 60\,\text{mi/h}$ but $V \neq 0$.

2.7 An EV racer will attempt to jump five city buses as shown in Figure P2.7. The vehicle will start at rest at a position 100 m from the beginning of the take-off ramp. The vehicle will accelerate *uniformly* until it reaches the end of the take-off ramp at which time it will be traveling at 100 mi/h. The vehicle has the following parameter values:

$$m = 692\,\text{kg},\ C_D = 0.2,\ A_F = 2\,\text{m}^2,\ C_0 = 0.009,\ C_1 = 1.75 \times 10^{-6}\,\text{s}^2/\text{m}^2.$$

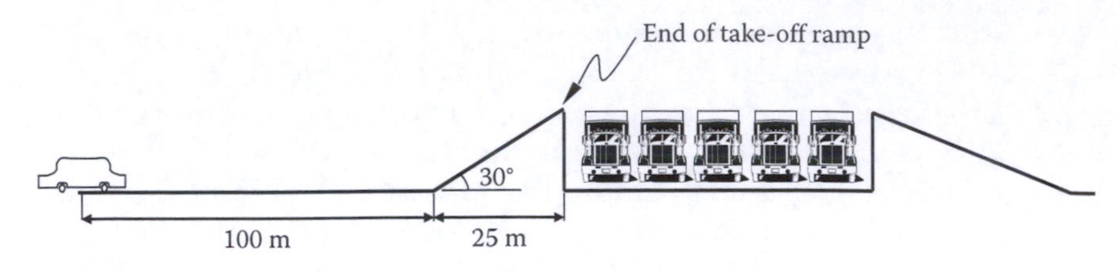

FIGURE P2.7

Also, take density of air $\rho = 1.16\,\text{kg/m}^3$, and acceleration due to gravity $g = 9.81\,\text{m/s}^2$.

(a) Sketch and label the velocity profile of the vehicle from the time it starts to the time it reaches the end of the take-off ramp. How much time does the vehicle take to reach the end of the take-off ramp?
(b) Calculate the change in gravitational potential energy, from start to the end of take-off ramp.
(c) Calculate energy loss, E_{loss}, from start to the end of take-off ramp, if $\Delta e_{TR} = 8.28 \times 10^5$ J during that period.

2.8 The parameters of a vehicle are given below:

Vehicle mass = 1800 kg
Driver/one passenger = 80 kg
Rolling resistance coefficient, $C_0 = 0.01$
Wheel radius, $r_{wh} = 0.3305$ m
Aerodynamic drag coefficient, $C_{AD} = 0.45$
Frontal area, $A_F = 2.5\,\text{m}^2$

The vehicle accelerates from 0 velocity to 21 m/s in 5 s on a 0.5% roadway grade when it reaches the maximum power limit of the propulsion unit. The vehicle then accelerates in the constant power mode for another 7 s. The maximum power limit is 145 kW

(a) Write the *dv/dt* equation for constant power acceleration for the given conditions.
(b) What is the velocity after a total time of 10 s.
(c) What is the velocity at 12 s if the roadway grade changes to 4% at 10 s?

2.9 (a) Find the wheel speed of the vehicle given in Problem 2.8 when the vehicle velocity is 60 mph and the wheel slip is 15%.
(b) The coefficient of friction as a function of wheel slip is given by

$$\mu(s) = \mu_{pk}\left[a\left(1 - e^{-bs}\right) - cs\right]$$

The parameters for the coefficient of friction on a dry pavement are:

$$\mu_{pk} = 0.85,\ a = 1.1,\ b = 20,\ \text{and}\ c = .0035$$

Calculate the traction force limit for slip = 15% on a 1% slope with a driver and a passenger for the vehicle in Problem 2.8.

References

1. J.Y. Wong, *Theory of Ground Vehicles*, John Wiley & Sons, Inc., Hoboken, NJ, 2008.
2. W.E. Meyer and H.W. Kummer, *Mechanism of Force Transmission Between Tire and Road*, SAE Publication No. 490A, National Automobile Week, March 1962.
3. T.D. Gillespie, *Fundamentals of Vehicle Dynamics*, SAE International, Warrendale, PA, 1992.

3

Alternative Vehicle Architectures

Advanced vehicle technologies alternative to the conventional IC engine vehicles are referred to as alternative vehicles. Electric vehicles (EV), hybrid electric vehicles (HEV), plug-in hybrid vehicles (PHEV), and fuel cell electric vehicles (FCEV) are all examples of alternative vehicles. This chapter presents the architectures and design of these alternative vehicle technologies. The chapter starts with the components used and their sizing in EV. The classification of hybrid architectures and component sizing of HEV will be discussed. The features of plug-in and fuel cell hybrid vehicles are also included in this chapter. Further details on fuel cells and fuel cell vehicles appear in Chapter 5.

The fuels in alternative vehicles can be derived from sources alternative to the fossil fuels. Many of these vehicles also use a combination of alternative fuels and conventional fuels. The EV may appear to be operating solely on alternative fuels by the use of modern battery technology, but the ultimate source may be the conventional fossil fuels depending on how the electricity to recharge the batteries was generated. EV with batteries charged from renewable energy sources, such as wind, solar, and hydro, provide the best scenario of alternative energy transportation.

3.1 Electric Vehicles

EV, as defined in Chapter 1, have portable energy source and the tractive effort comes solely from one or more electric machines. The battery electric vehicle powered by only one or more electric machines has the most straightforward architecture without the need for power blending. The detailed structure of an electric vehicle system along with the interaction among its various components is shown in Figure 3.1. The primary components of an electric vehicle system are the motor, controller, power source, and transmission. The figure also shows the choices available for each of the subsystem components. Electrochemical batteries have been the traditional source of energy in EV. Lead/acid batteries were used in the first commercially available EV Saturn EV1 in 1996, but since then the technology has progressed toward NiMH and Li-ion batteries. The batteries need a charger to restore the stored energy level once its available energy is near depletion due to usage. The limited range problem of battery-driven EV prompted

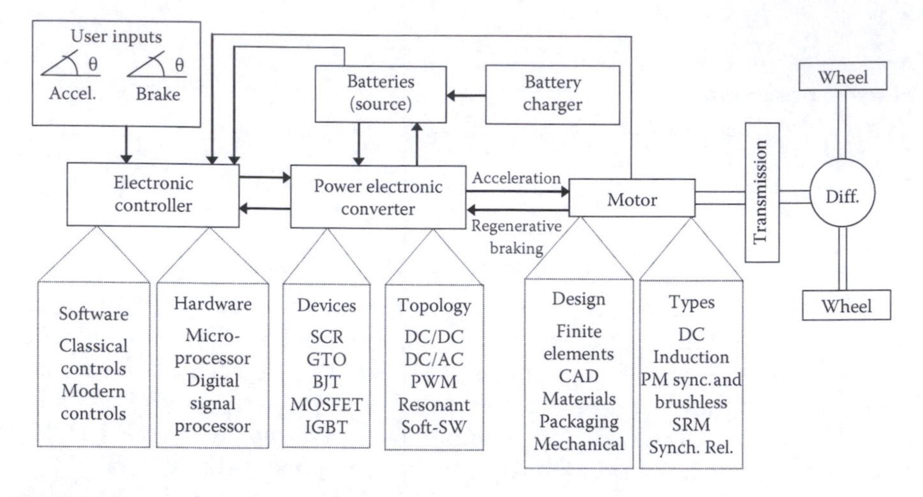

FIGURE 3.1
Major electrical components and choices for an electric vehicle system.

the search for alternative energy sources, such as fuel cells and flywheels. Prototypes and fleet vehicles have been developed with fuel cells over the last decade, but at the same time battery technology has also progressed significantly. Several different types of Li-ion battery chemistry have emerged that are being used in both EV and hybrid vehicles.

The majority of EV developed so far are based on DC machines, induction machines, or permanent magnet machines. The disadvantages of DC machines turned electric vehicle developers to look into various types of AC machines. The maintenance-free, low-cost induction machines became an appealing alternative to many developers. High-speed operation of induction machines comes with a penalty in size and weight, although recent studies on re-rating of induction motors for EV are opening new perspectives even in the high-speed range. The excellent performance together with high power density features of permanent magnet machines makes them an attractive solution for electric vehicle applications, although the cost of permanent magnets can become prohibitive in the future. The high power density and potentially low production cost of switched reluctance machines make it ideally suitable for electric vehicle applications. However, the acoustic noise problem has so far been the deterrent for the use of switched reluctance machines in EV. The electric motor design includes not only the electromagnetic aspects of the machine, but also the thermal and mechanical considerations. The motor design tasks of today are supported by the finite element studies and various computer-aided design tools making the design process highly efficient.

The electric motor is driven by a power electronics based power processing unit that converts the fixed DC voltage available from the source into a variable voltage, variable frequency source controlled to maintain the desired operating point of the vehicle. The power electronics circuit comprises of power semiconductor devices that saw a tremendous development over the past three decades. The enabling technology of power electronics is a key driving force in developing efficient and high-performance powertrain unit for EV. The advances in power solid state devices and very large scale integration (VLSI) technology are responsible for the development of efficient and compact power electronics circuits and electronic control units. High power devices in compact packaging are available today enabling the development of lightweight and efficient power processing units known as power electronic motor drives. The developments in high-speed digital signal processors or microprocessors enable complex control algorithm implementation with high degree of accuracy. The controller includes algorithms for both the motor drive in the inner loop as well as system level control in the outer loop.

3.2 Hybrid Electric Vehicles

The term hybrid vehicles in general usage refer to vehicles with two or three different type of sources delivering power to the wheels for propulsion. The most common hybrid vehicles have an IC engine and one or more electric machines for vehicle propulsion. The IC engine can be used to generate electric energy "on board" to power the electric machines. An energy storage device buffers the electrical energy flow between the electric machine operated as a generator and the electric machine operated as a motor. An electric machine can be operated both as a motor and as a generator. Vehicle architectures with only one electric machine use it both for electric power generation and for vehicle propulsion power, the constraint being that electric power generation and power delivery through electric propulsion is not possible simultaneously. In hybrid vehicles, the traction electric motors can operate independently or in association with the IC engine to power the wheels depending on the type of vehicle architecture. The subject of treatment in this section is the basic architectures of hybrid vehicles.

There are several ways of classifying hybrid vehicles; the most common approach is based on the path of energy flow from its store to the wheels through the power transmission paths. In an HEV, the propulsion power is transmitted to the wheels through either the mechanical power transmission path or the electric power transmission path, or the combination of the two. The mechanical path is made of an IC engine and a transmission, whereas the electrical path consists of an energy storage system, a generator, a propulsion motor, and a transmission. The vehicle powertrain is designed to meet the vehicle base load requirements as well as the peak load requirements during acceleration and starting. The arrangement of the hybrid powertrain components in the power transmission path gives rise to the architecture-based hybrids known as *series, parallel,* and *series–parallel* hybrids.

The hybrid vehicles can also be classified based on the degree of hybridization into *mild, power,* and *energy* hybrids. This classification is based on the powertrain size deviation from a conventional vehicle. The progression from mild to energy hybrids is related to the degree of downsizing the engine and upsizing the electrical and energy storage components.

The hybrid vehicles are also classified as *charge-depleting* or *charge-sustaining* hybrids depending on whether or not the energy storage device needs to be charged from an external source or is self-sustaining with its on-board electricity-generation capability. The battery EV are an extreme example of charge-depleting vehicle, which does not have any on-board electricity generation capability. PHEV are also examples of charge-depleting hybrids; these vehicles operate in the battery-only electric mode for certain distances and then as regular hybrid vehicle for longer distances. The plug-in hybrids are essentially hybrid vehicles but with a large enough energy storage system that will get depleted of its charge and will need to be plugged in to restore operation in the electric only mode. The charge-sustaining vehicles, such as the hybrids available commercially, have a smaller capacity energy storage system; the on-board IC engine and electric generator are sufficient to restore the charge in its energy storage device. The charge-sustaining hybrids never need to be plugged-in.

3.2.1 Hybrids Based on Architecture

3.2.1.1 Series and Parallel Architectures

The HEV evolved out of two basic configurations: series and parallel. A series hybrid is one in which only one energy converter can provide propulsion power. The IC engine acts as a prime mover in this configuration to drive an electric generator that delivers power to the battery or energy storage link and the propulsion motor. The components' arrangement of a series HEV is shown in Figure 3.2.

A parallel hybrid is one in which more than one energy conversion device can deliver propulsion power to the wheels. The IC engine and the electric motor are configured in parallel with a mechanical coupling that blends the torque coming from the two sources. The components' arrangement of a parallel hybrid is shown in Figure 3.3.

Series HEV is the simpler type where only the electric motor provides all the propulsion power to the wheels. A downsized IC engine drives a generator, which supplements the batteries and can charge them when they fall below a certain state of charge. The power required to propel the vehicle is provided solely by the electric motor. Beyond the IC engine and the

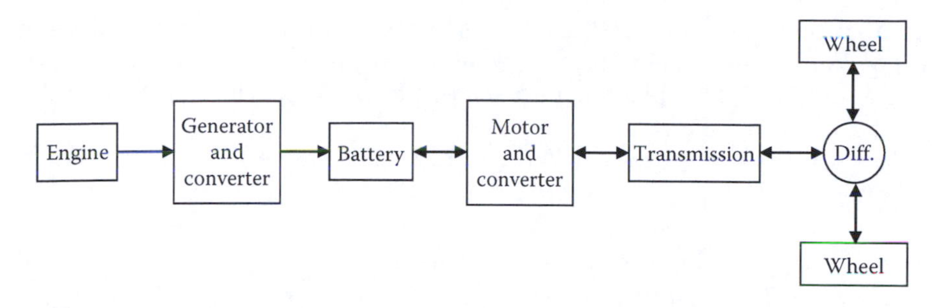

FIGURE 3.2
Series HEV powertrain.

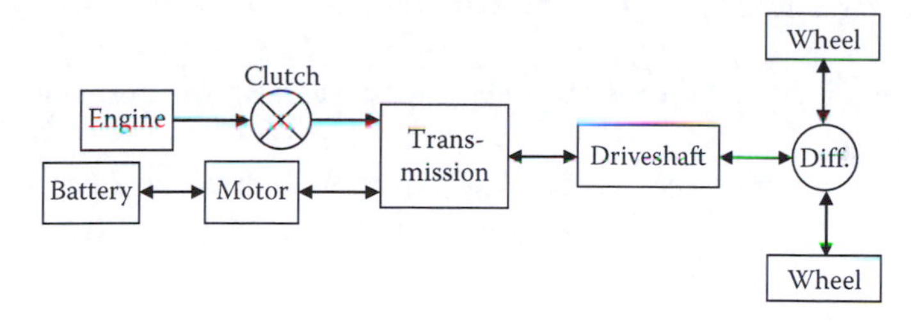

FIGURE 3.3
Parallel HEV powertrain.

generator, the propulsion system is exactly the same as that in an electric vehicle; the electric motor power requirements are exactly the same as that in an electric vehicle.

The drawback of the series configuration is the size of the electrical traction motor, which has to be rated for the maximum power requirement of the vehicle. The drawback can be removed if the engine is used in parallel with the electric machine to supply power to the wheels. This is called a parallel architecture, where both the IC engine and the electric motor are connected to the driveshaft through transmission and clutch. In the parallel HEV, the power requirements of the electric motor in the parallel hybrid is lower than that of an electric vehicle or series hybrid, since the IC engine complements for the total power demand of the vehicle. The propulsion power may be supplied by the IC engine alone, by the battery-motor set, or by the two systems in combination.

Both series and parallel hybrids come in a variety of types. The mission of the vehicle and the optimum design for that mission dictates the choice. If the HEV is to be basically an electric vehicle with an IC engine assist for achieving acceptable range, then the choice could be series hybrids with the IC engine ensuring that the batteries remain charged all the time. On the other hand, if the hybrid vehicle is to be basically a vehicle with almost all the performance characteristics and comforts of an IC engine vehicle, but with lower emission and fuel usage standards, then the choice could be a parallel configuration. Parallel HEV have been built with performance that is equal, in all aspects of normal operation, to a conventional car. However, some series hybrid vehicles have also been built that perform as well as the IC engine vehicles.

The advantages of a series hybrid architecture can be summarized as the following:

- Flexibility of location of engine-generator set
- Simplicity of drivetrain
- Suitable for short trips with stop and go traffic

The disadvantages of a series hybrid architecture are as follows:

- It needs three propulsion components: IC engine, generator, and motor.
- The motor must be designed for the maximum sustained power that the vehicle may require, such as when climbing a high grade. However, the vehicle operates below the maximum power for most of the time.
- All three drivetrain components need to be sized for maximum power for long-distance sustained, high-speed driving. This is required since the batteries will exhaust fairly quickly leaving IC engine to supply all the power through the generator.

The advantages of a parallel hybrid architecture are the following:

- It needs only needs propulsion components: IC engine and motor/generator. In parallel HEV, motor can be used as generator and vice versa.
- A smaller engine and a smaller motor can be used to get the same performance, until batteries are depleted. For short-trip missions, both can be rated at half the maximum power to provide the total power, assuming that the batteries are never depleted. For long-distance trips, engine may be rated for the maximum power, while the motor/generator may still be rated to half the maximum power or even smaller.

The disadvantages of a parallel hybrid architecture are the following:

- The control complexity increases significantly, since power flow has to be regulated and blended from two parallel sources.
- The power blending from the IC engine and the motor necessitates a complex mechanical device.

A simpler architecture with reduced components capable of delivering good acceleration performance is the "through-the-road" parallel architecture. This architecture has one engine powering the front wheels and one electric machine powering the rear wheels. The engine is the primary propulsion unit of the vehicle, while the electric machine is used for load leveling supplying the propulsion power during acceleration and capturing vehicle kinetic energy through regeneration during deceleration. When the charge in the energy storage device falls below a set level, the engine is commanded to deliver torque in excess of what is required to meet the driver demand. The additional energy supplied by the engine is used to charge the energy storage device by operating the rear electric machine as a generator. Power is transferred from the engine to the electric machine through the road, and hence the name *through-the-road* parallel. This charge sustaining mechanism may not be that efficient, but the architecture offers great simplicity without the need for a coupling device to blend the torque from the engine and the electric motor. The vehicle can deliver good acceleration performance using the parallel operation of two propulsion units; the fuel efficiency can also be improved by about 20%–30% compared to a similarly sized IC engine vehicle. Additionally, the architecture gives four-wheel drive capability.

3.2.1.2 Series–Parallel Architecture

The parallel hybrid architecture is better suited to passenger vehicles where the electric motor and IC engine can be operated in parallel to deliver high performance when higher power is required [1]. On the other hand, the series hybrid architecture is equipped with a small output engine that can always be operated in its most efficient operating region to generate electrical power. The output electric power can be used directly to drive the propulsion motor or charge the energy storage system. Series hybrid vehicles tend to be heavy and typically have difficulty meeting acceleration requirements since the powertrain components need to be sized for the maximum continuous output power for charge sustaining operation.

The advanced hybrids combine the benefits of series and parallel architectures into a series–parallel hybrid architecture with charge sustaining capability [1,2]. In these combination hybrids, the IC engine is also used to charge the battery. The architecture is relatively more complicated, involving additional mechanical links and controls compared to the series hybrid, and an additional generator compared to the parallel hybrid. The vehicle is primarily a parallel HEV, but with a small series element added to the architecture. The small series element ensures that the battery charge is sustained in prolonged wait periods such as in traffic lights or in a traffic jam. The controller for the series–parallel architectures effectively utilizes the IC engine and electric motors to deliver up to their maximum capabilities through flexible adaptation with driving conditions.

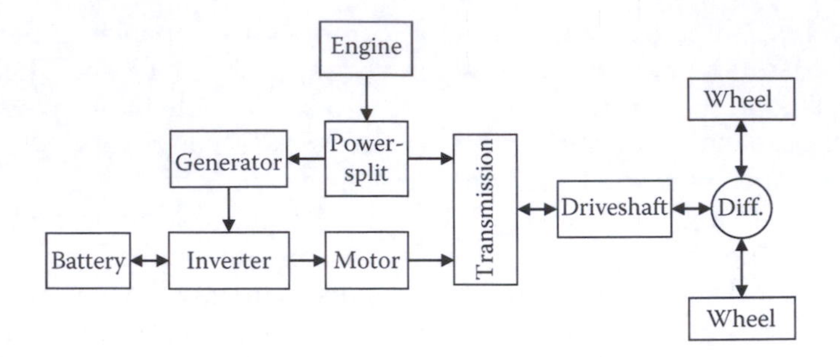

FIGURE 3.4
Series–parallel combination HEV.

The series–parallel architecture is the one that has been used in the first-ever commercially available hybrid vehicle, the Toyota Prius. The vehicle architecture uses a mechanical power-split device that was developed by the Japanese researchers from Equos Research [3,4]. The compact transaxle design in the powertrain integrates two electric motors to simplify the manufacture of both conventional vehicles and hybrid vehicle in the same production plant without significant alteration of the same assembly line. The series–parallel passenger hybrid vehicles available in the market today have all three power plants (the engine and two electric machines) mounted in the front transaxle. The power-split device divides the output from the engine into mechanical and electrical transmission paths using a planetary gear set. The components' arrangement in the series–parallel architecture based on the Toyota Prius hybrid design is shown in Figure 3.4. The power split device allocates power from the IC engine to the front wheels through the driveshaft and the electric generator depending on the driving condition. The power through the generator is used to recharge the batteries. The electric motor can also deliver power to the front wheels in parallel to the IC engine. The inverter is bidirectional and is used to either charge the batteries from the generator or to condition the power for the electric motor. For short bursts of acceleration, power can be delivered to the driveshaft from both the internal combustion engine and the electric motor. A central control unit regulates the power flow for the system using multiple feedback signals from the various sensors.

The series–parallel hybrids are capable of providing continuous high output power compared to either a series or a parallel hybrid for similarly sized powertrain components. The series–parallel vehicle can operated in all three modes of series, parallel, and power-split mode. There is significantly greater flexibility in the control strategy design for a series–parallel hybrid vehicle that leads to better fuel economy and lower emissions.

3.2.1.3 Series–Parallel 2 × 2 Architecture

A series–parallel 2 × 2 vehicle architecture with two electric machines, one IC engine and a battery energy storage system is shown in Figure 3.5. The

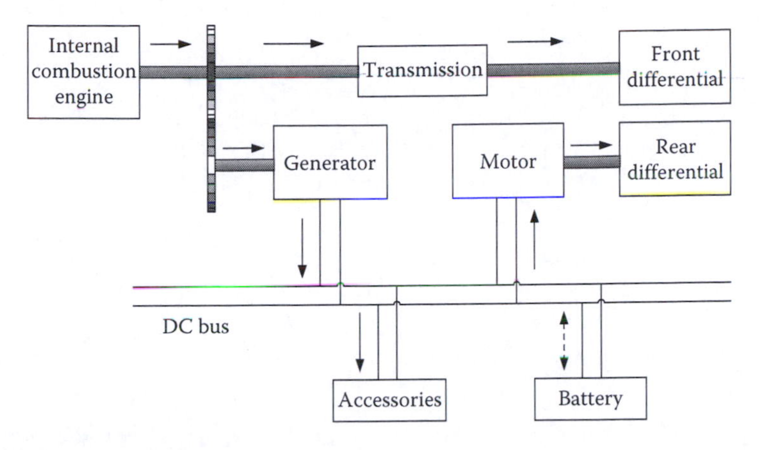

FIGURE 3.5
Series–parallel 2 × 2 vehicle architecture.

architecture offers the same attractive features of series–parallel architecture described in Section 3.2.1.2, but with an inherent four-wheel drive capability. The engine is coupled to one set of wheels, typically the front wheels, through a transmission. One electric machine is coupled to the engine mechanically while the other electric machine is coupled to the other set of wheels, typically the rear-wheels. The front-mounted electric machine or generator is mostly used for generation and starting. The rear-mounted motor is used for regenerative braking and traction. The engine is used for traction and also for supplying power to the generator. During acceleration, the engine powers the front wheels while the rear-mounted electric machine powers the rear wheels. However, during peak acceleration demand, the front electric machine can be operated as a motor to add torque at the front axle. The torque blending between the front axle and rear axles is through electronic controls and does not require any mechanical power-split device like that in the Prius series–parallel architecture. The downside to the series–parallel 2 × 2 is the complexity of the control algorithm and the mounting requirements of powertrain components in both the front and rear axle.

3.2.2 Hybrids Based on Transmission Assembly

Hybrid vehicles can be classified as pre- and post-transmission hybrids depending on the location of the mechanical transmission with respect to the electric drive. In the pre-transmission configuration, the output shafts of the electric motor and the IC engine are connected through a mechanical coupling before the mechanical transmission gearbox. The transmission matches the combined output of the electric drive and IC engine with the vehicle speed. The parallel hybrid architecture shown in Figure 3.3 is a pre-transmission configuration. The series–parallel architecture shown in Figure 3.4 is also a pre-transmission configuration. Figure 3.6 shows the

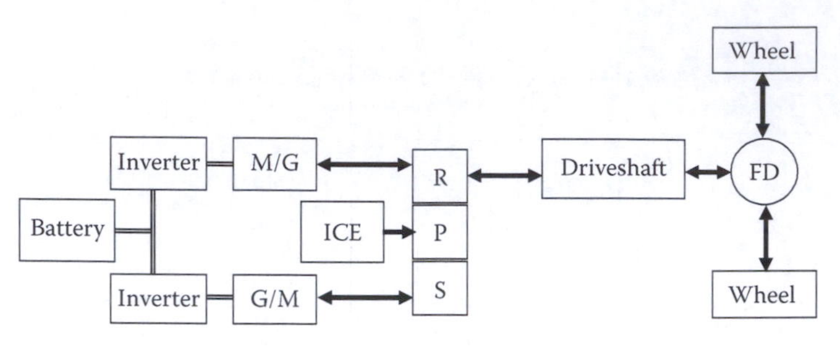

FIGURE 3.6
Power-split pre-transmission hybrid configuration.

series–parallel architecture using the epicyclic or planetary gear set: "R," "P," and "S" denote the "ring," "planet," and "sun" gears of the gear set to which the three powertrain components are connected. The planetary gearset output shaft delivers power to the wheels through the driveshaft and the final drive. The mechanical transmission component including the gearbox is located in between the final drive and the propulsion components of IC engine, electric motor, and generator, which makes the architecture a pre-transmission hybrid configuration. This nonshifting, clutchless, pre-transmission configuration with planetary gearset is the most popular passenger hybrid configuration. The planetary gearset torque–speed relationships will be presented later in Chapter 10.

In the post-transmission hybrid configuration, the electric motor drive is coupled to the output shaft of the transmission. A gearbox may be used to match the transmission output speed, which varies over the entire vehicle speed range as shown in Figure 3.7. The electric motor drive can be operated at higher speeds compared to the vehicle speed with the gear coupling, but electric motor drive must span the entire speed range of the vehicle. Consequently, motor drives for post-transmission configurations are required to have a wide constant power speed region (CPSR), preferably with 6:1 or higher ratio between top speed and base speed. The configuration poses

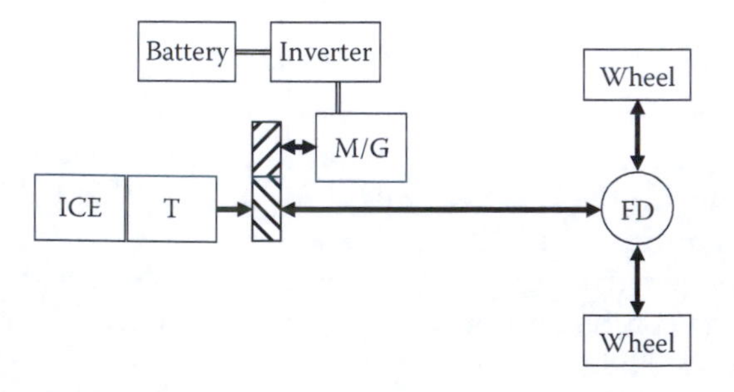

FIGURE 3.7
Parallel post-transmission hybrid.

packaging challenges and connection issues with the final drive; continuous engagement with the wheel results in no load spin losses with certain types of electric machines. Hybrid vehicles with wheel or hub motors are of similar post-transmission configuration. High torque, low-speed motors are required for hub motors, which are directly mounted on the wheels without any matching transmission or gear. Hub motor results in higher unsprung mass; the motors are subjected to higher levels of vibration and temperature and are also prone to environmental effects such as water, sand, dust, salt, gravel, etc.

3.2.3 Hybrids Based on Degree of Hybridization

The classification based on the degree of hybridization corresponds to the mission that the vehicle is designed to achieve [5]. This is a more consumer-oriented classification method used by the automotive industry. There are three "mission-based" classes: mild hybrids, power hybrids, and energy hybrids. The "mild" hybrids have the lowest degree of hybridization with a moderate effect on fuel economy and emissions. Typical electrical rating of a mild hybrid would be in the range of 5–10 kW, with energy capacity in the range of 1–3 kW h. The "power" hybrids have a larger electric propulsion component, with an electrical rating as high as 40 kW; these hybrids allow significant amount of power transfer between battery and motor drive system, although the battery storage is designed with relatively low energy capacity (3–4 kW h). The power hybrids have a greater potential to provide fuel economy improvements. These also have better engine-out emissions due to a more focused engine duty schedule. Power hybrids, like the mild hybrids, are charge-sustaining type receiving all motive energy from the on-board combustion of the fossil fuel. The "energy" hybrid employs a high-energy battery system capable of propelling the vehicle for a significant range without engine operation. The electrical rating and battery capacity are typically in the ranges of 70–100 kW and 15–20 kW h [5]. A zero emission range of 50 mi would meet the daily commute range of the majority of population based on average driving habits of about 12,000 mi/year in the United States. The energy hybrids are obviously charge-depleting type with provisions to recharge the batteries at home electrically. These vehicles are the subject of treatment in the following section.

3.3 Plug-In Hybrid Electric Vehicle

The PHEVs are similar to the charge-sustaining hybrids except that they have a higher capacity energy storage system with a power electronic interface for connection to the grid. The energy storage system in a PHEV can be charged on-board and also from an electrical outlet. The vehicle can operate in a battery-only mode for a much longer period than the charge-sustaining hybrid vehicles. The PHEV is intended to operate as a pure battery electric vehicle for the design specified distances during the daily commute. The IC engine is used to provide additional power and range for long distance driving. This type of vehicle is also sometimes known as a "range extender." The energy obtained from the external power grid in the PHEV displaces the energy that would otherwise be obtained by burning fuel in the vehicle's IC engine. This has the potential of higher usage of alternative fuels in comparison to other hybrid vehicles where all of the energy comes from the fossil fuel.

The architecture choices for PHEV are the same as those charge-sustaining hybrids. The series architecture is the simplest architecture and is highly suitable for PHEV. The IC engine and electric generator need not be sized to match the peak rating of the traction electric motor due to the availability of high energy capacity battery-pack. An on-board power electronic interface similar to that of a battery electric vehicle is necessary for grid connection. The architecture of the series PHEV is shown in Figure 3.8. Parallel and series–parallel architectures similar to the regular hybrids but with the grid connectivity are also possible with PHEV.

A plug-in hybrid is generally rated based on the zero-emission distance traveled; it is designated as PHEV'X' where "X" is the distance traveled in miles using off-board electrical energy. This range of travel where the IC engine is not used is known as the zero-emission vehicle (ZEV) range. A PHEV40 is a plug-in hybrid with a useable energy storage capacity equivalent to 40 mi of driving energy on a reference driving cycle. The PHEV40 can displace petroleum energy equivalent to 40 mi of driving on the reference cycle with off-board electricity.

The PHEV uses the battery-pack storage energy for most of its driving, thereby reducing the vehicle emissions, and air and noise pollution. The

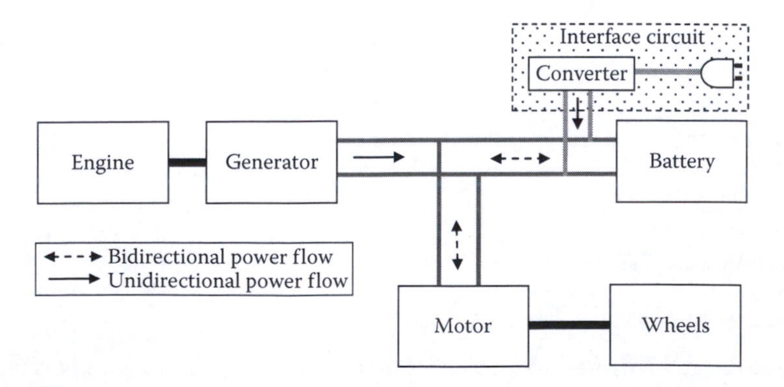

FIGURE 3.8
Block diagram of a series PHEV.

maintenance cost of a PHEV is low since the vehicle is primarily electric. A PHEV can also be used to even out electricity demands during peak load demand on the grid. Excess battery charge from the plug-in hybrid can be used to send power back to the grid during peak periods. The concept is known as "Vehicle-to-Grid" power or V2G. The vehicle can be recharged again during off-peak periods.

The major drawback of the PHEV is the poor well-to-wheel efficiency compared to conventional gasoline and regular hybrid vehicles. This is because the PHEV has the poor efficiency segments of both the IC engine vehicle and the EV in the well-to-wheel energy conversion pathway. However, the PHEV has the potentials to be both more efficient and environmentally friendlier if the electricity to recharge the batteries can be obtained from renewable sources.

The higher capacity battery in the PHEV increases the mass of the vehicle. The added mass affects the performance; components have to be sized for higher power rating to achieve good performance. The batteries are also quite expensive, and as a result the cost of the vehicle is also high.

The zero-emission range of the PHEV depends on the way the vehicle is used. The batteries used in the vehicle must be charged using grid electrical power to achieve maximum efficiency and cost reductions. The benefits of the PHEV decrease if the recharging capability through plug-in is unavailable. The use of the engine to charge the batteries should be minimized to maximize efficiency. Energy is always lost during charging and discharging of the battery-pack. The desired mode of operation when the IC engine has to be used is the series mode.

3.4 Powertrain Component Sizing

The complete design of a vehicle is complex involving numerous variables, constraints, considerations, and judgment. The full treatment of a vehicle design is beyond the scope of this book, but from the roadway fundamentals, the vehicle dynamics, and the architectures presented thus far, we can address the fundamental calculations involved in sizing of the powertrain components. The calculations provide the design data for starting computer modeling and simulation for detailed analysis of a complex HEV system. The computer modeling and simulation continues with design iterations, and subsystem resizing and controller updates until simulation results meet the specifications. The basic equations and principles that lay the foundation for the design of electric and hybrid vehicles are presented next. Design considerations will be highlighted as and when appropriate.

The primary design specifications related to powertrain sizing are (1) the initial acceleration, (2) rated velocity on a given slope, (3) maximum %

grade, and (4) maximum steady state velocity. The zero-emission range is the most important specification for sizing the capacity of the battery-pack. The energy required for a given acceleration and constant steady-state velocity can be used for the sizing of the energy system.

This section first presents the sizing of the propulsion component for an electric vehicle. Then the sizing of powertrain components for a hybrid vehicle is presented followed by a design example.

3.4.1 EV Powertrain Sizing

The major components in the powertrain of an electric vehicle are the electric motor and the energy storage system. The sizing of the electric motor involves finding the power rating of the motor and the operating speed range. The voltage rating and the dimensional packaging constraints must also be specified at the system level design stage. The design of the machine itself is handled by the electric machine designed to meet the requirements set at the system level. The discussion below addresses the power rating of the electric machine to meet vehicle performance requirements. The sizing of the energy storage device will be addressed later in the hybrid vehicle design section.

Electric motors have three major segments in its torque-speed characteristics: (1) constant torque region, (2) constant power region, and (3) natural mode region. The envelope of the electric motor torque-speed characteristics is shown in Figure 3.9. The motor delivers rated torque up to the base speed or rated speed of the motor when it reaches its rated power condition. The motor rated speed is defined as the speed at which the motor can deliver rated torque at rated power. The motor operates in a constant power mode beyond the rated speed where torque falls off steadily at a rate that is inversely proportional to speed. Electric motor can operate at speeds higher than rated using field weakening in the constant power region. There is a

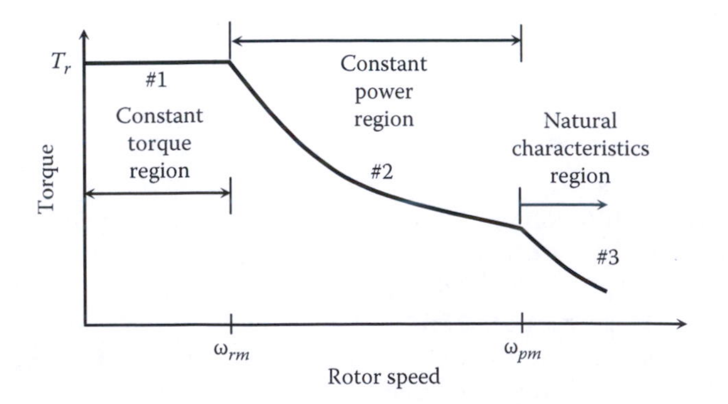

FIGURE 3.9
Electric motor torque-speed envelope.

third natural mode region for very high motor speeds where the torque falls off rapidly being inversely proportional to the square of the speed. The natural characteristic region can be an important part of the overall torque speed curve of certain motors that can be used to reduce the power rating of the motor. However, in most cases, the maximum speed of the vehicles is considered to be at the end of the constant power region. Note that the curves in Figure 3.9 show the envelope, i.e., the operating torque and speed limits in different regions. The electric motor can operate at any point within the envelope through the feed from a power electronics based motor drive component. The salient feature of wide operating speed range characteristics of an electric motor makes it possible to eliminate multiple gear ratios and the clutch in electric vehicle and other applications. A single gear ratio transmission is sufficient for linking the electric motor with the driveshaft. Electric motors with extended constant power region characteristics are needed to minimize the gear size in EV.

The size of an electric motor depends on the maximum torque required from the machine. The higher the maximum torque required, the larger will be the size of the motor. In order to minimize the size and weight, electric motors are designed for high-speed operation for a given power rating. Gears are used to match the higher speed of the electric motor with the lower speed of the wheels. Typical motor speeds can be in the vicinity of 15,000 rpm for typical wheel speeds of around 1,000 rpm for lightweight passenger vehicles. The transmission gear achieves this speed reduction in the range of ~10–15:1 typically in two stages of 3–4:1 of speed reduction. The gear sizing depends on whether the low speed or the high speed performance of the electric vehicle is more important based on the power rating determined for the electric vehicle.

The tractive force vs. speed characteristics of the propulsion system can be widely different for two gear ratios as shown in Figure 3.10. Note that

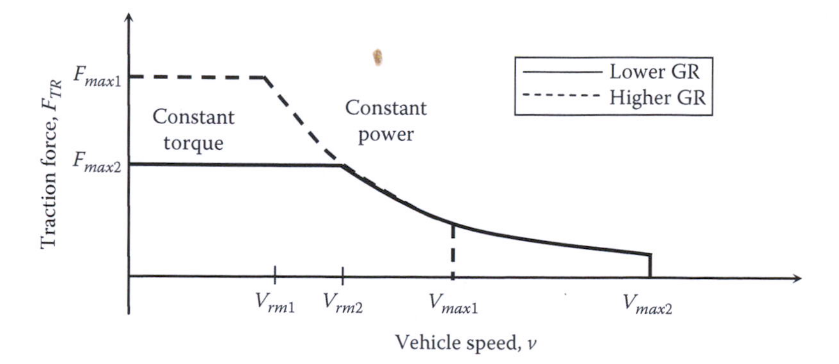

FIGURE 3.10
Electric motor torque-speed characteristics in terms of traction force and vehicle speed for two gear ratios.

the rated speeds shown are for the drivetrain unit comprising of the electric motor and transmission system, and that the electric motor rated speed is different from these values. The electric motor speed can first be converted into drivetrain unit speed or vehicle wheel speed using the gear ratio as shown earlier. The motor rated speed at the wheel $\omega_{rm,wh}$ in rad/s can be converted to linear vehicle speed using $v_{rm} = \omega_{rm,wh} \cdot r_{wh}$, where r_{wh} is the radius of the wheel. The symbol v_{rm} is referred to here as the equivalent electric motor linear speed after accounting for the transmission gear and the wheel radius. A higher wheel speed or vehicle speed can be attained with a smaller gear ratio, but the peak traction force that the drivetrain can deliver will be smaller. The smaller traction force will limit the initial acceleration and maximum gradability capabilities of the vehicle. On the other hand, if a high gear ratio is used in the transmission for the same electric motor, the peak traction force available at the wheels will be higher, but the maximum vehicle speed v_{max} will be limited. Therefore, we can conclude that the gear ratio depends on the motor rated speed, vehicle rated speed, vehicle maximum speed, wheel radius, and maximum gradability. It must be observed that a higher gear ratio entails a larger gear size. Therefore, the gear ratio and the electric motor rated speed must be selected simultaneously to optimize the overall size and performance requirements.

In the following, the design of the electric motor will be discussed in view of the specified requirements of the initial acceleration, the rated velocity on a given slope, the maximum steady state velocity, and the maximum gradability. In the process the following parameters will be used: rated motor power P_m, rated motor speed, ω_{rm}, rated wheel speed ω_{fwh}, rated vehicle speed v_f, vehicle total mass m, vehicle frontal area A_F, rolling resistance coefficients C_0 and C_1, and aerodynamic drag coefficient C_D. The design process starts with a set of known parameters and some educated guess and ends with the final design values that meet the requirements after several iterations.

3.4.1.1 Initial Acceleration

The initial acceleration is specified as 0 to v_f in t_f seconds. v_f is the vehicle rated speed obtained from $v_f = \omega_{fwh} \cdot r_{wh}$. The design problem is to solve for F_{TR} starting with a set of parameters including vehicle mass, rolling resistance, aerodynamic drag coefficient, percent grade, wheel radius etc., some of which are known, while others have to be assumed. The acceleration of the vehicle in terms of these variables is given by (2.24), repeated here for convenience

$$a = \frac{dv}{dt} = \frac{F_{TR} - F_{RL}}{m}$$

The motor power rating can be obtained by solving the above differential equation for a given force–velocity profile, such as one of the two shown in Figure 3.10, and the following boundary conditions:

At $t=0$, vehicle velocity $v=0$.

At $t=t_f$, vehicle velocity $v=v_f$.

Integrating the differential equation within the interval $t=0$ to $t=t_f$ for velocities 0 to v_f

$$m\int_0^{v_f}\frac{dv}{F_{TR}-F_{RL}(v)}=\int_0^{t_f}dt$$

The vehicle-rated velocity is higher than the motor-rated velocity and lies in the constant power region of motor torque-speed characteristics. Splitting the integral on the left side into two velocity regions of 0–v_{rm} for the constant torque mode and of v_{rm}–v_f for the constant power mode, one can write

$$m\int_0^{v_{rm}}\frac{dv}{(P_m/v_{rm})-F_{RL}(v)}+m\int_{v_{rm}}^{v_f}\frac{dv}{(P_m/v)-F_{RL}(v)}=t_f \tag{3.1}$$

The road load resistance force F_{RL} can be expressed as a function of velocity as shown in Chapter 2 for given values of rolling resistance, aerodynamic drag force, and roadway slope. Equation 3.1 can then be solved for motor power rating P_m for specified vehicle rated velocity v_f and rated motor speed. Note that Equation 3.1 is a transcendental equation with F_{RL} being a function of velocity and can be solved numerically to find the motor power rating P_m. In fact, extensive computer computation and simulation aids a practical design to derive the required motor power rating and gear ratio of the powertrain.

An interesting analysis has been presented in Ref. [6] to stress the importance of extended constant power region of motor torque-speed characteristics. It has been shown that for $F_{RL}=0$, the motor power rating is

$$P_m=\frac{m}{2t_f}\left(v_{rm}^2+v_f^2\right) \tag{3.2}$$

Equation 3.2 shows that the motor power rating will be minimum when $v_{rm}=0$, which means that the electric motor that operates entirely in the constant power mode is the smallest motor to satisfy the requirements. In the other extreme case, the motor power will be the double that of the smallest case if the motor operates entirely in the constant torque mode with $v_{rm}=v_f$. Of course, eliminating the constant torque region and operating entirely in the constant power region is not practically realizable. In a practical setting, the electric motor should be designed with a low base speed or rated speed and a wide constant power region.

3.4.1.2 Rated Vehicle Velocity

The drivetrain designed to accelerate the vehicle from 0 to rated velocity will always have the sufficient power to cruise the vehicle at rated speed, provided the roadway slope specified for initial acceleration has not been raised for rated velocity cruising condition.

3.4.1.3 Maximum Velocity

The traction power required to cruise the vehicle at maximum vehicle velocity v_{max} is

$$P_{TR,max} = mgv_{max}\sin\beta + \left[mgC_1 + \frac{P}{2}A_F C_D\right]v_{max}^3 + mgv_{max}C_0 \tag{3.3}$$

The dominant resistance force at high speeds is the aerodynamic drag force with the power requirement to overcome it increasing at a cubic rate. For vehicles designed with fast acceleration characteristics, P_m is likely to be greater than $P_{TR,max}$. If $P_{TR,max} > P_m$ derived earlier to meet the initial acceleration requirement, then $P_{TR,max}$ will define the electric motor power rating. The natural mode region of electric motors can be used to meet very high maximum vehicle velocity requirements to minimize the motor size.

3.4.1.4 Maximum Gradability

The maximum gradability of a vehicle for a given motor and gear ratio can be derived from

$$\text{Max. \% grade} = \frac{100F_{TR}}{\sqrt{(mg)^2 - F_{TR}{}^2}}$$

The maximum traction force F_{TR} available from the preliminary motor design can be plugged into the above equation to check whether the vehicle maximum gradability conditions are met or not. If the maximum electric motor power derived for acceleration or maximum vehicle velocity is not enough to meet the maximum gradability requirement of the vehicle, then either the motor power rating or the gear ratio has to be increased. Care must be taken not to violate the maximum vehicle velocity requirement when increasing the gear ratio. The gear ratio and motor power are decided in a coordinated manner to meet both the requirements, while maintaining a reasonable size for both the electric motor and the gear.

3.4.2 HEV Powertrain Sizing

The HEV powertrain architecture and control technique depends on the desired requirements including, but not limited to, performance, range, and emission. The performance requirements of initial acceleration, cruising velocity, maximum velocity, and gradability dictate the power requirements of the IC engine and motor. The power requirements can also be specified in terms of multiple driving schedules that have the worst case demands embedded in them. The energy required by the drivetrain to meet the range specification dictates the design of the energy storage system, which can be a battery-pack or a combination of battery and ultracapacitors. Meeting the emission standard depend solely on the IC engine emission characteristics, since the electric motor has zero emission.

The power required in a hybrid vehicle comes from a combination of the electric motor and IC engine outputs. The mission of the vehicle plays a big role in apportioning the power requirement between the electric motor and the heat engine. A hybrid vehicle designed for urban commute-type transportation will have a different combination of drivetrain subsystems than a family sedan designed for both urban and highway travel. The power requirement of the electric motor and the IC engine for an urban commute vehicle will definitely be lower than that required for cars designed for highway travel. For urban vehicles that are used solely for daily commutes of less than 100 mi, the designer must consider the battery electric vehicle instead of a PHEV. The design engineer evaluates the design trade-offs based on the mission and specifications, and selects one of the series or parallel configurations discussed earlier. The subsystems of the drivetrain have their individual control units, and the components are coordinated through a supervisory control unit. The mission of the vehicle also dictates the type of control to be employed for the vehicle. For example, the highway vehicles require a control strategy designed to maximize the fuel economy. The designed system must be capable of handling all real world situations within the limits of the design requirements. Appropriate safety measures must also be incorporated in the controller to handle situations when certain subsystems fail or underperform.

The sizing of the components of both the electrical system and the mechanical system starts once the drivetrain architecture is laid out based on the mission of the vehicle. In a series hybrid vehicle, the electrical system design is the same as that of an electric vehicle. The IC engine size is specified for keeping the batteries charged. The sizing of the components of a parallel hybrid vehicle is much more complex. If the vehicle is designed with heavier biasing on the IC engine, then the batteries can be downsized and reconfigured for maximum specific power instead of maximum specific energy. The battery and the motor serves to supply peak power demands during acceleration and overtaking without being completely discharged. The battery also acts as a reservoir for the regenerative braking energy. Ultracapacitors can be used instead of batteries provided they meet the requirements during peak power demands. If the vehicle is to be more battery biased, then the system is configured such that the battery energy is depleted as much as about 80% from its fully charged capacity at the end of the longest trip. Once the power requirements of the electrical and mechanical system are apportioned for the parallel hybrid vehicle, the electrical components are designed based on the power designated for the electrical system using the same design philosophy as that used for the electric vehicle components. A philosophy parallel to that used for electric vehicle can be used to design the mechanical system, where the components are sized based on initial acceleration, rated cruising velocity maximum velocity, and maximum gradability [7]. The gear ratio between the IC engine and the wheel shaft of a parallel hybrid vehicle can be obtained by matching the maximum speed of the IC engine to the maximum speed of the driveshaft. A single gear transmission is desired to minimize complexity. The sizing of the components of a parallel hybrid vehicle is discussed below.

3.4.2.1 Rated Vehicle Velocity

In a hybrid vehicle, the electric motor primarily serves to meet the acceleration requirements, while the IC engine delivers the power for cruising at rated velocity assuming that the battery energy is not sufficient to provide the required power throughout the desired range. Therefore, the IC engine size is determined by the vehicle cruising power requirement at its rated velocity independent of the electric motor power capacity. Thus, the IC engine size should be determined first in the case of hybrid vehicle; its size can be used to reduce the power requirement of the electric motor responsible for vehicle acceleration characteristics.

The road load characteristics developed in Chapter 2 and the force–velocity characteristics of the IC engine (derived from the torque speed characteristics of an IC engine) are useful in sizing the IC engine. Figure 3.11 shows example curves of IC engine characteristics with engine displacement as a parameter along with the road load characteristics for an assumed grade and vehicle parameters. The correct IC engine size is determined from the intersection of the worst case road load characteristics with the IC engine force–velocity

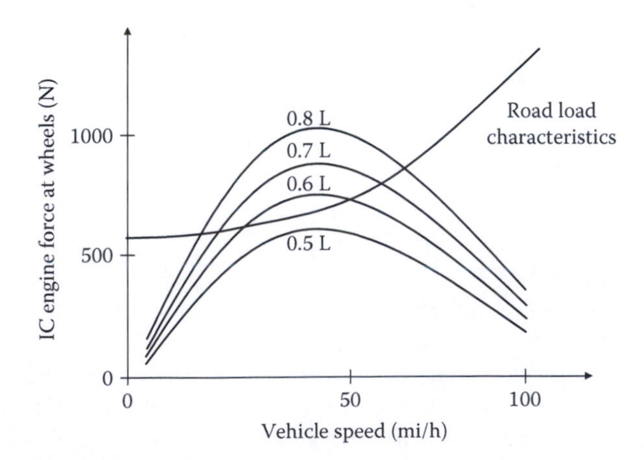

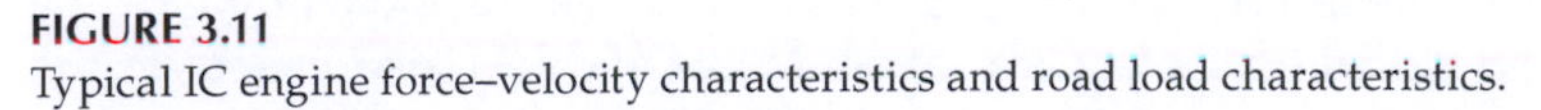
FIGURE 3.11
Typical IC engine force–velocity characteristics and road load characteristics.

profile at rated velocity, plus allowing a nominal 10% margin for the battery-pack recharging [7]. The exact amount of margin needed is the subject of a more complicated analysis, involving vehicle driving cycles, battery capacity, battery charge/discharge characteristics, and generator characteristics.

3.4.2.2 Initial Acceleration

The electric motor with its higher peak power capabilities is more heavily used during initial acceleration. The mechanical power available from the IC engine can be blended with the electric motor power for acceleration, thereby reducing the power requirement of the electric motor. The power required from the motor depends on the velocity at which torque blending from the two propulsion units starts. Figure 3.12 shows the effect of torque blending on the electric motor rated power requirement during initial acceleration with a single gear transmission for the IC engine. The figure shows that there will be very little power contribution from the engine until a minimum critical velocity v_{cr} mi/h of the vehicle is reached due to its poor low speed torque capability. Therefore, torque blending should start after the vehicle has attained the critical velocity with single gear transmission avoiding the use of engine for initial acceleration as much as possible without significantly increasing the rating of the electric motor. The power requirement from the electric motor increases nonlinearly with speed if IC engine torque blending is delayed beyond v_{cr} mi/h. The qualitative figure shows two curves with the extent of the constant power region as a parameter, x being an integer number. The lower curve has much wider constant power region than the upper curve. The curves emphasize once again the need for extended constant power region of operation of the electric motor to minimize its size. The power requirement of the IC engine determined from rated vehicle velocity

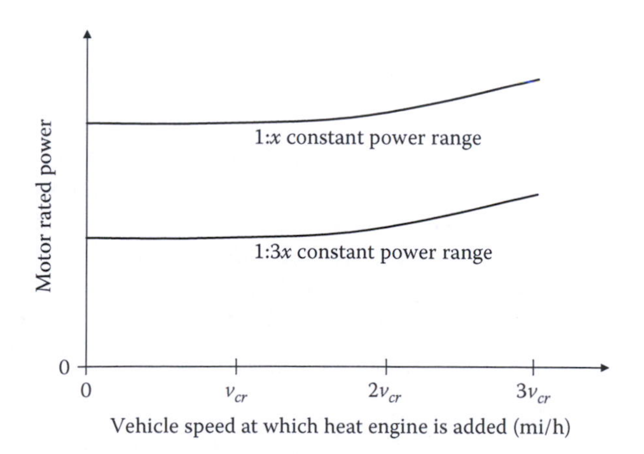

FIGURE 3.12
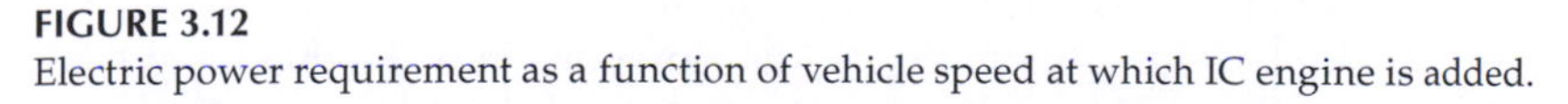
Electric power requirement as a function of vehicle speed at which IC engine is added.

condition would typically be enough to provide the initial acceleration in combination with the electric motor.

3.4.2.3 Maximum Velocity

The power requirement from the propulsion system at maximum velocity is $F_{TR} \cdot v_{max}$, which is supplied by a combination of the engine and the electric motor. This requirement is given by

$$P_{TR,max} = F_{TR} \cdot v_{max} = mgv_{max} \sin\beta + \left[mgC_1 + \frac{\rho}{2} A_F C_D \right] v_{max}^3 + mgv_{max} C_0$$

The electric motor power required to meet the maximum velocity conditions can be uniquely defined by subtracting the engine power determined for cruising at rated velocity from the maximum velocity power requirement. The electric motor power requirement calculated in this step would, in general, be less than the power requirement for the initial acceleration, unless the maximum velocity requirements are very stringent, such as very high speed on a steep grade.

3.4.2.4 Maximum Gradability

The maximum gradability condition must be checked once the sizing of the IC engine and the electric motor is done from the previous three requirements. The maximum gradability of a vehicle is given by

$$\text{Max. \% grade} = \frac{100 F_{TR}}{\sqrt{(mg)^2 - F_{TR}^2}}$$

If the condition is not met, then the size of either the engine or the motor or both must be increased or the gear ratio must be changed to meet the gradability requirements.

Although the design philosophy outlined above states that the IC engine sizing primary comes from the rated cruising velocity and the electric motor sizing comes from the initial acceleration, the practical design involves extensive computer simulation using various drive cycles, parameters of the vehicle and characteristics of chosen battery, motor, generator, and IC engine. As in all systems cases, the sizing and design of the components of electric and hybrid vehicles is an iterative process that ends when all the design requirements are met. The discussions presented provide the theoretical basis for initial estimates avoiding unnecessary oversizing of components.

3.4.3 HEV Powertrain Sizing Example

An HEV powertrain sizing example is presented below for a series–parallel 2×2 hybrid vehicle. The vehicle was re-engineered from a 2005 Chevrolet

TABLE 3.1

Akron Hybrid Vehicle Parameters and Requirements

Description	Requirements
Vehicle mass	4400 lb/1995 kg
Driver/one passenger	176 lb/80 kg
Trailering capacity	2500 lb/1133 kg
Rolling resistance coefficient, C_0	0.009
Wheel radius, r_{wh}	0.3305 m
Aerodynamic drag coefficient, C_{AD}	0.45
Frontal area, A_F	2.686 m^2
0–60 mi/h	9.0 s
50–70 mi/h	6.8 s

Equinox for the national collegiate student vehicle design competition called Challenge X. This vehicle is termed as Akron hybrid vehicle in this book. The vehicle parameters and performance requirements for the design are given in Table 3.1. The overall design goal for the vehicle is increased fuel economy and reduced emissions without sacrificing customer amenities. The gains in fuel economy with vehicle hybridization can be achieved on the basis of downsizing the engine, reduction in engine idling operation, recovery of braking energy losses, and flexibility in choosing engine operating point to enhance energy efficiency. These considerations form the basis of component sizing and selection, and control strategy development for the Akron hybrid vehicle.

The design considers initial acceleration without the trailer, but includes the mass of the driver. The vehicle cruising speed is 75 mi/h on a 3% grade, and against a 20 mi/h headwind at high engine fuel efficiency as well as high electric drive efficiency. The Akron hybrid vehicle architecture is the same as that shown in Figure 3.5. The vehicle components have been sized to meet the initial acceleration demand through a combination of power delivered by the electric motor and the IC engine, and to maintain charge sustaining operation at rated cruising velocity with the IC engine only. The electric motor, with its higher peak power capabilities, will be more heavily used during initial acceleration. The battery will be sized for peak power handling only, and not for continuous electric-only mode of operation. The IC engine and electric motor will be operated in the series mode for urban driving conditions. The electric generator will be sized to generate sufficient power with continuous operation during urban driving.

3.4.3.1 Total Power Required: Initial Acceleration

Power required during acceleration is usually sufficient to meet the maximum velocity and towing requirements. Therefore, the acceleration power, which comes from the combined efforts of electric drive motor and IC engine, is calculated first, but later it has to be verified that the other power requirements are met by the two drive components.

The initial acceleration requirements are met by operating the IC engine and electric motor at their peak torque capabilities until the power limit of the two are reached. Therefore, the drivetrain can be assumed to be operating with constant torque (i.e., constant force F_{TR}) acceleration initially, and then with constant power acceleration.

In order to estimate the constant force required initially and the peak power requirement, three velocity profiles are considered and analyzed. These velocity profiles are with constant force F_{TR} acceleration, uniform acceleration, and a third one termed smooth acceleration.

The velocity profile with constant force F_{TR} acceleration is given by

$$v(t) = \sqrt{\frac{K_1}{K_2}} \tanh\left(\sqrt{K_1 K_2}\, t\right) \tag{3.4}$$

where

$$K_1 = \frac{F_{TR}}{m} - gC_0 - mg\sin\beta > 0$$

$$K_2 = \frac{\rho}{2m} C_D A_F + gC_1 > 0$$

The velocity profile for constant acceleration is

$$v(t) = at \tag{3.5}$$

where a is the constant acceleration. For 0–60 mi/h in 9 s, the constant acceleration is

$$a = 2.98\ \text{m/s}^2$$

The velocity profile for smooth acceleration is [8]

$$v(t) = v_f \left(\frac{t}{t_f}\right)^x \tag{3.6}$$

The range of the exponent x is between 0.47 and 0.53 with lower values used for slow accelerating vehicles, and higher values used for fast accelerating vehicles. We will take $x = 0.5$ for our initial analysis.

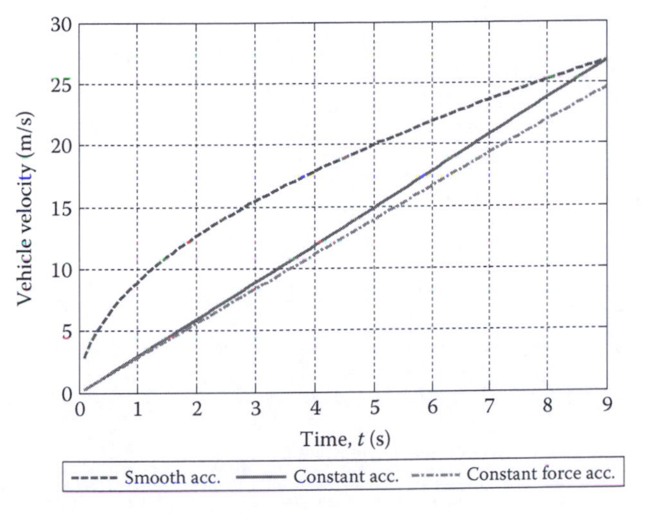

FIGURE 3.13
Velocity profile for three types of initial acceleration.

The 0–60mi/h velocity profiles for the three types of acceleration are shown in Figure 3.13. A force of F_{TR}=6000N has been assumed for the constant force acceleration. The velocity profile of Equation 3.6 requires significantly large force or torque initially, while the constant acceleration profile requires high power at the end of the initial acceleration period. The power and force required with constant acceleration at t=9s are 6899N and 184kW. The practical approach will be to accelerate faster than the constant acceleration profile, but less than the smooth acceleration profile, and transition to constant power acceleration approximately midway through the initial acceleration. The constant force for initial acceleration up to about 4s obviously needs to be greater than 6000N.

The total power required to meet the acceleration requirements prior to accounting for the drivetrain losses is calculated through design iterations using computer simulation. The numbers given in Table 3.2 are arrived at after a few design iterations with different peak tractive force, peak tractive power, and the time for transition of the control algorithm from constant force mode to constant power mode. The tractive force of 7500N is supplied by the combination of the electric motor and the IC engine. Furthermore,

TABLE 3.2
Final Numbers Following Initial Sizing

$F_{TR,peak}$	$P_{TR,peak}$	Time and Velocity of Mode Transition	Time for 0–60mi/h	Time for 50–70mi/h
7500N	120kW	4.6s/16m/s	9s	5s

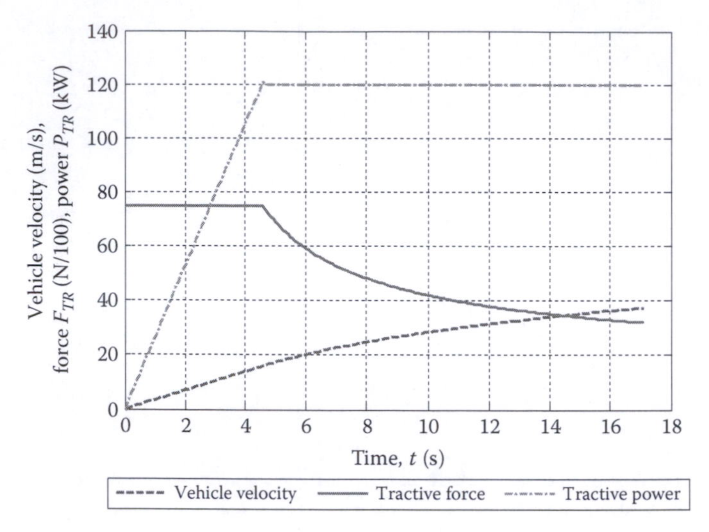

FIGURE 3.14
Tractive power and force requirements with a combination of constant force and constant power modes of acceleration.

there will be a gear between the electric motor and the drivetrain, which will bring the torque requirements of electric motor, but increase its speed requirement. The F_{TR} vs. velocity profile of Figure 3.14 matches well with the torque speed characteristics achievable from electric motors. The velocity ratio of the force characteristics between the peak vehicle velocity (60 m/s) to the onset of the constant power mode (16 m/s) is approximately 4:1, which can be obtained from well-designed electric motor drives. Therefore, assuming the drivetrain efficiency to be about 90%, the minimum combined power required is 132 kW.

3.4.3.2 IC Engine Power: Cruising Speed

Once the initial acceleration power requirements are calculated, the sizes of the IC engine and the electric traction motor can be determined. The IC engine size can be determined by the vehicle cruising power requirement at its rated velocity independent of the electric motor power capacity. With this approach, the IC engine size should be determined first for the Akron hybrid vehicle, and then the remainder of the 130 kW power required can be met by an electric motor.

The IC engine power can be calculated from

$$P_{TR}(t) = V\left(mg\sin\beta + \left[mgC_1 + \frac{\rho}{2}A_F C_D\right]V^2 + mgC_0\right) \tag{3.7}$$

where V is the rated velocity or cruising speed.

The hybrid vehicle IC engine can be sized for the engine power required to maintain a steady expressway velocity of x mi/h (say 75 mi/h), up a y% grade (say 3%), and against a z mi/h (say 20) headwind at high engine fuel efficiency. The engine power at highway speeds should also have good fuel efficiency; an IC engine's best fuel efficiency operating point is not necessarily at its maximum power capability point. Therefore, the IC engine can be selected with highway power capability less than the engine's maximum power rating for acceptable fuel efficiency.

3.4.3.3 Maximum Velocity

The power requirement from the propulsion system at maximum velocity is $F_{TR} \cdot v_{max}$, which is supplied by a combination of the IC engine and the electric motor. The power requirement for maximum velocity is

$$P_{TR,max} = F_{TR} \cdot v_{max} = mgv_{max} \sin\beta + \left[mgC_1 + \frac{\rho}{2} A_F C_D \right] v_{max}^3 + mgv_{max} C_0 \quad (3.8)$$

With the IC engine and electric motor sized to meet the acceleration and cruising speed requirements, it has to be verified that the combination of the two can meet the vehicle maximum velocity requirements. If the combined power is less than the power required for maximum velocity, then the size of one of the power plants should be increased.

3.4.3.4 Generator Sizing

The generator is sized to maintain series operation of the vehicle for typical urban driving conditions. The calculations showed that cruising at constant velocity of 40 mi/h with a driver and a passenger on a 1% grade and 10% drivetrain losses requires 12 kW. Allowing for power to recharge the batteries, the generator can be sized to be 20 kW for the Akron hybrid vehicle.

3.4.3.5 Battery Sizing

The energy required by the electric motor for initial acceleration and a minimum zero emission range with constant velocity can be used to estimate the size of the battery. The energy for initial acceleration, calculated by integrating the power required for acceleration between 0 and 9 s, is 1.62 MJ or 450 W h. The energy required for cruising at a constant velocity of 40 mi/h with a driver and a passenger on a 1% grade and 10% drivetrain losses is 3 kW h for a 10 mi (15 min) battery-only range or 1.5 kW h for a 5 mi (7.5 min) battery-only range. For a 300 V battery-pack, this requires a 10 A h or 5 A h battery-pack, respectively. Adding 10% margin, battery-packs with 11 A h or 5.5 A h can be used for 10 or 5 mi ZEV range, respectively. The estimated masses for three types of battery chemistry to meet these requirements are

TABLE 3.3

Battery Energy Requirements and Mass Estimates

Battery Type	Peak Acceleration and 10 mi ZEV Range at 40 mi/h Urban Driving			Peak Acceleration and 5 mi ZEV Range at 40 mi/h Urban Driving		
	Energy (W h)	Mass (kg)	Capacity (A h)	Energy (W h)	Mass (kg)	Capacity (A h)
Lead acid	3.5	94.3	11	2	47.2	5.5
NiMH	3.5	51.2	11	2	25.6	5.5
Li-ion	3.5	36.8	11	2	18.2	5.5

given in Table 3.3 based on nominal energy densities. The battery pack voltage is considered to be 300 V DC.

3.5 Mass Analysis and Packaging

Vehicle mass budget must be evaluated during the design of a hybrid vehicle whether it is made from ground up or re-engineered from an IC engine vehicle. The vehicle fuel economy and emissions will improve as vehicle mass decreases. Hybrids tend to be a little heavier compared to a similarly sized IC engine vehicle due to the use of both electrical and mechanical components in the powertrain. However, a good design can yield comparable masses for similarly rated hybrids and IC engine vehicles. The downsized engine in the hybrid vehicle can more than make up for the masses added due to hybridization, which includes masses for electric machines, their controllers, and cooling systems, and traction battery and its accessories. Mass advantages are also to be gained in hybrids by reducing fuel system sizes and removing the 12 V battery.

A mass comparison of crossover vehicles between the Chevrolet Equinox and the converted Akron hybrid is given in Table 3.4. The engine downsizing from a 3.6 L V6 gasoline engine to a 1.9 L diesel engine did not give significant mass advantage since diesel engines tend be larger and heavier than gasoline engines for comparable sizes. The exhaust system mass increased due to the addition of diesel aftertreatment components. The fuel tank size was reduced by 50%, which provided some mass savings for the hybrid. The masses of the electrical machines, controllers, traction battery, and thermal management systems depend on the components chosen. The removal of the alternator and downsizing the 12 V accessories battery provided small mass savings. The cabin climate control system mass is somewhat higher for the hybrid due to the switching to an electric motor driven compressor pump. The example analysis in Table 3.4 shows an increase of 82% in the powertrain mass for the Akron hybrid vehicle. The mass budget could have been restricted significantly had a downsized gasoline engine been used, although that would have adversely affected fuel economy.

TABLE 3.4
Mass Analysis of ICEV and HEV

Component	ICE Vehicle Mass (kg)	Hybrid Vehicle Mass (kg)
Engine and transmission	147	125
Exhaust system	40	50
Fuel system (tank and lines)	13	9
Fuel mass	38	15
Electric drive motor		108
Starter/generator	6	22
Starter/generator controller		26
Electrical thermal management		15
Traction battery		75
Battery hardware, cooling		28
12 V battery	14	6
Alternator	5	
Total powertrain	263	479
Climate control and accessory	26	30
Glider with Chassis subsystems	1445	1449
Total Curb mass	1734	1954

Packaging of components also has to be carefully evaluated during the design of a hybrid vehicle. A component packaging diagram must be developed to ensure that all the system components fit in the vehicle without compromise to safety and customer amenities. A simplified layout diagram based on the Akron hybrid is shown in Figure 3.15; the fuel lines and exhaust

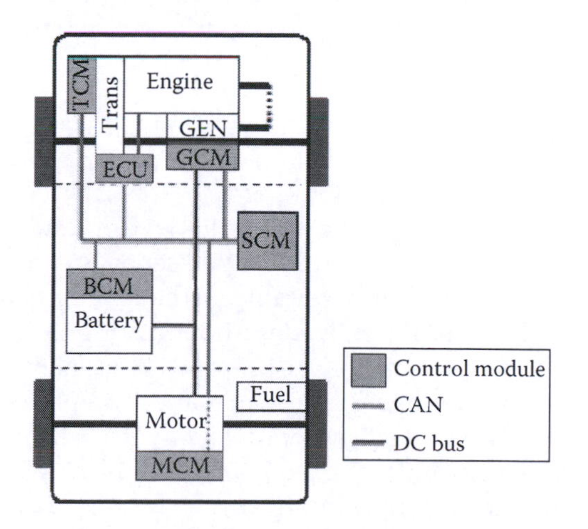

FIGURE 3.15
Component packaging in a hybrid vehicle.

piping are not shown here for keeping the figure simple. The packaging of the fuel and exhaust components can become cumbersome, especially when an IC engine vehicle is being re-engineered into a hybrid vehicle. Adequate spaces under the vehicle must be available for routing the fuel lines and the exhaust piping. A high voltage electrical circuit diagram has to be developed alongside fuel line and exhaust system drawing. While analyzing the packaging and mounting of the hybrid components, front to rear weight ratios must be maintained at 60:40 or less for acceptable drivability, ride performance, and dynamic braking.

In conclusion, the mass savings in engine and fuel system downsizing in a hybrid vehicle will more or less be consumed by the additional components added, and the packaging issues are much more complicated for re-engineered vehicles. It would always be advantageous to design a hybrid from ground up rather than retrofit and re-engineer an existing IC engine vehicle.

3.6 Vehicle Simulation

The simulation model of a vehicle system is an essential tool for the design and performance evaluation of the vehicle. With numerous choices available on hybrid architectures, and on powertrain component technologies, simulation saves time and cost in predicting tank-to-wheel energy usage as well as in-depth vehicle performance outputs. The simulation models can be developed with varying degrees of sophistication as required by the user. At the global system level, a simulation model can predict fuel economy, emissions, or wheel torque for a regulated drive cycle; at the subsystem level, component parameters such as electric machine or IC engine torque are available. The simulation model can be used to optimize and tune the supervisory control system of a hybrid vehicle.

3.6.1 Simulation Model

For vehicle simulation, a forward looking model is desired where a vehicle model responds to driver input commands to develop and deliver torque to the wheels similar to what it is in a real world scenario. A modular structure of the simulation model is highly desirable so that components and configurations can be easily added or removed without having to start from scratch. This facilitates comparisons among different component technologies or components of different ratings. The performance comparison of alternative hybrid architectures is also easily accomplished with the modular structure.

The models are developed using both physical principles and empirical data. The vehicle dynamics and roadway fundamentals presented in Chapter 2 are the physical principles based on which the vehicle model can be developed. The driver behavior is modeled using various drive cycles, which

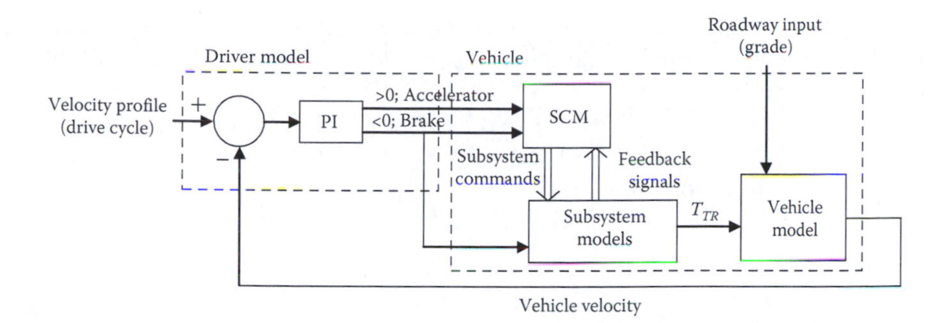

FIGURE 3.16
Vehicle simulation block diagram.

are standardized speed and road grade profiles used to evaluate a vehicle's performance. For the components, either simplified models or look-up tables based on empirical data are used in the simulation. Detailed dynamic behavior of the components is often not required, since the time responses of many of the subsystems (such as the electric machine) are much faster than the vehicle dynamics. Simpler models of subsystems minimize the simulation time, since otherwise it would be significant for such the highly complex vehicle system. With this approach, it is possible to run a large number of simulations in order to tune the control strategy to perform well under a wide range of driving conditions. Because there are both mechanical and electrical dynamics, a variable-step stiff system solver is required.

The simulation block diagram of a vehicle system showing the interactions among the driver, the supervisory controller, the subsystems, and the vehicle model is shown in Figure 3.16. The driver is modeled with a drive cycle and a PI regulator designed to minimize the error between the drive cycle velocity profile and the actual vehicle velocity. The speed profile of the drive cycle is assumed to be the vehicle speed desired by the driver. The PI regulator output creates a pedal position that is interpreted as either the accelerator pedal position or the brake pedal position. It is assumed that both pedals cannot be depressed simultaneously. The pedal position is read into the control strategy (labeled as SCM for supervisory control module) to generate the subsystem commands. The powertrain, modeled in the subsystem block, produces the tractive torque T_{TR}, which is sent to the vehicle model. The brake pedal position input is passed through directly into the brake subsystem model to address safety concerns; however, the brake pedal position is also required by the supervisor controller to determine the magnitude of the powertrain regeneration command.

3.6.2 Standard Drive Cycles

The standard drive cycles are used by the governmental agencies and automotive industry for performance evaluation of a vehicle, which includes certification of vehicle fuel economy. A drive cycle may have both speed and road grade components, although typically one is held constant while the other is varied.

Two commonly used standard drive cycles are the urban dynamometer driving schedule (UDDS) and the highway fuel economy test (HWFET), simulating urban and highway driving, respectively. The two drive cycles are shown in Figures 3.17 and 3.18. The UDDS drive cycle runs a distance of

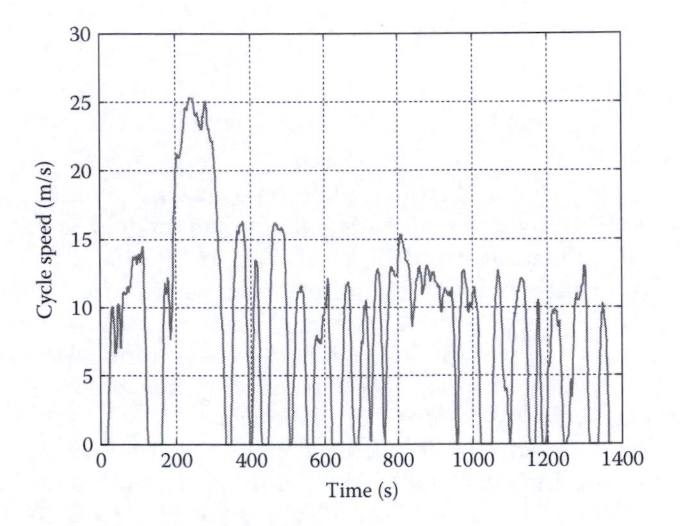

FIGURE 3.17
Urban dynamometer driving schedule (UDDS).

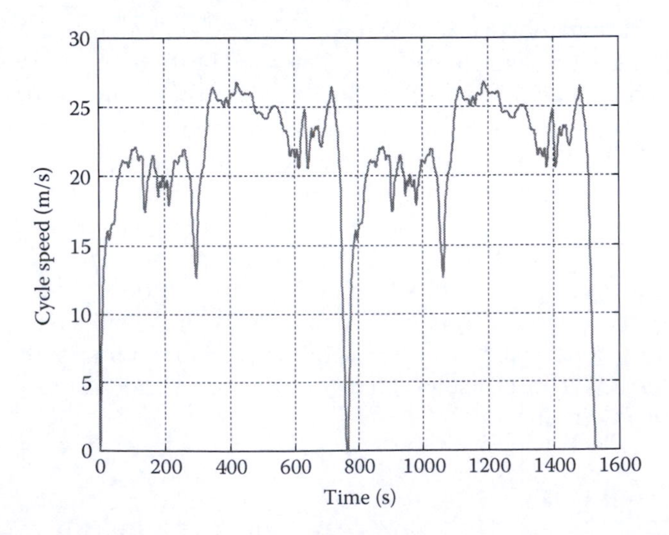

FIGURE 3.18
Highway fuel economy driving schedule (HWFET).

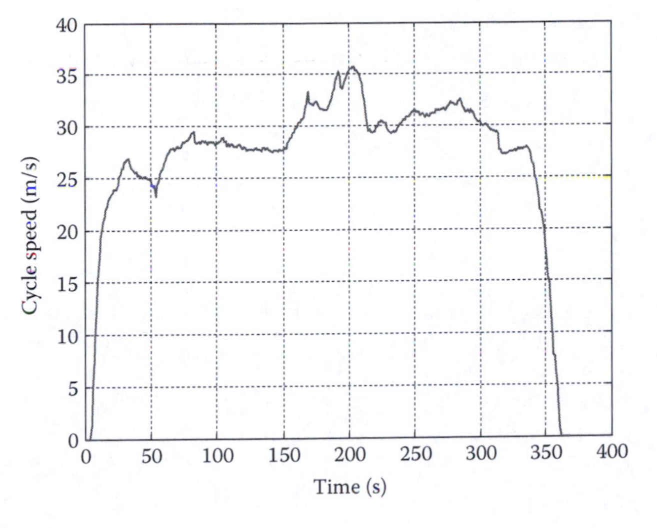

FIGURE 3.19
US06 standard drive cycle.

7.5 mi in 1369 s with frequent stops and has an average speed of 19.6 mi/h. The HWFET drive cycle runs a distance of 10.26 mi in 765 s and has an average speed of 48.3 mi/h.

Another standard drive cycle is the US06 drive cycle, which is used to test the effectiveness of the vehicle control strategy under extreme driving conditions. The drive cycle is shown in Figure 3.19. The US06 drive cycle is the most aggressive of the three drive cycles shown so far. It features rapid acceleration and hard braking, and has top speeds just over 80 mi/h. The Japanese technical standard drive cycle J1015, shown in Figure 3.20, is also useful in simulating urban driving with frequent stops.

The section will conclude describing the standard J227a driving cycle recommended by the Society of Automotive Engineers (SAE) to evaluate the performance of EV and energy sources. This short drive cycle is useful for hand calculations needed to develop the concepts of electric and hybrid vehicle fundamentals. The SAE J227a has three schedules designed to simulate the typical driving patterns of fixed-route urban, variable-route urban, and variable-route suburban travels. These three patterns are the SAE J227a driving schedules B, C, and D, respectively. Each schedule has five segments in the total driving period: (1) Acceleration time t_a to reach the maximum velocity from start-up, (2) cruise time t_{cr} at a constant speed, (3) coast time t_{co} when no energy is drawn from the source, (4) brake time t_b to bring the vehicle to stop, and (5) idle time t_i prior to the completion of the period. The driving cycle for J227a is shown in Figure 3.21 with the recommended times for each of the schedules given in Table 3.5. The figure drawn is slightly modified from the pattern recommended by the SAE. The J227a procedures specify only the cruise velocity and the time of transition from one mode to

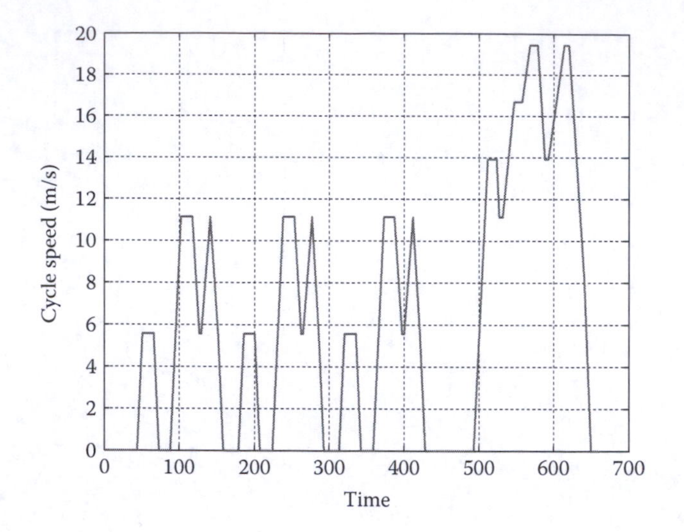

FIGURE 3.20
Japan1015 Japanese standard drive cycle.

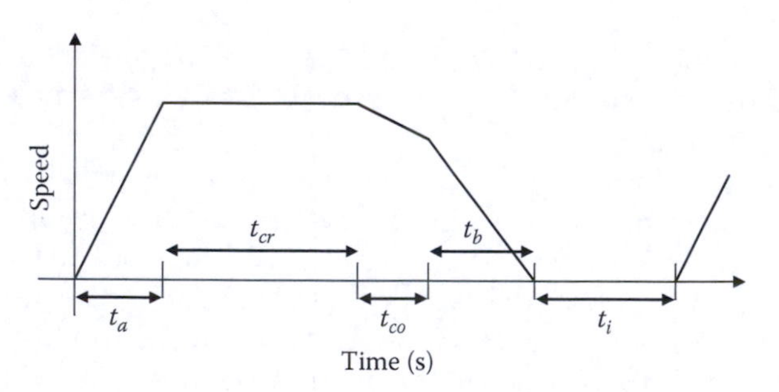

FIGURE 3.21
SAE J227a standard driving cycle.

TABLE 3.5
SAE J227a Standard Driving Schedules

	SAE J227a Schedules		
Test Parameter	**B**	**C**	**D**
Max. speed, km/h (mi/h)	32	48	72
Acceleration time, t_a (s)	19	18	28
Cruise time, t_{cr} (s)	19	20	50
Coast time, t_{co} (s)	4	8	10
Brake time, t_{br} (s)	5	9	9
Idle time, t_i (s)	25	25	25
Total time (s)	72	80	122
Approximate number of cycles per mile	4–5	3	1

the other. The velocity profile at segments other than the cruising part is not fixed, and hence the distance traversed during these other periods is also variable. In reality, the distances would depend on the acceleration capability of the vehicle under consideration. For simplicity, straight line approximations have been assumed for these schedules in this book.

Problems

3.1 The parameters of a parallel HEV are as follows:

Description	Parameters
Vehicle mass	1800 kg
Driver/one passenger	80 kg
Rolling resistance coefficient, C_0	0.01
Aerodynamic drag coefficient, C_{AD}	0.4
Frontal area, A_F	2.6 m^2

The vehicle is to accelerate uniformly (i.e., constant acceleration) from 0 to 60 mi/h in 10 s.

(a) Find an expression for traction force as a function of time $F_{TR}(t)$ for this initial acceleration period.

(b) Find an expression for traction power as a function of time $P_{TR}(t)$ for the initial acceleration period.

(c) Find the energy Δe_{TR} require for initial acceleration between 0 and 10 s.

3.2 The vehicle parameters and performance requirements of the vehicle are used for sizing calculations:

Description	Requirements
Vehicle mass	1800 kg
Driver/one passenger	176 lb/80 kg
Trailering capacity	1000 kg
Rolling resistance coefficient, C_0	0.01
Wheel radius, r_{wh}	0.3305 m
Aerodynamic drag coefficient, C_{AD}	0.45
Frontal area, A_F	2.5 m^2
0–60 mi/h	8.0 s
50–70 mi/h	6.3 s
Sustained cruising speed (with trailer)	55 mi/h at 7% grade
Sustained cruising speed (without trailer)	70 mi/h at 0.5% grade
Zero emission range (without trailer)	5 mi at 40 mi/h

The initial acceleration requirement is without the trailer, but the total mass must account for the driver and one passenger:

(a) Calculate the velocity and power at the end of 5 s for a constant force initial acceleration. The constant force is 7600 N.
(b) The vehicle accelerates in the constant power mode after the initial constant force acceleration period of part (a) with a constant power of 140 kW. Write the *dv/dt* equation for constant power acceleration for the given conditions. What is the velocity after a total time of 8 s? (You can solve the equation numerically).
(c) Calculate the power required for a steady state velocity of 55 mi/h at 7% with trailer.

References

1. C.C. Chan and K.T. Chau, *Modern Electric Vehicle Technology*, Oxford University Press, Oxford, U.K., 2001.
2. R. Hodkinson and J. Fentos, *Lightweight Electric/Hybrid Vehicle Design*, SAE International, Warrendale, PA, 2001.
3. SAE, *Strategies in Electric and Hybrid Vehicle Design*, SAE Publication SP-1156, SAE International, Warrendale, PA, 1996.
4. A. Nagasaka et al., *Development of Hybrid/Battery ECU for the Toyota Hybrid System, Technology for Electric and Hybrid Vehicles*, SAE Publication SP-1331, SAE International, Warrendale, PA, 1998.
5. J.F. Ronning and G.L. Grant, *Global Hybrid Electric Vehicle Markets and Missions*, Electric and Hybrid Electric Vehicles and Fuel Cell Technology, SAE Publication SP-1466, SAE International, Warrendale, PA, 1999.
6. M. Ehsani, K.M. Rahman, and H.A. Toliyat, Propulsion system design for electric and hybrid vehicles, *IEEE Transactions on Industrial Electronics*, 44(1), 19–27, February 1997.
7. J.R. Howell and R.O. Buckius, *Fundamentals of Engineering Thermodynamics*, 2nd edn., McGraw Hill, New York, 1992.
8. H.K. Ng, A.D. Vyas, and D.J. Santini, *The Prospects for Hybrid Electric Vehicles, 2000–2005: Results of a Delphi Study, Electric and Hybrid Electric Vehicles and Fuel Cell Technology*, SAE Publication SP-1466, SAE International, Warrendale, PA, 1999, pp. 95–106.

4

Battery Energy Storage

Energy, as available in nature for conversion and usage by the society, is known as primary energy. The sources of primary energy are in the wells or in other natural environments such as in the wind or in the sun. Examples of primary energy sources are fossil fuels, wind, solar, hydro, wave, nuclear, geothermal, and biomass. Energy is extracted from the primary source either by burning it or by transforming the source energy through an energy conversion system. When primary energy is transformed by one or more energy conversion processes and/or devices, it is known as secondary energy; additional energy conversion devices use the secondary energy to deliver useful work. Several stages of energy conversion are necessary, first, for processing of the primary energy, and then for delivering it to the end user. Inefficiencies at varying degrees are associated in each of these conversion stages.

Primary sources have energy stored in them in chemical, heat, kinetic, or some other alternative form. For example, energy is stored in fossil fuels in chemical form; energy is extracted from it in an internal combustion (IC) engine vehicle by burning it in a heat engine. Wind energy is available in kinetic form, which can be converted to electrical energy using a wind turbine. Electrical energy is an example of secondary energy that can be converted to mechanical work by an electric machine. Electrical energy can be obtained from the primary source by burning fossil fuels in thermal power plants, or from renewable primary sources of water, wind, solar, or other primary sources. Electrochemical devices also produce electrical energy from chemical energy.

Energy derived from sources other than the burning of fossil fuels is known as alternative energy. Ideally, converters used for the processing of alternative energy must avoid the usage of fossil fuels during any stage of the energy conversion process. For example, the ideal solution for recharging batteries in electric vehicles is to use electricity derived from renewable energy sources such as wind or solar. If electricity from coal-fired power plants is used for recharging, then the environmental problems are only being shifted from one location to another.

The energy content of various raw fuels or materials refers to the energy that can be extracted from it for useful work. The parameter for energy content evaluation is *specific energy* or *energy density*. Specific energy is the energy per unit mass of the energy source, and its unit is W h/kg. For fossil fuels, the energy content refers to the calorific or thermal energy that can be extracted from it by burning. Energy content for other materials is similarly

TABLE 4.1
Nominal Energy Densities of Sources

Energy Source	Nominal Specific Energy (W h/kg)
Gasoline	12,500
Diesel	12,000
Biodiesel	10,900
Natural gas	9,350
Ethanol	8,300
Methanol	6,050
Hydrogen	33,000
Coal (bituminous)	8,200
Lead-acid battery	35
Li-polymer battery	200
Flywheel (carbon-fiber)	200

evaluated in terms of specific energy for a level comparison. The specific energies of several energy sources are given in Table 4.1. These energies are shown without taking containment into consideration. The specific energies of hydrogen and natural gas would be significantly lower than that of gasoline when containment is considered.

One of the challenges of energy is to store it in a convenient form so that it can be utilized when needed. In terms of storage, fossil fuels have the biggest advantage since these can be conveniently stored in a container. On the other hand, electrical energy is very useful in delivering work on demand using a highly efficient electromechanical device; however, energy storage in electrical form is not simple.

The mechanism for energy delivery must also be considered for the system under consideration. The energy transportation system has to be secure, efficient, and environmentally friendly. Energy transportation over long distances using electrical transmission and distribution systems is highly efficient. Once an infrastructure is in place, energy converted to electrical form at one end (which is the generation side) can be delivered to industrial and residential units through the transmission and distribution system. On the other hand, fossil fuels are transported long distances using pipelines, ocean liners, and land transportation before being finally dispersed at the gas stations.

The electrochemical devices are the most promising alternative technologies to the conventional fossil fuel–burning power plants, both in vehicle and utility power stations. The electrochemical energy conversion processes have the advantages of high conversion efficiency, large enough power output, and the availability of a wide selection of fuels. Batteries and fuel cells are energy storage and power generating devices that are suitable for both portable and stationary applications. The topics covered in this chapter include the fundamentals of battery structure and operation, and battery applications in electric and hybrid vehicles. The presentations will help a design engineer size the energy storage system, and then select the appropriate battery technology for an application. Additional alternative energy storage devices are covered in the next chapter.

4.1 Batteries in Electric and Hybrid Vehicles

The basic requirement for purely electric vehicles is a portable supply of electrical energy, which is converted to mechanical energy in the electric motor for vehicle propulsion. The electrical energy is typically obtained through the conversion of chemical energy stored in devices such as batteries and fuel cells. The portable electrical energy storage presents the biggest obstacle in the commercialization of electric vehicles. One solution for minimizing the environmental pollution problem due to the lack of a suitable, high-energy density energy storage device for electric vehicles is the hybrid electric vehicles that combine the propulsion efforts from gasoline engines and electric motors.

Among the available choices of portable energy sources, batteries have been the most popular choice of energy source for electric vehicles since the beginning of research and development programs in these vehicles. The electric vehicles and hybrid electric vehicles that are available commercially today use batteries as the electrical energy source. The desirable features of batteries for electric and hybrid electric vehicle applications are high specific power, high specific energy, high charge acceptance rate for both recharging and regenerative braking, and long calendar and cycle lifes. Additional technical issues include methods and designs to balance the battery segments or packs electrically and thermally, accurate techniques to determine a battery's state of charge, and recycling facilities for battery components. Above all, the cost of batteries must be reasonable for electric and hybrid vehicles to be commercially viable.

There are two basic types of batteries: a primary battery and a secondary battery. Batteries that cannot be recharged and are designed for a single discharge are known as primary batteries. Examples of these are the lithium batteries used in watches, calculators, cameras, etc., and the manganese dioxide batteries used to power toys, radios, flashlights, etc., Batteries that can be recharged by flowing current in the direction opposite to that during discharge are known as secondary batteries. The chemical reaction process during cell charge operation when electrical energy is converted into chemical energy is the reverse of that during discharge. The batteries needed and used for electric and hybrid vehicles are all secondary batteries, since they all are recharged either during a mode of vehicle operation or during the recharging cycle in the stopped condition using a charger. All the batteries discussed in this book are examples of secondary batteries.

The lead-acid type of battery has the longest development history of all battery technology, particularly for their need and heavy use in the industrial electric vehicles, such as golf cars in sports, passenger cars in airports, forklifts in storage facilities and supermarkets. The power and energy densities of lead-acid battery are lower compared to several other battery technologies. The research and development for alternative batteries picked up momentum following the resurgence of interest in electric vehicles and hybrid vehicles in the late 1960s and early 1970s. The sodium-sulfur batteries showed great promise in the 1980s with high energy and power densities, but safety and manufacturing difficulties led to the abandoning of the technology. The development of battery technology for low power applications, such as cell phones and calculators opened the possibilities of scaling the energy and power of nickel-cadmium and lithium-ion-type batteries for electric and hybrid vehicle applications. The major types of rechargeable batteries used or being considered for electric and hybrid vehicle applications are

- Nickel-metal-hydride (NiMH)
- Lithium-ion (Li-ion)
- Lithium-polymer (Li-poly)
- Sodium-sulfur

The Li-ion battery technology is the most promising among the four battery chemistries mentioned above. It must also be stated that there are several different types of Li-ion battery technologies that are being developed for electric and hybrid vehicles.

The development of batteries is directed toward overcoming the significant practical and manufacturing difficulties. The theoretical predictions are difficult to match in manufactured products due to practical limitations. The theoretical and practical specific energies of several batteries are given in Table 4.2 for comparison.

The battery technology has gone through extensive research and development efforts over the past 30 years, yet there is currently no battery that can deliver an acceptable combination of power, energy, and life cycle at a reasonable cost for high-volume production urban usage electric vehicles. However, the extensive research and interest in alternative vehicles have resulted in several promising battery technologies. Pure electric vehicles are available commercially, but the cost is prohibitive since these require large capacity batteries. Hybrid electric vehicles minimize the battery capacity through using the combination of IC engine and electric machines. Even though

TABLE 4.2

Specific Energies of Batteries

	Specific Energy (W h/kg)	
Battery	**Theoretical**	**Practical**
Lead-acid	108	50
Nickel-cadmium		20–30
Nickel-zinc		90
Nickel-iron		60
Zinc-chlorine		90
Zinc-bromide		70
Silver-zinc	500	100
Sodium-sulfur	770	150–300
Aluminum-air		300
Nickel metal		70
Hydride		150
Li-ion		

the hybrid technology is much more complex than either a conventional or electric vehicle alone, the advantages gained in fuel economy through the addition of a smaller electric powertrain component are significant. Hybrid electric vehicles are commercially available for more than 10 years. Plug-in hybrid vehicles are the next step forward where larger capacity batteries are required. Several automotive manufacturers have announced the production plans of plug-in hybrid vehicles in the coming years. These steps along with societal interest and governmental incentives promote innovations that will ultimately drive the cost down. The next step toward the use of battery electric vehicles for urban transportation will be well within reach as traction battery cost comes down and technology improves. The move toward a battery electric vehicle becomes easier if consumers can accept vehicles with a range of 50 mi instead of 300 mi. This is certainly possible since the average daily commute is 32.8 mi/day in the United States, assuming that an average vehicle is driven 12,000 mi/year. The average number of miles driven per day is much less in other countries of the world.

The fundamentals of a battery are in a unit cell, and include the electrodes and electrolytes used, the chemical reactions involved, and the cell potential developed. In the following section, cell structure and chemical reactions are discussed with examples of some of the common cell chemistries. This will enable us to define the parameters of a battery, which are needed for the macroscopic point of view. We will then dig deeper into the theoretical aspects of a battery cell for analyzing, evaluating, and modeling different types of batteries. The chapter will then present cell chemistries of some of the promising battery technologies for electric and hybrid vehicle applications.

Finally in this chapter, battery-pack design for electric and hybrid vehicles' applications will be addressed.

4.2 Battery Basics

The batteries are made of unit cells containing the stored chemical energy that can be converted to electrical energy. One or more of these electrochemical cells are connected in series to form one battery. The grouped cells are enclosed in a casing to form a battery module. A battery pack is a collection of these individual battery modules connected in a series and/or parallel combination to deliver the desired voltage and energy to the power electronic drive system.

4.2.1 Battery Cell Structure

The energy stored in a battery is the difference in free energy between chemical components in the charged and discharged states. This available chemical energy in a cell is converted into electrical energy only on demand using the basic components of a unit cell; these components are the positive and negative electrodes, the separators, and the electrolytes. The electrochemically active ingredient of the positive or negative electrode is called the active material. Chemical reactions take place at the two electrodes, one of which releases electrons while the other consumes those. The electrodes must be electronically conducting and are located at different sites separated from each other by a separator, as shown in Figure 4.1. The connection points between the electrodes and the external circuit are called the battery terminals. The external circuit ensures that stored chemical energy is released only on demand and utilized as electrical energy.

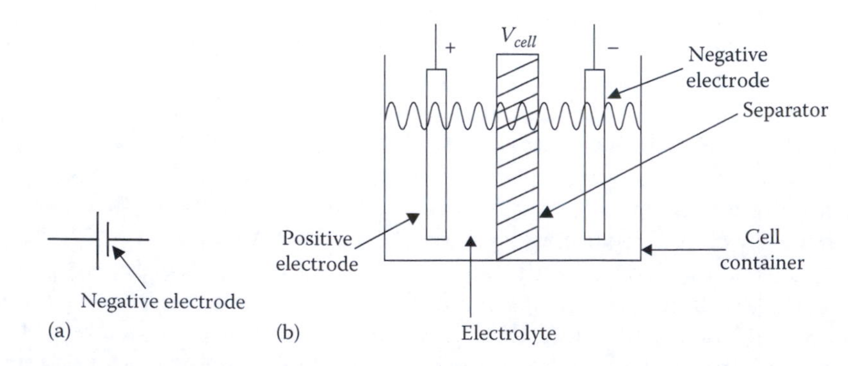

FIGURE 4.1
Components of a battery cell. (a) Cell circuit symbol; (b) cell cross-section.

The components of the battery cell are described as follows:

1. Positive electrode: The positive electrode is an oxide or sulfide or some other compound that is capable of being reduced during cell discharge. This electrode consumes electrons from the external circuit during cell discharge. Examples of positive electrodes are lead oxide (PbO_2) and nickel oxyhydroxide (NiOOH). The electrode materials are in the solid state.
2. Negative electrode: The negative electrode is a metal or an alloy that is capable of being oxidized during cell discharge. This electrode releases electrons to the external circuit during cell discharge. The examples of negative electrodes are lead (Pb) and cadmium (Cd). The negative electrode materials are in the solid state within the battery cell.
3. Electrolyte: The electrolyte is the medium that permits ionic conduction between positive and negative electrodes of a cell. The electrolyte must have high and selective conductivity for the ions that take part in electrode reactions, but must be a nonconductor for electrons in order to avoid self-discharge of batteries. The electrolyte may be liquid, gel, or solid material. Also, the electrolyte can be acidic or alkaline, depending on the type of battery. Traditional batteries such as lead-acid and nickel-cadmium use liquid electrolytes. In lead-acid batteries, the electrolyte is the aqueous solution of sulfuric acid (H_2SO_4(aq)). Advanced batteries for electric vehicles, such as sealed lead-acid, NiMH, and lithium-ion batteries use an electrolyte that is gel, paste, or resin. Lithium-polymer batteries use a solid electrolyte.
4. Separator: The separator is the electrically insulating layer of material, which physically separates electrodes of opposite polarity. Separators must be permeable to the ions of the electrolyte and may also have the function of storing or immobilizing the electrolyte. Present day separators are made from synthetic polymers.

4.2.2 Chemical Reactions

During battery operation, chemical reactions at each of the electrodes are sustainable only if electrons generated at the electrodes are able to flow through an external electrical circuit that connects the two electrodes. When a passive electrical circuit element is connected to the electrode terminals of a battery, electrons are released from the negative electrode and consumed at the positive electrode, resulting in current flow into the external circuit. In this process, the battery gets discharged. The supply of electrons is due to the chemical reactions at the electrode surfaces inside the battery cell, which are collectively known as reduction and oxidation (redox) reactions. During battery discharge, the positive electrode gets chemically reduced as it absorbs electrons from the external circuit; the negative electrode gets oxidized as it releases electrons to the external circuit. For battery charging, a source with voltage higher than the battery terminal voltage has to be applied so that current can flow into the battery in the opposite direction. During charging, electrons are released at the positive electrode and consumed at the negative electrode; consequently, the positive electrode is oxidized and negative electrode is reduced.

Regardless of the battery cell chemistry, redox reactions take place at the electrodes during both cell charging and discharging for the release and absorption of electrons at the terminals. The generalized redox reactions are given by [1]

$$aA \underset{\text{Discharge}}{\overset{\text{Charge}}{\rightleftarrows}} cC + nE^{+} + ne^{-} \tag{4.1}$$

for the positive electrode, and

$$bB + nE^{+} + ne^{-} \underset{\text{Discharge}}{\overset{\text{Charge}}{\rightleftarrows}} dD \tag{4.2}$$

for the negative electrode. The combined chemical reaction is

$$aA + bB \underset{\text{Discharge}}{\overset{\text{Charge}}{\rightleftarrows}} cC + dD \tag{4.3}$$

Chemical reactions 4.1 and 4.2 illustrate that electrons are released and absorbed during any redox reaction. The positive electrode reaction 4.1 shows that during cell charging, species A within the electrode is oxidized and becomes energized species C, releasing electron(s) into the external circuit and positive ion(s) into the electrolyte. Similarly, the negative electrode reaction 4.2 shows that species B at the electrode combines with positive ion(s) from the electrolyte and electron(s) from the external circuit to form energized species D. The converse is true at the two electrodes during cell discharging. The coefficients *a*, *b*, *c*, and *d* represent the numbers of moles associated with the species in the reactions; the coefficient *n* represents the number of electrons and ions involved in the redox reactions.

In electric traction applications, battery cell operation is that of cell discharging when the energy is supplied from the battery to the electric motor for propulsion power and of cell charging when energy is supplied from an external source to store energy in the battery. In conventional vehicles, battery cells supply power to electrical accessories while discharging, and accept energy from an external device to replenish the stored energy during charging. We will next review the redox reactions during cell charging and discharging in a few battery chemistries, starting with the lead-acid battery

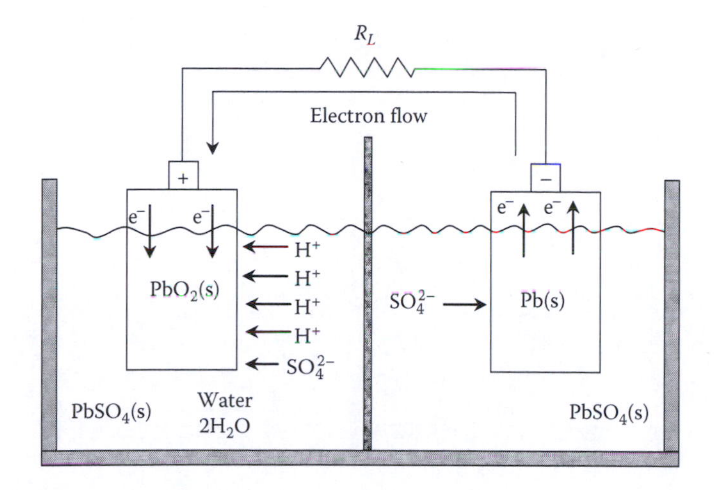

FIGURE 4.2
Lead-acid battery: cell discharge operation.

cell. Lead-acid is still the battery choice for powering electrical accessories in conventional, electric, and hybrid electric vehicles.

Figure 4.2 shows the cell discharge operation of a lead-acid battery cell into a passive resistive element. The positive electrode made of lead-oxide (PbO_2) is reduced by consuming electrons and ions. The electron supply is through the external circuit which originates at the negative electrode. The current flow is therefore out of the positive electrode into the electrical load with the battery acting as the source. The positive electrode reaction is given by

$$PbO_2(s) + 4H^+(aq) + SO_4^{2-}(aq) + 2e^- \rightarrow PbSO_4 + 2H_2O(l)$$

A highly porous structure is used for the positive electrode to increase $PbO_2(s)$/electrolyte contact area. A porous electrode structure results in higher current densities since PbO_2 is converted to $PbSO_4(s)$ during cell discharge. As discharge proceeds, the internal resistance of the cell rises due to $PbSO_4$ formation and decreases the electrolyte conductivity as H_2SO_4 is consumed. $PbSO_4(s)$ deposited on either electrode in a dense, fine-grain form can lead to sulfation. The discharge reaction is largely inhibited by the buildup of $PbSO_4$, which reduces cell capacity significantly from the theoretical capacity.

The negative electrode is made of solid lead (Pb); during cell discharge lead is oxidized, releasing electrons into the external circuit. The negative electrode reaction during cell discharge is

$$Pb(s) + SO_4^{2-}(aq) \rightarrow PbSO_4 + 2e^-$$

The production of $PbSO_4(s)$ can degrade battery performance by making the negative electrode more passive.

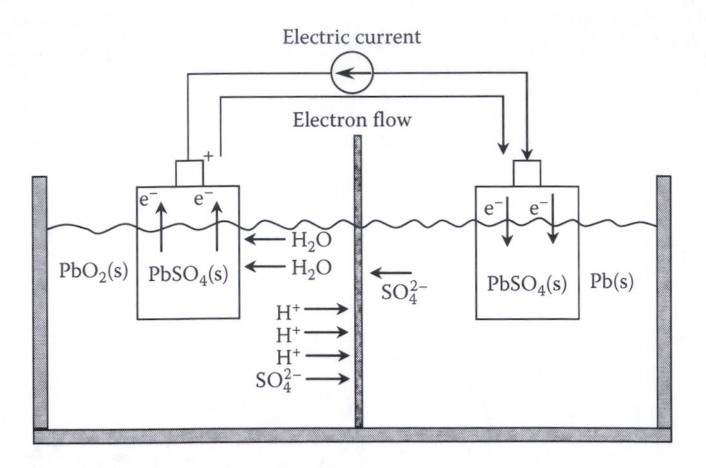

FIGURE 4.3
Lead-acid battery: cell charge operation.

The overall cell discharge chemical reaction is

$$Pb(s) + PbO_2(s) + 2H_2SO_4(aq) \rightarrow 2PbSO_4 + 2H_2O(l)$$

The cell charge operation is the reverse of the cell discharge operation. An external electrical source supplies current into the battery to reverse the chemical reactions as shown in Figure 4.3. During cell charging, the lead sulfate is converted back to the reactant states of lead and lead oxide. The electrons are consumed from the external source at the negative electrode, while the positive electrode produces the electrons. The current flows into the positive electrode from the external source, thereby delivering electrical energy into the cell where it gets converted into chemical energy. The positive electrode is oxidized, releasing electrons during cell charging as follows:

$$PbSO_4(s) + 2H_2O(L) \rightarrow PbO_2(s) + 4H^+(aq) + SO_4^{2-}(aq) + 2e^-$$

The negative electrode is reduced during cell charging absorbing electrons, and the chemical reaction is

$$PbSO_4(s) + 2e^- \rightarrow Pb(S) + SO_4^{2-}(aq)$$

The overall chemical reaction during cell charging is

$$2PbSO_4(s) + 2H_2O(l) \rightarrow Pb(s) + PbO_2(s) + 2H_2SO_4(aq)$$

The nickel-cadmium (NiCd) and NiMH batteries are examples of alkaline batteries where electrical energy is derived from the chemical reaction of a metal with oxygen in an alkaline electrolyte medium. The specific energy of alkaline batteries is lowered due to the mass addition of the carrier metal. The NiCd battery employs a nickel-oxide positive electrode and a metallic-cadmium negative electrode. The reactions take place in the presence of potassium hydroxide (KOH) electrolyte. The positive electrode chemical reaction is

$$NiOOH + H_2O + e^- \underset{\text{Charge}}{\overset{\text{Discharge}}{\rightleftarrows}} Ni(OH)_2 + OH^-$$

The negative electrode chemical reaction is

$$Cd + 2OH^- \underset{\text{Charge}}{\overset{\text{Discharge}}{\rightleftarrows}} Cd(OH)_2 + 2e^-$$

The overall chemical reaction is

$$Cd + 2NiOOH + 2H_2O \underset{\text{Charge}}{\overset{\text{Discharge}}{\rightleftarrows}} 2Ni(OH)_2 + Cd(OH)_2$$

In NiMH batteries, the positive electrode is a nickel oxide similar to that used in a NiCd battery, while the negative electrode has been replaced by a metal hydride which stores hydrogen atoms. The concept of NiMH batteries is based on the fact that fine particles of certain metallic alloys, when exposed to hydrogen at certain pressures and temperatures, absorb large quantities of the gas to form the metal hydride compounds. Furthermore, the metal hydrides are able to absorb and release hydrogen many times without deterioration. The two electrode chemical reactions in a NiMH battery are as follows:

At the positive electrode,

$$NiOOH + H_2O + e^- \underset{\text{Charge}}{\overset{\text{Discharge}}{\rightleftarrows}} Ni(OH)_2 + OH^-$$

At the negative electrode,

$$MH_x + OH^- \underset{\text{Charge}}{\overset{\text{Discharge}}{\rightleftarrows}} MH_{x-1} + H_2O + e^-$$

where

M stands for metallic alloy, which takes up hydrogen at ambient temperature to form the metal hydride MH_x

x is the number of hydrogen atoms absorbed

The overall chemical reaction is

$$NiOOH + MH_x \underset{\text{Charge}}{\overset{\text{Discharge}}{\rightleftarrows}} Ni(OH)_2 + MH_{x-1}$$

4.3 Battery Parameters

In this section, the various battery parameters including capacity and state of charge (SoC) are defined. The parameters mostly relate to the terminal characteristics of the battery that have the practical implications for use in an application.

4.3.1 Battery Capacity

The amount of charge released by the energized material at an electrode associated with complete discharge of a battery is called the *battery capacity*. The capacity is measured in A h (1 A h = 3600 C or Coulomb, where 1 C is the charge transferred in 1 s by 1 A current in the SI unit of charge).

The theoretical capacity of a battery can be obtained by Faraday's law of electrolysis, which states that the mass of the elemental material altered at an electrode is directly proportional to the element's equivalent weight for a given quantity of electrical charge. The equivalent weight of the elemental material is given by the molar mass divided by the number of electrons transferred per ion for the reaction undergone by the material. This number is known as the valency number of ions for the substance. Mathematically, Faraday's law can be written as

$$m_R = \frac{Q}{F}\frac{M_m}{n} \tag{4.4}$$

where

m_R is the mass of the limiting reactant material altered at an electrode
Q is the total amount of electric charge passing through the material
F is the Faraday number or Faraday constant
M_m is the molar mass
n is the number of electrons per ion produced at an electrode

M_m/n is the equivalent weight of the reactant substance. The Faraday number is given by the amount of electric charge carried by one mole of electrons. The number of molecules or atoms in a mole is given by the Avogadro number N_A which is equal to 6.022045×10^{23} mol^{-1}. The amount of charge in one electron, which is the elemental charge, is equal to $e_0 = 1.6021892 \times 10^{-19}$ C. Therefore, the Faraday number is equal to $F = N_A e_0 = 96{,}485$ C/mol. The number of Faradays required to produce one mole of substance at an electrode depends on the way in which the substance is oxidized or reduced.

Therefore, the theoretical capacity of a battery (in Coulomb) can be obtained from Equation 4.4 as

$$Q_T = xnF \text{ C} \tag{4.5}$$

where x is the number of moles of limiting reactant associated with complete discharge of battery, and is given by

$$x = \frac{m_R}{M_m}$$

Here

m_R is the mass of the reactant material in kg
M_m is the molar mass of that material in g/mol

The theoretical capacity in A h is

$$Q_T = 0.278F \frac{m_R n}{M_m} \text{ A h} \tag{4.6}$$

The cells in a battery are connected in series and the capacity of the battery is dictated by the smallest cell capacity. Therefore, $Q_{Tbattery} = Q_{Tcell}$. Six battery cells connected in series to form a battery are shown in Figure 4.4.

4.3.2 Open Circuit Voltage

The battery in its simplest form can be represented by an internal voltage E_v and a series resistance R_i, as shown in Figure 4.5a. More representative but complex battery models are discussed later in the chapter. The battery internal voltage appears at the battery terminals as open circuit voltage when there is no load connected to it. The internal voltage or the open circuit voltage (OCV) depends on the state of charge of the battery, temperature, and past discharge/charge history (memory effects) among other factors. The

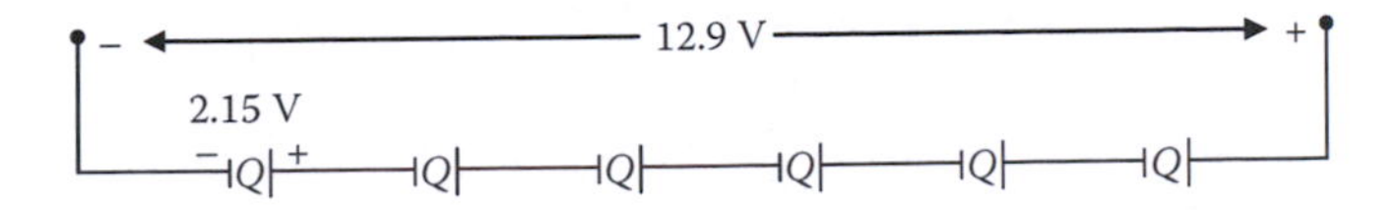

FIGURE 4.4
Battery cells connected in series.

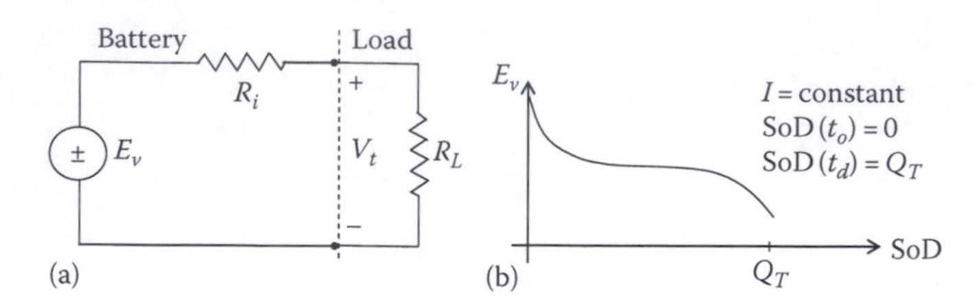

FIGURE 4.5
(a) Steady-state battery equivalent circuit. (b) Battery open circuit voltage characteristics.

open circuit voltage characteristics are shown in Figure 4.5b. As the battery is gradually discharged, the internal voltage decreases, while the internal resistance increases. The open circuit voltage characteristics have a fairly extended plateau of linear characteristics with a slope close to zero. The open circuit voltage is not a good indicator of the state of charge; state of charge of a battery pack needs to be calculated considering discharge current characteristics, battery chemistry, temperature effects, and number of charge/discharge cycles. Once the battery is completely discharged, the open circuit voltage decreases sharply with more discharge.

4.3.3 Terminal Voltage

Battery terminal voltage V_t is the voltage available at the terminals when a load is connected to the battery. The terminal voltage is at its full charge voltage V_{FC} when the battery is fully charged. For example, with lead-acid battery it means that there is no more $PbSO_4$ available to react with H_2O to produce active material. V_{cut} is the battery cut-off voltage, where discharge of the battery must be terminated. The battery terminal voltage characteristic in relationship with the state of discharge (SoD) is shown in Figure 4.6.

4.3.4 Practical Capacity

The *practical capacity* C_P of a battery is the actual charge released by the energized material at an electrode associated with complete discharge of the battery. The symbol C_P is used for the practical capacity, since the more

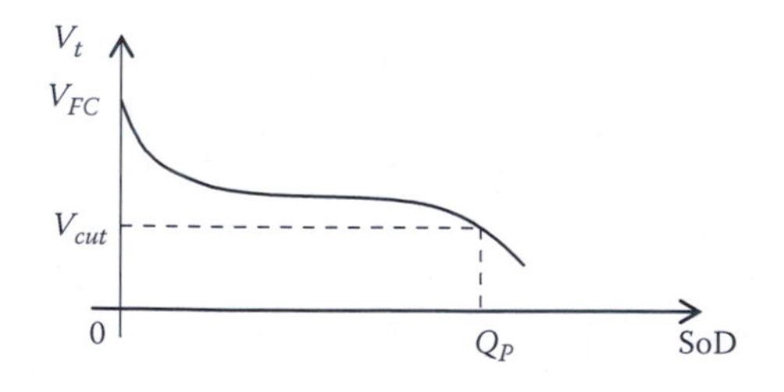

FIGURE 4.6
Battery terminal voltage.

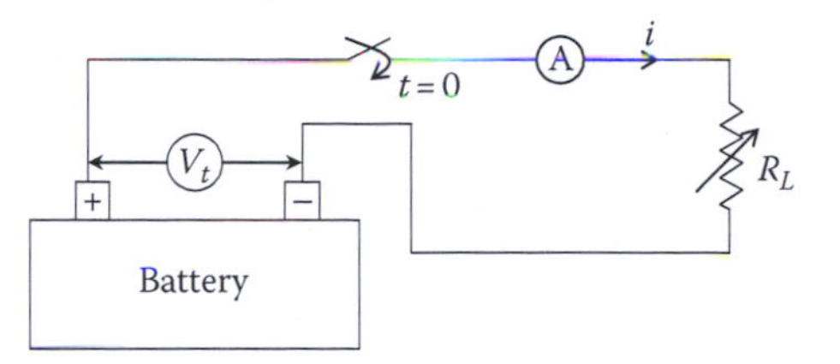

FIGURE 4.7
Battery capacity measurement.

commonly used symbol for battery capacity is *C*. The practical capacity is always much lower compared to the theoretical capacity Q_T due to practical limitations. The practical capacity of a battery is given as

$$C_P = \int_{t_o}^{t_{cut}} i(t)dt \tag{4.7}$$

where
t_o is the time at which battery is fully charged
t_{cut} is the time at which battery terminal voltage is at V_{cut}

Therefore, $V_t(t_{cut}) = V_{cut}$.

The practical capacity of a battery is defined in the industry by a convenient and approximate approach of A h instead of Coulomb under constant discharge current characteristics. Let us consider the experiment shown in Figure 4.7, where the battery is discharged at constant current starting from time $t = 0$. The ammeters and voltmeters measure the discharge current and the battery terminal voltage. The current is maintained constant by varying the load resistance R_L until the terminal voltage reaches the cut-off voltage V_{cut}. The qualitative graphs of two constant current discharge characteristics at two different current levels are shown in Figure 4.8. The following data are obtained from the experiment:

$I = 80\text{ A}$: Capacity $C_{80\text{ A}} = (80\text{ A})t_{cut} = 80 \times 1.8 = 144\text{ A h}$
$I = 50\text{ A}$: Capacity $C_{50\text{ A}} = (50\text{ A})t_{cut} = 50 \times 3.1 = 155\text{ A h}$
$I = 30\text{ A}$: Capacity $C_{30\text{ A}} = (30\text{ A})t_{cut} = 30 \times 5.7 = 171\text{ A h}$

The results show that the capacity depends on the magnitude of discharge current. The smaller the magnitude of the discharge current, the higher the capacity of the battery is. To be accurate, when the capacity of a battery is stated, the constant discharge current magnitude must also be specified.

4.3.5 Discharge Rate

The *discharge rate* is the current at which a battery is discharged. The rate is expressed as *C/h* rate, where *C* is rated battery capacity and *h* is discharge

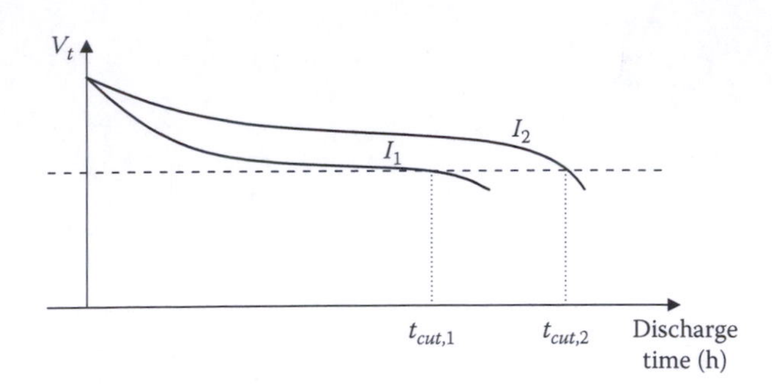

FIGURE 4.8
Constant current discharge curves.

time in hours. For a battery that has a capacity of C A h and is discharged over Δt, the discharge rate is $C/\Delta t$.
Example: Let the capacity of a battery be 1 C = 100 A h. (1 C usually denotes rated capacity of the battery.) Therefore,

$$\frac{\text{C}}{5}\text{ rate is }\frac{100\text{ A h}}{5\text{ h}} = 20\text{ A}$$

and

$$2\text{C rate is }\frac{100\text{ A h}}{0.5\text{ h}} = 200\text{ A}$$

4.3.6 State of Charge

The *state of charge* (SoC) represents the present capacity of the battery. It is the amount of capacity that remains after discharge from a top-of-charge condition. The current is the rate of change of charge given by

$$i(t) = \frac{dq}{dt}$$

where q is the charge moving through the circuit. The instantaneous theoretical state of charge $\text{SoC}_T(t)$ is the amount of equivalent charge remaining at the positive electrode and ready to be released by the energized material. If the state of charge is Q_T at the initial time t_o, then $\text{SoC}_T(t_o) = Q_T$. For a time interval dt

$$\begin{aligned} d\text{SoC}_T &= -dq \\ &= -i(i)dt \end{aligned}$$

Integrating from the initial time t_o to the final time t, the expression for instantaneous state of charge is obtained as

$$\mathrm{SoC}_T(t) = Q_T - \int_{t_o}^{t} i(\tau)d\tau \tag{4.8}$$

SoC_T is often expressed as a percentage of the capacity of the battery as follows:

$$\mathrm{SoC}_T(t) = \frac{Q_T - \int_{t_o}^{t} i(\tau)d\tau}{Q_T} \times 100\%$$

The state of charge will be increasing when a battery is being charged. If the state of charge is zero initially at $t=0$, the state of charge at time t expressed in percentage form is given by

$$\mathrm{SoC}_T(t) = \frac{\int_{0}^{t} i(\tau)d\tau}{Q_T} \times 100\%$$

4.3.7 State of Discharge

The *state of discharge* (SoD) is a measure of the charge that has been drawn from a battery during discharge. Mathematically, state of discharge is given as

$$\mathrm{SoD}_T(t) = \Delta q = \int_{t_o}^{t} i(\tau)d\tau$$

$$\Rightarrow \mathrm{SoC}_T(t) = Q_T - \mathrm{SoD}_T(t) \tag{4.9}$$

4.3.8 Depth of Discharge

The *depth of discharge* (DoD) is the percentage of battery rated capacity to which a battery is discharged. The depth of discharge is given by

$$\mathrm{DoD}(t) = \frac{Q_T - \mathrm{SoC}_T(t)}{Q_T} \times 100\%$$

$$= \frac{\int_{t_o}^{t} i(\tau)d\tau}{Q_T} \times 100\% \tag{4.10}$$

The withdrawal of at least 80% of battery (rated) capacity is referred to as deep discharge.

4.3.9 Battery Energy

Energy of a battery is measured in terms of the capacity and the discharge voltage. To calculate the energy, the capacity of the battery must be expressed in coulombs. 1 A h is equivalent to 3600 C, while 1 V refers to 1 J (J for joule) of work required to move 1 C charge from the negative to positive electrode. Therefore, the stored electrical potential energy in a 12 V, 100 A h battery is $(12)(3.6 \times 10^5)$ J = 4.32 MJ. In general, the theoretical stored energy

$$E_T = V_{bat} Q_T$$

where

V_{bat} is the nominal no load terminal voltage
Q_T is the theoretical capacity in C units

Therefore, using Equation 4.6, the theoretical energy is

$$E_T = \left[\frac{1000Fn}{M_m} m_R \right] V_{bat} = 9.65 \times 10^7 \frac{nm_R}{M_m} V_{batt} \text{ J} \tag{4.11}$$

The practical available energy is

$$E_p = \int_{t_o}^{t_{cut}} v(t) i(t) dt \text{ W h} \tag{4.12}$$

where

t_o is the time at which battery is fully charged
t_{cut} is the time in hours at which battery terminal voltage is at V_{cut}
v is the battery terminal voltage
i is the battery discharge current

E_P is dependent on the manner in which the battery is discharged. Practical energy in Watt-hours (W h) multiplied by 3600 gives the energy in Joules, i.e., Watt-seconds.

4.3.10 Specific Energy

The *specific energy* of a battery in terms of discharge energy related to complete discharge from fully charged condition is given by

$$\text{SE} = \frac{\text{Discharge energy}}{\text{Total battery mass}} = \frac{E}{M_B}$$

The unit for specific energy is W h/kg. The theoretical specific energy of a battery using Equation 4.9 is

$$\mathrm{SE}_T = 9.65 \times 10^7 \times \frac{nV_{bat}}{M_m}\frac{m_R}{m_B}\ \mathrm{W\,h/kg} \quad (4.13)$$

If the mass of the battery M_B is proportional to the mass of the limiting reactant of the battery m_R, then SE_T is independent of mass. The specific energy of lead-acid battery is 35–50 W h/kg at $C/3$ rate. Since practical energy E_P varies with discharge rate, the practical specific energy SE_P is also variable.

The term energy density is also used in the literature to quantify the quality of a battery or other energy sources. The term *energy density* refers to the energy per unit volume of a battery. The unit for energy density is W h/L.

4.3.11 Battery Power

The instantaneous battery terminal power is

$$p(t) = V_{bat}i \quad (4.14)$$

where
- V_{bat} is the battery terminal voltage
- i is the battery discharge current

Using Kirchhoff's voltage law for the battery equivalent circuit of Figure 4.5a,

$$V_t = E_v - R_i i \quad (4.15)$$

Substituting Equation 4.15 into 4.14 yields

$$p(t) = E_v i - R_i i^2 \quad (4.16)$$

Power versus current characteristic is shown in Figure 4.9. Using the maximum power transfer theorem in electric circuits, the battery can deliver maximum power to a DC load when the load impedance matches the battery internal impedance. The maximum power is

$$P_{max} = \frac{E_v^2}{4R_i} \quad (4.17)$$

Since E_v and R_i vary with the state of charge, P_{max} also varies accordingly.

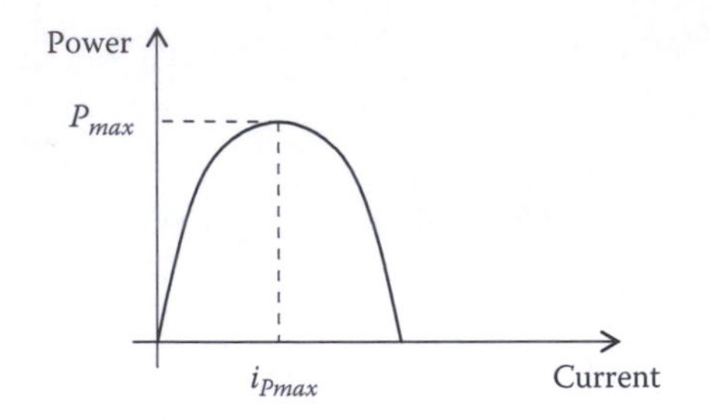

FIGURE 4.9
Battery power characteristics.

Maximum power output is needed from the battery in fast discharge conditions in vehicle applications, which occur when the electric motor is heavily loaded. Acceleration on a slope is such a condition, when the motor draws a lot of current to deliver maximum power required for traction.

The performance of batteries to meet acceleration and hill climbing requirements can be evaluated with the help of rated power specifications, which are based on the ability of the battery to dissipate heat. The *rated continuous power* is the maximum power that the battery can deliver over prolonged discharge intervals without damage to the battery. These do not necessarily correspond to P_{max} on p–i curve of battery characteristics. The *rated instantaneous power* is the maximum power that the battery can deliver over a very short discharge interval without damage to the battery.

4.3.12 Specific Power

The *specific power* of a battery is

$$SP = \frac{P}{M_B} (\text{units}: \text{W/kg}) \tag{4.18}$$

where

P is the power delivered by battery
M_B is the mass of battery

Typically, lead-acid battery's maximum specific power is around 280 W/kg (which corresponds to P_{max}) at DoD = 80%. Similar to specific energy and energy density, the term *power density* is used to refer to the power of the battery per unit volume with units of W/L.

4.3.13 Ragone Plots

In electrochemical batteries, there is a decrease in charge capacity (excluding voltage effects) with increasing currents. This is often referred to as the

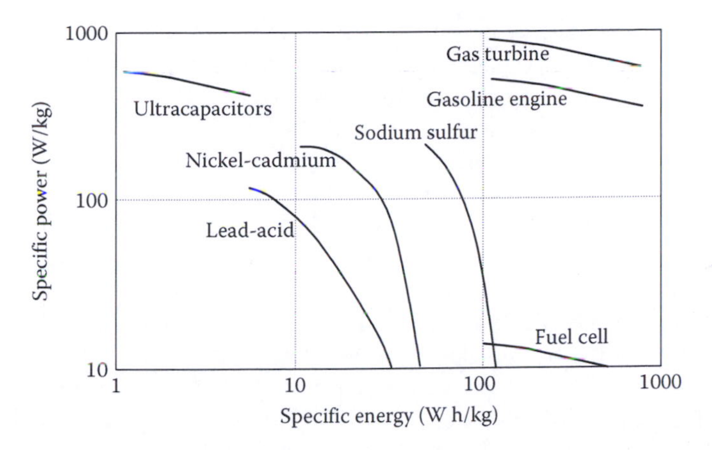

FIGURE 4.10
Specific power vs. specific energy (Ragone plots) of several batteries, a gasoline engine, and a fuel cell.

Ragone relationship and is described by *Ragone plots*. Ragone plots are usually obtained from constant power discharge tests or constant current discharge plots. Let us consider the experiment of Figure 4.6 again, but this time the current i is adjusted by varying R_L such that the power output at the battery terminals is kept constant. The experiment stops when the battery terminal voltage reaches the cut-off voltage, i.e., $V_t = V_{cut}$. We assume that the battery is fully charged at $t=0$. The experiment is performed at several power levels and the following data are recorded: (1) power $p(t) = V_t i = P$, (2) time to cut-off voltage t_{cut}, and (3) practical energy $E_P = Pt_{cut}$. The plot of SP vs. SE on a log-log scale is known as the Ragone plot. The Ragone plots of several batteries along with alternative energy sources and IC engines are given in Figure 4.10 to give an idea about the relative power and energy capacities of these different devices.

4.4 Electrochemical Cell Fundamentals

The *electrochemical cells* use the oxidation and reduction reactions for the flow of electric currents that are known as faradaic currents. These cells form the basis of batteries and fuel cells that supply electrical energy from stored chemical energy using the electrochemical reactions. The electrochemical cells can be classified as either *galvanic cells* or *electrolytic cells*. The galvanic cells are those where the substances react spontaneously when an external electrical load is connected to the cell. In the galvanic cell, electrons flow through the external circuit and ions transfer from one electrode to another within the cell resulting in work done. The galvanic cells are also known as voltaic cells since volts or potentials are established due to the spontaneous chemical reactions. The electrolytic cells are those in which reactions are nonspontaneous, and effected by the imposition of external voltage greater than the open circuit voltage of the cell. Electrical energy is expended by the external source and work is done on the electrochemical cell. When a battery cell is being charged to restore chemical energy within, it functions as an electrolytic cell. The battery cells will be referred to as electrochemical cells in general, since both discharging and charging are associated with the secondary battery cells used in electric and hybrid vehicles.

The fundamentals of a battery cell, or for that matter, of any other electrochemical cell, is in the physics and chemistry involved in the energy conversion process. In this section, we will analyze the fundamentals to gain sufficient understanding of the process so that electrochemical cell models can be developed. The models are useful not only for analyzing the electrochemical cells, but also for evaluating systems that include the electrochemical cell-based components for energy storage.

The fundamentals governing the principles and operation of an electrochemical cell are chemical thermodynamics, electrochemical reaction rates, electrode kinetics, and mass transport [1–7]. Each one of these mechanisms influences the energy conversion from stored chemical energy into electrical energy in an electrochemical cell. The mechanisms are treated in detail in this section.

4.4.1 Thermodynamic Voltage

Electrochemical thermodynamics determine the electrical potential difference between the electrode and the electrolyte of an electrochemical cell even in the absence of any external electrical connection to the electrodes when there is no external electric current flow. The thermodynamic physics of the electrochemical cell explains the mechanism of electric potential buildup at an electrode and establishes the theoretical limits of work, conversion efficiency, and potential of the cell.

The first law of thermodynamics states that energy is conserved during any process; energy can be transferred from a system to its surroundings or vice versa, but it cannot be created or destroyed. In chemical processes, system refers to the chemical species involved and the associated reactions; the container where the reaction takes place serves as the boundary of the chemical reaction. The three forms of energy for chemical processes are the internal energy of each substance, work done due to changes in pressure and volume or due to electrical current flow, and the heat transfer with the surroundings. In all chemical reactions, heat is either absorbed from or released to the surroundings. The amount of heat absorbed or released by the chemical reaction is equal to the change in internal energy only if there is no work done on or by the surroundings. However, if work done is not kept zero, some of the heat energy will be converted to work, and the total of heat absorbed or released is no longer equal to the change in internal energy. To represent the combined activity due to heat absorption or release, the term enthalpy representing the heat content of a substance or system has been introduced for convenience. By thermodynamic definition, enthalpy H is given by

$$H = U + pV$$

where
U is the internal energy
p is the pressure
V is the volume of the system

The product of pressure and volume (pV) is related to the work of expansion.

In galvanic devices, stable substances react spontaneously to form new substances, which mean that the reaction can only go from a higher energy state to a lower one according to the first law of thermodynamics. The lower energy state that the substances will assume depends on another measure known as the entropy of a substance. *Entropy* is a measure of the disorder level for the particles (molecules, atoms, etc.) that make up a substance. Entropy is a property that is specified for every equilibrium state of a substance. Since entropy is a property, the change in entropy in going from one state to another is the same for all processes. The SI unit for entropy is J/K.

The second law of thermodynamics states that a spontaneous reaction in a closed environment proceeds until the maximum entropy is reached for the substances involved. Based on the two laws of thermodynamics, part of the enthalpy change has to be reflected in the form of entropy change ($T\Delta S$), where T is the temperature in Kelvin and ΔS is the change of entropy. Therefore, the maximum available work from a chemical reaction or any process is [8,9]

$$\Delta G = \Delta H - T\Delta S$$

In chemical reactions, G is the Gibbs-free energy, which determines the relative importance between the enthalpy and entropy terms driving a particular reaction. A chemical reaction is favorable or spontaneous if $\Delta H < 0$, and unfavorable or nonspontaneous if $\Delta H > 0$. Again, a reaction is spontaneous if $\Delta S > 0$, and nonspontaneous if $\Delta S < 0$. The Gibbs-free energy of a system provides the net result when enthalpy and entropy forces of a reaction drive the system in opposite directions. A reaction is spontaneous if $\Delta G < 0$, and nonspontaneous if $\Delta G > 0$.

Gibbs-free energy can be calculated from the enthalpy and entropy changes between the products and reactants of the chemical reaction as

$$\Delta G = \Delta H^o - T\Delta S^o \tag{4.19}$$

where

$$\Delta H^o = \sum H_f^o(\text{products}) - \sum H_f^o(\text{reactants})$$

$$\Delta S^o = \sum S^o(\text{products}) - \sum S^o(\text{reactants})$$

The superscript *o* denotes standard-state conditions of 25°C and one atmospheric pressure, while the subscript *f* denotes the standard-state free energies of formation. The standard enthalpies of formation and standard entropies are given in thermodynamic tables. The standard enthalpy of formation of an element is defined as zero.

Alternatively, when the free energies of formation for each substance involved in a chemical reaction are known, the Gibbs-free energy change of reaction is

$$\Delta G = \Delta G_f^o(\text{products}) - \Delta G_f^o(\text{reactants})$$

Similar to the definition of enthalpies of formation, the standard free energy of formation of an element is defined as zero.

Let us consider the chemical reaction involved in sliver-zinc battery to illustrate the concept of energy available in an electrochemical cell. The chemical reaction is [8]

$$\text{Zn} + 2\text{AgCl} \rightarrow \text{Zn}^{++} + 2\text{Ag} + 2\text{Cl}^-$$

The amount of heat content liberated from the chemical reaction between metallic zinc and silver chloride solution when mixed at standard-state conditions is −233 kJ/mol of Zn reacted, i.e., $\Delta H^o = -233$ kJ/mol. For the chemical reaction, $T\Delta S = -43$ kJ/mol. The Gibbs-free energy in this example is

$$\Delta G = \Delta H - T\Delta S = -233 + 43 = -190 \text{ kJ/mol}$$

The example illustrates that all of the enthalpy change cannot be converted to work; a minimum amount of energy has to be consumed by the entropy change which is usually reflected in the generated heat during chemical reactions.

Gibbs-free energy is also useful in determining the maximum thermodynamic conversion efficiency of a galvanic cell, which is given by the ratio of Gibbs-free energy and the total enthalpy change:

$$\eta_{EC} = \frac{\Delta G}{\Delta H} = 1 - \frac{T\Delta S}{\Delta H} \tag{4.20}$$

For the given example of silver-zinc battery cell, $\eta_{EC} = 81.5\%$. The inherent high conversion efficiency compared to the thermodynamic upper limit of efficiencies in heat engines is an advantage of galvanic devices. However, just like the heat engines, the practical efficiency of galvanic devices during normal operation is much lower than the theoretical efficiency. The decrease in efficiency is directly related to the practical currents required in practical systems.

Gibbs-free energy released in the chemical process of the electrochemical cell imposes the theoretical limit on the maximum work that can be done by the cell. This work is the charge transferred per mole under the force of the open-circuit voltage (OCV) of the cell. With E being the electrode potential difference at equilibrium, the work done by the cell can be expressed as

$$\Delta G = \text{Charge transferred per mole} \times \text{OCV}$$
$$= -nFE^o \tag{4.21}$$

The theoretical upper limit of a galvanic cell potential can be obtained from the above as

$$E^o = -\frac{\Delta G}{nF} \tag{4.22}$$

We will use the chemical reaction in a NiCd battery cell to calculate the equilibrium cell potential. The chemical reaction in a NiCd battery cell is given by

$$\text{Cd} + 2\text{NiOOH} \rightarrow \text{Cd(OH)}_2 + 2\text{Ni(OH)}_2$$

The Gibbs-free energy change for the NiCd cell is

$$\Delta G = \Delta G_f^o(\text{products}) - \Delta G_f^o(\text{reactants})$$
$$= \left[(-470.25) + 2(-452.7)\right] - \left[0 + 2(-541.3)\right]$$
$$= -293.05 \text{ kJ/mol}$$

For the NiCd battery cell, two electrons are involved in the chemical reaction. The theoretical cell potential is thus

$$E^o = -\frac{-293{,}050}{2 \times 96{,}485} = 1.52 \text{ V}$$

The theoretical cell potential is never achievable in practice due to various electrochemical phenomena in the cell, which are described in the Sections

4.4.3 through 4.4.7. The nominal practical cell voltage in a NiCd cell is 1.3 V. This is true for all electrochemical cells. The thermodynamic cell potential only gives the theoretical upper value of cell potential for a battery chemistry.

The potential of an electrode is the potential difference between the electrode and the electrolyte that it is in contact with. The electrode potential is determined with respect to a reference electrode, since an absolute potential value cannot be obtained. Both chemical and electrical processes contribute to the electrode potential difference [6]. The environment at the vicinity of an electrode is changed due to chemical activities between the electrode and the electrolyte regardless of the electric potential difference at the solid-liquid phase boundary. The measure of the work done to bring a particle to its assumed potential is the chemical potential. Again, regardless of the changes in the chemical environment, the transfer across the electric potential is accomplished by electric work done in its original sense. Although one cannot separate these two components for single species experimentally, the differences in the scales of the two environments make it possible to separate them mathematically [7,8]. The resultant potential for these two kinds of energy change is the electrochemical potential or simply the electric potential.

4.4.2 Electrolysis and Faradaic Current

The process of *electrolysis* is the transfer of electrons between an electrode and a chemical species in solution, resulting in an oxidation or a reduction reaction. An external electrical circuit is necessary for the redox reactions at the electrodes of the electrochemical cell. In order to understand the electrochemical phenomenon in a cell, and subsequently utilize that knowledge to develop simple models for these cells, the nature of current and potential in the cell must be evaluated [8]. This section describes how current controls the reaction using Faraday's law of electrolysis and the relationship between charge and current.

The chemical reactions 4.1 and 4.2 represent both a chemical process and an electrical process. The reaction rates can be completely determined using the electrical process by seeing that the chemical conversion can only occur if electrons are either arriving or leaving at the electrodes. Therefore, the chemical conversion rates are controlled or measured by the electrical current passing through a given electrode. Faraday's law of electrolysis relates the mass of a substance altered at an electrode to the quantity of electrical charge transferred at that electrode. The electrolysis rate in terms of moles electrolyzed is given by Faraday's law as (see Equation 4.5)

$$x = \frac{Q}{nF} \tag{4.23}$$

As was defined previously, n is the number of electrons per ion released at an electrode, F is the Faraday constant, and x is the number of moles electrolyzed. Again, by definition, electric current is the number of coulombs of electric charge flowing per second, i.e.,

$$i = \frac{dQ}{dt} \tag{4.24}$$

Equations 4.23 and 4.24 relate the faradaic current in an electrochemical cell with the reaction rate. Combining the two equations, we can write the reaction rate as

$$\frac{dx}{dt} = \frac{i}{nF} \text{ mol/s}$$

The reaction rate is typically expressed in mol/s per unit area since electrode reactions are a heterogeneous process occurring only at the electrode–electrolyte interface. The heterogeneous reaction rate depends on the mass transfer to the electrodes and various surface effects in addition to the electrode kinetics. The reaction rate per unit area is given by

$$\frac{dx}{dt} = \frac{i}{nFA} = \frac{j}{nF} \text{ mol/s cm}^2 \tag{4.25}$$

where
- A is the area in cm^2
- j is the current density in A/cm^2

4.4.3 Electrode Kinetics

The study of electrode kinetics includes the processes that govern the electrode reaction rates or the faradaic currents flowing in an electrochemical cell. A number of process rates dictate the electrode kinetics of which the two most common ones are

1. The rate of electron transfer at the electrode surface between the electrodes and species in solution.
2. Mass transport of the active materials from the bulk solution to the electrode interface.

The rate of electron transfer at the electrode surface is governed by the Faraday's law of electrolysis presented in Section 4.4.2. In this section, the additional fundamental principles of electrode kinetics are described to establish the terminal voltage–current relationship in an electrochemical cell. The mass transport is another fundamental mechanism for the continuity of electrochemical reactions and will be addressed in Section 4.4.4.

It must be mentioned that the study of electrochemistry is vast and a detailed treatment is beyond the scope of this book. Only an overview of the fundamental theory for electrode kinetics is given here. The readers are referred to [8–10] for further details on this topic of electrochemistry.

When a redox couple is present at each electrode and there are no contributions from liquid junctions, then the open circuit potential is also the equilibrium potential. However, in general there is always ongoing activity at the electrode–electrolyte interface for many electrochemical cells. The critical potential at which the electrode reactions occur is known as standard potential E^0 for the specified chemical substances in the system. The electrode potential deviates from its so-called equilibrium state when there is external electric current flow. The relationship between the energy flow and the current is more complicated in the electrochemical cell than that for the conduction in a solid, since the current flow and chemical reactions are heterogeneous processes. Consequently, the relationships are nonlinear, and approximations are often used to arrive at simpler mathematical expressions.

The electrode voltage–current relationship can be obtained using the forward and reverse reactions between the electrode and electrolytes. These currents can be obtained by relating reaction rates to the rate constants and the concentration of the reactants. Let us consider a general case where n electrons are transferred between two species (O) and (R) at an electrode–electrolyte interface. The general electrode reaction is

$$\mathrm{O} + ne^- \underset{k_r}{\overset{k_f}{\rightleftharpoons}} \mathrm{R}$$

where k_f and k_r are the forward and reverse rate constants, respectively. The rate constants are the proportionality factors linking the concentration of the species to the reaction rates. The concentration of species undergoing oxidation at a distance x from the surface and at time t will be denoted as $C_O(x,t)$; hence, the surface concentration is $C_O(0,t)$. Similarly, the surface concentration for the species undergoing reduction is $C_R(0,t)$.

The reaction rate obtained from the product of rate constant and species concentration can be equated to the reaction rate given by Equation 4.25 to establish the relationship between the species concentrations and faradaic current. Therefore, for the forward and reverse currents, we have

$$k_f C_O(0,t) = \frac{i_f}{nFA}$$

and

$$k_r C_R(0,t) = \frac{i_r}{nFA}$$

The net current flow at the electrode is the difference between the forward and reverse currents

$$i = i_f - i_r = nFA\left[k_f C_O(0,t) - k_r C_R(0,t)\right] \tag{4.26}$$

For the sake of simplicity in analyzing the electrode process, we will assume a single electron transfer (i.e., $n = 1$) at the electrode–electrolyte interface. In this case, the rate constants can be related to the electrical potential across the electrode–electrolyte interface using free energy considerations [8]. For the standard potential E^0, the forward and reverse rate constants are equal; this constant is known as the standard rate constant and is given the symbol k^0. The rate constants at other potentials are given in terms of the standard rate constant as

$$k_f = k^0 e^{(F/RT)(-\alpha)(E-E^0)}$$

$$k_r = k^0 e^{(F/RT)(1-\alpha)(E-E^0)}$$

Inserting these relations into Equation 4.26 gives the complete current–voltage characteristics at the electrode–electrolyte interface

$$i = FAk^0\left(C_O(0,t)e^{(F/RT)(-\alpha)(E-E^0)} - C_R(0,t)e^{(F/RT)(1-\alpha)(E-E^0)}\right) \tag{4.27}$$

The approach used is referred to as the Butler–Volmer approach for analyzing electrode kinetics [8].

Although this equation describes the electrode kinetics quite accurately, it is generally impossible to express voltage in terms of current which would be the more useful form for modeling electrochemical cells. One approach is the Nernst solution which assumes that the current is so small that it can be neglected. The Nernst equation is given by

$$E = E^o + \frac{RT}{nF}\ln\frac{C_O(0,t)}{C_R(0,t)} \tag{4.28}$$

Although Nernst defined the equation independently, it can be derived from the Butler–Volmer equation assuming that the system is in equilibrium and the net current is zero. At equilibrium, the electrodes adopt a potential based on the bulk concentrations, as dictated by Nernst, and the bulk concentrations of O and R are also found at the surface. Using $i(t) = 0$ and $E = E_{eq}$ in Equation 4.27, we have

$$C_O(0,t)e^{(F/RT)(-\alpha)(E_{eq}-E^0)} = C_R(0,t)e^{(F/RT)(1-\alpha)(E_{eq}-E^0)}$$

This equation takes the Nernst relation form as

$$E_{eq} = E^o + \frac{RT}{F} \ln \frac{C_O^*}{C_R^*} \tag{4.29}$$

where

E_{eq} is the equilibrium potential

C_O^* and C_R^* are the bulk concentrations of the oxidation and reduction reactants, respectively

Another approach for estimating the terminal potential is the Tafel solution which assumes that the current is large in one direction or the other. The approximation means that one of the two exponential terms in the Butler–Volmer expression of Equation 4.27 is negligible. The Tafel solution is given by

$$E(t) = E^0 + \frac{RT}{\alpha nF} \ln(i_0) - \frac{RT}{\alpha nF} \ln(i(t)) \tag{4.30}$$

where i_0 is the exchange current obtained from the equilibrium condition. Although the net current is zero at equilibrium, there is balanced faradaic activity with equal forward and reverse currents. The exchange current is equal to these faradaic currents and is given by

$$i_0 = i_f = i_r = FAk^0 C_0^* e^{-\alpha(F/RT)(1-\alpha)(E_{eq}-E^0)} \tag{4.31}$$

Although the Tafel relationship was originally derived from experimental data, it can also be deduced from the Butler–Volmer expression.

The reactant activities at the electrode–electrolyte interface cause the open circuit voltage E to deviate from the standard state voltage E^0. The electrode voltage difference between that at the equilibrium condition and when there is current flow due to charge transfer at the electrode–electrolyte interface is often referred to as the activation or charge transfer polarization overpotential. The charge transfer polarization is reflected as a voltage drop from the equilibrium position during cell discharge and will be denoted by E_{ct}.

4.4.4 Mass Transport

A complete electrochemical cell is formed when two electrodes are immersed in the same electrolyte. When an electrical circuit is completed by connecting an external load to the two electrodes, current flows through the external circuit. Electrode reactions and mass transport are two mechanisms for supporting continuous current flow. Current flow is maintained inside the cell through the mass transport of ions in the electrolyte. The dominant process for mass transport is the diffusion process where molecules transfer from a location with higher concentration to one with lower concentration. The mass transport can also occur through convection and migration. Convection is the mechanical movement of particles, which does not occur in a battery cell. In fuel cells, the pressure of fuel supply does cause some mass transport due to the convection effect. Migration is ion movement under the influence of an electric field where positively charged ions will migrate toward the negative electrode, while the negatively charged ions will move toward the positive electrode. The ion movement due to migration may be in the same direction as that due to the diffusion process or in the opposite direction. Since the diffusion process dominates over the convection and migration process, only the diffusion process is addressed below.

The diffusion process for an active species in an electrochemical cell can be described by Fick's second law as

$$\frac{\partial C(x,t)}{\partial t} = D\frac{\partial^2 C(x,t)}{\partial x^2}$$

where

$C(x,t)$ is the active species concentration

D is the diffusion coefficient of the electrochemical cell

x and t are the space and time variables, respectively

Let the species concentration at the electrode responsible for chemical reactions to maintain current flow be C_d. This concentration is less than the bulk concentration of the electrolyte C_{bulk}. A linear relationship for the diffusing current derived from Fick's law is given as [6]

$$C_d(t) = C_{bulk} - \frac{\delta}{nFAD} i(t) \tag{4.32}$$

where

δ is the diffusion layer thickness

A is the surface area

Porous electrodes are almost invariably used in electrochemical cells to decrease the activation potential. The active materials penetrate into the pores of the electrode to reach the reaction site. The increased surface area due to the pores results in parallel diffusion processes, which is termed as branched diffusion process. The pores and increased surface area complicates the analysis of the diffusion process. The behavior of the branched diffusion has been shown to follow the pattern described by the constant phase element (CPE) instead of Fick's second law [10]. For a CPE, the phase angle of the frequency response remains the same for all frequencies. The CPE transfer function used to represent the overall diffusion process is given by

$$H(s) = \frac{C_d(s)}{i(s)} = \frac{K}{s^q}, \quad 0 < q < 1$$

The time response of the CPE is easier to solve for simpler operations such as constant current discharge. That time response is

$$C_d(t) = C_{bulk} - Kit^q \tag{4.33}$$

The diffusion process described serves the purpose of representing both the energy storage and the impedance. The energy within a battery is stored or spatially distributed in the electrolyte in terms of the concentration of the active material. The movement of active material during cell reactions is controlled by the inherent impedance of the electrolyte. Both mechanisms have been represented by the diffusion process and are represented as a coupled mechanism from the electrical perspective. However, for certain analysis, it is desirable to separate the source and impedance. This separation is desirable for certain applications. For example, the fuel cell-type electrochemical device does not store any energy; the materials for chemical reactions in a fuel cell are supplied from external fuel. The electrochemical process is more accurately modeled by separated energy source (the fuel) and the impedance to the source. Another important need is for the prediction of how much energy is left in the battery, which is essentially calculating the SoC of the battery.

The separation of energy storage and impedance enables an improved modeling of pulsed discharge characteristics of batteries, which is essentially what takes place in electric and hybrid vehicles. For the complicated discharge currents in such applications, it is difficult to obtain analytical solutions of the CPE, and often one has to resort to numerical solutions.

4.4.5 Electrical Double Layer

The electrical potential difference between the electrode and the electrolyte is due to the excess charges residing at the electrode–electrolyte interface. The excess charges at the electrode surface have to be counterbalanced by the charges of opposite polarity in the electrolyte. The two parallel layers formed with charged particles have a structure similar to that of a capacitor. This electrode interface layer is called electrical double-layer or simply a double layer.

The electrical double layer has a thin but finite thickness in the range of a few Angstroms where all the excess charges reside. There is a strong electrostatic field at the surface, since there is no dielectric material other than the charged particles in the electrical double layer. The important properties of the electrical double layer are its capacitance and the variation of electric potential and ion concentrations. Experimental results showed that the behavior of the capacitance is nonlinear, depending on the interface potential [8].

In the electrochemical models, this double layer capacitive behavior can be represented by an equivalent nonlinear capacitor C_{dbl}. In batteries, the double layer capacitor does not a play a significant role in energy storage and can be neglected in simpler equivalent circuit models. However, this electrical double layer concept forms the basis for making nonfaradaic electrochemical devices known an supercapacitors or ultracapacitors. The ultracapacitors are discussed in Chapter 5.

4.4.6 Ohmic Resistance

The voltage drop due to migration in the electrochemical cell is caused by the ohmic resistance of the electrolyte. The electrical resistance of the electrode materials, bulk electrolyte, and the electrode–electrolyte contact areas contribute to the ohmic voltage drop. The contact resistance gradually increases as the cell is being discharged during the formation of nonconducting film during cell reactions. A linear resistance is typically added in equivalent circuit models to represent the ohmic voltage drop.

A secondary electrolyte, known as supporting electrolyte or inert electrolyte, is often added in an electrochemical cell to reduce the ohmic voltage drop due to migration. The supporting electrolyte increases the conductivity of ions in the electrolyte. Additionally, the electrolyte reduces the electric field substantially which reduces the migration of the active species. The current conduction due to diffusion or migration reduces significantly with the addition of the supporting electrolyte. The inert ions in the cell are primarily responsible for the current conduction within the cell. The diffusion process still remains the dominant mechanism of supplying reactant materials to the electrodes.

4.4.7 Concentration Polarization

In an electrochemical cell, there are species that do not participate in the chemical reactions at the electrode, but contribute to the conduction of current. These species may be the ions from the supporting electrolyte or from the composition of the primary electrolyte. These species have to accumulate near the electrodes to aid in the flow of current, but do not react with the electrodes. Positive ions accumulate near the negative electrode, while the negative ions gather around the positive electrode. The result of this accumulation is a voltage drop caused by the electric field formed by the accumulation of the unreacted but charged particles. This is known as concentration polarization. This polarization follows a Nernst equation like relationship using the concentration of the inert ions [5]

$$E_C = \frac{RT}{nF} \ln \frac{C_{electrode,i}}{C_{bulk,i}} \tag{4.34}$$

where $C_{electrode,i}$ and $C_{bulk,i}$ are the concentrations of the inert ions at the electrode and the bulk solution, respectively. Polarization, regardless of its origin, is reflected at cell terminals as a voltage reduction from the open circuit voltage.

4.5 Battery Modeling

Batteries and other electrochemical cells can be modeled at various levels, depending on the use of the model. Battery models are useful for battery design, performance evaluation, and system simulation at the application level. Modeling aids research on device design, construction, and materials through understanding the factors that affect the energy conversion process. Models also help research on the performance of the device in an application, which can be utilized for improved design and better utilization.

At the most complex level, the fundamental physics- and chemistry-based theories are used to develop *theoretical models* of electrochemical cells. These models reflect material properties and design factors on the device performance. The fundamental mechanisms of electrical power generation are characterized in these models in terms of both macroscopic behavior (terminal voltage and current characteristics) and microscopic (internal material and reactant behavioral processes) information. The models are very useful for the design and performance evaluation of a particular type of battery. The strength of these models is in the information obtained on the effect of design variables on performance during the design stage. These models characterize the physical and chemical relationships applied to each element of the device [5–8]. Numerical simulation techniques such as finite element analysis or computational fluid dynamics are also sometimes used to develop the analytical models. The drawbacks of the theoretical models are their complexity; often the models cannot be used to represent the device as a component of a larger system. Device parameters are not always available to the end user. The models could be specific for a particular chemistry and design. Dynamic response such as that of the battery SoC is difficult, if not impossible, to analyze with most of these models.

Battery models that emphasize the macroscopic behavior are more useful for the performance evaluation at a system level (such as the electric or hybrid vehicle systems) and for the design of these systems. For example, a simplified battery model can be used for the dynamic simulation of a hybrid vehicle to predict the powertrain characteristics as well as the range on electric-only operating mode. Depending on the simulation objective, the models can be represented by a set of electrical circuit components or by a simple empirical relationship of two parameters. These two types of models are the *electric circuit models* and *empirical models,* which are presented in this section. The electric circuit-based models are somewhat more complex than the empirical models, but are extremely useful for vehicle system level analysis. On the other hand, the empirical models allow a quick evaluation on the range of a vehicle based on the capacity or energy density.

The energy storage device models presented in this section are useful not just for electric and hybrid vehicle applications, but also for utility power system applications. Distributed power systems require energy storage devices with similar features as those required for electric and hybrid vehicles.

4.5.1 Electric Circuit Models

The equivalent electrical circuit-based models use lumped parameters that make them suitable for integration in the simulation model of a larger system. The models use a combination of circuit elements (resistors, capacitors, and inductors) and dependent sources to give a circuit representation of the behavior and the functionality of the electrochemical cell. The model parameters are extracted from response data of the device, eliminating the need to know the chemical processes and the design details. The electric circuit models range from a simple linear resistive model to a fairly complex one that characterizes the chemical processes in terms of lumped parameters. The accuracy of these models is in between those of the theoretical models and the empirical models; yet the circuit models are very useful for both simulation and design of a system. The application aspects of the battery can be evaluated effectively with insights into operation of the device as well as that of the system. Some of the more complex circuit models can be used to study the dynamic response such as the effect of pulse discharge which is a characteristic of hybrid and electric vehicles [7].

The primary electrochemical activities in the electrochemical cell are governed by two fundamental relationships: (1) Butler–Volmer relationship characterizing the electron exchange at the electrode–electrolyte interface and (2) Faraday's law of electrolysis, which states that current controls the reaction. Relating these relationships with the stored charge and the diffusing charge in the electrochemical cell enables us to develop an electric circuit model whose parameters can be obtained from experimental data.

In developing the battery models, it is more convenient to consider the stored and diffusion charges at a surface rather than the effective species concentration or surface activities. Let $q_s(t)$ and $q_d(t)$ be the instantaneous stored charge and the instantaneous diffusion charge in the vicinity of the electrode representing surface activities. If Q is the total capacity of a cell, then the charge in the nonenergized species can be represented as $Q - q_s(t)$.

As was mentioned previously, the difficulty is in finding an inverse for the Butler–Volmer equation so that terminal voltage can be represented in terms of electrode current. The Nernst and Tafel equations are approximations with limitations on the terminal current. One simplified approximation is the Unnewehr universal model [11] given by

$$E(t) = E_0 + R_\Omega i(t) + k_1 q_s(t)$$

where

E_0 is the initial voltage of the cell
R_Ω is the series resistance
k_1 is a constant parameter

A generalized form to represent the solution to the Butler–Volmer equation is presented by Hartley and Jannette [12]:

$$E(t) = E_0 + R_\Omega i(t) + k_1 \ln\left(1 + |i|\right)\text{sgn}(i) + k_2 \ln\left(1 + |q_d|\right)\text{sgn}(i) + k_3 \ln(1 - q_s)$$

The constants E_0, R_Ω, k_1, k_2, and k_3 depend on the properties of the electrochemical cell and can be determined from experimental data. While the Hartley model gives a mathematical representation of the terminal voltage, it is often convenient to find an equivalent electric circuit model for simulation and analysis of a battery cell. In the following, several such electric circuit models representing an electrochemical cell are discussed, starting with a basic model derived from the Hartley model.

4.5.1.1 Basic Battery Model

Let us begin with a simple electrical equivalent circuit model that incorporates the fundamental principles, yet simple enough for characterization based on cell discharge data is shown in Figure 4.11. One of the key dynamics that has to be modeled is the diffusion process. While complex representations using a CPE or Warburg impedance can be used, an approximate solution to the change in diffusing charge has the same form as that of a voltage across an RC circuit element. Therefore, the effect on the terminal voltage due to diffusion charge will be represented by the following first-order differential equation,

$$\frac{dv_d(t)}{dt} = \frac{1}{C_d} i(t) - \frac{1}{C_d R_d} v_d(t)$$

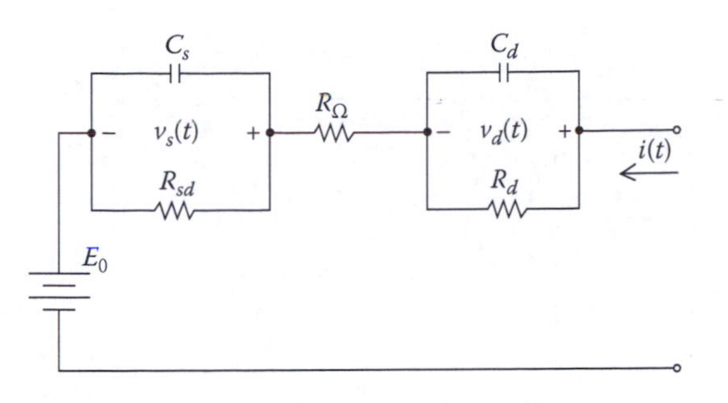

FIGURE 4.11
Electric equivalent circuit battery model.

where $v_d(t)$ is the voltage dropped across the R_dC_d parallel circuit that is proportional to the diffusion charge $q_d(t)$. Additional RC circuit elements can be added to represent the diffusion charge, but we will keep it as a single RC time constant for our simple model shown in Figure 4.11.

Another key cell dynamic that needs to be modeled is the effect of SoC on the terminal voltage of the cell. Figure 4.7 showed how the battery terminal voltage decreases as the cell is being discharged. In the middle of the characteristics, the terminal voltage decrease is approximately linear which can be modeled by a series capacitor C_s to represent the stored charge in the cell. The voltage across this storage capacitor C_s is proportional to the stored charge $q_s(t)$. As the SoC of the cell increases or decreases during charging or discharging, the voltage across the capacitor will increase or decrease, respectively. Additionally, an electrochemical cell loses charge while it is at rest. A resistor can be added in parallel to the storage capacitor to account for this loss of charge. This resistor R_{sd} represents the self-discharge of the cell. The C_sR_{sd} circuit elements representing the storage capacitor and self-discharge resistor are shown in Figure 4.11 in series with the diffusion parameters. The mathematical representation of this segment of the circuit model in relation to the terminal current is

$$\frac{dq_s(t)}{dt} = i(t) - \frac{1}{R_{sd}} q_s(t)$$

The two other parameters that need to be added to complete the electrochemical cell equivalent circuit is a voltage source in series with a resistor representing the ohmic resistance drop described in Section 4.4.6. The voltage source is taken to be the open circuit voltage of the cell E_0, and R_Ω is the ohmic resistance, both of which are shown in Figure 4.11 in series with the storage and diffusion parameters. This completes the simple equivalent circuit model of an electrochemical cell. The values of these circuit elements can be determined experimentally by applying a step change in battery current. The procedure for obtaining the parameters of this cell is given in Example 4.1.

Example 4.1

A step discharge current of 15 A is applied to a three-cell generic battery to calculate its parameters for the model shown in Figure 4.11. The data collected from the experiment is shown graphically in Figure 4.12. The step command of 15 A constant current discharge is applied at 3150 s and removed at 4370 s. After the discharge, the battery terminal voltage settles to a lower voltage level of 5.873 V compared to its initial no-load voltage due to the reduction in the state of charge.

Solution

ΔV_d, ΔV_{Cs}, $\Delta V_{R\Omega}$ are the voltage differences that need to be calculated from the test data to obtain the diffusion, storage, and series resistance parameters, respectively. The time to reach 63% of ΔV_d is 100 s. Neglecting the self-discharge of the cells, calculate the battery equivalent circuit parameters.

Let us first calculate the equivalent series resistance of the battery. The voltage drop for the series resistance shows up in the output voltage characteristics as an instantaneous increase or decrease of the terminal voltage due to the step change in current. The voltage increase due to the 15 A step change in current is $\Delta V_{R\Omega} = 5.775 - 5.58 = 0.195$ V. Therefore, the series resistance value is

$$R_\Omega = \frac{\Delta V_{R\Omega}}{\Delta I} = \frac{0.195}{15} = 0.013\ \Omega$$

The resistance for the diffusion component R_d is

$$R_d = \frac{\Delta V_d}{\Delta I} = \frac{0.098}{15} = 0.00653\ \Omega$$

The RC time constant for the diffusion parameters is 100 s. Therefore, the diffusion capacitor C_d can be calculated as

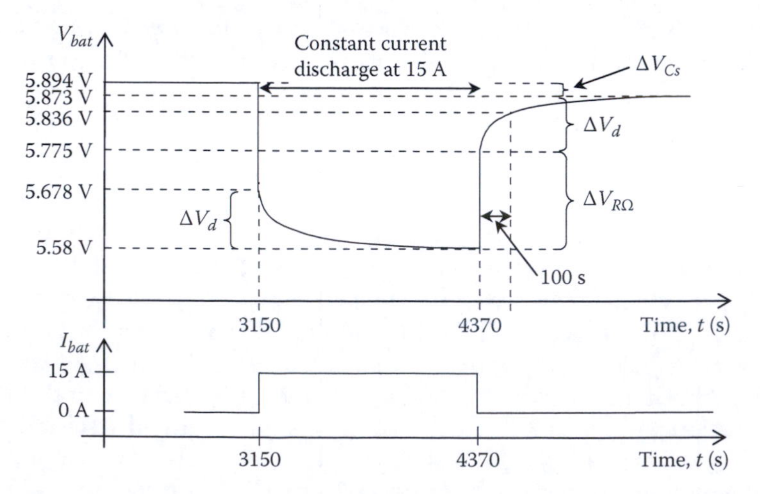

FIGURE 4.12
Test data for a battery to calculate equivalent circuit parameters.

$$C_d = \frac{100}{0.00653} = 15,306\ \text{F}$$

The storage capacitor C_s can be calculated from the voltage change due to the constant current discharge ΔV_{Cs} and the change in stored charge. This is calculated as follows:

$$C_s = \frac{\Delta Q}{\Delta V_{Cs}} = \frac{15(4370 - 3150)}{5.894 - 5.873} = 871,428.6\ \text{F}$$

4.5.1.2 Run-Time Battery Model

The Thevenin-type circuit model shown in Figure 4.11 with a constant open circuit voltage does not allow prediction of the battery terminal voltage V_t variations (i.e., DC response) and run-time information. The prediction of SoC, transient response, terminal voltage, run-time, and temperature effects is possible with run-time models. A run-time model capable of practicing the capacity of battery has been developed by Chen and Mora [13]. The circuit model, shown in Figure 4.13, has dependent current and voltage sources in addition to several passive components. The terminal voltage–current characteristics segment of the model is similar to that of Figure 4.11, except that the open circuit voltage depends on the capacity or SoC of the battery.

The capacitor $C_{capacity}$ and a current-controlled current source model the capacity, SoC, and run-time of the battery. The two RC networks simulate the voltage–current transient response characteristics. The SoC is calculated based on the current drawn out of the cell and the initial capacity in the run-time segment of the model. The value of the capacitor $C_{capacity}$ is given by

$$C_{capacity} = 3600 \cdot Q_C \cdot k_1 \cdot k_2$$

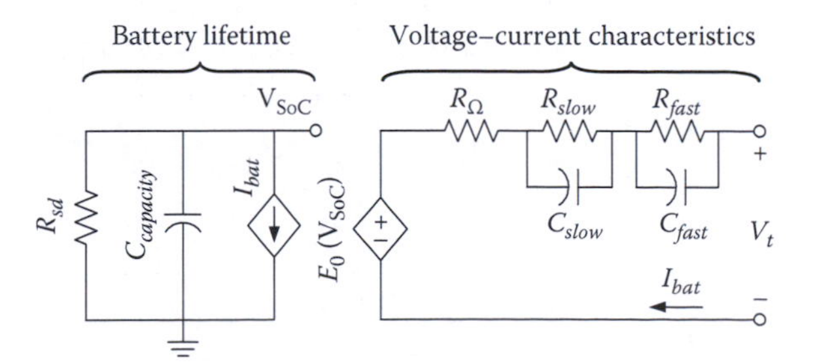

FIGURE 4.13
Run-time battery model proposed by Mora et al. [13].

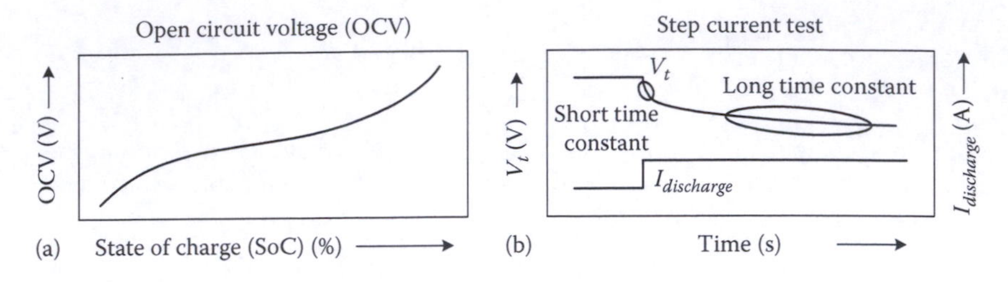

FIGURE 4.14
Example experimental curves to obtain run-time model parameters: (a) SoC vs. open circuit voltage characteristics; (b) discharge plot for calculating RC time constants.

where
- Q_c is the battery capacity in A h
- k_1 an k_2 are cycle number and temperature-dependent correction parameters, respectively

The initial voltage across $C_{capacity}$ is set to 1 or lower, depending on whether the battery is initially fully charged or not. A value of "1" represents 100% SoC. Similarly, a value of "0" would indicate that the battery is fully discharged, i.e., SoC is 0%.

SoC is bridged to the open circuit voltage through a voltage-controlled voltage source. The relationship between SoC and open circuit voltage is nonlinear and has to be represented from experimentally obtained data for this model. However, the collection of the open circuit voltage versus SoC data is extremely time consuming [14]. An example SoC versus open circuit voltage characteristic and the discharge profile to calculate the RC time constants are shown in Figure 4.14.

4.5.1.3 Impedance-Based Model

Another type of battery equivalent circuit-based model is the *impedance model*. Electrochemical impedance spectroscopy is applied to develop equivalent AC impedance-based circuit representation of the battery characteristics. The battery model based on impedance spectroscopy is shown in Figure 4.15. Impedance-based models are less intuitive, and are applicable only for

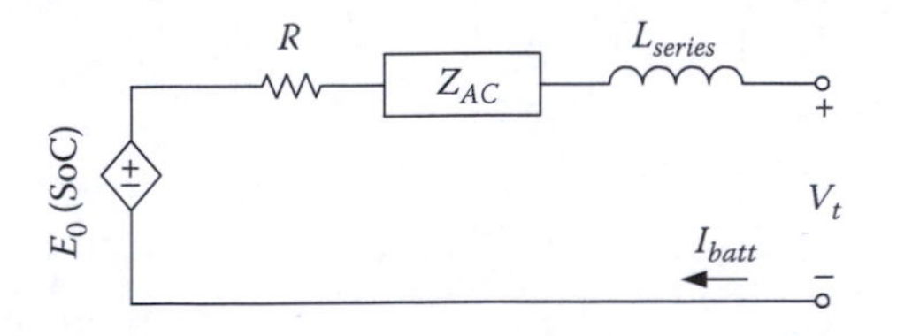

FIGURE 4.15
Impedance-based equivalent electric circuit battery model.

a fixed SoC and temperature; prediction of DC response and run-time of a battery are difficult with these models.

4.5.1.4 First Principle Model

An interesting equivalent circuit model based on the fundamental electrochemical principles has been developed by Lei Xia [7]; the model is called the *first principle model*. While this is not one of the simpler electric equivalent circuit models, it isolates and relates the physical and chemical fundamentals of an electrochemical cell to an equivalent circuit parameter. The model has discrete, lumped parameter representation of all the electrochemical processes within the cell. The first principle model, shown in Figure 4.16, incorporates the following phenomena within an electrochemical cell:

- Electrochemical energy conversion
- Diffusion process
- Charge transfer polarization
- Concentration polarization
- Electric double layer
- Ohmic resistance
- Self-discharge

In the equivalent circuit, the diffusion process has been described by a general CPE; open circuit voltage and concentration polarization have been represented by Nernst equations; charge transfer polarization has been represented by Tafel equation; ohmic voltage drop has been represented by resistance R_Ω; electric double layer has been represented by capacitance C_{dbl}; and resistance R_{sd} represents self-discharge of the cell.

The first principle model is construction and chemistry independent. The parameters of the model can be derived from experimental response data of the device, which eliminates the need for the knowledge of electrochemical

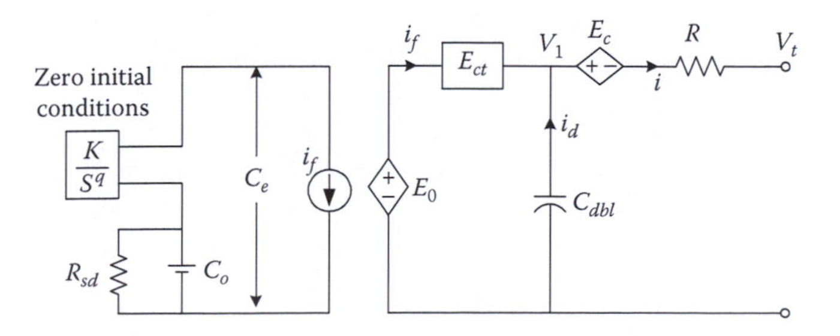

FIGURE 4.16
First principle battery model with constant current source. (From Xia, L., Behavioral modeling and analysis of galvanic devices, PhD dissertation, University of Akron, Akron, OH, 2000.)

properties and the design details. The model is based on the fact that before the discharge of any current, the internal voltage E_0, the double-layer capacitor voltage V_1, and the terminal voltage V_t (the variables are shown in Figure 4.16) are all the same. The charge transfer potential and the concentration polarization potential are zeroes for this condition. When a load is connected to the terminals, initially the discharge current is almost entirely supplied by the double-layer capacitor. As the double-layer capacitor discharges and V_1 decreases, the charge transfer potential is established and the faradaic current i_f starts to increase. When current i_f increases to a point where E_{ct} does not change appreciably, i_d becomes minimal. In this situation, the faradaic current i_f supplies the majority of the load current. The potential drop in this region is primarily due to the ohmic resistance.

As an example, the parameters for a generic battery cell are given below [7]:

Bulk electrolyte concentration	$C_0 = 2.616$ (dimensionless, but represents numerical value of the concentration)
Diffusion process parameters	$C_d(t) = C_0 - Ki_f(t)t^q; \quad K = \frac{1}{227.5}, \quad q = 0.68$
Open circuit voltage (Nernst equation)	$E(t) = 1.95 + 0.052 \ln C_d(t)$
Charge transfer polarization (Tafel equation)	$E_{ct} = 0.118 + 0.28 \ln (i_f)$
Ohmic resistance	$R_\Omega = 0.05\,\Omega$
Double-layer capacitor	$C_{dbl} = 50\,\mathrm{F}$
Concentration polarization	$E_c(t) = 0.04 \ln \frac{C_d(t)}{C_0}$

4.5.2 Empirical Models

The empirical models are the simplest of all models developed primarily for simple input–output relationships of the electrochemical devices. These models describe the performance of the device using arbitrary mathematical relationships matched with experimental or theoretical model data. The mathematical or empirical relationships are established by curve fitting with experimental data. The physical reasons for the behavior are not as important as the terminal relationship between certain parameters of the device. The physical basis for device functionalities is nonexistent in these models. The effects of design variations are impossible to analyze with the empirical models. Often only a subset of behaviors of the device is described such as the constant current discharge characteristics of a battery. These models do not provide the terminal i–v characteristics of the device which is necessary in circuit simulation and analysis for hybrid and electric vehicles. However, the empirical models are very useful for a quick estimation of the range of an electric vehicle for a particular type of battery pack.

One of the widely used empirical battery model is based on the Peukert's equation relating discharge current with the battery capacity. The model is based on constant current discharge characteristics of the battery. A series of constant current discharge experiments give the I vs. t_{cut} data for different constant current levels; t_{cut} is the time when the terminal voltage reaches the cut-off voltage limit V_{cut} during constant current discharge. The data obtained is used to fit Peukert's equation to develop the model as

$$I^n t_{cut} = \lambda \tag{4.35}$$

where

I is the constant discharge current
n and λ are curve fitting constants of a particular battery

n is a number between 1 and 2 with the value approaching 1 for smaller currents, but tends toward 2 for larger currents. The model does not specify the initial capacity, nor does it model the voltage variation or temperature, and aging factors. Peukert's model does not give any terminal i–v information.

Example 4.2

Find the curve fitting constants n and λ for Peukert's equation for the two measurements available from a constant current discharge experiment of a battery: (1) $(t_1,I_1)=(10,18)$ and (2) $(t_2,I_2)=(1,110)$.

Solution

Equation 4.35 is the Peukert's empirical formula using the constant current discharge approach. Taking logarithm of both sides of Equation 4.35

$$\text{Log}_{10}\left(I^n \times t_{cut}\right) = \text{Log}_{10}(\lambda)$$

$$=> \text{Log}_{10}(I) = \frac{1}{n}\text{Log}_{10}(t_{cut}) + \frac{1}{n}\text{Log}_{10}(\lambda)$$

Comparing with the equation for a straight line, $y=mx+b$; I versus t_{cut} curve is linear on a log-log plot, as shown in Figure 4.17.

The slope of the straight line is

$$m = \frac{\Delta y}{\Delta x} = \frac{\log(I_1)-\log(I_2)}{\log(t_1)-\log(t_2)} = \frac{\log(I_1/I_2)}{\log(t_1/t_2)}$$

Therefore, $n = -\dfrac{\log(t_1/t_2)}{\log(I_1/I_2)}$.

For the graph shown, $n = \dfrac{-1}{18/110} = 1.27 \quad [\because t_1 = 10t_2]$

The other constant can now be calculated from Peukert's equation

$$\lambda = 110^{1.27} \times 1 = 391.4\ \text{A h}$$

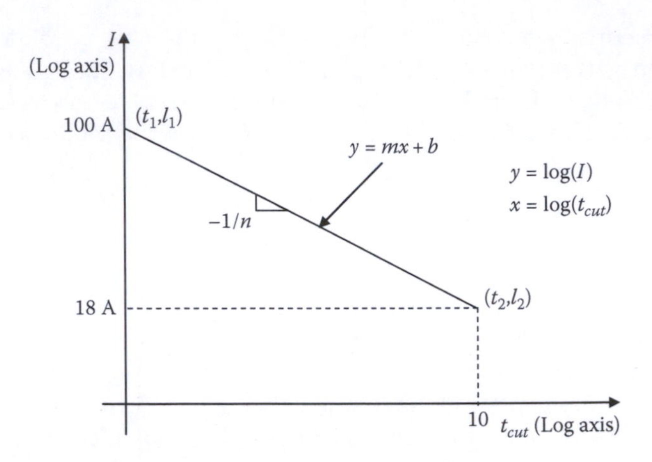

FIGURE 4.17
Plot of Peukert's equation using constant current discharge.

or

$$\lambda = 18^{1.27} \times 10 = 392.8 \text{ A h}$$

Another well-known and more general empirical model of a battery is based on the Shepherd equation [15]. The simplest form of Shepherd model is

$$E = E_0 - iR - \left(\frac{\lambda}{\lambda - it}\right) Ki$$

where
- E is the battery voltage
- i is the current
- t is the time

The parameters of the model are E_0, R, K, and λ representing battery reference voltage, internal resistance, polarization constant, and reference capacity, respectively. The model parameters have some physical meaning and relate the electrochemical behavior with the terminal i–v characteristics of the battery. The output response is expressed as a function of time in this model; however, the model is difficult to use for discharge patterns other than constant current discharge.

4.5.2.1 Range Prediction with Constant Current Discharge

Peukert's equation with constant current discharge characteristics can be used to develop a *fractional depletion model* (FDM) of batteries. FDM of a battery can be used to predict the range of an electric vehicle. Using Peukert's equation, we can establish the relationship between Q and I. The practical capacity of a battery is

$$Q = I \times t_{cut}$$

$$=> t_{cut} = \frac{Q}{I}$$

Substituting into Peukert's equation

$$I^n \left(\frac{Q}{I} \right) = \lambda$$

$$=> Q = \frac{\lambda}{I^{n-1}}$$

Since $0 < n - 1 < 1$, for $I > 1$, Q decreases as I increases.

From Section 4.3.7, we know that

$$\text{SoD} = \int i(\tau) d\tau$$

and

$$\text{DoD} = \frac{\text{SoD}}{Q(i)}$$

SoD is the amount of charge that the battery generates to the circuit. Assume that at $t = t_0$, the battery is fully charged. Let us consider a small interval of time dt. Therefore,

$$d(\text{DoD}) = \frac{d(\text{SoD})}{Q(i)}, \quad \text{where } d(\text{SoD}) = i(t)dt$$

We know that $Q = \lambda / I^{n-1}$ for constant current discharge. Let, $Q = \lambda / i^{n-1}$ for time-varying current as well, for the lack of anything better.

Therefore,

$$d(\text{DoD}) = \frac{idt}{\lambda / i^{n-1}} = \frac{i^n}{\lambda} dt$$

Integrating, we obtain,

$$\int_{t_0}^{t} d(\text{DoD}) = \int_{t_0}^{t} \frac{i^n}{\lambda} dt$$

$$=> \text{DoD}(t) - \text{DoD}(t_0) = \int_{t_0}^{t} \frac{i^n}{\lambda} dt$$

$\text{DoD}(t_0) = 0$, if the battery is fully charged at $t = t_0$.

The *fractional depletion model* is thus obtained as

$$\text{DoD}(t) = \left[\int_{t_0}^{t} \frac{i^n}{\lambda} dt \right] \times 100\% \quad (4.36)$$

The FDM based on current discharge requires knowledge of the discharge current $i(t)$. Therefore, this model to predict the electric vehicle range should be used when $i(t)$ is known.

Example 4.3

The constant current discharge characteristics of the battery pack used in an electric vehicle are

$$\ln I = 4.787 - 0.74 \ln t_{cut} - 0.0482(\ln t_{cut})^2$$

The current drawn from the battery during test drives of the electric vehicle for the SAE schedule J227a has the profile shown in Figure 4.18. The current magnitudes for the three SAE schedules are given in Table 4.3.

Find the range of the electric vehicle for each of the three schedules.

Solution

Apply the FDM (Equation 4.36) to find the number of driving cycles for DoD = 100%. From FDM

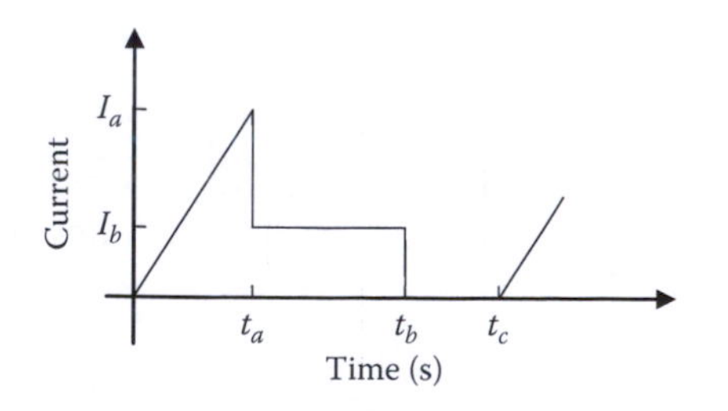

FIGURE 4.18
Pattern of current drawn from the battery.

$$1 = \int_{t_0}^{t_{100\%}} \frac{i^n}{\lambda} dt$$

First, we need to determine λ and n from the given battery characteristics

$$\frac{-1}{n} = -0.74 \Rightarrow n = 1.35$$

$$\frac{1}{n}\ln(\lambda) = 4.787 \Rightarrow \lambda = 645 \times 3600\,\text{As}$$

TABLE 4.3

Current Data for the Driving Schedules

Schedule J227a	I_a(A)	I_b(A)
B	100	35
C	216	54.6
D	375	88.7

Therefore,

$$1 = \int_{0}^{t_{100\%}} \frac{i^{1.35}}{645 \times 3600} dt$$

For schedule B, fraction depleted over 1 cycle

$$\text{DoD for 1 cycle} \Rightarrow f_{cyc} = \int_0^{72} \frac{i^{1.35}}{645 \times 3600} dt$$

$$\Rightarrow f_{cyc} = \int_0^{72} \frac{i^{1.35}}{645 \times 3600} dt = 4.31 \times 10^{-7} \left[\int_0^{19} \left(\frac{100t}{19} \right)^{1.35} dt + \int_{19}^{38} (35)^{1.35} dt \right]$$

$$= 4.31 \times 10^{-7} \left[9.41 \left(\frac{1}{2.35} \right) 19^{2.35} + 121.5(38 - 19) \right]$$

$$\Rightarrow f_{cyc} = 2.74 \times 10^{-3}$$

Let N = # of cycles required for 100% DoD,

$$\therefore 1 = N \times f_{cyc} \Rightarrow N = \frac{1}{f_{cyc}}$$

$$\therefore N = \frac{1}{2.74 \times 10^{-3}} = 365\ \text{cycles}$$

From Table 3.5, the EV goes 1 mi in about 4 cycles for schedule B.

Therefore,

$$\text{EV range} = 365/4 = 91\,\text{mi for } \textit{schedule B}$$

$$\text{Measured } N = 369 \Rightarrow \text{error} = 1.08\%$$

J 227a schedule C: From FDM, $N = 152$; EV range $= 152/3 = 51$ mi.
Measured, $N = 184$ => error $= 17.4\%$
J 227a schedule D: FDM, $N = 41$; EV range $= 41/1 = 41$ mi.
Measured, $N = 49$ => error $= 16.3\%$.

4.5.2.2 Range Prediction with Power Density Approach

An alternative approach for using Peukert's equation to develop a battery model is through the use of its Ragone relationship which is the specific power vs. specific energy characteristics. Ragone relationship and the corresponding plots are linear on the log-log scale to a first-order approximation. Battery model in terms of specific power and specific energy is

$$(\mathrm{SP})^n(\mathrm{SE}) = \lambda \tag{4.37}$$

where n and λ are curve-fitting constants.

Example 4.4

The data given in Table 4.4 is collected from an experiment on a battery with mass 15 kg. The data is used to draw the Ragone plot shown in Figure 4.19. Using the data points (8,110) and (67.5,10), calculate the constants of Peukert's equation n and λ.

Given a battery terminal power profile $p(t)$, the specific power SP(t) profile can be obtained by diving the power profile $p(t)$ by the total vehicle mass m_V (Figure 4.20). The battery is assumed to be fully charged at $t = 0$.

Let, $f_r(t)$ = fraction of available energy provided by battery from 0 to t, where $f_r(0) = 0$, since SoD(0) = 0. Now, consider the time interval dt over which a fraction of available energy df_r is provided by the battery

$$df_r = \frac{dE}{E_{avail}} = \frac{dE/m_V}{dE_{avail}/m_V} = \frac{d(\mathrm{SE})}{\mathrm{SE}_{avail}}.$$

TABLE 4.4

Data from Constant Power Discharge Test

P (W) (Measured)	t_{cut} (h) (Measured)	E_P (W h) (Calculated)	SP (W/kg) (Calculated)	SE (W h/kg) (Calculated)
150	6.75	(150)(6.75) = 1013	150/15 = 10	1013/15 = 67.5
450	0.85	381	30	25.4
900	0.23	206	60	13.7
1650	0.073	120	110	8

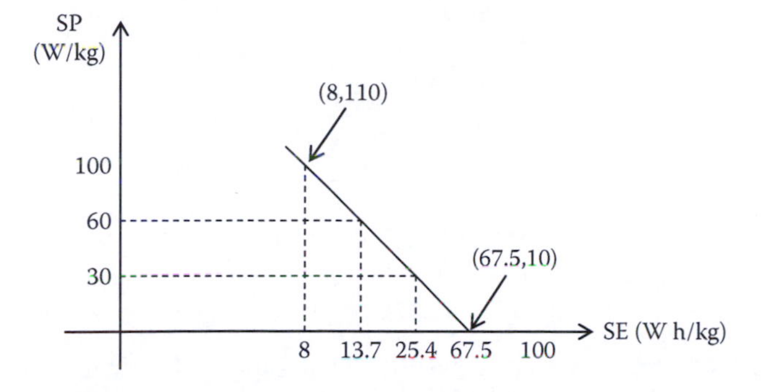

FIGURE 4.19
Ragone plot for Example 4.4.

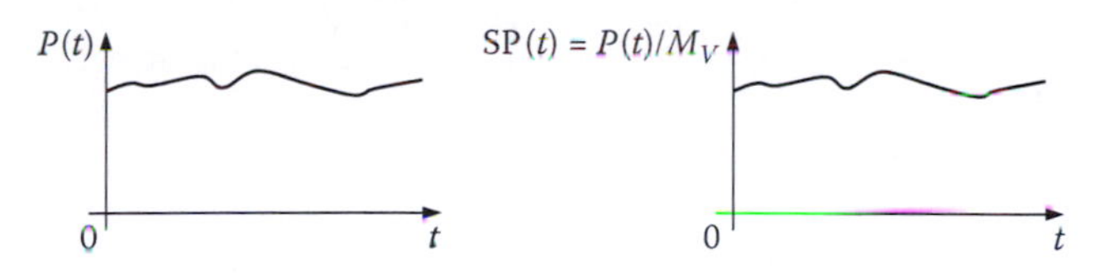

FIGURE 4.20
Power and specific power.

If dE is the energy provided by battery to the electrical circuit over dt and E_{avail} is the total available energy, then

$$dE = pdt$$

Now E_{avail} is a function of instantaneous power and we know that,

$$d(\text{SE}) = (\text{SP})dt$$

Therefore,

$$\text{SE}_{avail} = f(\text{SP})$$

We will use Peukert's equation to relate specific power and specific energy as follows:

$$(\text{SP})^n * \text{SE}_{avail} = \lambda$$

Therefore,

$$df_r = \frac{\text{SP}}{\lambda/(\text{SP})^n}dt = \frac{(\text{SP})^{n+1}}{\lambda}dt$$

Integrating,

$$\int_{f_r(0)}^{f_r(t)} df_r = \int_0^t \frac{(SP)^{n+1}}{\lambda} d\tau$$

$$=> f_r(t) = \int_0^t \frac{(SP)^{n+1}}{\lambda} d\tau \quad (4.38)$$

Equation 4.38 is the FDM using power density approach. If t = time at which x% of available energy has been used, then

$$\frac{x}{100} = \int_0^t \frac{(SP)^{n+1}}{\lambda} d\tau$$

Note that $1 = \int_0^{t_{100\%}} \frac{(SP)^{n+1}}{\lambda} d\tau$

At $t_{100\%}$, 100% i.e., all the available energy has been used by the system.

4.6 Traction Batteries

Lead-acid batteries that have served the automotive industry over the past 100 years for powering electrical accessories in conventional IC engine vehicles do not have the power and energy density required in electric vehicles and hybrid vehicles. The push for zero-emission vehicles led to numerous research and development efforts on advanced batteries activities in the United States, Europe, and Japan. Desirable features sought after in alternative battery technologies are high power and energy density, faster charge time, and long cycle life. The research and development progressed slowly until recent years due to the lack of market penetration of electric vehicles. In the meantime, the growth in the electronics industry over the past several decades has led to tremendous advancements in alternative batteries, such as NiCd, NiMH, and Li-based batteries technologies. The rechargeable Li-ion battery is the technology of choice for cell phones and laptop computers. Further research on scaling of these battery technologies led to the development of several battery technologies for electric and hybrid vehicle applications. NiMH battery packs are currently used in commercially available hybrid electric vehicles, while the Li-ion battery pack is used in the electric vehicle Tesla roadster. Emerging plug-in hybrid vehicles are also likely to use the Li-ion battery technology. While the NiMH and Li-ion batteries are the frontrunners today for electric and hybrid electric vehicles

TABLE 4.5

Properties of Electric and Hybrid Electric Vehicles Batteries

Battery Type	Specific Energy (W h/kg)	Specific Power (W/kg)	Energy Efficiency (%)	Cycle Life
Lead-acid	35–50	150–400	80	500–1000
Nickel-cadmium	30–50	100–150	75	1000–2000
Nickel-metal hydride	60–80	200–400	70	1000
Aluminum-air	200–300	100	<50	Not available
Zinc-air	100–220	30–80	60	500
Sodium-sulfur	150–240	230	85	1000
Sodium-nickel-chloride	90–120	130–160	80	1000
Li-polymer	150–200	350	Not available	1000
Li-ion	90–160	200–350	>90	>1000

applications, several other battery technologies have been used in various prototype vehicles. In this section, we will review not only the promising battery technologies, but also those that have been tried in various prototype electric vehicles.

The future of the battery technologies for electric and hybrid vehicle applications depends on factors including system cost, availability of raw materials, mass production capabilities, and life cycle characteristics. One must note that the electric and hybrid vehicles industry covers a wide spectrum and is not just limited to road vehicles. Some technologies may be more suitable for certain applications for various reasons. The representative properties of the promising batteries technologies along with that of lead-acid battery are summarized in Table 4.5 with information obtained from various literatures. The chemistry and additional information on the alternative battery technologies will then be presented in this chapter.

4.6.1 Lead–Acid Battery

The lead-acid batteries have been the most popular choice of batteries for electric vehicles during the initial development stages. The lead-acid battery has a long history that dates back to the middle of the nineteenth century and is currently a very mature technology. The first lead-acid battery was produced as early as in 1859. In the early 1980s, over 100 million lead-acid batteries were produced per year. The long existence of the lead acid battery is due to

- Relatively low cost
- Easy availability of raw materials (lead, sulfur)
- Ease of manufacture
- Favorable electromechanical characteristics

Lead-acid batteries can be designed to be of high power and are inexpensive, safe, and reliable. A recycling infrastructure is in place for them. However, low specific energy, poor cold temperature performance, and short calendar and cycle life are among the obstacles to their use in electric vehicles and hybrid electric vehicles.

Conventionally, lead-acid batteries are of flooded-electrolyte cells, where free acid covers all the plates. This imposes the constraint of maintaining an upright position for the battery, which is difficult in certain portable situations. Efforts in developing hermetically sealed batteries faced the problem of buildup of an explosive mixture of hydrogen and oxygen on approaching the top-of-charge or overcharge condition during cell recharging. The problem is addressed in the valve-regulated-lead-acid (VRLA) batteries by providing a path for the oxygen, liberated at the positive electrode, to reach the negative electrode where it recombines to form lead sulfate. There are two mechanisms for making the sealed VRLA batteries, the gel battery and the AGM (absorptive glass microfiber) battery. Both types are based on immobilizing the sulfuric acid electrolyte in the separator and the active materials leaving sufficient porosity for the oxygen to diffuse through the separator to the negative plate [16].

The construction of a typical battery consists of positive and negative electrode groups (elements) interleaved to form a cell. The through partition connection in the battery is illustrated in Figure 4.21. The positive plate is made of stiff paste of the active material on a lattice type grid, which is shown in Figure 4.22. The grid made of a suitably selected lead alloy is the framework of a portable battery to hold the active material. The positive plates can be configured as flat pasted or in tubular fashion. The negative plates are always manufactured as pasted types.

4.6.2 Nickel–Cadmium Battery

The advantages of NiCd batteries are superior low-temperature performance compared to the lead-acid battery, flat discharge voltage, long life, and excellent reliability. The maintenance requirements of the batteries are also low.

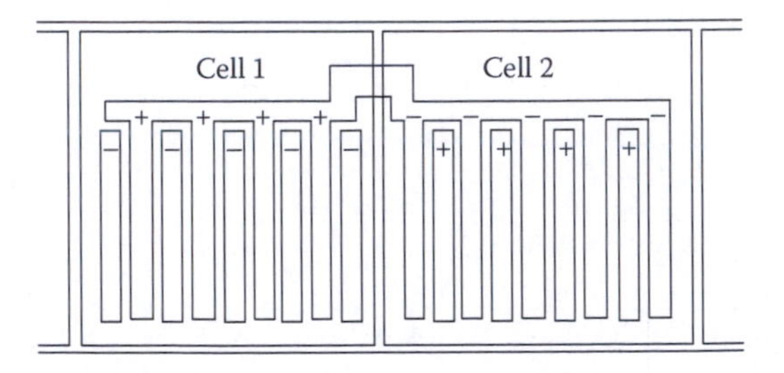

FIGURE 4.21
Schematic diagram of lead-acid battery showing through-partition connection.

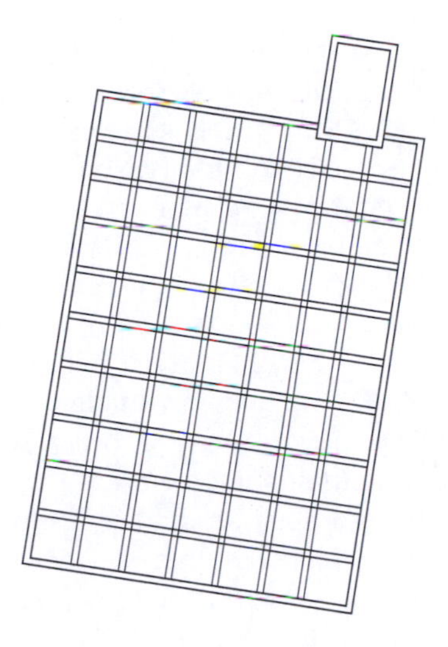

FIGURE 4.22
A lead-acid battery grid.

The lower practical cell voltage between 1.2 and 1.3 V means that more cells have to be connected in series to get the desired voltage. The specific energy of NiCd batteries is 30–50 W h/kg, which is similar to that of lead-acid batteries.

The biggest drawbacks of NiCd batteries are the high cost and the toxicity contained in cadmium. The environmental concerns may be overcome in the long run through efficient recycling, but the insufficient power delivered by the NiCd batteries is another important reason for not considering these batteries for electric and hybrid electric vehicles applications. The drawbacks of the NiCd batteries led to the rapid development of NiMH batteries, which are deemed more suitable for electric and hybrid electric vehicle applications.

4.6.3 Nickel–Metal–Hydride Battery

The NiMH battery is a successor to the nickel-hydrogen battery, and is already in use in the production of hybrid electric vehicles. The positive electrode in a NiMH battery cell is nickel hydroxide ($Ni(OH)_2$) and the negative electrode is metal hydride. The chemical reactions of the NiMH battery cell have already been presented in Section 4.2. The negative electrode consists of a compressed mass of fine metal particles. The metallic alloy can absorb a large number of hydrogen molecules under certain temperature and pressure to form the metal hydride. This can be thought of as an alternative approach of storing hydrogen. The proprietary alloy formulations used in NiMH are known as AB_5 and AB_2 alloys. In the AB_5 alloy, A is the mixture of rare earth elements and B is partially substituted nickel. In the AB_2 alloy,

A is titanium and/or zirconium and B is again partially substituted nickel. The AB_2 alloy has higher capacity for hydrogen storage and less costly. The operating voltage of NiMH is almost the same as that of NiCd with flat discharge characteristics. The capacity of the NiMH is significantly higher than that of NiCd, with specific energy ranging from 60 to 80 W h/kg. The specific power of NiMH batteries can be as high as 250 W/kg.

The NiMH batteries have penetrated the market in recent years at an exceptional rate. NiMH battery pack was used in Chrysler "EPIC" minivans, which give a range of 150 km. NiMH battery packs are exclusively used in the commercially available Toyota hybrid vehicles.

The components of NiMH are recyclable, but a recycling infrastructure is not yet in place. NiMH batteries have a much longer life cycle than lead-acid batteries and are safe and abuse-tolerant. The disadvantages of NiMH batteries are the relatively high cost, higher self-discharge rate compared to NiCd, poor charge acceptance capability at elevated temperatures, and low cell efficiency. NiMH is likely to survive as the leading rechargeable battery in the future for traction applications with strong challenge coming only from Li-ion batteries.

4.6.4 Li–Ion Battery

The lithium metal has high electrochemical reduction potential relative to that of hydrogen (3.045 V) and the lowest atomic mass (6.94), which shows promise for a battery of 3 V cell potential when combined with a suitable positive electrode. The interest in secondary lithium cells soared soon after the advent of lithium primary cells in the 1970s, but the major difficulty was the highly reactive nature of the lithium metal with moisture that restricted the use of liquid electrolytes. The discovery in late 1970s by researchers at Oxford University that lithium can be intercalated (absorbed) into the crystal lattice of cobalt or nickel to form $LiCoO_2$ or $LiNiO_2$ paved the way toward the development of Li-ion batteries [17]. The use of metallic-Li is bypassed in Li-ion batteries by using lithium intercalated (absorbed) carbons (Li_xC) in the form of graphite or coke as the negative electrode along with the lithium metallic oxides as the positive electrode. The graphite is capable of hosting lithium up to a composition of LiC_6. The majority of the Li-ion batteries use either a layered oxide or iron phosphates of lithium as the positive electrode. The layered positive electrodes of cobalt oxide are expensive, but proved to be the most satisfactory. Nickel-oxide $LiNiO_2$, which costs less, can also be used, but is structurally more complex. The performance is similar to that of cobalt-oxide electrodes. The manganese oxide-based positive electrodes ($LiMn_2O_4$ or $LiMnO_2$) are also used since manganese is cheaper, widely available, and less toxic. Alternative positive electrode material is the lithium-iron-phosphate ($LiFePO_4$) which can deliver stable and good performance at lower costs.

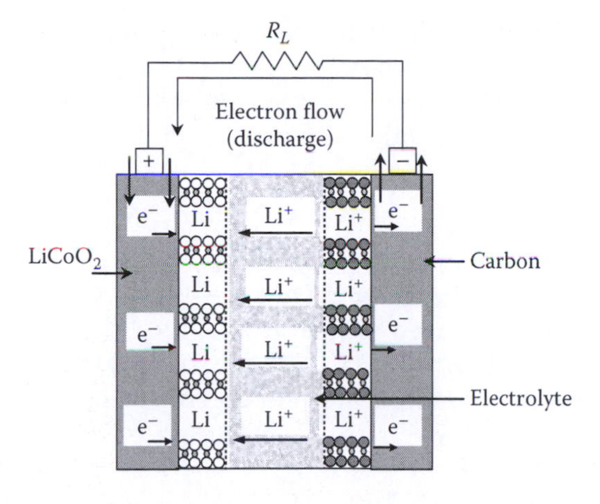

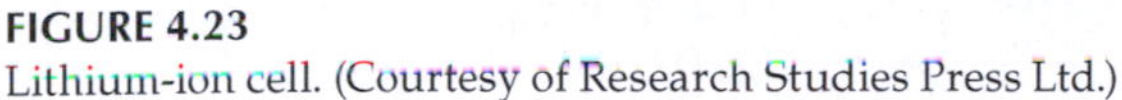

FIGURE 4.23
Lithium-ion cell. (Courtesy of Research Studies Press Ltd.)

The cell discharge operation in a Li-ion cell using $LiCoO_2$ is illustrated in Figure 4.23. During cell discharge, lithium ions (Li^+) are released from the negative electrode that travels through an organic electrolyte toward the positive electrode. In the positive electrode, the lithium ions are quickly incorporated into the lithium compound material. The process is completely reversible. The chemical reactions at the electrodes are

At the negative electrode,

$$Li_xC_6 \underset{\text{Charge}}{\overset{\text{Discharge}}{\rightleftarrows}} 6C + xLi^+ + xe^- \quad \text{where } 0 < x < 1$$

At the positive electrode,

$$xLi^+ + xe^- + Li_{(1-x)}CoO_2 \underset{\text{Charge}}{\overset{\text{Discharge}}{\rightleftarrows}} LiCoO_2$$

During cell charge operation, the lithium ions move in the opposite direction from the positive electrode to the negative electrode. The nominal cell voltage for a Li-ion battery is 3.6 V, which is equivalent to three NiMH or NiCd battery cells.

The lithium ion batteries have high specific energy, high specific power, high energy efficiency, good high-temperature performance, and low self-discharge. The components of Li-ion batteries are also recyclable. These characteristics make Li-ion batteries highly suitable for electric and hybrid vehicles and other applications of rechargeable batteries. The main drawback of Li-ion batteries is that these are very sensitive to overvoltages and overdischarges. The overvoltage of Li-ion cell positive electrode results in solvent oxidation and the exothermic decomposition of the active material. Overvoltage and overdischarge can result in irreversible cell damage possibly accompanied by cell ignition [18].

4.6.5 Li–Polymer Battery

The Li-polymer evolved out of the development of solid-state electrolytes, i.e., solids capable of conducting ions but are electron insulators. The solid-state electrolytes resulted from the research in the 1970s on ionic conduction in polymers. These batteries are considered solid-state batteries since their electrolytes are solids. The most common polymer electrolyte is the polyethylene oxide compounded with an appropriate electrolyte salt.

The most promising positive electrode material for Li-polymer batteries is vanadium oxide V_6O_{13} [16]. This oxide interlaces up to 8 lithium atoms per oxide molecule with the following positive electrode reaction:

$$Li_x + V_6O_{13} + xe^- \underset{\text{Charge}}{\overset{\text{Discharge}}{\rightleftarrows}} Li_xV_6O_{13} \quad \text{where } 0 < x < 8$$

The Li-polymer batteries have the potentials for the highest specific energy and power. The solid polymers, replacing the more flammable liquid electrolytes in other type of batteries, can conduct ions at temperatures above 60°C. The use of solid polymers also has a great safety advantage in case of electric and hybrid electric vehicles accidents. Since the lithium is intercalated into carbon electrodes, the lithium is in ionic form and is less reactive than pure lithium metal. The thin Li-polymer cell gives the added advantage of forming a battery of any size or shape to suit the available space within the electric and hybrid electric vehicles chassis. The main disadvantage of the Li-polymer battery is the need to operate the battery cell in the temperature range of 80°C–120°C. Li-polymer batteries with high specific energy, initially developed for electric vehicle applications, also have the potential to provide high specific power for hybrid electric vehicle applications. The other key characteristics of the Li-polymer are good cycle and calendar life.

4.6.6 Zinc–Air Battery

The zinc-air batteries have a gaseous positive electrode of oxygen and a sacrificial negative electrode of metallic zinc. The practical zinc-air battery is only mechanically rechargeable by replacing the discharged product, zinc hydroxide with fresh zinc electrodes. The discharged electrode and the potassium hydroxide electrolyte are sent to a recycling facility. In a way, the zinc-air battery is analogous to a fuel cell with the fuel being the zinc metal. A module of zinc air batteries tested in German Mercedes Benz postal vans had a specific energy of 200 W h/kg, but only a modest specific power of

100 W/kg at 80% DoD (see Sections 4.3.8 and 4.3.12 for definition of depth of discharge and specific power, respectively). With the present-day technology, the range of zinc-air batteries can be between 300 and 600 km between recharges.

Other metal air systems have also been investigated but the work has been discontinued due to severe drawbacks in the technologies. These batteries include iron-air and aluminum-air batteries where iron and aluminum are respectively used as the mechanically recyclable negative electrode.

The practical metal-air batteries have two very attractive positive features: (1) The positive electrode can be optimized for discharge characteristics, since the batteries are recharged outside the battery and (2) the recharging time is rapid with a suitable infrastructure.

4.6.7 Sodium–Sulfur Battery

Sodium, similar to lithium, has a high electrochemical reduction potential (2.71 V) and low atomic mass (23.0), making it an attractive negative electrode element for batteries. Moreover, sodium is abundant in nature available at a very low cost. Sulfur, which is a possible choice for the positive electrode, is also a readily available and another low cost material. The use of aqueous electrolytes is not possible due to the highly reactive nature of sodium and solid polymers, like those used for lithium batteries are not known. The solution of electrolyte came from the discovery of beta-alumina by scientists in Ford Motor Company in 1966. Beta-alumina is a sodium aluminum oxide with a complex crystal structure.

Despite the several attractive features of NaS batteries, there are several practical limitations. The cell operating temperature in NaS batteries is around 300°C, which requires adequate insulation as well as a thermal control unit. The requirement forces a certain minimum size of the battery limiting the development of the battery for only electric vehicles, a market for which is not yet established. Another disadvantage of NaS batteries is the absence of an overcharge mechanism. At the top-of-charge one or more cells can develop a high resistance, which pulls down the entire voltage of the series-connected battery cells. Yet another major concern is the safety issue, since the chemical reaction between molten sodium and sulfur can cause excessive heat or explosion in the case of accident. The safety issues were addressed through efficient design, and manufactured NaS batteries have been shown to be safe.

The practical limitations and manufacturing difficulty of NaS batteries have led to the discontinuation of its development programs, especially when the simpler concept of sodium-metal chloride batteries was developed.

4.6.8 Sodium–Metal–Chloride Battery

The sodium-metal-chloride battery is a derivative of sodium-sulfur battery with intrinsic provisions of overcharge and overdischarge. The construction

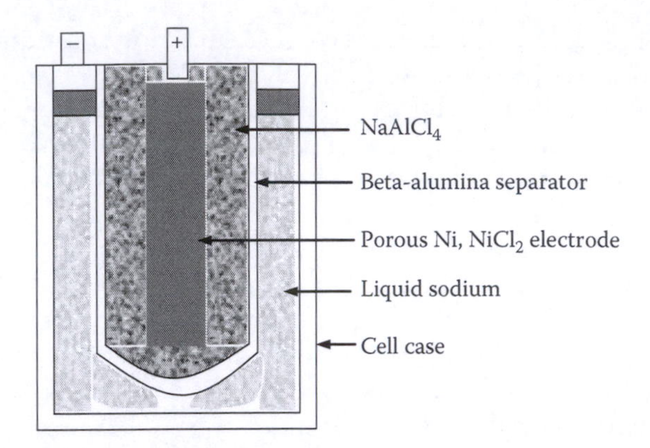

FIGURE 4.24
A sodium-nickel-chloride cell.

is similar to that of NaS battery, but the positive sulfur electrode is replaced by nickel chloride ($NiCl_2$) or a mixture of nickel chloride and ferrous chloride ($FeCl_2$). The negative electrode and the electrolyte are the same as in NaS battery. The schematic diagram of a $NaNiCl_2$ cell is shown in Figure 4.24. In order to provide good ionic contact between the positive electrode and the electrolyte, both of which are solids, a second electrolyte of sodium chloraluminate ($NaAlCl_4$) is introduced in a layer between $NiCl_2$ and beta-alumina. The $NaAlCl_4$ electrolyte is a vital component of the battery, although it reduces the specific energy of the battery by about 10% [17]. The operating temperature is again high, similar to that of NaS battery. The basic cell reactions for the nickel chloride and ferrous chloride positive electrodes are

$$NiCl_2 + 2Na \underset{\text{Charge}}{\overset{\text{Discharge}}{\rightleftarrows}} Ni + 2NaCl \ (2.58 \text{ V})$$

$$FeCl_2 + 2Na \underset{\text{Charge}}{\overset{\text{Discharge}}{\rightleftarrows}} Fe + 2NaCl \ (2.35 \text{ V})$$

The cells in a sodium-metal-chloride battery are assembled in a discharged state. The positive electrode is prefabricated from a mixture of Ni or Fe powder and NaCl (common salt). On charging after assembly, the positive electrode compartment is formed of the respective metal and the negative electrode compartment is formed of sodium. This procedure has two significant advantages, (1) pure sodium is manufactured in situ through diffusion in beta-alumina and (2) the raw materials for the battery (common salt and metal powder) are inexpensive. Although iron is cheaper than nickel, the latter is more attractive as the metallic component because of fewer complications and wider operating temperature range.

The sodium chloride batteries are commonly known as the ZEBRA batteries, which originally resulted from a research collaboration between scientists from the United Kingdom and South Africa in the early 1980s. The ZEBRA batteries have been shown to be safe under all conditions of use. The ZEBRA batteries have high potentials for being used as batteries for electric vehicles and hybrid electric vehicles. There are several test programs performed with the ZEBRA batteries.

4.6.9 Goals for Advanced Batteries

The California legislative mandates in the early 1990s led to the formation of the U.S. Advanced Battery Consortium (USABC) to oversee the development of power sources for electric vehicles. USABC is within the U.S. Council of Automotive Research (USCAR), which is an umbrella organization of U.S. auto manufacturers formed in 1992 to strengthen the automotive technology base through collaborative research and development. The USABC addresses factors to continue the development of high energy density and high power density energy storage technologies. USABC establishes the goals for energy storage developments to support electric, hybrid, and fuel cell vehicles. Goals are set for long term development as well as long-term commercialization. The purpose of the commercialization goals is to develop batteries with a reasonable goal, while the long-term criteria was set to develop batteries for electric vehicles, which would be directly competitive with the IC engine vehicles. The specific power and specific energy long-term goals have been set aggressively at 400 W/kg and 200 W h/kg to promote research and development. A subset of the goals set by the USABC for advanced electric vehicle batteries is listed in Table 4.6. The calendar life for these batteries is targeted for 10 years, while cycle life has been set for 1000 cycles at 80% DoD for both commercialization and long-term goals.

TABLE 4.6

USABC Objectives for EV Advanced Battery Packs

Parameter	Minimum Goals for Long-Term Commercialization	Long-Term Goals
Specific energy (W h/kg) (C/3 discharge rate)	150	200
Specific power (W/kg) (80% DoD per 30 s)	300	400
Specific power (W/kg), Regen. (20% DoD per 10 s)	150	200
Recharge time (h) (20% → 100% SoC)	4–6	3–6
Cost, U.S. $/kW h	150	100

TABLE 4.7

USABC Goals for HEV Advanced Energy Storage Systems

Parameter	Power Assist (Minimum)	Power Assist (Maximum)
Pulse discharge power, 10 s (kW)	25	40
Peak regenerative pulse power, 10 s (kW)	20	35
Total available energy (kW h)	0.3 at C/1 rate	0.5 at C/1 rate
Maximum weight (kg)	40	60
Maximum volume (L)	32	45
Cost, @1,000,000 units/year (U.S. $)	500	800

The USABC has also set goals for hybrid electric vehicles at two levels of power-assist, one at the 25 kW level and the other for the 40 kW level. A subset of the goals set by the USABC for power-assist hybrid electric vehicle energy storage system is listed in Table 4.7. The calendar life for these batteries is targeted for 15 years, while cycle life has been set for 300,000 cycles for specified SoC increments. The roundtrip energy efficiency has been set to 90% for both 25 and 40 kW power-assist hybrid electric vehicles.

The USABC has also specified goals for two main PHEV battery types: a high power/energy ratio battery providing 10 mi of all-electric range (PHEV-10), and a low power/energy ratio battery providing 40 mi of all-electric range (PHEV-40). PHEV-10 goals are set for a "crossover utility vehicle" weighing 1950 kg and the PHEV-40 goals are set for a midsize sedan weighing 1600 kg. Few of these important goals set for PHEV development are listed in Table 4.8. The calendar life for these batteries is also targeted for 15 years, and roundtrip energy efficiency has been set to 90% for both PHEV-10 and PHEV-40. All the specified goals for energy storage systems for

TABLE 4.8

USABC Goals for PHEV Energy Storage Systems

Parameter	PHEV-10	PHEV-40
Pulse discharge power, 10 s (kW)	45	38
Peak regenerative pulse power, 10 s (kW)	3,025	35
Available energy for charge depleting mode (kW h)	3.4	11.6
Available energy for charge sustaining mode (kW h)	0.3	0.5
Charge depleting life/discharge throughput (cycles/MW h)	5,000/17	5,000/58
Charge sustaining life cycle (cycles)	300,000	300,000
Maximum weight (kg)	60	120
Maximum volume (L)	40	80
Cost, @100,000 units/year (U.S. $)	1,700	3,400

electric and hybrid electric vehicles are listed in the USABC Web site under USCAR at *www.uscar.org*.

4.7 Battery Pack Management

Batteries can be configured in series or in parallel or in a combination thereof. The battery pack, i.e., the energy storage device in an electric and hybrid vehicle, consists of a number of individual electrochemical cells connected in a series string to deliver the required voltage. The strings of series-connected cells can be connected in parallel to increase the capacity of the storage system. The battery pack also includes electronics, which is typically located outside the battery pack. The electronic circuit of a battery pack controls charging, discharging, and balanced utilization of the battery cells. The electronic circuit along with its controller hardware and software algorithms is responsible for managing the battery pack and protecting the cells within the pack. The primary function of the battery management system is to protect the cells from operating outside the safe region. This ensures longer life of the battery and minimizes replacement costs.

The battery pack management techniques are general and suitable for any energy storage-based system, such as electric and hybrid vehicles, distributed power generation units, and portable consumer electronics. Of all the applications, the most rigorous usage of energy storage systems is in hybrid vehicles where it goes through pulsed charge/discharge cycles. Hence, battery pack management is required to be of the most advanced type. The essentials of battery management systems, SoC measurement techniques, cell balancing methods, and battery charging methods are covered in the Sections 4.7.1 through 4.7.4.

The battery management systems and methods of cell balancing presented are equally applicable to an ultracapacitor bank replicating an energy storage device. The ultracapacitor cells are also electrochemical cells, which are connected in a series string to form the energy storage system to supply power at the desired voltage level.

4.7.1 Battery Management System

The battery management system (BMS) consists of a set of algorithms based on voltage, current, and temperature measurements to calculate essential battery parameters and determine charge/discharge power limits at a given time. Depending on the level of sophistication in the BMS, measurements can be from individual cells or group of cells or from the entire pack. The BMS is also responsible for generating command signals for cell equalization circuits if used in a battery pack. BMS ensures reliability and protection against overcharge, overdischarge, short circuits, and thermal abuse. The

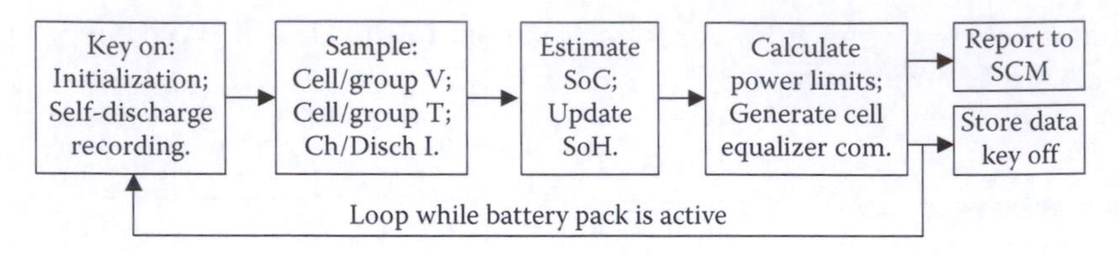

FIGURE 4.25
Parameter estimations and pack management in a BMS.

BMS for an energy storage system is designed to have all or some of the following features:

- State-of-charge (SoC) estimation
- State-of-health (SoH) monitoring for cell and pack protection
- Temperature control
- Charge/discharge power control
- Cell equalization
- Data logging

The measurements, parameter estimations, and outputs generated in a BMS are shown in Figure 4.25 [19]. The BMS initializes once the system is powered, which happens when with key on in a hybrid electric vehicle. The only function during initialization is to record the self-discharge during the system off period. If the self-discharge is excessive, then it is reported to the SoH monitoring algorithm. The other parameters for battery management are estimated during each measurement cycle while the pack is on.

SoC provides information on the available capacity of a battery. This is necessary not only for the protection of the battery pack, but also for vehicle powertrain controls. The SoC should also be maintained within a certain band to enhance the life of the battery. In sophisticated management systems, the SoC of individual cells or group of cells in a pack are determined to verify the uniform distribution of SoC among the cells. The SoC is typically expressed as a percentage of the rated capacity instead of the maximum available capacity, which could be less due to aging and environmental effects. However, the SoC could also be calculated based on the maximum available capacity. This SoC calculation can be used for cell equalization, since all cells in the series-string generally experience the same environment.

SoH is the working condition of the pack and measures the pack's ability to deliver power compared to a new pack. The fading of a cell capacity and power compared to other cells in the pack with aging indicates the deteriorating health of the cell. The cell capacity and other parameters are used in an algorithm to estimate the SoH of a battery pack. The SoH information is useful for battery safety and for delivering power up to its maximum capability. The SoH estimation algorithms are based on comparing measured and estimated cell parameters with references or neighboring cells. The voltages and SoC anomalies of one cell compared to the nearby cells is indicative of poor health of that cell. Similarly, the excessive self-discharge of the pack raises a flag, and is compared with preset limits to estimate the pack SoH. The SoH information can be used to replace damaged cells in a pack instead of replacing the entire pack.

The temperature is the primary environmental factor that affects the SoC of an energy storage system. Imbalances in temperature among the various cells in a pack will result in imbalances in the SoC. Additionally, temperature affects the self-discharge rates. The thermal management in a pack is part of the cooling system design for the pack, but the temperature information of the cells should be effectively utilized for protection and health monitoring of the cells.

The maximum power available from the battery at a given time is calculated in the BMS based on the SoC and terminal voltages, ensuring that operating voltage, current, SoC, and other design limits are not violated. The BMS sets the power limits during charging and discharging for battery protection. Batteries could get severely damaged due to inappropriate charging. The limits are reported to the supervisory controller for powertrain controls in electric and hybrid vehicles.

There are three levels of management systems: pack-level management, modular-pack-level management, and cell-level management. Pack-level management is the most basic one where overall pack voltage and SoC are monitored, whereas the most complete cell equalization and balancing is possible when individual cell parameters of voltage, current, and temperature are monitored. The charging and discharging power managements at the pack level leaves individual cells vulnerable to damage. In modular-pack-level management, groups of cells are treated as a module for cell balancing and equalization; the BMS algorithms depend on group voltage, current, and temperature measurements rather than pack or individual cell measurements. For packs employing cell equalizer circuits, the BMS generates command signals for cell voltage equalization based on its measurements and estimations. The circuits act on these signals to balance the cells or groups of cells.

Data logging is another important function of the energy storage management system. The data for voltage, current, temperature, SoC, and number of charge/discharge cycles could be stored as a function of time for SoH monitoring, diagnostics, and fault analysis.

4.7.2 SoC Measurement

The SoC of the energy storage system is calculated using measurements of a physical parameter that varies with the SoC. The SoC varies with voltage, charge/discharge rate, self-discharge rate, temperature, and aging. Depending on the parameters monitored, the SoC calculation can be either

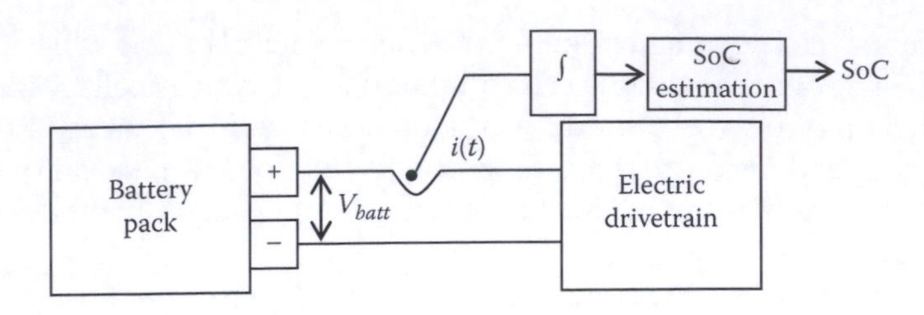

FIGURE 4.26
Battery SoC measurement.

a voltage-based method or a current-based method. The more accurate SoC calculations use both voltage and current measurements in an observer-based method.

The voltage-based SoC measurement is applicable to cell chemistries where the voltages are directly related to the SoC. The relation between open circuit voltage and SoC must be known a priori for a good estimation of the SoC. The voltage-based measurement is not at all suitable for Li-ion cells, since the voltage for these cells is fairly steady over most of the charge/discharge cycle.

In current-based estimation, the SoC is obtained from the integration of current using the fundamental definition of charge $q = \int_0^t i(t)dt$. A simple SoC measurement diagram is shown in Figure 4.26. The charge and discharge currents out of and into the storage device are measured directly using a current sensor. The integration of measured current gives the SoD when the initial condition is the fully charged condition of the battery. Knowing the initial capacity C_p of the battery, SoC is calculated from

$$\text{SoC}(t) = C_P - \text{SoD}(t)$$

The method is also known as Coulomb counting. The method tends to accumulate errors if it is solely based on current information. A method of improving the SoC estimation is to incorporate the directly measurable parameters (voltage and current) into a mathematical model of the storage system to implement an observer-based SoC estimation method. This is a closed-loop Coulomb counting method as opposed to the open-loop method of Figure 4.26. The closed-loop method is based on feedbacks that can be empirically designed or generated using Kalman filters [19].

4.7.3 Battery Cell Balancing

The individual cells in the connected string of the battery pack are the unit battery cells. The available energy stored in a battery cell is $E_{avail} = qV$, which states that both charge and voltage need to be balanced in a series string to maximize the output of a pack. When a series-string of electrochemical cells is charged as a pack, slight parameter mismatches in individual cells and temperature differences result in charge and voltage imbalances. The imbalances adversely affect the vehicle performance by reducing the throughput of the battery pack.

The chemical reactions in an electrochemical cell depend on the temperature and pressure. The temperature differences among the cells change the self-discharge rates, causing imbalances in the charge of the cells. A low cell temperature reduces chemical activity which increases the cell's internal impedance. The increased internal resistance reduces the terminal voltage, and thus the cell capacity. In addition, manufacturing differences and different aging characteristics result in parameter mismatches among individual cells, which cause voltage and capacity imbalances [20,21].

The charge imbalance also shows up as voltage differences. The imbalances tend to grow as the pack goes through repeated charge/discharge cycles. The weaker cells tend to charge slower and the stronger cells charge faster. The process shortens the pack life and reduces its utilization. The number of charge/discharge cycles affects some battery chemistry more than the others. For example, Li-ion batteries are highly sensitive to overvoltages and undervoltages. Li-ion batteries are recommended to limit the charge/discharge rates to no more than 2C, and also to keep the cells charged to at least 40% SoC to minimize aging.

The maximum throughput of the pack can be ensured by balancing the voltage and charge of individual cells. The cell balancing methods utilize electronic circuits and control to even out the voltages and SoC of a series-string of electrochemical cells. The simplest strategy adopted for charging a series-string of cells is to monitor the cell voltages and discontinue charging when one of the cells (strongest cell) reaches the voltage limit for individual cells. Extended charging is another option where charging is continued even after the strongest cell has reached its capacity to bring the weaker cells up to capacity. When charging continues to bring the weaker cells to the maximum voltage, overvoltage results in the stronger cells. Overcharging is not at all an option with certain battery chemistry, while in others the process vents hydrogen gas (known as *gassing*) and removes water from the overcharged cells. Increased gassing in the cell at elevated temperatures shortens the cell life.

The overcharging in the stronger cells can be avoided if there is a path to shunt the charging currents once they reach the voltage limit. Similarly, the simplest protection during discharging of a pack is to shut down when the first cell reaches the minimum voltage limit. This cell is consequently the weakest cell in the series-string and is limiting the capacity of the pack. If discharging is continued to extract energy from the stronger cells, then the weaker cell voltage will fall below the minimum voltage level, possibly causing damage to the cells.

The simple cell balancing strategies result in underutilization of the battery pack. Improved cell balancing circuits provide a path to bypass the weaker cells once they reach the minimum voltage, provided the pack voltage level is still above the minimum voltage level of the system. Power electronic converter circuits are used to divert charging currents to boost the weaker cells or deplete charge from stronger cells for cell voltage equalization. The circuit topologies for cell balancing are presented in Chapter 7 after the power electronic devices and concepts are introduced.

4.7.4 Battery Charging

The charging of secondary batteries is accomplished in several phases using different charging currents. The phases are structured based on the battery chemistry to minimize the damage on the cells. The initial charging phase is the *bulk charging phase* when the cells are charged with the maximum current to replenish most of the charge lost during discharge. The last few percentages of SoC are replenished with the *absorption charging phase*. The charging current in this phase is kept low to prevent any damage to the cells. An *equalization phase* can also be used to fully charge and balance all the cells in the battery pack. The *float charge phase* starts once the battery is fully charged to compensate for energy lost over time due to self-discharge. Microprocessor controllers are used to set the charging profile based on an algorithm to tune the charging for a particular type of battery chemistry.

The battery charging circuits can apply either a constant current or a constant voltage or any combination of the two to design the charge profile. In the constant current charging method, known as I-charging, a current regulator in the battery charger maintains the set current level. The charging current levels are adjusted by the current regulator for the different phases of charging. The charging current can also be applied in the form of pulses by pulse width modulation (PWM) control of the output voltage. The charging rate is controlled by adjusting the pulse width. The short durations between the pulses allow the chemical reactions within the cells to stabilize. Excessive chemical reactions that could lead to gassing are avoided by using pulse charging. An example of multistep charging profile with I-charging and PWM control is shown in Figure 4.27 [22].

For constant voltage charging, a voltage greater than the battery upper limit voltage is applied by the charger for bulk charging. A constant voltage charge is also usually applied during the absorption charging phase. During the float charging phase, the charger applies a DC voltage slightly lower than the battery upper limit voltage across the battery. A slight drop in the battery voltage results in charge being replenished in the battery. This form of charging is also known as trickle charging used to compensate for self-discharge in the cells.

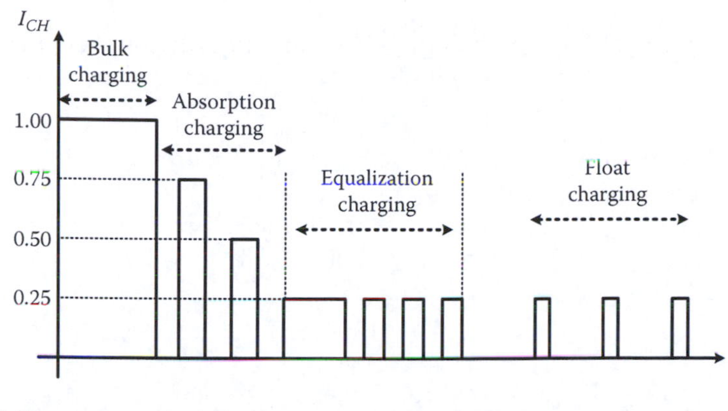

FIGURE 4.27
Multistep I-charging with PWM control.

Problems

4.1 Estimate the weight of a 12 V, 100 A h lead-acid battery. Do this by calculating the reactant masses participating in the overall chemical reaction. Also, assume that the mass of H_2O in the electrolyte is twice the mass of H_2SO_4. Neglect battery casing mass, electrode grid mass, separator mass, and current bus mass. (Note that $n=2$ for Pb and PbO_2 and $n=1$ for H_2SO_4.)

4.2 In the nickel-cadmium cell, nickel oxyhydroxide, NiOOH is the active material in the charged positive plate. During discharge it reduces to the lower valence state, nickel hydroxide $Ni(OH)_2$, by accepting electrons from the external circuit:

$$2NiOOH + 2H_2O + 2e^- \underset{\text{Charge}}{\overset{\text{Discharge}}{\rightleftarrows}} 2Ni(OH)_2 + 2OH^- (0.49\ V)$$

Cadmium metal is the active material in the charged negative plate. During discharge, it oxidizes to cadmium hydroxide, $Cd(OH)_2$, and releases electrons to the external circuit:

$$Cd + 2OH^- \underset{\text{Charge}}{\overset{\text{Discharge}}{\rightleftarrows}} Cd(OH)_2 + 2e^- (0.809\ V)$$

The net reaction occurring in the potassium hydroxide (KOH) electrolyte is:

$$\text{Cd} + 2\text{NiOOH} + 2\text{H}_2\text{O} \underset{\text{Charge}}{\overset{\text{Discharge}}{\rightleftarrows}} 2\text{Ni(OH)}_2 + \text{Cd(OH)}_2^- (1.299\text{ V})$$

Estimate the weight of a 11.7 V, 100 A h Ni-Cd battery. Neglect the mass KOH component of the electrolyte.

4.3 A 12 V battery is connected to a series RL load, as shown in Figure P4.3. The battery has a rated capacity of 80 A h. At $t = 0$, the switch is closed and the battery begins to discharge.

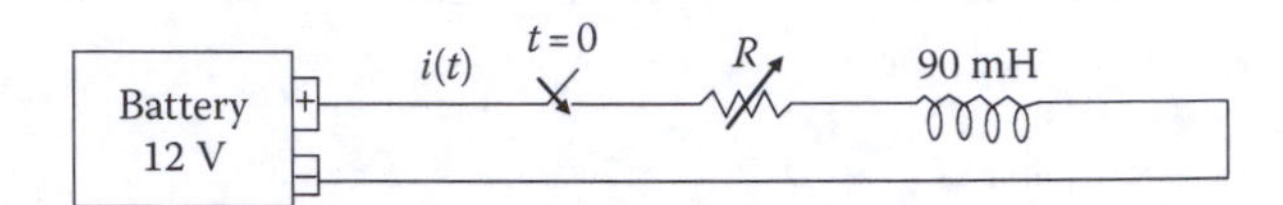

FIGURE P4.3

(a) Calculate and plot the battery discharge current, $i(t)$, if the steady-state discharge rate is C/2. Neglect battery voltage drop.
(b) Calculate and plot SoD(t) in A h for $0 < t < 2$ h.
(c) Calculate and plot SoC(t) assuming that at $t = 0$, the battery is charged to rated capacity. Assume also that the rated capacity is the practical capacity.
(d) Calculate the time corresponding to 80% DoD.

4.4 Given below are constant power discharge characteristics of a 12 V lead-acid battery:

SP (W/kg)	SE (W h/kg)
10	67.5
110	8

The battery characteristics are to be expressed in terms of Peukert's equation, which has the following form:

$$(\text{SP})^n(\text{SE}) = \lambda\left(n \text{ and } l \text{ are curve fitting constants}\right)$$

(a) Derive the constants n and λ, assuming a linear relationship between log (SP) and log (SE).
(b) Find the capacity Q_T of the battery if the theoretical energy density is $\text{SE}_T = 67.5$ W h/kg, given the battery mass of 15 kg.

4.5 An EV battery pack consists of four parallel sets of 6 series connected 12 V, 100 A h lead-acid batteries. One steady-state motoring (discharge) cycle of battery current is shown in Figure P4.5a. The steady-state regenerative braking (charge) cycle of battery is shown in Figure P4.5b.

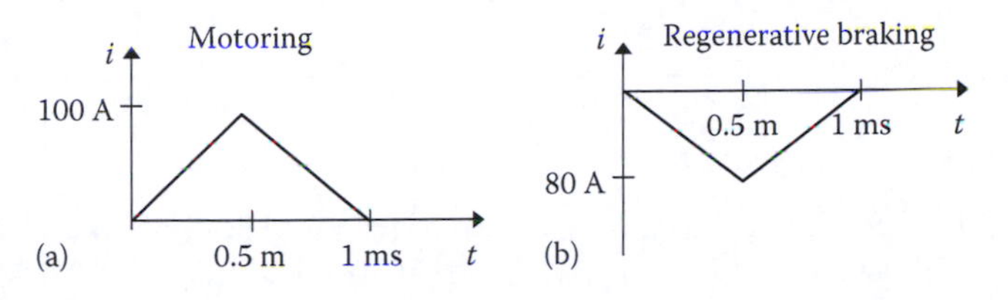

FIGURE P4.5

(a) Suppose no regenerative braking is employed. How much time does it take to reach 80% DoD?
(b) If regenerative braking is employed such that for every 50 motoring cycles there is 1 regenerative braking cycle, how much time does it take to reach 80% DoD?

(Note: In this problem, neglect variation of capacity with discharge rate. Assume that the practical capacity is equal to the rated capacity.)

4.6 Given a lead-acid battery having the following empirical characteristics:

$$(SP)^{.9}(SE) = 216E4$$

where SP = specific power and SE = specific energy. The EV parameters are:

$m = 700\,kg$, $M_B = 150\,kg$, $C_D = 0.2$, $A_F = 2\,m^2$, $C_0 = 0.009$, $C_1 = 0$.
Also, take $\rho = 1.16\,kg/m^3$ and $g = 9.81\,m/s^2$.

(a) Derive and plot $F_{TR}(t)$ vs. t. (Assume level road.)
(b) Derive and plot $P_{TR}(t)$ vs. t.

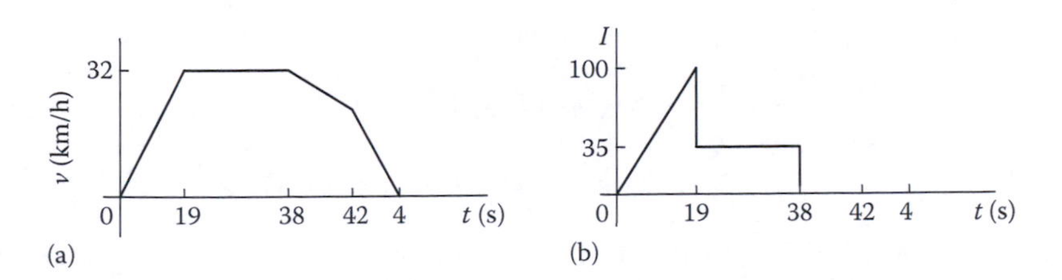

FIGURE P4.6

Calculate the EV range based on SAE J227a schedule B driving cycle using the power density approach of fractional depletion model (FDM). The SAE J227a driving cycle and the current profile of the EV are given in Figure P4.6a and b. (Assume no regenerative braking.)

References

1. D. Linden, *Handbook of Batteries*, 2nd edn., McGraw-Hill Inc., New York, 1995.
2. C. Mantell, *Batteries and Energy Systems*, McGraw-Hill Inc., New York, 1983.
3. K. Kordesch and G. Simadar, *Fuel Cells and Their Applications*, John Wiley & Sons Inc., Chichester, U.K., 1996.
4. J.O.M. Bockris, *Fuel Cells: Their Electrochemistry*, McGraw-Hill Inc., New York, 1969.
5. J.S. Newman, *Electrochemical Systems*, Prentice Hall, Englewood Cliffs, NJ, 1991.
6. T. Erdey-Gruz, *Kinetics of Electrode Process*, John Wiley & Sons Inc., New York, 1972.
7. L. Xia, Behavioral modeling and analysis of galvanic devices, PhD dissertation, University of Akron, Akron, OH, 2000.
8. A.J. Bard and L.R. Faulkner, *Electroanalytical Methods: Fundamentals and Applications*, John Wiley & Sons Inc., New York, 1996.
9. J.R. Elliott and C.T. Lira, *Introductory Chemical Engineering Thermodynamics*, Prentice Hall, Upper Saddle River, NJ, 1999.
10. T.T. Hartley, Insights into the fractional order initial value problem via semi-infinite systems, NASA TM-1998-208407, November 1998.
11. R.L. Hartman II, *An Aging Model for Lithium-ion Cells*, PhD dissertation, University of Akron, Akron, OH, 2008.
12. T.T. Hartley and A.G. Jannette, A first principles model for nickel-hydrogen batteries, AIAA (American Institute of Aeronautics and Astronautics), in *3rd International Energy Conversion Engineering Conference*, San Francisco, CA, August 2005.
13. M. Chen and G.A. Rincon-Mora, Accurate electrical battery model capable of predicting runtime and I–V performance, *IEEE Transactions on Energy Conversion*, 21(2), 504–511, June 2006.
14. S. Abu-Sharkh and D. Doerffel, Rapid test and non-linear model characterization of solid-state lithium-ion batteries, *Journal of Power Sources*, 130, 266–274, 2004.
15. C.M. Shephard, Design of primary and secondary cells, *Journal of the Electrochemical Society*, 112, 657–664, July 1965.
16. D.A.J. Rand, R. Woods, and R.M. Dell, *Batteries for Electric Vehicles*, John Wiley & Sons Inc., New York, 1998.
17. R.M. Dell and D.A.J. Rand, *Understanding Batteries*, Royal Society of Chemistry, Cambridge, U.K., 2001.
18. M.J. Isaacson, R.P. Hollandsworth, P.J. Giampaoli, F.A. Linkowsky, A. Salim, and V.L. Teofilo, Advanced lithium ion battery charger, in *The Fifteenth Annual IEEE Battery Conference on Applications and Advances*, Long Beach, CA, January 2000, pp. 193–198.

19. G.L. Plett, Extended Kalman filtering for battery managements systems of LiPB-based HEV battery packs: Part 1. Background, *Journal of Power Sources*, 134, 252–261, 2004.
20. W.F. Bentley, Cell balancing considerations for lithium-ion battery systems, in *Twelfth Annual IEEE Battery Conference on Applications and Advances*, New York, January 1997, pp. 223–226.
21. R.M. Laidig and W.J. Wurst, Technology implementation of stationary battery failure prediction, in *Proceedings of the Ninth Annual IEEE Battery Conference on Applications and Advances*, Long Beach, CA, January 1994, pp. 168–172.
22. N.H. Kutkut and M.D. Divan, Dynamic equalization techniques for series battery stacks, in *IEEE Telecommunications Energy Conference*, Boston, MA, October 1996, pp. 514–521.

5

Alternative Energy Storage

Alternatives to batteries as the portable energy storage device for electric and hybrid electric vehicles are the fuel cells, ultracapacitors, compressed air tanks, and flywheels. Many of these energy storage devices are equally useful for stationary power generation. The alternatives are to be evaluated based on the technological challenges, energy conversion efficiencies, and fuel sources. Fuel cell is powered by hydrogen that has to be derived from primary energy sources. Electricity produced by the fuel cell using hydrogen as the fuel propels the electric powertrain of a fuel cell electric vehicle. Hydrogen fuel delivery method needs to be in place in addition to the development of fuel cell electric vehicles. One possible infrastructure is to establish hydrogen-filling stations where hydrogen will be produced and stored in tanks using electricity supplied through the transmission grid. The alternative to this is to produce hydrogen on board using the reformer technology.

Ultracapacitor, similar to battery, is another electrochemical device where energy can be stored and used on demand by an electric powertrain. The ultracapacitor technology has advanced tremendously in recent years, although it is unlikely to achieve specific energy levels high enough to serve as the sole energy storage device of a vehicle. However, ultracapacitors in conjunction with a battery or fuel cell have the possibility of providing an excellent portable energy storage system with sufficient specific energy and specific power for next-generation vehicles.

Compressed air presents another type of energy storage concept that has been utilized to develop compressed air vehicles. The compressed air vehicles have recently gained attention since the well-to-wheel efficiencies are comparable to those of fuel cell electric vehicles, but with a much simpler fuel chain. The fuel infrastructure requirement for compressed air vehicle is similar to that of fuel cell vehicles; electricity from the grid would be used to compress air at local filling stations, which would be dispensed to the air tanks of compressed air vehicles.

Flywheel is another storage device where energy is stored in mechanical form as kinetic energy. The energy is stored in a rotating disk and released on demand. Once again, electrical energy is the source for storing energy. Flywheel technology is not yet competitive enough with the alternatives discussed.

Technological challenges have to be overcome for the alternative energy storage devices before they can supply energy to mass-produced alternative vehicles; some of the device technologies are well advanced compared to others. For all of the alternative energy storage devices, the energy source is electricity, which is secondary in nature and has to be derived from either unsustainable or renewable sources. Again on the end-user side, the powertrain is electric for all the alternative vehicles, except for the compressed air vehicles.

The alternative energy storage devices and the source of their fuels are covered in this chapter. The principles of operation and modeling of the fuel cell and the ultracapacitor devices are the focus of this chapter; compressed air and flywheels will be covered in brief.

5.1 Fuel Cells

A "fuel cell" is an electrochemical device that produces electricity by means of a chemical reaction, much like a battery. The major difference between batteries and fuel cells is that the latter can produce electricity as long as fuel is supplied, while batteries produce electricity from stored chemical energy and hence require frequent recharging.

The basic structure of a fuel cell (Figure 5.1) consists of an anode and a cathode similar to a battery. The fuel that is supplied to the cell is hydrogen and oxygen. The concept of fuel cell is the opposite of electrolysis of water where hydrogen and oxygen are combined to form electricity and water. The hydrogen fuel supplied to the fuel cell consists of two hydrogen atoms per molecule chemically bonded together in the form H_2. This molecule includes two separate nuclei, each containing one proton, while sharing two electrons. The fuel cell breaks apart these hydrogen molecules to produce electricity. The exact nature of accomplishing the task depends on the fuel cell type, although what remains the same for all fuel cells is that this reaction takes place at the anode. The hydrogen molecule breaks into four parts at the anode due to the chemical reaction releasing hydrogen ions and electrons. A catalyst speeds the reaction and an electrolyte allows the two hydrogen ions,

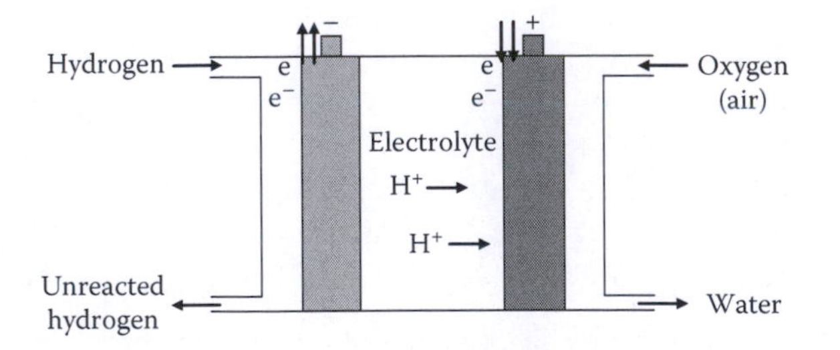

FIGURE 5.1
Basic fuel cell structure.

which essentially are two single protons, to move to the cathode through the electrolyte placed between the two electrodes. The flow of electrons from the anode to the cathode through the external circuit is what produces electricity. For the overall cell reaction to complete, oxygen or air must be passed over the cathode. The cathode reaction takes place in two stages. First, the bond between the two oxygen atoms in the molecule breaks and then each ionized oxygen atom grabs two electrons coming from the anode through the external circuit to become negatively charged. The negatively charged oxygen atoms are balanced by the positively charged hydrogen atoms at the cathode and the combination produces H_2O, commonly known as water. The chemical reaction taking place in a fuel cell is

$$\text{Anode: } H_2 \rightarrow 2H^+ + 2e^-$$

$$\text{Cathode: } 2e^- + 2H^+ + \frac{1}{2}(O_2) \rightarrow H_2O$$

$$\text{Cell: } H_2 + \frac{1}{2}O_2 \rightarrow H_2O$$

The first rudimentary version of a fuel cell was built by Sir William Robert Grove in 1845. Fuel cell found its first application in space when NASA developed the device for an alternative power source. The source was first used in a moon buggy and is still used in NASA's space shuttles. There has been a tremendous interest in fuel cells in recent years for applications in other areas, such as electric vehicles and stationary power systems. The research sponsored by several U.S. research agencies and corporations has attempted to improve cell performance with two primary goals: the desire for higher power cells, which can be achieved through higher rates of reaction, and the desire for fuel cells that can internally reform hydrocarbons and are more tolerant of contaminants in the reactant streams. For this reason, the searches have concentrated on finding new materials for electrodes and electrolytes. There are several different types of fuel cells having their own strengths and weaknesses. Low operating temperature is desirable for the vehicle applications, despite the fact that higher temperatures result in higher reaction rates. Rapid operation and cogeneration capabilities are desirable for the stationary applications. *Cogeneration* refers to the capability of utilizing the waste heat of a fuel cell to generate electricity using conventional means.

5.1.1 Fuel Cell Characteristics

Theoretically, fuel cells operate isothermally meaning that all free energy in a fuel cell chemical reaction should convert into electrical energy. The hydrogen "fuel" in the fuel cell does not burn as in the internal combustion (IC) engines, bypassing the thermal to mechanical conversion. Also, since the operation is isothermal, the efficiency of such direct electrochemical converters is not subject to the limitation of Carnot cycle efficiency imposed on heat engines. The fuel cell converts the Gibbs-free energy of a chemical reaction into electrical energy in the form of electrons under isothermal conditions. The maximum electrical energy for a fuel cell operating at constant temperature and pressure is given by the change in Gibbs-free energy

$$W_{el} = -\Delta G = nFE \tag{5.1}$$

where

n is the number of electrons produced by the anode reaction
F is Faraday's constant = 96,485 C/mol
E is the reversible potential

The Gibbs-free energy change for the reaction $H_2(g) + (1/2)O_2\ (g) \rightarrow H_2O(l)$ at standard condition of 1 atmospheric pressure and 25°C is −236 kJ/mol or −118 MJ/kg. With $n = 2$, the maximum reversible potential under the same conditions is $E_0 = 1.23$ V, using (5.1). The maximum reversible potential under actual operating conditions for the hydrogen–oxygen fuel cell is given by the Nernst equation as follows [1]:

$$E = E_0 + \left(\frac{RT}{nF}\right)\ln\left[\frac{C_H \cdot C_O^{1/2}}{C_{H_2O}}\right] \tag{5.2}$$

where

T is the temperature in K
R is the specific gas constant
C_H, C_O, and C_{H_2O} are the concentrations or pressures of the reactants and products, respectively

The voltage–current output characteristic of a hydrogen–oxygen cell is illustrated in Figure 5.2. The higher potentials around 1 V per cell are

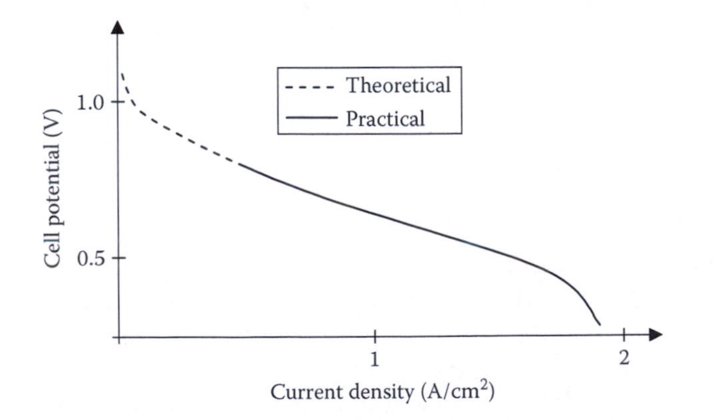

FIGURE 5.2
Voltage–current relationship of a hydrogen–oxygen cell.

theoretical predictions that are not achievable in a practical cell. The linear region where the reduction in cell potential is due to ohmic losses is where a practical fuel cell operates. The resistive components in the cell limit the practical achievable efficiency of a fuel cell. The working voltage of the cell falls with an increasing current drain, a knowledge that is very important in designing fuel cell–powered electric and hybrid vehicles. Since the cell potential is small, several cells are stacked in series to achieve the desired voltage. The major advantage of fuel cells is the lower sensitivity to scaling, which means that fuel cells have very similar overall system efficiencies from those in the kW range to those in the MW range.

5.1.2 Fuel Cell Types

The six major types of fuel cells are the alkaline, proton exchange membrane (PEM), direct methanol, phosphoric acid, molten carbonate, and solid oxide fuel cells. A short description of the relevant characteristics of each type in the context of vehicular and stationary applications is given below [2,3].

5.1.2.1 Alkaline Fuel Cell

In an alkaline fuel cell (AFC), an aqueous solution of potassium hydroxide (KOH) is used as the electrolyte. Compared to some other fuel cells where acidic electrolytes are used, the performance of the alkaline electrolyte is as good as the acid electrolytes, while being significantly less corrosive toward the electrodes. Alkaline fuel cells have been in actual use for a long time, delivering electrical efficiencies of up to 60%. They require pure hydrogen as fuel and operate at low temperatures (at 80°C), and are, therefore, suitable for vehicle applications. Residual heat can be used for heating, but the cell temperature is not sufficiently high to generate steam that can be used for cogeneration.

5.1.2.2 Proton Exchange Membrane Fuel Cell

The PEM fuel cells use solid electrolytes and operate at low temperatures (around 80°C). Nafion is an example of solid polymer electrolyte. These fuel cells are also known as solid polymer membrane fuel cells. The electrical efficiency of PEM fuel cells is lower than that of the alkaline cells (about 40%). However, the rugged and simple construction makes these types of fuel cells highly suitable for vehicle applications. The PEM fuel cell and the alkaline fuel cell are the two types that are currently being considered for vehicle applications. The advantage of PEM cells is that they can tolerate impurity in the fuel as compared to pure hydrogen needed in alkaline fuel cells.

5.1.2.3 Direct Methanol Fuel Cell

The direct methanol fuel cell (DMFC) is a result of the research on using methanol as the fuel that can be carried on board a vehicle and reformed to supply hydrogen to the fuel cell. A DMFC works on the same principle as the PEM, except that the temperature is increased to the range of 90°C–120°C such that internal reformation of methanol into hydrogen is possible. The electrical efficiency of DMFC is quite low at about 30%. This type of fuel cell is still in the design stages, since the search for a good electro-catalyst to both reform the methanol efficiently and to reduce oxygen in the presence of methanol is ongoing.

5.1.2.4 Phosphoric Acid Fuel Cell

Phosphoric acid fuel cells (PAFC) are the oldest type whose origin extends back to the creation of the fuel cell concept. The electrolyte used is phosphoric acid and the cell operating temperature is about 200°C, which makes some cogeneration possible. The electrical efficiency of this cell is reasonable at about 40%. These types of fuel cells are considered too bulky for transportation applications, while higher efficiency designs exist for stationary applications.

5.1.2.5 Molten Carbonate Fuel Cell

Molten carbonate fuel cells (MCFC), originally developed to operate directly from coal, operate at 600°C and require CO or CO_2 on the cathode side and hydrogen on the anode. The cells use carbonate as the electrolyte. The electrical efficiency of these fuel cells is high at about 50%, but the excess heat can be used for cogeneration for improved efficiency. The high temperatures required make these fuel cells not particularly suitable for vehicular applications, but can be used for stationary power generation.

5.1.2.6 Solid Oxide Fuel Cell

Solid oxide fuel cells (SOFC, ITSOFC) use a solid ionic conductor as the electrolyte rather than a solution or a polymer, which reduces corrosion problems. However, to achieve adequate ionic conductivity in such a ceramic, the system must operate at very high temperatures. The original designs, using yttria-stabilized zirconia as the electrolyte, required temperatures as high as 1000°C to operate, but the search for materials capable of serving as the electrolyte at lower temperatures resulted in the "intermediate temperature solid oxide fuel cell." This fuel cell also has high electrical efficiency of 50%–60%, and residual heat can also be used for cogeneration. Although not a good choice for vehicle applications, it is at present the best option for stationary power generation.

The several fuel cell features described above are summarized in Table 5.1. The usable energy and relative cost of various fuels used in fuel cells are listed in Table 5.2. The selection of fuel cells as the primary energy source in electric and hybrid vehicles depends on a number of issues ranging from fuel

TABLE 5.1

Fuel Cell Types

Fuel Cell Variety	Fuel	Electrolyte	Operating Temperature	Efficiency	Applications
Phosphoric acid	H_2, reformate (LNG, methanol)	Phosphoric acid	~200°C	40%–50%	Stationary (>250 kW)
Alkaline	H_2	Potassium hydroxide solution	~80°C	40%–50%	Mobile
Proton exchange membrane	H_2, reformate (LNG, methanol)	Polymer ion exchange film	~80°C	40%–50%	EV/HEV, industrial up to ~80 kW
Direct methanol	Methanol, ethanol	Solid polymer	90°C–100°C	~30%	EV/HEVs, small portable devices (1 W–70 kW)
Molten carbonate	H_2, CO (coal gas, LNG, methanol)	Carbonate	600°C–700°C	50%–60%	Stationary (>250 kW)
Solid oxide	H_2, CO (coal gas, LNG, methanol)	Yttria-stabilized zirconia	~1000°C	50%–65%	Stationary

TABLE 5.2

Usable Energy and Cost of Fuels

Fuel	Usable Energy (MJ/kg)	Relative Cost (MJ)
Hydrogen:		
95% pure at plant	118.3	1.0
99% pure in cylinders	120	7.4
LPG (propane)	47.4	0.5
Gasoline	45.1	0.8
Methanol	21.8	3.3
Ammonia	20.9	3.6

cell technology itself to infrastructure to support the system. Based on the discussion in this section, the choice of fuel cell for the vehicular application is an alkaline or proton exchange design, while for stationary applications it will be the solid oxide fuel cell. The size, cost, efficiency, and startup transient times of fuel cells are yet to be at an acceptable stage for electric and hybrid vehicle applications. The complexity of the controller required for fuel cell operation is another aspect that needs further attention. Although its viability has been well-proven in the space program, as well as in prototype vehicles, the immature status make it a longer-term enabling technology for electric and hybrid vehicles.

5.1.3 Fuel Cell Model

A fuel cell is an electrochemical device similar to that of a battery where all the physical processes are essentially the same. The only difference between a fuel cell and battery is that the energy source in the former is external, whereas it is internally stored in a battery. A lumped parameter model for fuel cell has been developed by Lei Xia using the fundamentals of electrochemistry [4]. The structure of the fuel cell model is shown in Figure 5.3.

The parameters of the model are

Fuel concentration	$C_0=10$
Diffusion process parameters	$R_{TL}=0.001\ \Omega$, $C_{TL}=200$ F
Open circuit voltage (Nernst equation)	$E=1.4+0.052(2p_0)\ln C_d$ (p_0 is the normalized fuel flow rate)
Charge transfer polarization (Tafel equation)	$E_{ct}=0.1+0.26\ln(i_f)$
Ohmic resistance	$R_\Omega=0.002\ \Omega$
Double-layer capacitor	$C_{dbl}=3000$ F
Concentration polarization	$E_c=0.06\ln[1-(i/100)]$

The given parameters are for a single cell. Several cells are stacked in series to make a fuel cell device for an application. The model is useful for predicting the output *i–v* characteristics for a given flow rate.

The major difference between the lumped parameter models of the fuel cell and the battery is the representation of the diffusion process parameters. The diffusion process in a fuel cell is entirely different from that of a battery. In a battery, the electrolyte not only provides the medium for mass transport of ions, but also stores the energy. Therefore, the relative volume or physical size of the electrolyte needs to be relatively large in a battery. This difference in the amount of electrolyte used in battery can be attributed to its much faster response characteristics compared to that of a fuel cell. In a fuel cell, the electrolyte functions only as the current conduction medium

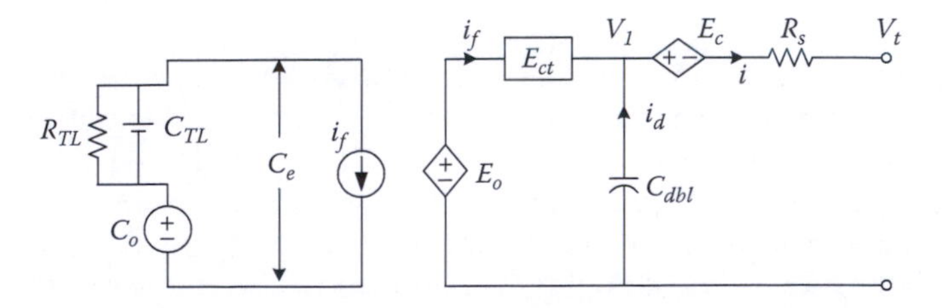

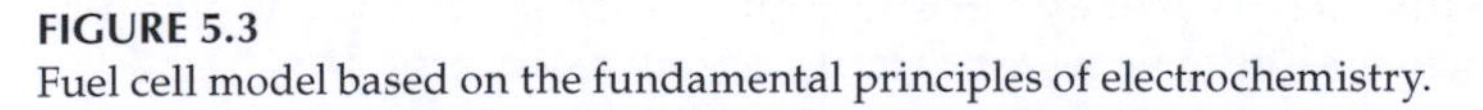
FIGURE 5.3
Fuel cell model based on the fundamental principles of electrochemistry.

through the diffusion process. The electrolyte layer in a fuel is designed to be fairly thin, and just enough to provide the electrical insulation between the electrodes. The different nature of diffusion process in a fuel cell is represented in the model with an RC network, primarily to represent the transient response. The function for the concentration polarization is also represented differently in the fuel cell model; it is modeled as a function of the discharge current.

5.1.4 Hydrogen Storage Systems

The options for storage of hydrogen play a critical role in the future development of infrastructure for fuel cell–powered electric and hybrid vehicles. The hydrogen gas at atmospheric pressure has a fairly low energy density and is not a suitable fuel for storage. Hydrogen could be stored as compressed or liquefied gas, or in more advanced manner by using metal hydrides or carbon nanotubes. Gas storage in compressed form is an option that has been in use for a long time. In this method, a large amount of energy is required to compress the gas to a level that will make storage viable, usually at a pressure of several hundred atmospheres [5]. The generation of liquid H_2 requires further compression along with refrigeration to cryogenic temperatures, and is not likely to become a viable means of storage for vehicle applications.

Advanced methods for H_2 storage include the use of metal hydrides or carbon nanotubes. Here, the gas is compressed to a lower pressure level (a few to a few tens of atmospheres) and fed into a container filled with a material that can absorb and release H_2 as a function of the pressure, temperature, and amount of stored hydrogen in the system. The use of metal hydrides reduces the volumetric and pressure requirements for storage, since when fully loaded, these metal hydrides can actually contain twice as many hydrogen atoms than an equivalent volume of liquid hydrogen. The sole problem is that it is much heavier than the other solutions. However, current efforts are underway by several automakers to include this in the structure of the vehicle, which may result in an overall acceptable vehicle weight. The prospect of using carbon nanotube-based materials for hydrogen storage is highly exciting, since it could eliminate most of the weight penalty. However, it should be noted that the properties of carbon nanotubes regarding their usefulness as H_2 storage materials is still highly controversial.

One of the myths that must be overcome to popularize fuel cell EVs is the safety of carrying pressurized hydrogen on board. The safety of hydrogen handling has been explored by both commercial entities as well as public institutions, such as Air Products and Chemicals, Inc. [6] and Sandia National Laboratories [7]. The recommendations for its safe handling have been issued [6]. In addition, the Ford report suggests that with proper engineering, the safety of a hydrogen vehicle could be better than that of a propane or gasoline vehicle [5].

5.1.5 Reformers

Many automotive industries have been exploring the use of methanol, ethanol, or gasoline as a fuel and reforming it on board into hydrogen for the fuel cell. The reformer is the fuel processor that breaks down a hydrocarbon, such as methanol, into hydrogen and other by-products. The advantage of the approach is the ease of handling that of hydrocarbon fuel compared to hydrogen gas, substantiated by the difficulty in storage and generation of pure hydrogen.

The accepted methods of reforming technique for vehicular fuel cells are steam reforming, partial oxidation, and auto-thermal processing. The two types of steam reformers available use methanol and natural gas as the fuel. Gasoline can also be used as the fuel, but reforming it is an expensive and complex process. Methanol is the most promising fuel for reformers, since it reforms fairly easily into hydrogen and is liquid at room temperature. A brief description of how a methanol steam reformer works is given in the following.

The first step in a methanol (CH_3OH) steam reformer is to combine the methanol with water. This methanol–water mixture is then heated to about 250°C–300°C and reacted with a catalyst at low pressure. Platinum is the most widely used catalyst in reformers, but other metals such as zinc and copper can also be used. The chemical reactions taking place in the methanol steam reformer are

$$CH_3OH \rightarrow CO + 2H_2$$

$$2H_2O(\text{gas}) \rightarrow 2H_2 + O_2$$

$$H_2 + O_2 + CO \rightarrow CO_2 + H_2O$$

The catalyst first splits the methanol into its two constituents, carbon monoxide and hydrogen gas. The heat generated splits the water into hydrogen and oxygen. The leftover oxygen from the decomposition of water combines with the pollutant carbon monoxide to produce another greenhouse gas carbon dioxide. Therefore, the major pollutant released as exhaust from the methanol steam reformer is CO_2, although the concentration is minimal compared to the exhaust of IC engines.

The argument used for using reformers is that the infrastructure for the production and distribution of such fuel is already in place, although widespread conversion to methanol systems is not straightforward for methanol fuels due to its high corrosivity [3]. While hydrogen gas would lead to true zero emission vehicles, it should be noted that reforming hydrocarbon fuels, including methanol and other possible biomass fuels, only shifts the source of emissions to the reformer plant. Other factors to consider are safety of methanol versus hydrogen handling, including the fact that methanol is violently toxic, whereas hydrogen is innocuous. The methanol vapors tend to accumulate in enclosed spaces like those of gasoline, leading to the formation of potentially explosive mixtures, whereas hydrogen will easily escape even from poorly ventilated areas. The overall efficiency from the well to the wheel of methanol-based transportation will be comparable or even lower than that can be achieved today from gasoline-based IC engine vehicles.

5.1.6 Fuel Cell Electric Vehicle

A fuel cell electric vehicle consists of a fuel storage system that is likely to include a fuel processor to reform raw fuel to hydrogen, fuel cell stack and its control unit, power processing unit and its controller, and the propulsion unit consisting of the electric machine and drive train. The fuel cell has current source-type characteristics and the output voltage of a cell is low. Several fuel cells have to be stacked in series to obtain a higher voltage level, and then the output voltage needs to be boosted up in order to interface with the DC/AC inverter driving an AC propulsion motor, assuming that an AC motor is used for higher power density. The block diagram of a fuel cell electric vehicle system is shown in Figure 5.4. Fuel cell output voltage is relatively low; a DC/DC converter is used to boost and regulate the voltage before being fed to the electric motor drive. The power electronic interface circuit between the fuel cell and electric motor includes the DC/DC converter for voltage boost, DC/AC inverter to supply an AC motor, microprocessor/digital signal processor for controls, and battery/capacitors for energy storage. The time constant of the fuel cell stack is much slower than that of the electrical load dynamics. A battery storage system is necessary to supply the power during transient and overload conditions and also to absorb the reverse flow of energy due to regenerative braking. The battery pack voltage rating must be high in order to interface directly with the high-voltage DC link, which means a large number of series batteries will be needed. Alternatively, a bidirectional DC/DC converter link can interface a lower voltage battery pack and the high-voltage DC bus. The battery pack can be

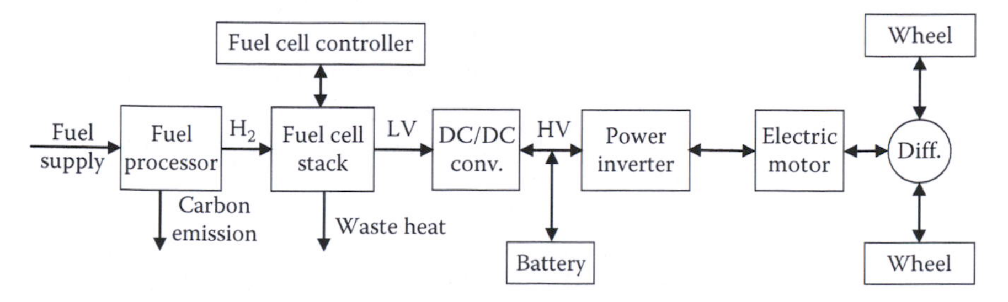

FIGURE 5.4
Fuel cell–based electric vehicle.

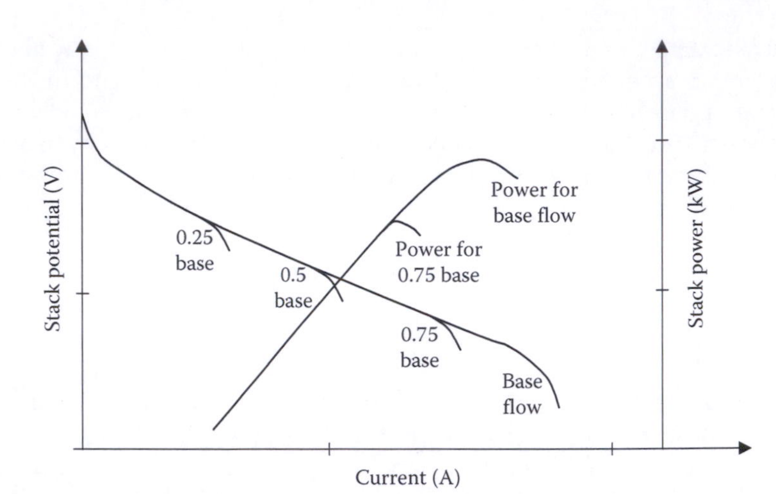

FIGURE 5.5
Fuel cell characteristics as a function of flow rate.

replaced by ultracapacitors in a fuel cell electric vehicle, although the technology is not yet ready to replace batteries.

Fuel cell performance is sensitive to the load variations because of the low-voltage and high-current output characteristics. The fuel cell controller using voltage and current feedback information regulates the flow of hydrogen into the fuel cell stack to achieve a reaction rate that delivers the required electrical power with minimum excess hydrogen vented. Attempts to draw more power out of the fuel cell without changing the flow rate depletes the concentration of hydrogen, which reduces the output voltage and may lead to damage to the fuel cell membrane [8]. The fuel cell characteristic curves as a function of flow rate are shown in Figure 5.5. When the hydrogen utilization rate approaches 100%, the cell goes into the current limit mode when it is dominated by high internal losses. The fuel cell controller must avoid operation in the current limit regime in order to maintain a decent efficiency of operation. The output power deliverability of the fuel cell stack reduces with reduced flow rate of hydrogen, but if lower power is required for traction, then operating the fuel cell at reduced flow rate minimizes wasted fuel. The ideal controller delivers fuel to the cell at exactly the same rate at which it is consumed by the cell to generate the electricity for the desired propulsion power. However, due to the slow response characteristics of the fuel cell, a reserve of energy is required to provide uninterrupted operation.

The by-product of the fuel cell reaction is water in the form of steam that exits the cell along with any excess hydrogen. The water vapor can be used for heating the inside of the vehicle, but the hydrogen that is vented out is a waste for the system.

Example 5.1

The current drawn by an electric motor of a fuel cell EV for a SAE schedule D J227A driving cycle is

$$I = \begin{cases} 9.36t + 1.482e - 3t^3 \text{ A} & \text{for } 0 < t < 28 \\ 61.42 \text{ A} & \text{for } 28 < t < 78 \\ 0 \text{ A} & \text{otherwise} \end{cases}$$

The fuel flow rate for PEM fuel cell used in the vehicle is

$$N_f = \frac{405I}{nF} \text{ g/s}$$

(a) Calculate the amount of fuel (hydrogen) needed for 1 cycle of schedule D.
(b) Calculate the range of the vehicle using schedule D for the storage capacity of 5 kg of hydrogen.

Solution

(a) The fuel flow rate from the given equation is

$$N_f = \begin{cases} 0.01964t + 3.11\text{e}{-6}t^3 & 0 < t < 28 \\ 0.1289 & 28 < t < 78 \\ 0 & \text{otherwise} \end{cases}$$

The amount of fuel (hydrogen) needed in 1 cycle can be obtained by integrating the $N_f(t)$ for $t = 0$ to $t = 78$ s.
Using numerical integration, we can show that

$$\int_0^{78} N_f(t)dt = 15.10 \text{ g}$$

(b) For schedule D, 1 cycle is equivalent to 1 mi. Therefore, 5 kg of hydrogen will give a range of 5000/15.1 = 331 mi.

5.2 Ultracapacitors

Ultracapacitors and supercapacitors are derivatives of the conventional capacitors where energy density has been increased at the expense of power density to make the devices function more like a battery. The terms ultracapacitors and supercapacitors are often interchangeably used. There are two types of ultracapacitors: symmetrical and asymmetrical ultracapacitors. In the symmetrical ultracapacitors, there is no electrochemical reaction and the process is completely non-faradaic. These ultracapacitors use two identical polarizable carbon electrodes, and are symmetrically designed. In this book, we will refer to these types as simply ultracapacitors. Asymmetrical ultracapacitors are designed for both faradaic and non-faradaic processes to take place simultaneously, which improves the energy density of the device. The asymmetrical ultracapacitors are more commonly referred to as supercapacitors.

5.2.1 Symmetrical Ultracapacitors

Capacitors are devices that store energy by the separation of equal positive and negative electrostatic charges. The basic structure of a capacitor consists of two conductors, known as plates separated by a dielectric, which is an insulator. The power density of the conventional capacitors are extremely high ($\sim 10^{12}$ W/m^3), but the energy density is very low (~ 50 W h/m^3) [9]. These conventional capacitors, commonly known as *electrolytic capacitors*, are widely used in electrical circuits as intermediate energy storage elements for time constants that are of a completely different domain and are of much smaller order compared to the energy storage devices that are to serve as the primary energy source for electric vehicles. The capacitors are described in terms of capacitance, which is directly proportional to the dielectric constant of the insulating material and inversely proportional to the space between the two conducting plates. The capacitance of a structure in terms of dimensional parameters is given by

$$C = \frac{\varepsilon A}{d}$$

where
- ε is the permittivity of the dielectric material
- A is the area
- d is the separation distance of the charges

The capacitance of a conventional capacitor is established by wrapping metal foil plates separated by a dielectric film. The thickness of the dielectric film separates the charges built in the metal electrodes. The voltage rating of the capacitor is determined by the dielectric strength, given by *V/m*, and thickness of the film.

In terms of physical parameters, the capacitance is measured by the ratio of the charge magnitude between either plate and the potential difference between them, i.e.,

$$C = \frac{q}{V}$$

where
 q is the charge between the parallel plates
 V is the voltage between them

The structure of an ultracapacitor is completely different from that of a conventional capacitor, although the physics of establishing the capacitance is the same. Ultracapacitors contain an electrolyte that enables the storage of electrostatic charge in the form of ions in addition to the conventional energy storage in electrostatic charges like in an electrolytic capacitor. The internal functions in an ultracapacitor do not involve any electrochemical reaction. The electrodes in ultracapacitors are made of porous carbon with high internal surface area to help absorb the ions and provide a much higher charge density than is possible in a conventional capacitor. The ions move much more slowly than electrons, enabling a much longer time constant for charging and discharging compared to electrolytic capacitors. There is no ion or electron transfer within an ultracapacitor; the process is simply charge separation or polarization. Therefore, the chemical process in the cell is entirely non-faradaic.

Instead of using a conventional dielectric, the ultracapacitors use activated carbon plates that are of two layers of the same substrate, facilitating charge separation distances of an ion diameter, which is about 1 nm. The structure of an ultracapacitor is shown in Figure 5.6a. The carbon plates are made of extremely small powder particles, which form a mushy structure facilitating large surface areas. The carbon mush or matrix is impregnated with conductive electrolytes. The effective charge separation across extremely small distances and large surface areas form the electric double layer (EDL), giving rise to extremely high capacitances. Positive and negative carbon matrixes in the ultracapacitor are separated by the electronic separator that is porous to ions. The EDL arises at the interfaces between the porous activated carbon and the electrolyte. The double layer includes a compact layer and a diffused layer [10]. The compact layer is formed at the interface of the solid

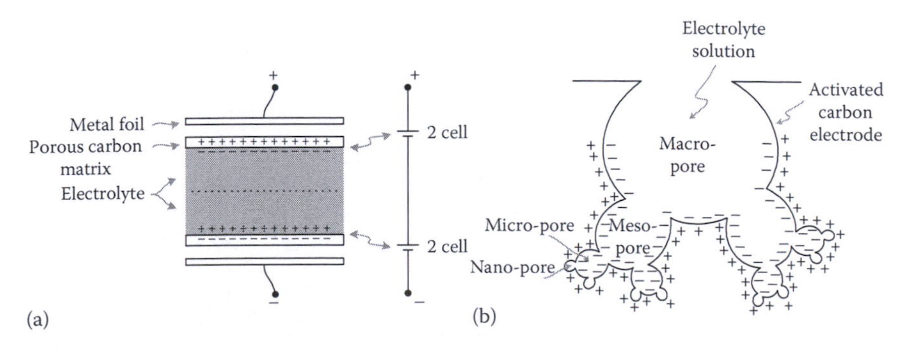

FIGURE 5.6
Ultracapacitor structure: (a) electric double-layer capacitance; (b) carbon matrix double layer.

electrode and the electrolyte contacts. The diffused layer penetrates into the electrolyte. Metal foils or electrodes for carrying currents to external circuits are placed against the positive and negative carbon matrixes. Each of the carbon matrixes forms a capacitor; hence, each ultracapacitor is essentially the series combination of two capacitors. The drawback of using the carbon matrix double layers instead of a solid dielectric is the low voltage ratings that are in the range of 2–3 V.

The porosity of the electric double-layer structure is shown in further detail in Figure 5.6b. The carbon matrix has three types of pores: macro-, meso-, and micropores. The micropores are as small as 1.5–2 times an ion diameter, which means that no more than one complete ion would fit inside the pore. Pores smaller than an ion diameter termed as submicropores exist, but ions cannot enter into those. Ions accumulate in layers in the macro- and meso-pores, resulting in the electric field within the electrolyte. As the capacitor gets charged, the electrolyte slowly becomes depleted of ions and the charging slows down.

The power density and energy density of ultracapacitors are of the order of 10^6 W/m^3 and 10^4 W h/m^3, respectively. The energy density is much lower compared to those of batteries (~5–25 × 10^4 W h/m^3), but the discharge times are much faster (1–10 s compared to ~5 × 10^3 s of batteries) and the cycle life is much more (~10^5 compared to 100–1000 of batteries) [9,11,12].

Current research and development aims to create ultracapacitors with capabilities in the vicinity of 4000 W/kg and 15 W h/kg. The possibility of using ultracapacitors as the primary energy source is quite far reaching, although it is likely that these can be improved to provide sufficient energy storage all by itself in hybrid vehicles. On the other hand, the ultracapacitors with high specific power are highly suitable as an intermediate energy transfer device in conjunction with batteries or fuel cells in electric and hybrid vehicles to provide the sudden transient power demand, such as during acceleration and hill climbing. The devices can also be used highly efficiently to capture the recovered energy during regenerative braking.

5.2.2 Asymmetrical Ultracapacitors

The symmetrical ultracapacitors with aqueous electrolyte first appeared in the late 1970s. The designs of asymmetrical ultracapacitors started in the 1980s. The asymmetrical devices use one polarizable carbon electrode and one nonpolarizable electrode and a nonaqueous electrolyte. The chemical activities include both faradaic as well as non-faradaic processes, which help increase the cell operating voltage. The asymmetrical ultracapacitor, which can also be termed a pseudo-battery, has higher energy density than a symmetrical ultracapacitor and is also capable of very high pulse power.

One design of asymmetrical ultracapacitor by ESMA in Russia uses a negative electrode of activated carbon like that in a symmetrical ultracapacitor and a positive nonpolarizable, faradaic electrode [13]. The positive electrode

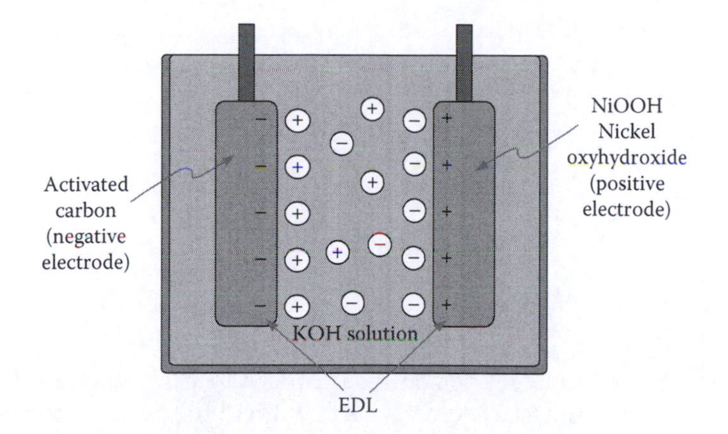

FIGURE 5.7
Asymmetrical ultracapacitor.

is made of nickel oxyhydroxide. The electrolyte is an aqueous solution of potassium hydroxide (KOH), which is used in alkaline batteries. The cell structure is shown in Figure 5.7. The capacitance of this positive electrode is significantly greater than the capacitance of a negative electrode having the same dimensions, which is different from the symmetrical ultracapacitors. The total capacitance of the asymmetrical ultracapacitor is approximately equal to the capacitance of the polarizable electrode.

5.2.3 Ultracapacitor Modeling

Ultracapacitors rely on polarization of the electrolyte at carbon electrodes. The huge number of pores of the activated carbon increases the surface area, allowing many electrons to accumulate. The number of electrons stored in the electrodes is proportional to the ultracapacitor capacitance. Since the size of pores in the activated carbon is not uniform, the capacitance depends on frequency. The ions in the electrolyte can charge the largest of the pores, the macro-pores, at highest frequency since these are closest to the current collectors and offers the least resistance path. The contribution to the total capacitance due to the macro-pores represents the high frequency behavior of the ultracapacitor with the fastest time constant. At medium frequencies, the ions are able to migrate to the meso-pores as well. The capacitance for this part is lower, but the spreading resistance is higher compared to the macro-pores. At very low frequencies, ions are able to penetrate into the micropores, establishing their contribution to the double-layer capacitance.

The behavioral models of ultracapacitor use a distributed RC network with multiple time constants. A three-branch nonlinear model developed by Zubieta et al. is shown in Figure 5.8 [14]. The first RC branch with the elements of R_{fast}, C_{fast0}, and C_{fast1} models the charge accumulation behavior in the macro-pores. The components represent the highest frequency behavior in

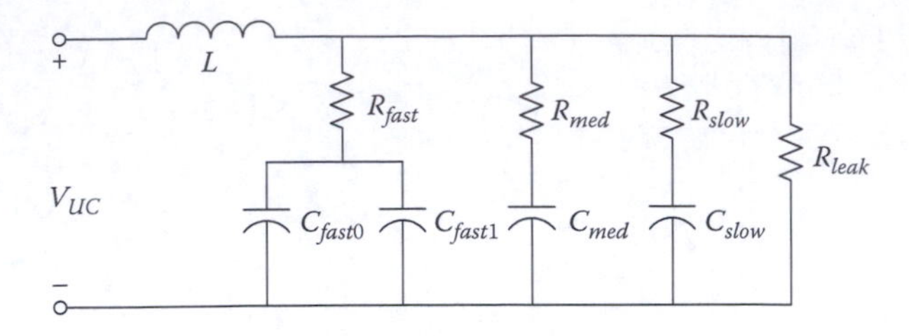

FIGURE 5.8
Toronto model.

the time frame of seconds. The nonlinear capacitance C_{fast1} models the variation in the effective surface geometry as ions fill in the pores as a function of voltage. The second branch of R_{med}, and C_{med} models the behavior around the meso-pores in the time frame of seconds. The third branch with the parameters of R_{slow}, and C_{slow} is the slowest branch, representing the activity around the micropores and has the slowest time constant. The model also includes a parallel resistor to represent leakage and a series inductor for the terminals and electrodes. The model was developed to represent the behavior of ultracapacitor intervals of 30 min or less, which is sufficient for motor drive systems in electric and hybrid vehicles.

A three-time constant model developed at MIT and based on fast, medium, and slow branches is shown in Figure 5.9 [15]. The model is a simplified version of the Toronto model where the nonlinear dependence on voltage has been eliminated. The model describes the behavior of the ultracapacitor in the shorter-term time frames.

Ultracapacitor models can also be developed based on frequency response data obtained using electrochemical image spectroscopy (EIS). EIS models the dynamic behavior of the system very well, but the derived model does not accurately predict the voltage-dependent behavior unless different DC biases are considered. The parameters are identified using the algorithm embedded in the EIS system. An EIS-based model, developed at Aachen, is shown in Figure 5.10 [16]. The series resistance R_s in the model represents the contact and electrolyte resistances while the series inductance L_s represents the terminal and electrode inductance. The parallel RC network represents

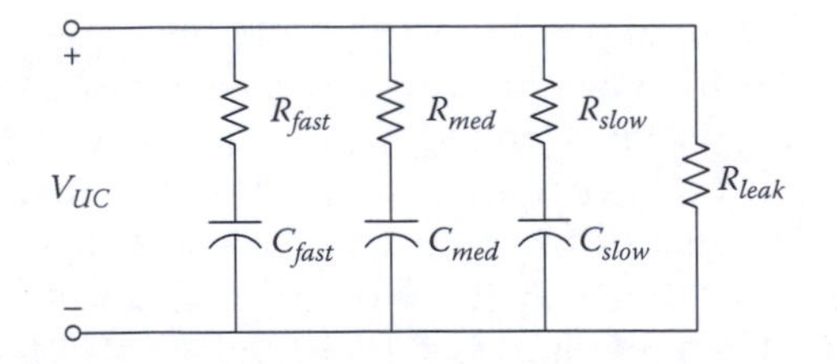

FIGURE 5.9
Short-term three-time constant model.

the complex pore behavior. The model does not include any leakage resistance to represent the longer-term behavior of ultracapacitors. The frequency domain impedance of the model is given by

$$Z(s) = R_s + sL_s + \frac{\tau}{\sqrt{s\tau}\,\tanh(\sqrt{s\tau})}$$

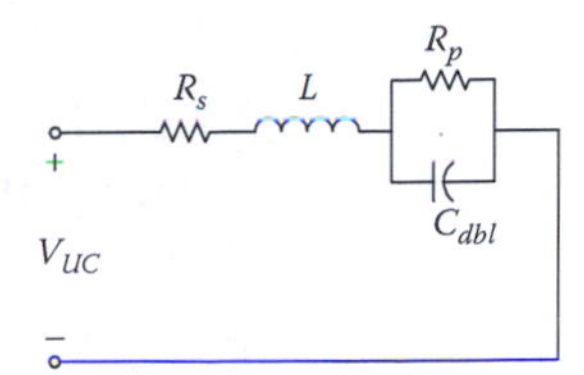

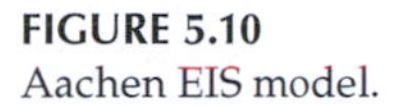
FIGURE 5.10
Aachen EIS model.

5.3 Compressed Air Storage

The compressed air storage is another alternative approach of energy storage where air is compressed and stored in a tank using electrical energy. The compressed air in the tank is expanded in a turbine, converting stored energy into work done for the propulsion of a vehicle. The concept could be used solely for the propulsion of an alternative vehicle or it could be combined with an IC engine or electric motor in a hybrid vehicle. The first compressed air vehicles were built in Paris in the mid-1800s, but the concept never transformed into commercialization even after repeated attempts to develop a viable prototype vehicle. In recent years, French developer MDI demonstrated advanced compressed air vehicles, but the commercial production of such vehicles is yet to take place. At present, the low energy content of the compressed air in an acceptable tank size limits the range of the vehicle. However, the technology has merits as a viable alternative for clean and efficient local transportation, and further research and development is an attractive option. Some auto industries are pursuing this technology.

The components and the airflow for a compressed air vehicle is shown in Figure 5.11. The air compressor shown in Figure 5.11 is to be located in a filling station for the air vehicle where an electric motor compresses the air in a compressor, and passes the gas through an air cooler before transferring it to the vehicle. Air compression produces substantial amount of heat, and hence, the working fluid is passed though heat exchanger for it to return to ambient temperatures. The vehicle itself has a tank that is filled with the compressed air at high pressures, typically around 30 MPa or 300 bars. The compressed air drives two expansion turbines under suitable thermodynamic conditions for vehicle propulsion. The air coming out of the expansion turbine is extremely cold due to the thermodynamics involved in rapid expansion of the working fluid.

The efficiency of a compressed air vehicle is related to the thermodynamics of the compression and expansion processes. The thermodynamic analysis of compressed air energy storage showed that significant advantages are to be gained with much improved overall energy utilization when both the

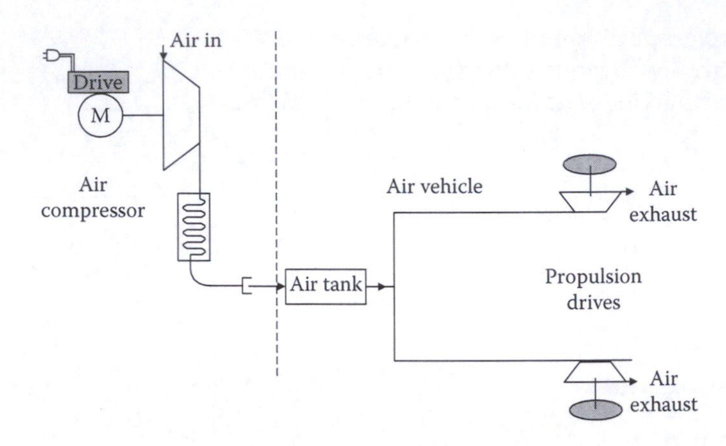

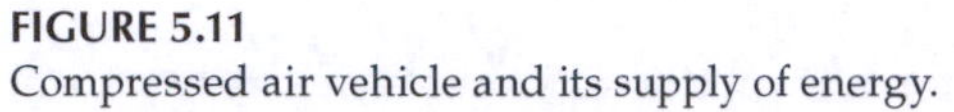
FIGURE 5.11
Compressed air vehicle and its supply of energy.

compression and expansion processes are distributed over several stages with integrated interstage cooling [17]. In the compression process, the initial air volume is compressed to the vehicle tank volume followed by heat removal to bring the air temperature back to the original ambient temperature. The energy efficiency of compression is much higher when a multistage process is used with interstage cooling. The overall energy utilization can be improved if heat produced during air compression is otherwise used for domestic water and space heating. Similarly, the expansion process is also much more efficient when it is distributed over several stages with integrated interstage cooling. In a multistage turbine, the air is reheated using the ambient air at each stage before the working fluid enters the next turbine for further expansion. The multistage turbine is essentially extracting heat from the ambient which had been released during compression, although full recovery will never be possible. Both the compression and expansion processes must proceed as close as possible to the ideal isothermal limit; this can be done in practice with multistage polytropic processes with heat addition and removal during the intermediate stages. The efficiency improvements with multistage processes are significant compared to single-stage processes.

The exhaust for the compressed air vehicle is air which is clean, and hence, these vehicles will get certified as zero-emission vehicles. However, the source of energy for compressing the air at the filling station is electricity, and the well-to-wheel energy efficiency depends on how electrical energy is generated from a primary source. The ideal scenario is, of course, when electricity is produced from renewable sources. A wind-to-wheel energy analysis showed that the efficiency of compressed air vehicles is in between those of battery electric vehicles and fuel cell electric vehicles, with the latter being the worse [18]. There is justification in pursuing the development of this compressed air alternative vehicle that uses an unconventional energy storage method. The engineering hurdles to overcome are associated with ice formation after expansion and interstage heat exchanger efficiency.

5.4 Flywheels

The flywheel is the kind of energy supply unit that stores energy in a mechanical form. Flywheels store kinetic energy within a rotating wheel-like rotor or disk made of composite materials. Flywheels have a long history of usage in automobiles being routinely used in all of today's IC engines to store energy and smoothen the power delivered by the abrupt pulses of the engine. However, the amount of energy storage required in flywheels of IC engines is small and is limited by the need of the vehicle to accelerate rapidly. The flywheel is currently being looked into for use in a number of different capacities. Flywheels can be used in hybrid vehicles with a standard IC engine as a power-assist device. Alternatively, the flywheels can be used to replace chemical batteries in electric vehicles to serve as the primary energy source or could be used in conjunction with batteries. However, technological breakthroughs in increasing the specific energy of flywheels are necessary before they can be considered as the energy source for electric and hybrid vehicles. The flywheels of today are quite complex, large, and heavy. Safety is also a concern with flywheels.

The flywheel design objective is to maximize the energy density. The energy U stored in the flywheel is given by

$$U = \frac{1}{2} J\omega^2$$

where

J is the polar moment of inertia
ω is the angular velocity

The energy storage is increased by spinning at higher velocities without increasing the inertia, which is directly proportional to mass. Increasing the angular velocity ω in turn increases the centrifugal stress, which must not exceed the failure stress with a given factor of safety. The stored energy per unit mass can be expressed as

$$\frac{U}{m} = k\frac{\sigma}{\rho}$$

where

k is a constant depending on the geometry
σ is the tensile strength
ρ is the density of the material

Therefore, the material to be used in flywheel must be lightweight with high tensile strength, conditions that are well satisfied by composite materials.

The flywheels have several advantages as the energy source, the most important one of which is the high specific power. Theoretically, the specific power of flywheels has been shown to be of the order of 5–10 kW/kg, with 2 kW/kg easily achievable without exceeding safe working stresses. The other performance features that make the flywheels attractive can be attributed to their mechanical nature. The flywheels are not affected by temperature extremes. There are no concerns with toxic chemical processing and disposal of waste materials, making flywheels environmentally friendlier than the chemical batteries. The flywheel energy storage is reliable that possesses excellent controllability and repeatability characteristics. The state of charge in flywheels is precisely known at all times through measurement of the rotational speed. The energy conversion process to and from the flywheel approaches 98% compared to 75%–80% of batteries. The service life of a flywheel is many times than that of battery with very little maintenance requirement. The charging of flywheels is a fraction of that required by batteries, and can be less than 10 min for full recharge in a flywheel charging station. The ability to absorb or release high amount of power in a short period of time also aids the regenerative braking process.

Despite the several advantages, there are still a number of significant drawbacks with flywheels. The major difficulty in implementing a flywheel energy storage system is in the extra equipment needed to operate and contain the device. The extras are particularly difficult in electric and hybrid vehicle applications where the extra weight and expense make a big difference. In order to reduce windage losses, the flywheel needs to be enclosed in a vacuum chamber. The vacuum condition adds additional constraints on the bearings, since liquid-lubricated bearings do not survive in vacuum. The alternative is to use magnetic bearings, which themselves are in a development stage. The biggest extra weight in flywheels comes from the safety containment vessel, which is required to protect from the dangerous release of sudden energy and material in the case of a burst failure.

The flywheels, similar to a battery, go through a charge/discharge process to store and extract energy, which earned it the name *electromechanical battery*. The rotor's shaft is coupled with a motor/generator, which, during charging, spins the rotor to store the kinetic energy and during discharging converts the stored energy into electric energy. Interface electronics is necessary to condition the power input and output and to monitor and control the flywheel. Modern flywheels are made of composite materials such as carbon fiber instead of steel to increase the energy density, which can be up to

200 W h/kg. The composite material flywheels have the additional advantage that these disintegrate in the form of a fluid as compared to large metallic pieces for steel-made flywheels in the case of a catastrophic burst.

Problems

5.1 The current drawn by an electric motor of a fuel cell EV for a SAE schedule D J227A driving cycle is

$$I = \begin{cases} 9.5t + 1.5\text{ A} & \text{for } 0 < t < 28 \\ 55\text{ A} & \text{for } 28 < t < 78 \\ 0\text{ A} & \text{otherwise} \end{cases}$$

The fuel flow rate for a PEM fuel cell used in the vehicle is

$$N_f = \frac{405I}{nF}\text{ g/s}$$

(a) Calculate the amount of fuel (hydrogen) needed for 1 cycle of schedule D.
(b) Calculate the amount of hydrogen needed for a range of 200 mi.

5.2 The efficiency of a fuel cell is related to the average voltage of each cell in the fuel cell stack V_{fc}. Referring to the lower heating value (LHV) of hydrogen, the efficiency of a fuel cell is given by

$$\eta_{fc} = \frac{V_{fc}}{E_0}$$

The practical average cell voltage has been used to determine the following equation for the rate of use of hydrogen in a PEM fuel cell:

$$H_2\text{ usage rate in kg/s} = 1.05\times10^{-8}\times\frac{P(t)}{V_{fc}}$$

The power required for a FCEV during a SAE J227A schedule B driving cycle is

$$P_{TR}(t) = \begin{cases} 222t + .03t^3 & \text{for } 0 < t \le 19 \\ 830 & \text{for } 19 < t \le 38 \end{cases}\text{ W}$$

There is no regenerative braking in the driving cycle. Calculate the range of the FCEV with 8 kg of hydrogen. Consider the PEM fuel cell efficiency operating with pure hydrogen as 40% and reversible potential $E_0 = 1.23$ V.

References

1. A.J. Appleby and F.R. Foulkes, *Fuel Cell Handbook*, Van Nostrand Reinhold, New York, 1989.
2. N. Andrews, Poised for growth: DG and ride through power, *Power Quality*, 13(1), 10–15, January/February 2002.
3. M.A. Laughton, Fuel cells, *Power Engineering Journal*, 16(1), 37–47, February 2002.
4. L. Xia, Behavioral modeling and analysis of galvanic devices, PhD dissertation, University of Akron, Akron, OH, 2000.
5. Ford Motor Co, Direct-hydrogen-fueled proton-exchange-membrane fuel cell system for transportation applications: Hydrogen vehicle safety, Report no. DOE/CE/50389-502, Directed Technologies Inc., Arlington, VA, May 1997.
6. R.E. Linney and J.G. Hansel, in T.N. Veziroglu et al. (eds.), *Hydrogen Energy Progress XI: Proceedings of the 11th World Hydrogen Energy Conference*, International Association for Hydrogen Energy, Coral Gables, FL, 1996.
7. J.T. Ringland et al., Safety issues for hydrogen powered vehicles, Report no. SAND-94-8226, UC407, Sandia National Laboratories, Albuquerque, NM, March 1994.
8. EC&G Services, Parson's Inc., *Fuel Cell Handbook*, 5th edn., U.S. Department of Energy, Office of Fossil Energy, Morgantown, WV, October 2000.
9. R.M. Dell and D.A.J. Rand, *Understanding Batteries*, Royal Society of Chemistry, Harwell, U.K., 2001.
10. F. Belhachemi, S. Rael, and B. Davat, A physical based model of power electronic double-layer supercapacitors, in *IEEE Industry Applications Society Conference Record*, Rome, Italy, October 2000, Vol. 5, pp. 3069–3076.
11. S. Dhameja, *Electric Vehicle Battery Systems*, Newnes, Boston, MA, 2002.
12. D.A.J. Rand, R. Woods, and R.M. Dell, *Batteries for Electric Vehicles*, John Wiley & Sons Inc., New York, 1998.
13. I.N. Varakin, A.D. Klementov, S.V. Litvinenko, N.F. Starodubsev, and A.B. Stepanov, Application of ultracapacitors as traction energy sources, in *7th International Seminar on Double Layer Capacitors and Similar Energy Storage Devices*, Deerfield Beach, FL, December 1997.
14. L. Zubieta and R. Bonert, Characterization of double layer capacitors for power electronic applications, *IEEE Transactions on Industry Applications*, 36(1), 199–205, January/February 2000.
15. J.M. Miller, *Propulsion Systems for Hybrid Vehicles*, Institution of Electrical Engineers (IEE), London, U.K., 2004.
16. S. Buller, E. Karden, D. Kok, and R. W. De Doncker, Modeling the dynamic behavior of supercapacitors using impedance spectroscopy, *IEEE Transactions on Industry Applications*, 38(6), 1622–1626, November/December 2002.

17. U. Bossel, Thermodynamic analysis of compressed air vehicle propulsion, in *Proceedings of the European Fuel Cell Forum* [Online]. Available: www.efcf.com/reports (E14).
18. P. Mazza and R. Hammerschlag, Wind-to-wheel energy assessment, in *Proceedings of the European Fuel Cell Forum* [Online]. Available: www.efcf.com/reports (E18).

6

Electric Machines

An *electric machine* is an electromechanical device used for energy conversion from electrical to mechanical and vice versa. In a vehicle system, the electric machine can be designed to process supplied energy and deliver power or torque to the transaxle for propulsion. The machine also processes the power flow in the reverse direction during regeneration when the vehicle is braking converting mechanical energy from the wheels into electrical energy. The term *motor* is used for the electric machine when energy is converted from electrical to mechanical, and the term *generator* is used when power flow is in the opposite direction with the machine converting mechanical energy into electrical energy. The braking mode in electric machines is referred to as *regenerative braking*. There are electrical, mechanical, and magnetic losses during the energy conversion process in either direction in an electric machine, which affect the conversion efficiency. Some energy is always lost from the system for any energy conversion process. However, the conversion efficiency of electric machines is typically quite high compared to that of other types of energy conversion devices.

In electric vehicles, the electric motor is the sole propulsion unit, while in hybrid vehicles, the electric motor and the internal combustion (IC) engine together in a series or parallel combination provide the propulsion power. In electric and hybrid vehicles, the electric traction motor converts electrical energy from the energy storage device to mechanical energy that drives the wheels of the vehicle. The major advantages of an electric motor over an IC engine is that the motor provides full torque at low speeds and the instantaneous power rating can be two or three times the rated power of the motor. These characteristics give the vehicle excellent acceleration with a nominally rated motor.

Electric motors can be DC type or AC type. The DC series motors were used in a number of prototype electric vehicles in the 1980s and prior to that due to its excellent match with the road load characteristics and ease of control. However, the size and maintenance requirements of DC motors are making their use obsolete not just in the automotive industry, but in all motor drive applications. The more recent electric and hybrid vehicles employ AC and brushless motors, which include induction motors, permanent magnet (PM) motors, and switched reluctance motors. The AC induction motor technology is quite mature, and significant research and development activities have taken place on induction motor drives over the past 50 years. The control of induction motors is much more complex than DC motors, but with the availability of fast digital processors, the computational complexity can be easily managed. Vector control techniques developed for sinusoidal machines make the control of AC motors similar to that of a DC motor through reference frame transformation techniques. The computational complexity arises from these reference frame transformations, but today's digital processors are capable of completing complex algorithms in a relative short time.

The competitor to induction motors is the PM motors. The PM AC motors have magnets on the rotor, while the stator construction is the same as that of the induction motor. The PM motors can be of surface-mounted type or the magnets can be inset within the rotor in the interior permanent magnet (IPM) motors. The PM motor can also be classified as sinusoidal type or trapezoidal type, depending on the flux distribution in the air gap. The trapezoidal motors have concentrated three-phase windings and are also known as brushless DC motors. The PM motors are driven by a six-switch inverter just like an induction motor, but the control is relatively simpler than that of the induction motor. The use of high-density, rare earth magnets in PM motors provides high power density, but at the same time the cost of magnets is on the negative side for these motors. For electric and hybrid vehicle applications, the motor size is relatively large compared to the other smaller power applications of PM motors, which amplifies the cost problem. However, the hybrid vehicle motors are much smaller than electric vehicle motors, and the performance and efficiency achievable from PM motors may be enough to overcome the cost problem. The IPM motors have excellent performance characteristics, much superior than the surface-mount PM motors, but the manufacturing complexity is one of the drawbacks of these motors.

Another candidate for traction motors is the switched reluctance (SR) motors. These motors have excellent fault tolerance characteristics and their construction is fairly simple. The SR motors have no windings, magnets, or cages on the rotor, which helps increase the torque/inertia ratio and allows higher rotor operating temperature. The constant power speed range is the widest possible in SR motors compared to other technologies, which makes it ideally suitable for traction applications. The two problems associated with SR motors are acoustic noise and torque ripple. There are well-developed techniques to address both; moreover, for several traction applications, noise and torque ripple are not of big concern.

6.1 Simple Electric Machines

The torque in electric machines is produced utilizing the basic principles of electromagnetic theory in one of two ways: (1) by the mutual interaction of two orthogonal magnetomotive forces (mmf) utilizing the Lorentz force principle, and (2) by using a varying reluctance flux path where the rotor moves toward attaining the minimum reluctance position. The DC and AC machines, including the PM machines, work on the first principle, while the switched and synchronous reluctance machines work on the latter principle. The fundamental machine phenomenon responsible for inducing voltage and producing torque is explained in this section followed by two simple machine configurations that operate based on the two different principles.

6.1.1 Fundamental Machine Phenomena

An electric machine converts electrical energy to mechanical energy in the motoring mode and mechanical energy to electrical energy in the generating mode. The same electric machine can be used to serve either as a motor or a generator with appropriate changes in the control algorithm. Two basic phenomena responsible for this electromechanical energy conversion occur simultaneously in electrical machines. These are

1. Voltage is induced when a conductor moves in an electric field.
2. When a current-carrying conductor is placed in an electric field, the conductor experiences a mechanical force.

6.1.1.1 Motional Voltage

When a conductor is moving in a uniform magnetic field $\vec{B}$ with a velocity $\vec{v}$, there will be voltage induced in the conductor given by

$$\vec{e} = (\vec{v} \times \vec{B}) \cdot \vec{l} \tag{6.1}$$

The voltage induced is known as motional voltage or speed voltage. The above is a mathematical definition of Faraday's law of electromagnetic induction, which is written as

$$e = -\frac{d\lambda}{dt} = -N\frac{d\phi}{dt} \tag{6.2}$$

where
λ is the flux linkage
ϕ is the flux
N is the number of turns

For the conductor of length l shown in Figure 6.1, which is moving with a velocity $\vec{v}$ in a magnetic field $\vec{B}$ that is perpendicularly directed into the plane of the paper, the induced voltage is

$$e = Blv \tag{6.3}$$

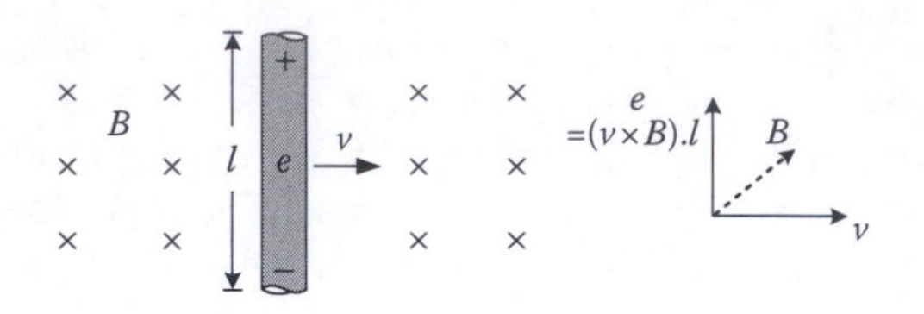

FIGURE 6.1
Induced voltage in a conductor.

6.1.1.2 Electromagnetic Force

A current-carrying conductor placed in a magnetic field experiences a force given by the Lorentz force equation. For a conductor-carrying current i in a uniform magnetic field $\vec{B}$, this electromagnetic force (emf) is given by

$$\vec{f} = i(\vec{l} \times \vec{B}) \tag{6.4}$$

Let us consider a magnetic field established by a pair of magnets and a conductor of length l carrying current i, as shown in Figure 6.2. The force in this case is

$$f = Bil \tag{6.5}$$

6.1.2 Simple DC Machine

Let us consider a simple one-turn loop placed in a magnetic field to understand the principles of voltage induced and torque developed. This magnetic field can be established by a magnet pole-pair directing flux from the North pole to the South pole, as shown in Figure 6.3a. The one-turn loop is placed in the gap between the pole-pair where it can freely rotate around a pivot axis under the influence of the magnetic field. The magnet pole-pair and the one-turn loop forms the simple electric machine that we will use to discuss the

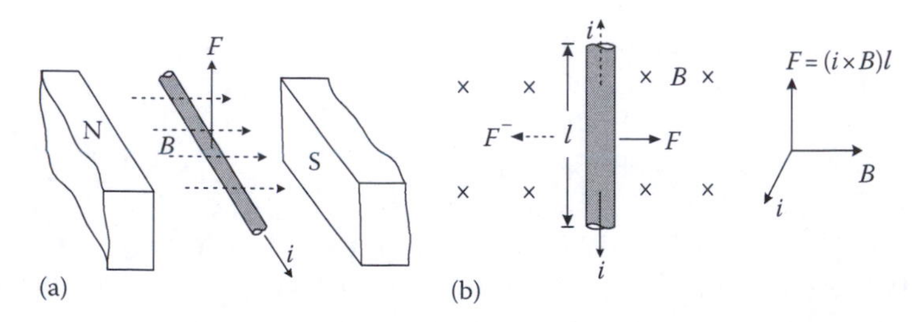

FIGURE 6.2
Force on a current-carrying conductor in a magnetic field; (a) magnets and conductor; (b) force, flux, and current directions.

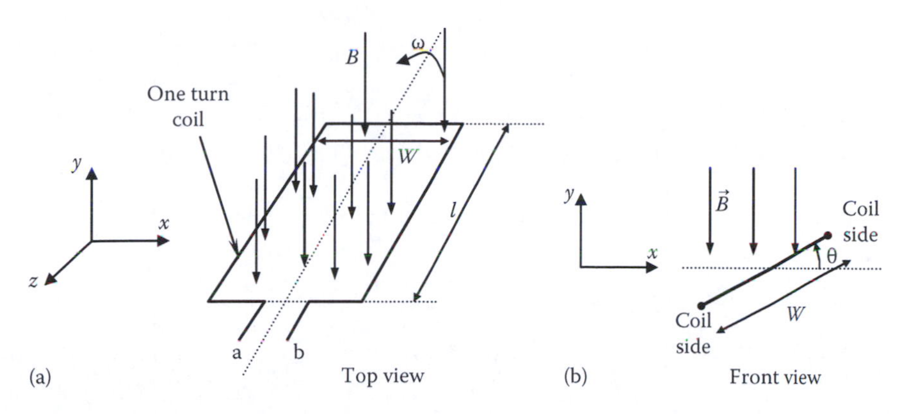

FIGURE 6.3
A one-turn coil in a magnetic field: (a) top view and (b) front view.

principles of electromagnetic energy conversion. Another view of the one-turn loop under the influence of the magnetic field is shown in Figure 6.3b. The one-turn coil cuts the magnetic field as it rotates around its pivot axis.

6.1.2.1 Induced Voltage

The one-turn rotating part of the simple machine is the rotor, while the permanent magnets (PMs) form the stator. The flux from the stator magnet pole-pair establish the magnetic field for the machine, which is known as the field flux. Let us assume that the flux established in between the magnet pole-pair is uniform, flowing from the North pole to South pole in the negative y-direction, as shown in Figure 6.3b. Now, if the single-turn conductor rotates in this uniform magnetic field at a constant angular velocity ω, a voltage will be induced in the sides of the coil according to Equation 6.1. The sides of the one-turn loop of length l, which passes under the pole faces, are called the *coil sides*. The side in the back of width W, which connects the coil sides, is known as the *end turn*. The one-turn coil rotating at an angular velocity of ω makes an angle θ with the magnetic field $\vec{B}$, as shown in the Figure 6.3a. The cross product $(\vec{v} \times \vec{B})$ for the coil sides at any instant of time is $vB\cos\theta$, and this resultant is always parallel with the length l of the coil sides. Therefore, the induced voltage in each of the coil sides is $vBl\cos\theta$ according to Equation 6.1. There will be no voltage induced on the end turn since $(\vec{v} \times \vec{B})$ is always perpendicular to the end turns that result in a zero dot product with the width W. Therefore, the total voltage induced in the one-turn loop accounting for the two coil sides and the end turn is

$$e_{ab} = 2vBl\cos\theta$$

Since the linear velocity v can be expressed in terms of the angular velocity ω as $v = \omega(W/2) = \omega r$, the voltage induced in the one-turn loop can be expressed as

$$e_{ab} = 2Blr\omega\cos\theta \tag{6.6}$$

The induced voltage can also be derived from the alternative form of Faraday's law ($e = -N(d\phi/dt)$) with the flux linkage for the one-turn coil expressed as

$$\phi(\theta) = -2Blr\sin\theta \tag{6.7}$$

The peak flux linking the coil is

$$\phi_P = 2Blr \tag{6.8}$$

The simple electric machine of Figure 6.3 will produce sinusoidal voltages when rotated at a constant speed in the uniform magnetic field. The alternating voltage generated at the terminals *ab* need to be rectified into unidirectional voltages to obtain a DC output. This is achieved with pairs of commutators and brushes attached at the ends of the one-turn conductor for the simple machine to produce a DC voltage (Figure 6.4). The commutators are connected to the ends of the turn and rotate with the coil. The brushes are stationary for connection with an external circuit. The two brushes are positioned in alignment with one of the stator magnetic poles such that they can collect the voltage created by that magnet pole and one of the coil sides through the commutator segment. The voltages seen by the brushes are thus unidirectional. The commutator and brush arrangement is a mechanical rectifier that converts the alternating voltage on the coil side e_{ab} into a unidirectional or DC voltage E_{12}. The field flux, the induced voltage, and the DC voltage are shown in Figure 6.5. The average DC voltage at the terminal of the brushes is

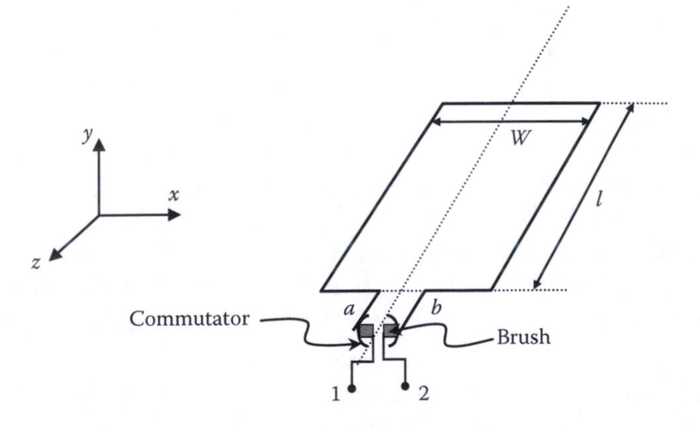

FIGURE 6.4
Commutators and brushes attached to the one-turn coil.

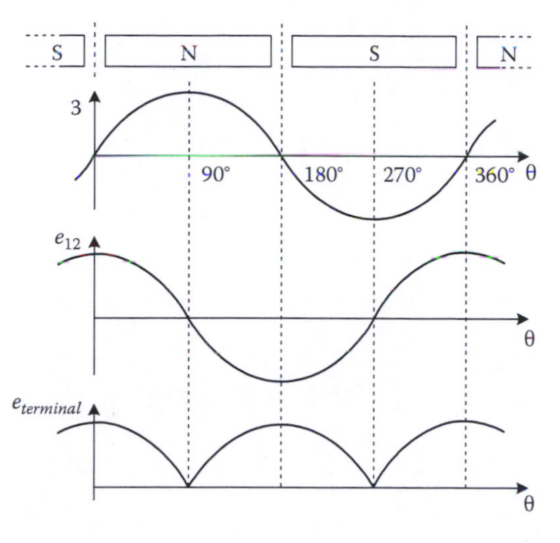

FIGURE 6.5
Field flux, induced voltage, and DC voltage.

$$E_{12} = \frac{2}{\pi} \int_{\pi/2}^{3\pi/2} \phi_P \omega \cos(\omega t) d(\omega t) = \frac{2}{\pi} \phi_P \omega \tag{6.9}$$

6.1.2.2 Force and Torque

In the one-turn coil of Figure 6.4, the circuit is open at the terminals 1–2. If a resistor is connected between the terminals 1–2, current starts to flow within the coil as long as there is motion of the conductor in the presence of the magnetic field. This current is the armature current in a DC machine; for the one-turn coil, the current for a resistance R is given by

$$I_a = \frac{e_{12}}{R} = \frac{2Blr}{R} \omega_m \cos\theta$$

The coil sides experience a force when current flows through the conductor according to Equation 6.5. The forces on the two coil sides, as shown in Figure 6.6, are given by

$$F_1 = F_2 = BI_a l$$

The electromagnetic torque of the machine can be derived from the two forces F_1 and F_2 forming a couple. The torque T_e on the armature turn about the axis-of-rotation is

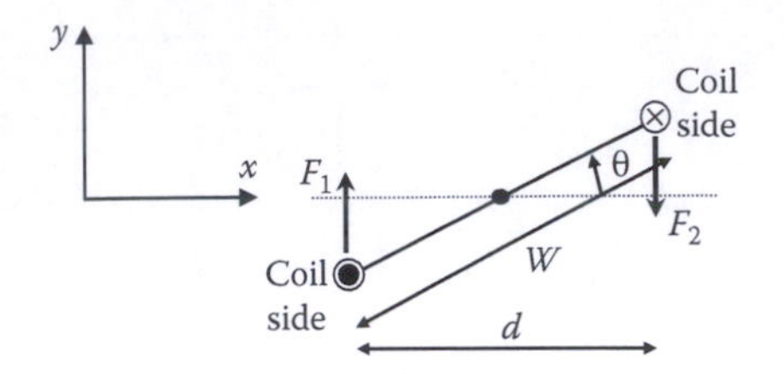

FIGURE 6.6
Forces on the coil sides due to current.

$$T_e = F_1 d = F_1 W \cos\theta$$

$$\Rightarrow T_e = \phi_P I_a \cos\theta$$

The average torque produced is

$$T_e = \frac{2}{\pi}\int_{\pi/2}^{3\pi/2} \phi_P I_a \cos(\omega t) d(\omega t) = \frac{2}{\pi}\phi_P I_a \tag{6.10}$$

6.1.2.3 DC Machine Back-EMF and Torque

In a practical machine, the coil would be wound around an iron structure which is the rotor of the electric machine. The rotor in the magnetic circuit is free to turn about the vertical axis. The reluctance of the iron is much smaller than the reluctance of the air. The rotor iron material must have much higher relative permeability than air to facilitate the flux flow from one stator pole to another through the rotor. The rotor iron, together with its curved shape, provides a constant width air gap between the stator and the rotor. The uniform air gap maintained between the stator and rotor poles allows the flux to be radially directed between the stator and the rotor.

The simple machine of Figure 6.3 is shown to have two poles only, one N-pole and one S-pole. The number "2" in Equation 6.9 represents those two poles in the machine. In a practical machine, there can be more than two poles, but always in pairs, for better utilization of the space around the stator and rotor circumference, and also for different torque–speed characteristics of the machine. In general, the higher speed machines will have few number of pole pairs, and higher torque machines will have more number of pole pairs. In addition, the machine geometry is better utilized when there are multiple turns in a coil instead of the single turn shown in Figure 6.3 for the simple machine. The multiple turns will increase the induced voltage. The multi-turn arrangement, known as the *conductor*, is shown in Figure 6.7. The sides of the turn along the *z*-direction are known as the *coil side* similar

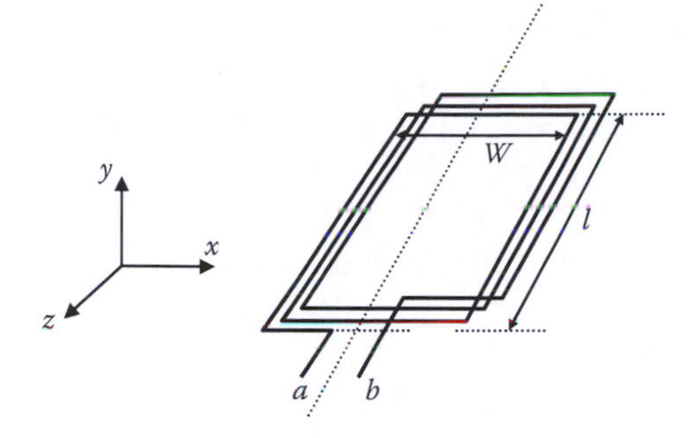

FIGURE 6.7
A multi-turn conductor.

to the one-turn coil. In the machine, many conductors are arranged radially along the rotor surface and connected in a series-parallel combination depending on the number of poles used. In DC machines, the conductor connectors are arranged in one of two winding patterns known as lap winding or wave winding. The induced voltage and torque equations for the simple machine can be generalized using the number of turns N_t in each conductor, the number of poles P, and the type of winding arrangement used. The equations then become

$$E_a = K_m \phi_P \omega \tag{6.11}$$

$$T_e = K_m \phi_P I_a \tag{6.12}$$

where K_m is a machine constant given by $K_m = N_t P/\pi a$, a is the number of parallel paths used in the conductor arrangement; $a = 2$ for wave winding, while $a = P$ for lap winding. E_a is the induced voltage or motional voltage, more commonly known as back-emf of the machine. Equation 6.12 shows that for the DC machine, the back-emf E_a is proportional to speed while the torque is proportional to current when the flux is held constant, since for a designed machine K_m is constant.

The end turns in a conductor, as was shown earlier, does not contribute to voltage generation or torque production, but contributes only to losses due to the resistances in that segment of the coil length. Therefore, one of the objectives in machine design is to minimize the end-turn lengths.

The flux in the simple machine shown is radially directed from the rotating component to the stationary component. The rotor rotates around the axis-of-rotation resulting in rotary motion. These machines are known as *radial flux machines*. Electric machines can also be constructed to deliver linear motion, which are known as *linear machines*. The movable component in a linear machine has linear or translator motion over the stationary structure.

6.1.3 Simple Reluctance Machine

The reluctance machines have a variable reluctance path for the magnetic flux as the rotor rotates about its axis-of-rotation. The saliencies in both the stator and the rotor poles provide a variable reluctance path for the magnetic flux as the rotor position changes. The windings are on the stationary member, while the rotor is a stack of laminations without any windings or magnets. A simple reluctance machine with only one phase is shown in Figure 6.8. The reluctance of the air gap varies as a function of the rotor angular displacement θ. The reluctance of the air gap is minimum when the rotor and the stator are in perfect alignment at θ=0°. The minimum reluctance position corresponds to the maximum inductance seen by the current flowing through the coil. The reluctance becomes maximum when the rotor attains the position with the largest air gap, which occurs when θ=90°. The inductance in this position is the minimum. The stator coil inductance variation as a function θ can be expressed by

$$L(\theta) = L_1 + L_2 \cos(2\theta)$$

The torque for the simple reluctance machine can be expressed as

$$T_e = \frac{1}{2} i^2 \frac{\partial L}{\partial \theta} \tag{6.13}$$

The single-phase simple reluctance machine operates as a synchronous machine when supplied with an AC voltage. The machine will produce torque as long as the stator excitation frequency is synchronized with the rotor speed. If the initial rotor position is δ, the rotor angular displacement for a rotor speed ω is given by

$$\theta = \omega t + \delta$$

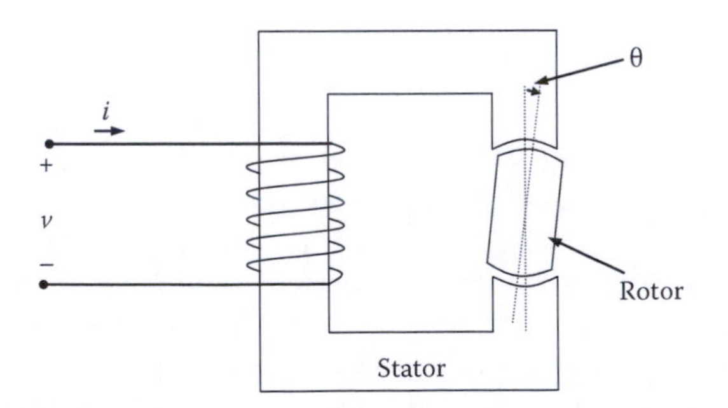

FIGURE 6.8
A simple reluctance machine.

The average torque produced for an excitation current of $i = I_m \cos \omega t$ at the synchronous speed is

$$T_e = -\frac{1}{4} L_2 I_m^2 \sin \delta$$

The simple reluctance machine will produce the average torque only at synchronous speed. The torque will be maximum when the initial position is 45° and zero when it is 0°. The machine will not be able to start rotation if the rotor is in the aligned position initially.

The reluctance machines used for high power applications are of multiple phases irrespective of whether they are of synchronous or switched types. Both the type of machines use power electronic inverters for controlling the torque and the speed of the machine. However, the fundamental principle for torque production is essentially the same as that described in this section with the simple reluctance machine. The principles and characteristics of the switched reluctance machines will be discussed further later in the chapter.

6.2 DC Machines

DC machines have two sets of windings, one in the rotor and the other in the stator, which establish the two fluxes; hence, the mmfs that interact with each other produce the torque. The orthogonality of the two mmfs, which is essential for maximum torque production, is maintained by a set of mechanical components called commutators and brushes. The winding in the rotor is called the *armature winding*, while the winding in the stationary part of the machine is called the *field winding*. Both the armature and the field windings are supplied with DC currents. The armature windings carry the bulk of the current, while the field windings carry a small field excitation current. The armature and the field currents in the respective windings establish the armature and field mmfs. The magnitude of the mmfs is the product of the number of turns in the windings and the current. Depending on the number of supply sources and the type of connection between the armature and field windings, there can be several types of DC machines. When the armature and field windings are supplied from independently controlled DC sources, then it is known as a separately excited DC machine. The separately excited DC machine offers the maximum flexibility of torque and speed control through independent control of the armature and field currents. The DC shunt machine has the similar parallel configuration of the armature and field windings as in the separately excited motor, except that the same DC source supplies both the armature and field windings. In the shunt motors, the simplicity in power supply is compromised for the reduced flexibility in control. In another type of DC machine, known as the series DC machine, the armature and the series windings are connected in series and the machine is supplied from a single source. Since the armature and the field windings carry the same current, the field is wound with a few turns of heavy gauge wires to deliver the same mmf or ampere-turns as in the separately excited machine. The greatest advantage of the series machine is the very high starting torque that helps achieve rapid acceleration. However, the control flexibility is lost due to the series connection of armature and field windings.

The positive attributes of DC machines are

- Ease of control due to linearity
- Capability for independent torque and flux control
- Established manufacturing technology

The disadvantages of DC machines are

- Brush wear that leads to high maintenance
- Low maximum speed
- Electromagnetic interference (EMI) due to commutator action
- Low power-to-weight ratio

The separately excited DC motor used in an electric or hybrid electric vehicle has two separate DC/DC converters supplying the armature and field windings from the same energy source, as shown in Figure 6.9. The DC/DC converters process the fixed supply voltage of the energy source to deliver a variable DC to the armature and field circuits. The power rating of the converter supplying the armature windings is much larger than that of the converter supplying the field winding. The control inputs to the converter circuits are the desired torque and speed of the motor, and the outputs of the converters are the voltages that are applied to the armature and field circuits of the DC motor.

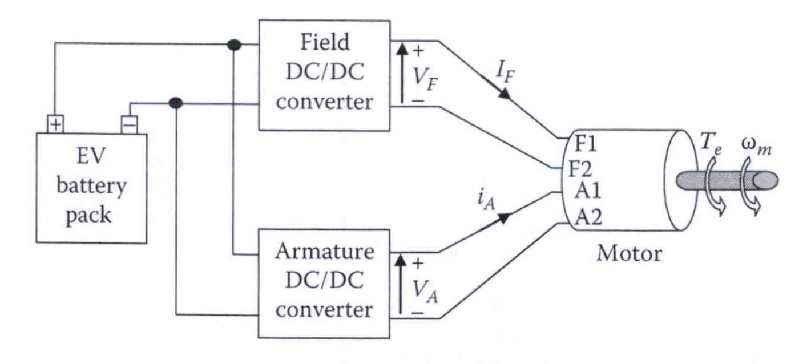

FIGURE 6.9
DC motor drive including the power electronics and battery source.

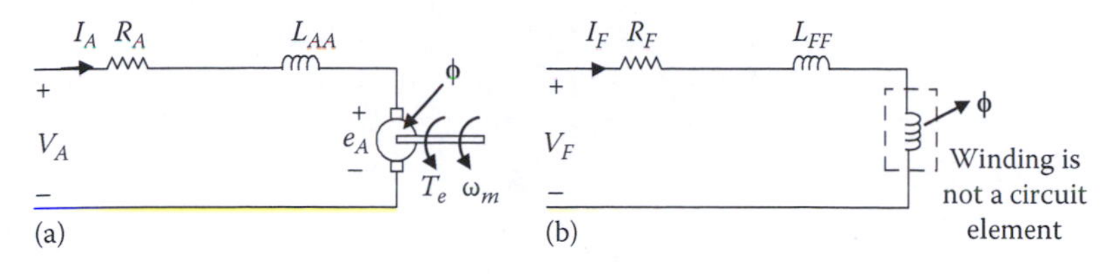

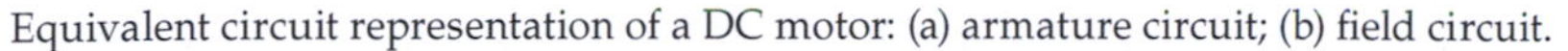

FIGURE 6.10
Equivalent circuit representation of a DC motor: (a) armature circuit; (b) field circuit.

The independent armature voltage and field current control, possible in separately excited DC machines or motors, offer the possibility of additional performance optimization in addition to meeting the torque–speed requirements of the machine. The indices used for measuring performance in motor drives include efficiency, torque per ampere, torque ripple, response time, etc. The weights on the individual performance indices depend on the application and the design requirements. The most critical performance index for electric and hybrid vehicle applications is the efficiency. The analysis to follow on DC motors, based on separately excited DC motors, is intended to set forth the premise for performance analysis of DC drives in the next chapter.

The armature equivalent circuit of a DC motor is shown in Figure 6.10a. The circuit consists of the armature winding resistance R_A, the self-inductance of armature winding L_{AA}, and the back-emf e_A. The variables shown in the figure are

V_A = armature voltage
I_A = armature current
T_e = developed motor torque
ω_m = shaft speed
ϕ = armature linking flux (primarily from field current)

Applying Kirchhoff's voltage law (KVL) around the armature circuit, the voltage balance equation is

$$V_A = R_A i_A + L_{AA}\frac{di_A}{dt} + e_A \tag{6.14}$$

where

$$e_A = K\phi\omega_m$$
$$T_e = K\phi i_A \tag{6.15}$$

where
e_A is known as the back-emf
K is a machine constant that depends on the machine construction, number of winding, and core material properties

The field equivalent circuit of the DC motor is shown in Figure 6.10b. The field circuit consists of the field winding resistance R_F and the self-inductance of the field winding L_{FF}. V_F is the voltage applied to the field.

The field circuit equation is

$$V_F = R_F i_F + L_{FF} \frac{di_F}{dt}$$

The resistance of the field winding in separately excited and shunt DC motors is very high, since there are a lot of turns in the winding. The transient response in the field circuit is thus much faster than the armature circuit. The field voltage is also typically not adjusted frequently, and for all practical purposes, a simple resistor fed from a DC source characterizes the electrical unit of the field circuit. The field current establishes the mutual flux or field flux, which is responsible for torque production in the machine. The field flux is a nonlinear function of field current and can be described by

$$\phi = f(i_F)$$

The electromagnetic properties of the machine core materials is defined by the relationship

$$B = \mu H$$

where

B is the magnetic flux density in T or Wb/m^2
H is the magnetic field intensity in A turn/m
μ is the permeability of the material

The permeability in turn is given by $\mu = \mu_0 \mu_r$ where $\mu_0 = 4\pi \times 10^{-7}$ H/m is the permeability of free space and μ_r is the relative permeability. The relative permeability of air is 1. The B–H relationship of magnetic materials is nonlinear and is difficult to describe by a mathematical function. Likewise, the field circuit of the DC machines is characterized by nonlinear electromagnetic properties of the core, which is made of ferromagnetic materials. The properties of core materials are often described graphically in terms of the B–H characteristics, as shown in Figure 6.11. The nonlinearity in the characteristics is due to the saturation of flux for higher currents and the hysteresis effects. When an external magnetization force is applied through the currents in the windings, the magnetic dipole moments tend to align to orient in a certain direction. This dipole orientation establishes a large magnetic flux, which would not exist without the external magnetization force applied on the core. The magnetic dipole moments relax toward their random orientation up on removal of the applied magnetic force, but few dipole moments retain their orientation in the direction of the previously existing magnetization force. The retention of direction phenomenon of the dipole moments is known as magnetic hysteresis. The hysteresis causes magnetic flux density B to be a multivalued function that depends on the direction of magnetization. The magnetic effect that remains in the core after the complete removal of the magnetization force is known as the residual magnetism (denoted by B_r in Figure 6.11). The direction of the residual flux, as mentioned previously, depends on the direction of field current change. The B–H characteristics can also be interpreted as the ϕ-i_F characteristics, since B is proportional to ϕ and H is proportional to i_F for a given machine. The saturation in the characteristics reflects the fact that more magnetic dipole moments remain to be oriented once sufficient magnetization force has been applied and the flux has reached the maximum or saturation level.

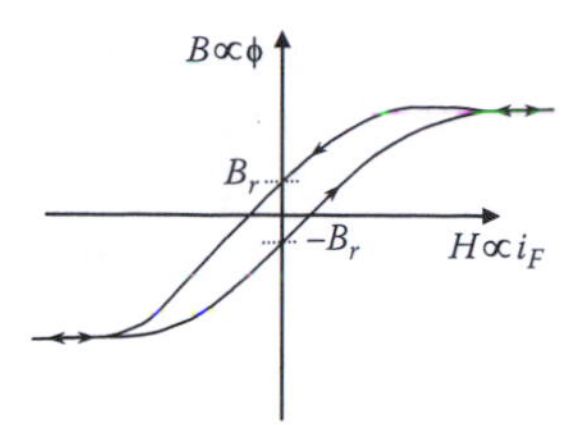

FIGURE 6.11
Typical DC motor magnetization characteristics.

The energy required to cause change in the magnetic orientations is wasted in the core material and is referred to as hysteresis loss. The area of the hysteresis loop in the magnetization characteristics is proportional to the hysteresis loss.

For most applications, it is sufficient to show the magnetic properties of core materials through a single-valued, yet nonlinear function, which is known as the DC magnetization curve. The magnetization curve of a DC machine is typically shown as a curve of open-circuit-induced voltage E_0 versus field current i_F at a particular speed. The open-circuit-induced voltage is nothing but the back-emf e_A, which is linearly proportional to the flux at a constant speed (refer to Equation 6.15). Therefore, the shape of this characteristic, shown in Figure 6.12, is similar to that of the magnetic characteristics of the core material.

The torque–speed relationship of a DC motor can be derived from Equations 6.14 and 6.15 and is given by

$$\omega_m = \frac{V_A}{K\phi} - \frac{R_A}{(K\phi)^2} T_e \qquad (6.16)$$

The torque–speed characteristic is shown in Figure 6.13. The positive torque axis represents the motoring characteristics, while the negative torque region represents the generating characteristics. The

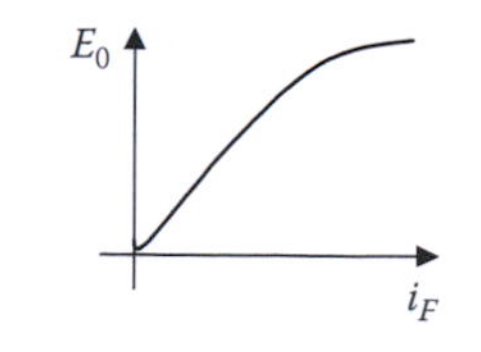

FIGURE 6.12
Magnetization characteristics of DC machines.

torque–speed characteristic is adjusted through either the armature voltage or the field current. For a given speed and torque (i.e., a point (T^*, ω_m^*) in the ω-T plane), there are an infinite number of corresponding armature voltage and field current, as shown in Figure 6.14a that would satisfy Equation 6.16. A possible control design is one that optimizes one or more performance indices and operates the motor on the optimized characteristic curve. To follow up on the concept, let us assume that the controller can set both the field current and the armature voltage and we are interested in minimizing the loss in the machine. The driver input commands set the desired torque and speed (T^*, ω_m^*) of the machine. Inserting the operating point in the armature voltage Equation 6.14, we have

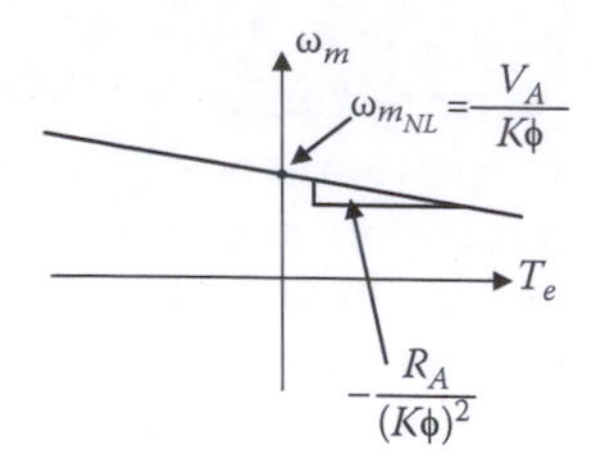

FIGURE 6.13
Torque–speed characteristics of a DC motor.

$$V_A = R_A I_A^* + K\phi\omega_m^*$$

$$\Rightarrow V_A = \frac{R_A}{K\phi}T^* + K\phi\omega_m^* \quad \text{or} \quad \omega_m^* K^2\phi^2 - V_A K\phi + R_A T^* = 0 \tag{6.17}$$

Equation 6.17 gives all the possible combinations of armature voltage and field flux that will give the same operating point (T^*, ω_m^*). The possible combinations are shown graphically in Figure 6.14b. The optimization algorithm will select the right combination of V_A and ϕ that will minimize the losses. The loss in DC machines is minimized when armature circuit-dependent losses equal the field circuit dependent losses [1]. Knowing the machine parameters, V_A and ϕ commands can be set such that the armature circuit losses equals the field circuit losses to minimize the overall loss, hence maximizing the efficiency.

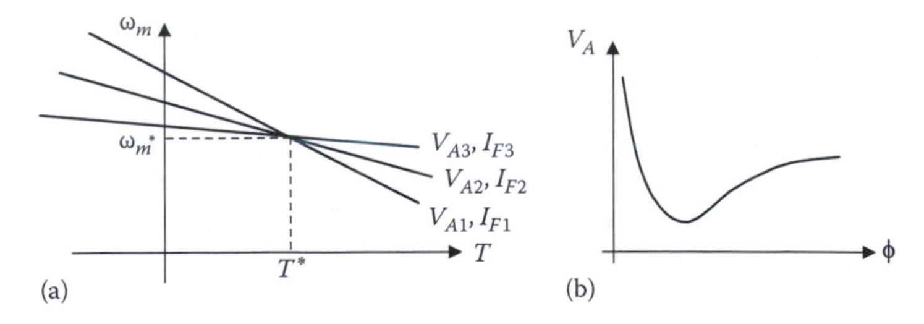

FIGURE 6.14
(a) Steady-state operating point. (b) Required armature voltage vs. flux at a fixed operating point.

6.3 Three-Phase AC Machines

The primary difference between AC and DC machines is that the armature circuit of the former is located in the stationary piece of the structure while for the latter it is in the rotor. The advantage gained in having the armature circuit in the stator is the elimination of the commutator and brushes of the DC machines. The AC machines require alternating supply that can be derived from a DC source using a DC/AC inverter. The machines can be single-phase or multiple-phase types. Single-phase AC machines are used for low-power appliance applications, while higher power machines are always of three-phase configuration. The second mmf required for torque production in AC machines (equivalent to the field mmf of DC machines) comes from the rotor circuit. Depending on the way the second mmf is established, the AC machines can be induction type or synchronous type. For either of the two types of AC machines, the stator windings are similar in configuration.

6.3.1 Sinusoidal Stator Windings

The three-phase stator windings of AC machines are sinusoidally distributed spatially along the stator circumference, as shown in Figure 6.15a to establish a sinusoidal mmf waveform. Although the windings are shown as concentrated in locations aa′, bb′, and cc′ for the three phases, the number of turns for each of the phase windings vary sinusoidally along the stator circumference. This space sinusoidal distribution of Phase-a winding is shown in Figure 6.15b, which has been represented by an equivalent concentrated winding a-a′ in Figure 6.15a. The equivalent distribution of Phase-a windings in a horizontally laid stator axis as if the stator cross section was split along the radius at $\theta=0$ and developed longitudinally is shown in Figure 6.16a. The current passing through these Phase-a stator windings causes a sinusoidal

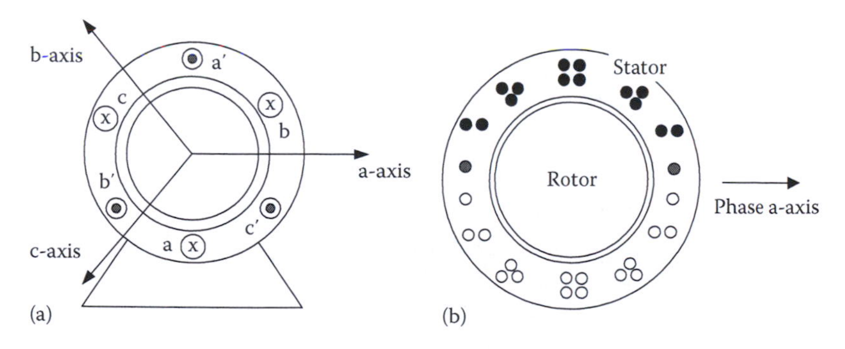

FIGURE 6.15
(a) Three-phase winding and magnetic axes of an AC machine. (b) Sinusoidal distribution of Phase-a winding along the stator circumference.

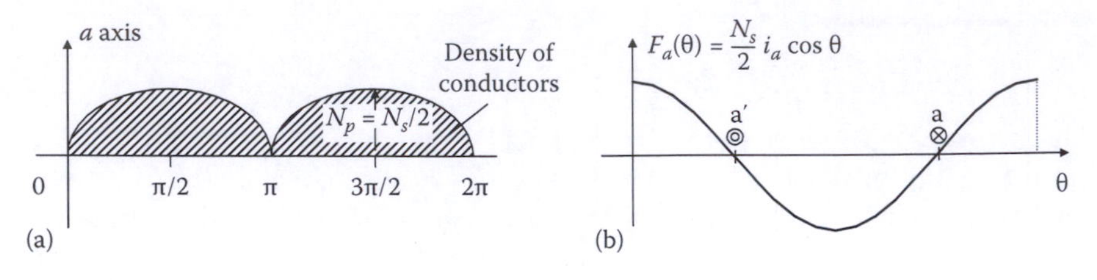

FIGURE 6.16
(a) Equivalent distribution of Phase-a winding; (b) mmf distribution of Phase a.

Phase-a mmf $F_a(\theta)$, which is shown in Figure 6.16b. The mmf primarily exists in the air gap due to the high permeability of the stator and rotor steel and tends to be radial in direction due to the short length of air gap relative to stator inside diameter. The sinusoidal distribution of the windings can be expressed as

$$\begin{aligned} n_{as}(\theta) &= N_P \sin\theta, \qquad 0 \le \theta \le \pi \\ &= -N_P \sin\theta, \qquad \pi \le \theta \le 2\pi \end{aligned}$$

where N_P is the maximum turns or conductor density expressed in turns per radian. Suppose Phase-a winding has an N_s equivalent number of turns (i.e., $2N_s$ conductors), which would give the same fundamental sinusoidal component as the actual winding distribution. Therefore, the integral of the conductor density in Figure 6.16b between 0 and π has a total of N_s conductors (accounting for half the turns in a winding-half), which is

$$N_s = \int_0^{\pi} N_P \sin\theta \, d\theta = 2N_P$$

$$\Rightarrow N_P = \frac{N_s}{2}$$

The sinusoidal conductor–density distribution in Phase-a winding is

$$n_s(\theta) = \frac{N_s}{2} \sin\theta, \quad 0 \le \theta \le \pi \tag{6.18}$$

The equivalent conductor–density is used to calculate the air gap magnetic field parameters, which consist of field intensity, flux density, and mmf. The basic relationship between magnetic field intensity H and current i is given by Ampere's law, which states that the line integral of H around a closed path is equal to the net current enclosed $\left(\oint H \cdot dl = \sum Ni\right)$. $\sum Ni$ is the ampere-turn product defining the net current enclosed and is known as the total mmf in magnetic circuit terms. The radial magnetic field intensity H_a in the AC machine under discussion is established in the air gap when current i_a flows through Phase-a windings, which can be derived using Ampere's law as

$$H_a(\theta) = \frac{N_s}{2l_g} i_a \cos\theta$$

where l_g is the length of the air gap. The flux density $B_a(\theta)$ and mmf $F_a(\theta)$ can be derived as

$$B_a(\theta) = \mu_0 H_a(\theta) = \frac{\mu_0 N_s}{2l_g} i_a \cos\theta$$

where μ_0 is the permeability of free space or air and

$$F_a(\theta) = l_g H_a(\theta) = \frac{N_s}{2} i_a \cos\theta \tag{6.19}$$

The mmf, flux intensity, and field intensity are all 90° phase shifted in space with respect to the winding distribution. The angle θ is measured in the counterclockwise direction with respect to the Phase-a magnetic axis. The field distribution shown in Figure 6.16b is for positive current. Irrespective of the direction of current, the peak of the mmf (positive or negative) will always appear along the Phase-a magnetic axis, which is the characteristic of mmf produced by a single-phase winding.

6.3.2 Number of Poles

The two equivalent Phase-a conductors in Figure 6.15a represent two poles of the machine. Electric machines are designed with multiple pairs of poles for the efficient utilization of the stator and rotor magnetic core material. In multiple pole-pair machines, the electrical and magnetic variables (such as induced voltages, mmfs, and flux densities) complete more cycles during one mechanical revolution of the motor. The electrical and mechanical angles of revolution and the corresponding speeds are related by

$$\begin{aligned} \theta_e &= \frac{P}{2}\theta_m \\ \omega_e &= \frac{P}{2}\omega_m \end{aligned} \tag{6.20}$$

where P is the number of poles. The four-pole machine cross section is represented, as shown in Figure 6.17a, while the Phase-a mmf F_a as a function of θ_e or θ_m is illustrated in Figure 6.17b. The mmf is mathematically represented as

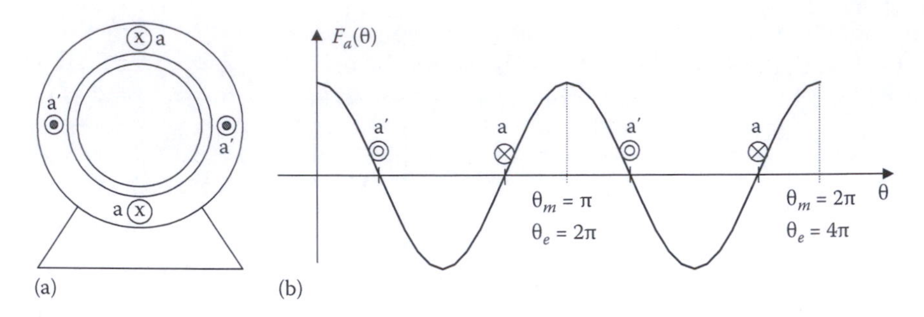

FIGURE 6.17
(a) Cross-section of a four-pole machine showing Phase-a windings only. (b) Phase-a mmf distribution.

$$F_a(\theta_e) = \frac{N_s}{P} i_a \cos(\theta_e) \tag{6.21}$$

6.3.3 Three-Phase Sinusoidal Windings

Phases b and c of the three-phase machine have identical sets of windings as the Phase-a winding described in Section 6.3.2, except that they are displaced 120° spatially with respect to each other, as shown in Figure 6.15a. The resulting mmfs due to currents in phases b and c can be expressed as

$$\begin{aligned} F_b(\theta) &= \frac{N_s}{2} i_b \cos\left(\theta - \frac{2\pi}{3}\right) \\ F_c(\theta) &= \frac{N_s}{2} i_c \cos\left(\theta + \frac{2\pi}{3}\right) \end{aligned} \tag{6.22}$$

6.3.4 Space Vector Representation

The extensive amount of coupling existing among the circuits of three-phase AC machines makes the analysis a formidable task. Axes transformations or reference frame theory is necessary to decouple the voltage expressions of the phases as well as to implement control algorithms that achieve the best performance. The *space vector representation* is a convenient method of expressing the equivalent resultant effect of the sinusoidally space-distributed electrical and magnetic variables in AC machines, in a way that is similar to the use of phasors in describing the sinusoidally time-varying voltages and currents in electrical circuits. The space vectors provide a very useful and compact form of representing the machine equations, which not only simplifies the representation of three-phase variables, but also facilitates the transformation between three-phase and two-phase variables. The two-phase system is an equivalent representation of the three-phase variables in a *dq* (two-axis) coordinate system, which is necessary for control implementation. The *dq* coordinate system will be elaborated in Section 9.2.

The concept of reference frame transformations originates from Park's transformations [2], which provided a revolutionary new approach of analyzing three-phase electric machines by transforming three-phase variables (voltages, currents, and flux linkages) into two-phase variables with the help of a set of two fictitious windings (known as *dq* windings) rotating with the rotor. The notations of space vector evolved later as a compact set of representation of the three-phase machine variables either in the three-phase *abc* reference frame or in the fictitious two-phase *dq* reference frame [3–5]. The space vectors are more complex than the phasors, since they represent time variation as well as space variation. The space vectors, just like any other vectors, have a magnitude and an angle, but the magnitude can be time varying. For example, the stator mmfs of the three phases in the AC machine can be represented by space vectors as

$$\begin{aligned} \vec{F}_a(t) &= \frac{N_S}{2} i_a(t)\angle 0^\circ \\ \vec{F}_b(t) &= \frac{N_S}{2} i_b(t)\angle 120^\circ \\ \vec{F}_c(t) &= \frac{N_S}{2} i_c(t)\angle 240^\circ \end{aligned} \tag{6.23}$$

Note that space vectors are complex numbers and "→" is used to denote the vector characteristic. The time dependence is also explicitly shown. The magnitude of the vector represents the positive peak of the sinusoidal spatial distribution, and the angle represents the location of the peak with respect to the Phase-a magnetic axis (chosen by convention). The space vectors of the individual phases can now be added conveniently by vector addition to give the resultant stator mmf as

$$\vec{F}_s(t) = \vec{F}_a(t) + \vec{F}_b(t) + \vec{F}_c(t) = \hat{F}_S \angle \theta_F \tag{6.24}$$

where

$\hat{F}_S$ is the stator mmf space vector amplitude

θ_F is the spatial orientation with respect to the Phase-a reference axis

In general, if f represents a variable (mmf, flux, voltage, or current) in a three-phase AC machine, the corresponding resultant space vector can be calculated as

$$\vec{f}_{abc}(t) = f_a(t) + af_b(t) + a^2 f_c(t)$$

where

$f_a(t)$, $f_b(t)$, and $f_c(t)$ are the magnitudes of the phase space vectors of the variables

a and a^2 are spatial operators that handle the 120° spatial distribution of the three windings, one with respect to the other along the stator circumference

The operators a and a^2 are $a = e^{j2\pi/3}$ and $a^2 = e^{j4\pi/3}$, and hence, the space vector can also be represented as

$$\vec{f}_{abc}(t) = f_a(t) + f_b(t)\angle 120 + f_c(t)\angle 240 \qquad (6.25)$$

The space vector can be used to represent any of the AC machine sinusoidal variables either in the stator circuit or in the rotor circuit. For example, the flux density, current, and voltage space vectors of the stator can be expressed as

$$\vec{B}_S(t) = \frac{\mu_0 N_S}{2l_g} i_a(t) + \frac{\mu_0 N_S}{2l_g} i_b(t)\angle 120 + \frac{\mu_0 N_S}{2l_g} i_c(t)\angle 240 = \hat{B}_S \angle \theta_B$$

$$\vec{i}_S(t) = i_a(t) + i_b(t)\angle 120 + i_c(t)\angle 240 = \hat{I}_S \angle \theta_I$$

$$\vec{v}_S(t) = v_a(t) + v_b(t)\angle 120 + v_c(t)\angle 240 = \hat{V}_S \angle \theta_V \qquad (6.26)$$

The space vector $\vec{f}_{abc}$ for a balanced set of three-phase variables f_a, f_b, and f_c has a magnitude 3/2 times greater than the magnitude of the phase variables; its spatial orientation is at an angle ωt at time t with respect to the reference Phase-a axis. Here, ω is the angular frequency of the phase variables. Therefore, the amplitude of the space vector is constant for a balanced set of variables, but the phase angle (i.e., the spatial orientation) is a function of time. The space vector at any instant of time can be obtained by the vector sum of the three-phase variables, as shown in Figure 6.18 for stator currents. Note that there is a unique set of phase variables that would sum up to give the resultant space vector $\vec{f}_{abc}$, since $f_a + f_b + f_c = 0$ for a balanced set of variables. Some examples with numerical values are given in the following to supplement the theory.

Example 6.1

The stator currents of a three-phase machine at $\omega t = 40°$ are

$$i_a = 10\cos 40° = 7.661\,\text{A}$$

$$i_b = 10\cos(40° - 120°) = 1.74\,\text{A}$$

$$i_c = 10\cos(40° - 240°) = -9.4\,\text{A}$$

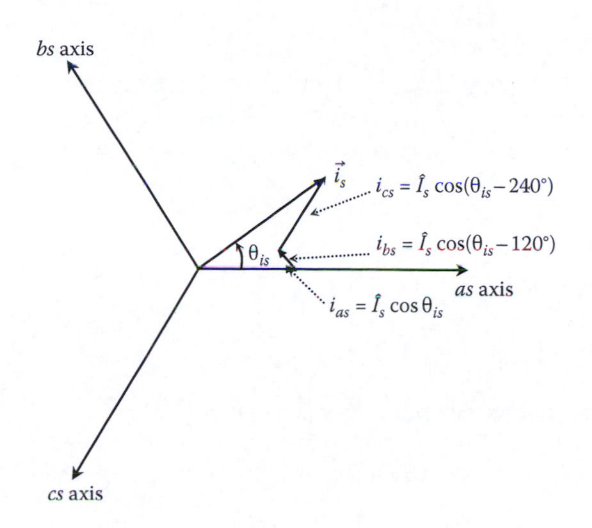

FIGURE 6.18
Space vector and its components in *abc* reference frame.

Calculate the resultant space vector.

Solution
The space vector at $\omega t = 40°$ is

$$\vec{i}_S = i_a(t) + i_b(t)\angle 120 + i_c(t)\angle 240$$

$$= 7.66 + 1.74\angle 120 + (-9.4)\angle 240$$

$$= \frac{3}{2}(7.66 + j6.43)$$

$$= \frac{3}{2} \cdot 10\angle 40 \text{ A}$$

The projection of $\vec{i}_s$ on the phase-a axis is 15 cos 40 = 11.49, which is 3/2 times i_a.

Example 6.2

(a) The phase voltage magnitudes of a three-phase AC machine at time $\omega t = 0$ are $v_a = 240\,\text{V}$, $v_b = -120\,\text{V}$, and $v_c = -120\,\text{V}$. Calculate the resulting space vector voltage.
(b) Recalculate the space vector at a different time when $v_a = 207.8\,\text{V}$, $v_b = 0\,\text{V}$, and $v_c = -207.8\,\text{V}$.
(c) Plot the space vector distribution in the air gap in the two cases.

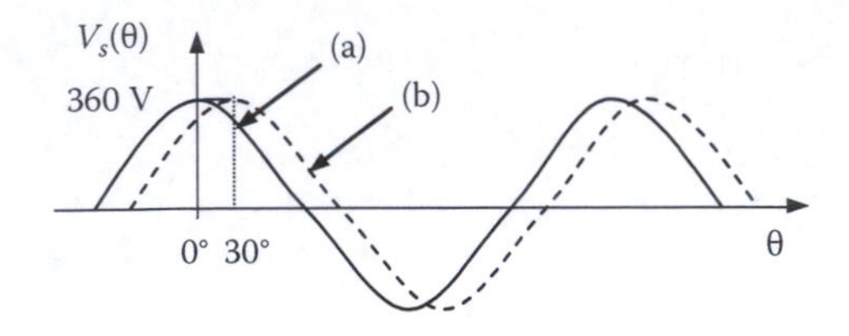

FIGURE 6.19
Plot of space vectors for Example 6.2 (a) and (b).

Solution

(a) The resulting space vector using Equation 6.25 is

$$\vec{v}_S(t) = v_a(t) + v_b(t)\angle 120 + v_c(t)\angle 240$$

$$\Rightarrow \vec{v}_S(t_0) = 240 + (-120)(\cos 120° + j\sin 120°) + (-120)(\cos 240° + j\sin 240°)$$

$$= 360\angle 0° \text{ V}$$

(b) The resulting space vector using Equation 6.25 is

$$\vec{v}_S(t) = v_a(t) + v_b(t)\angle 120 + v_c(t)\angle 240$$

$$\Rightarrow \vec{v}_S(t_0) = 207.8 + (0)(\cos 120° + j\sin 120°) + (-207.8)(\cos 240° + j\sin 240°)$$

$$= 360\angle 30° \text{ V}$$

(c) The plot is given in Figure 6.19.

Example 6.3

The phase voltage magnitudes of a three-phase AC machine at time t_0 are $v_a = 240\text{ V}$, $v_b = 50\text{ V}$, and $v_c = -240\text{ V}$. Calculate the resulting space vector voltage and plot the space vector distribution in the air gap.

Solution
The resulting space vector using Equation 6.25 is

$$\vec{v}_S(t) = v_a(t) + v_b(t)\angle 120 + v_c(t)\angle 240$$

$$\Rightarrow \vec{v}_S(t_0) = 240 + (50)(\cos 120° + j\sin 120°) + (-240)(\cos 240° + j\sin 240°)$$

$$= 418.69\angle 36.86° \text{ V}$$

The voltage space vector and its sinusoidal plot is shown in Figure 6.20.

The voltages given in Example 6.2 are a balanced set at two different times which correspond to $\omega t = 0$ and $\omega t = 30°$ for parts (a) and (b), respectively. The peak magnitude of the resulting voltage space vector remain the same in the two cases

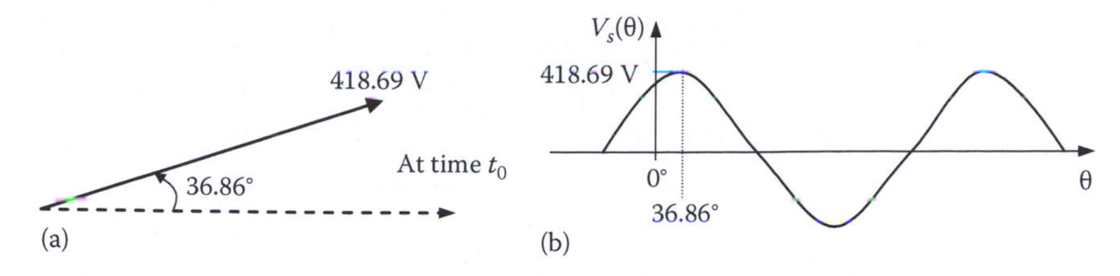

FIGURE 6.20
(a) Resultant voltage space vector; (b) voltage space vector sinusoidal distribution in the air gap at t_0.

and the location of these peaks along the machine axes are at $\theta = 0°$ and $\theta = 30°$, which corresponds to the time dependence of the voltages. This is not coincidental, but will be so for a balanced set of voltages. In Example 6.3, the voltages are unbalanced and the magnitude of the space vector depends on instantaneous values of the phase voltages.

6.3.4.1 *Interpretation of Space Vectors*

The space vectors, through one convenient and compact vector form, express the same resultant effect that the three individual phase variables would produce. For example, the stator mmf distribution in the air gap is a result of three phase currents i_a, i_b, and i_c, while the equivalent space vector current $\vec{i}_S(t)$ is developed in such a way that this resultant current flowing through an equivalent winding of N_s turns would produce the same resultant mmf distribution.

The relationships between electrical and magnetic quantities are conveniently expressed with the help of space vectors. Using Equations 6.23, 6.24, and 6.26, we can write

$$\vec{F}_S(t) = \frac{N_S}{2}\vec{i}_S(t) \tag{6.27}$$

The mmf and current vector magnitudes are related by the scalar constant $N_S/2$ and they have the same angular orientation.

The flux density can be similarly shown to be

$$\vec{B}_S(t) = \frac{\mu_0 N_S}{2l_g}\vec{i}_S(t) \tag{6.28}$$

6.3.4.2 *Inverse Relations*

The phase quantities can be derived from the space vectors through the inverse relations established using the complex variable mathematics. We know that

$$|A|\angle\theta = |A|\cos\theta + j|A|\sin\theta$$

Applying this to (6.25), we get

$$\vec{f}_{abc}(t) = f_a(t) - \frac{1}{2}\left(f_b(t) + f_c(t)\right) + j\frac{\sqrt{3}}{2}\left(f_b(t) - f_c(t)\right) = \frac{3}{2}f_a(t) + j\frac{\sqrt{3}}{2}\left(f_b(t) - f_c(t)\right)$$

since $f_a(t) + f_b(t) + f_c(t) = 0$ for balanced three-phase systems and for circuits without a neutral connection. Therefore, the inverse relation for Phase-a variable is

$$f_a(t) = \frac{2}{3}\mathrm{Re}\left[\vec{f}_{abc}(t)\right] \tag{6.29}$$

Similarly, it can be shown that

$$f_b(t) = \frac{2}{3}\mathrm{Re}\left[\vec{f}_{abc}(t)\angle 240^\circ\right] \tag{6.30}$$

and

$$f_c(t) = \frac{2}{3}\mathrm{Re}\left[\vec{f}_{abc}(t)\angle 120^\circ\right] \tag{6.31}$$

6.3.4.3 Resultant mmf in a Balanced System

In the typical operation of an AC machine, the stator windings are supplied with a balanced set of voltages, and since the windings are electrically symmetrical, a balanced set of currents flows through the windings. Let us assume that the rotor is open circuited and all the current that is flowing through the stator winding is the magnetizing current required to establish the stator mmf. The three-phase currents have the same magnitude and frequency, but are 120° shifted in time with respect to each other. The currents in the time domain can be expressed as

$$\begin{aligned} i_a(t) &= \hat{I}_M \cos\omega t \\ i_b(t) &= \hat{I}_M \cos(\omega t - 120^\circ) \\ i_c(t) &= \hat{I}_M \cos(\omega t - 240^\circ) \end{aligned} \tag{6.32}$$

The space vector for the above balanced set of currents is (using Equation 6.25)

$$\vec{i}_M(t) = \frac{3}{2}\hat{I}_M \angle\omega t \tag{6.33}$$

The resultant stator mmf space vector is

$$\vec{F}_{ms}(t) = \frac{N_S}{2}\vec{i}_M(t) = \frac{3}{2}\frac{N_S}{2}\hat{I}_M \angle\omega t = \hat{F}_{ms} \angle\omega t \tag{6.34}$$

The result shows that the stator mmf has a constant peak amplitude $\hat{F}_{ms}$ (since N_S and $\hat{I}_M$ are constants) that rotates around the stator circumference at a constant speed equal to the angular speed of the applied stator voltages. This speed is known as the synchronous speed. Unlike the single-phase stator mmf (shown in Figure 6.16b), the peak of the stator mmf resulting in the three-phase AC machine is rotating synchronously along the stator circumference with the peak always located at $\theta = \omega t$. The mmf peak position is time-varying for the three-phase winding, whereas the peak mmf position for the single-phase winding is *not* time-varying. The mmf wave is a sinusoidal function of the space angle θ. The wave has constant amplitude and a space-angle ωt, which is linear with respect to time. The angle ωt provides rotation of the entire wave around the air gap at a constant angular velocity ω. Thus, at a fixed time t_x, the wave is a sinusoid in space with its positive peak displaced ωt_x from the reference $\theta = 0$. The resultant space vector at three different times is shown in Figure 6.21. The polyphase windings excited by balanced polyphase currents produce the same general effect as that of spinning a PM about an axis perpendicular to the magnet, or as in the rotation of the DC-excited field poles.

The three-phase stator mmf is known as the rotating mmf, which can be equivalently viewed as a magnet rotating around the stator circumference at a constant speed. Note that with the vector sum of $F_a(\theta_e)$, $F_b(\theta_e)$, and $F_c(\theta_e)$, as described in Equations 6.21 and 6.22, with $i_a(t)$, $i_b(t)$, and $i_c(t)$ replaced by the balanced set of Equation 6.32, we will arrive at the same result.

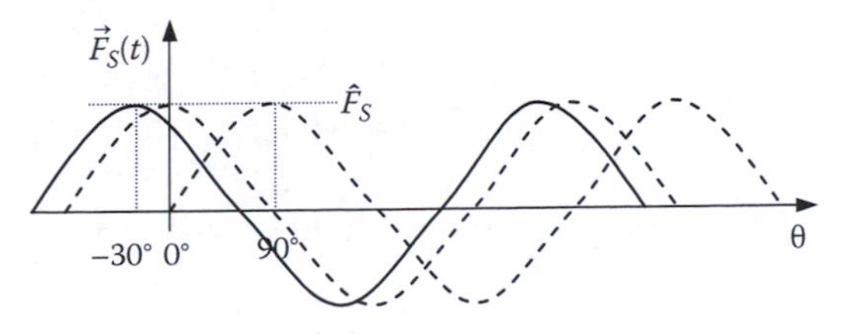

FIGURE 6.21
Resultant mmf space vector for $\omega t = -30°$, 0°, and 90°.

Exercise 6.1

Show that $F_a(\theta_e)+F_b(\theta_e)+F_c(\theta_e)=\frac{3}{2}\frac{N_S}{2}\hat{I}_M\angle\omega t$ with

$$i_a(t)=\hat{I}_M\cos\omega t$$

$$i_b(t)=\hat{I}_M\cos(\omega t-120°)$$

$$i_c(t)=\hat{I}_M\cos(\omega t-240°)$$

6.3.4.4 Mutual Inductance L_m and Induced Stator Voltage

In an ideal situation, the equivalent electrical circuit for the stator windings with no rotor existing consists of the applied stator voltage source and a set of winding that is represented by an inductance known as magnetizing or mutual inductance. The practical circuit extends on this ideal circuit by adding the stator winding resistance and the stator leakage inductance in series with the magnetizing inductance. The magnetizing inductance for the three-phase AC machine, including the effects of mutual coupling among the three phases, can be shown to be [4,5]

$$L_m=\frac{3}{2}\left[\frac{\pi\mu_0 rl}{l_g}\left(\frac{N_S}{2}\right)^2\right] \tag{6.35}$$

where
r is the radius to the air gap
l is the rotor axial length
l_g is the air gap length

Note that the form of Equation 6.35 is the same as that of a simple inductor given by $L=N^2/\mathfrak{R}$, where N is the number of turns and $\mathfrak{R}$ = Reluctance = (flux-path length)/($\mu\times$cross-sectional area). The voltage induced in the stator windings due to the magnetizing current flowing through L_m in space vector form is

$$\vec{e}_{ms}(t)=j\omega L_m\vec{i}_M(t) \tag{6.36}$$

The magnetizing flux density $\vec{B}_{ms}(t)$ established by the magnetizing current $\vec{i}_M(t)$ is (from Equation 6.28)

$$\vec{B}_{ms}(t)=\frac{\mu_0 N_S}{2l_g}\vec{i}_M(t)$$

which gives

$$\vec{i}_M(t) = \frac{2l_g}{\mu_0 N_S} \vec{B}_{ms}(t) \tag{6.37}$$

Using the expression for L_m from Equation 6.35 and the expression for $\vec{i}_M(t)$ in terms of $\vec{B}_{ms}(t)$, the induced voltage is

$$\vec{e}_{ms}(t) = j\omega \frac{3}{2} \pi r l \frac{N_S}{2} \vec{B}_{ms}(t)$$

The induced voltage can be interpreted as the back-emf induced by the flux density B_{ms}, which is rotating at the synchronous speed. For a P-pole machine, the expression for the induced voltage is

$$\vec{e}_{ms}(t) = j\omega \frac{3}{2} \pi r l \frac{N_S}{P} \vec{B}_{ms}(t) \tag{6.38}$$

6.3.5 Types of AC Machines

The second rotating mmf needed for torque production in AC machines is established by the rotor circuit. The interaction of the two rotating mmfs, essentially chasing each other at synchronous speed, is what produces torque. The method through which the rotor mmf is established differentiates the different types of AC machines. Broadly, the AC machines can be classified into two categories, synchronous machines and asynchronous machines. In synchronous machines, the rotor always rotates at synchronous speed. The rotor mmf is established either by using a PM or an electromagnet created by feeding DC currents into a rotor coil. The latter-type synchronous machines are typically the large machines that are used in electric power generating systems. The PM machines are more suitable for the electric and hybrid vehicle applications, since these offer higher power density and superior performance compared to the rotor-fed synchronous machines. The several types of PM AC machines will be discussed later in the chapter. The rotor-fed synchronous machines will not be discussed further in this book, since these are not of interest for the electric and hybrid vehicle applications. In the asynchronous-type AC machine, the rotor rotates at a speed that is close to but different from the synchronous speed. These machines are known as induction machines, which in the more common configurations are fed only from the stator. The voltages in the rotor circuit are induced from the stator, which in turn induces the rotor rotating mmf, and hence the name induction machines. The induction machines that are generally labeled as AC machines are the subject of treatment in the following section.

6.4 Induction Machines

The two types of induction machines are the squirrel cage induction machines and the wound rotor induction machines. The squirrel cage induction machines are the workhorse of the industry, because of their rugged construction and low cost. The rotor windings consist of short-circuited copper bars that form the shape of a squirrel cage. The squirrel cage of an induction motor is shown in Figure 6.22. The rotor winding terminals of the wound rotor induction machines are brought outside with the help of slip rings for external connections, which are used for speed control. The squirrel cage induction motors are of greater interest for electric and hybrid vehicles and most other general purpose applications, and hence, are discussed further.

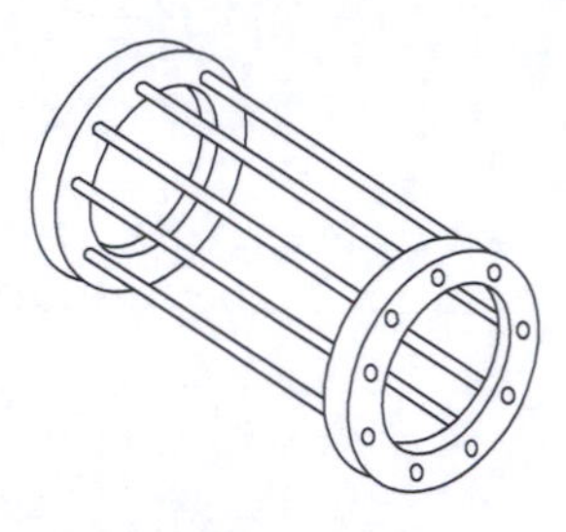

FIGURE 6.22
The squirrel cage of an induction motor.

The stator windings of the induction machines are exactly the same as that discussed in the previous section. The rotor, usually made of stacked laminations, has copper or aluminum rotor bars molded around the periphery in axial direction. The bars are short-circuited at the ends through electrically conducting end rings. The electrical equivalent circuit of a three phase induction machine along with the direction of Phase-a stator and rotor magnetic axes are shown in Figure 6.23. The rotor windings have been short-circuited and the angle between the rotor and stator axes is θ_r, which is the integral of the rotor speed ω_r.

When a balanced set of voltages is applied to the stator windings, a magnetic field is established which rotates at synchronous speed, as described in Section 6.3.4. By Faraday's law, as long as the rotor rotates at a speed other

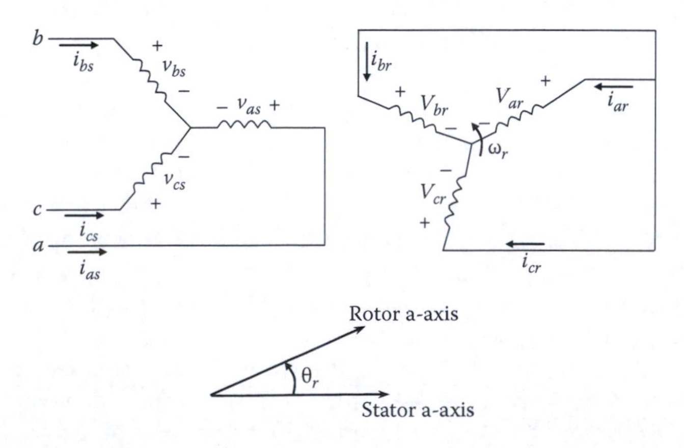

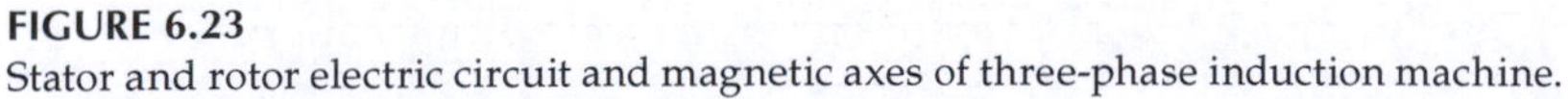

FIGURE 6.23
Stator and rotor electric circuit and magnetic axes of three-phase induction machine.

than the synchronous speed (even at zero rotor speed), the rotor conductors are cutting the stator magnetic field and there is a rate of change of flux in the rotor circuit, which will induce a voltage in the rotor bars. This is also analogous to transformer action where a time-varying AC flux established by the primary winding induces voltage in the secondary set of windings. The induced voltage will cause rotor currents to flow in the rotor circuit, since the rotor windings or bars are short-circuited in the induction machine. The induction machine can be thought of as a transformer with a short-circuited secondary or rotor windings. The rotor-induced voltages and the current have a sinusoidal space distribution, since these are created by the sinusoidally varying (space sinusoids) stator magnetic field. The resultant effect of the rotor bar currents is to produce a sinusoidally distributed rotor mmf acting on the air gap.

The difference between the rotor speed and the stator synchronous speed is the speed by which the rotor is slipping from the stator magnetic field, and is known as the slip speed:

$$\omega_{slip} = \omega_e - \omega_m \tag{6.39}$$

where

ω_e is the synchronous speed
ω_m is the motor or rotor speed

The slip speed expressed as a fraction of the synchronous speed is known as the slip:

$$s = \frac{\omega_e - \omega_m}{\omega_e} \tag{6.40}$$

The rotor bar voltage, current, and magnetic field are of the slip speed or slip frequency with respect to the rotor. The slip frequency is given by

$$f_{slip} = \frac{\omega_e - \omega_m}{2\pi} = sf, \quad \text{where } f = \frac{\omega_e}{2\pi} \tag{6.41}$$

From the stator perspective, the rotor voltages, currents, and rotor mmfs all have the synchronous frequency, since the rotor speed of ω_m is superimposed on the rotor variables' speed of ω_{slip}.

6.4.1 Per-Phase Equivalent Circuit

The steady-state analysis of induction motors is often carried out using the per-phase equivalent circuit. A single-phase equivalent circuit is used for the three-phase induction machine assuming a balanced set, as shown in Figure 6.24. The per-phase equivalent circuit consists of the stator loop and

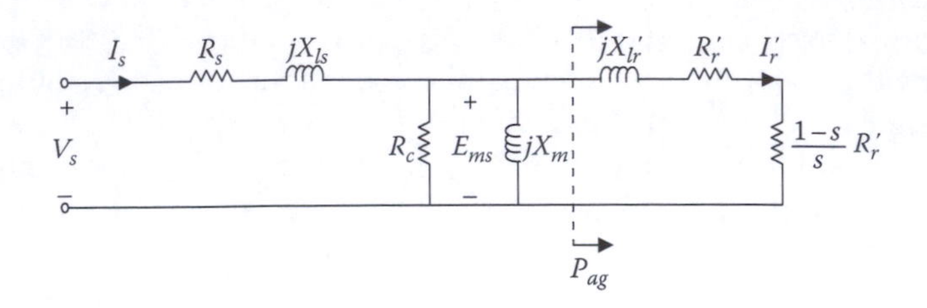

FIGURE 6.24
Steady-state per-phase equivalent circuit of an induction motor.

the rotor loop with the magnetic circuit parameters in the middle. The inductance representing the magnetization current path is in the middle of the circuit along with an equivalent core loss resistance. For the stator and rotor electrical parameters, the circuit includes the stator winding resistance and leakage reactance and the rotor winding resistance and leakage reactance. A slip-dependent equivalent resistance represents the mechanical power delivered at the shaft due to the energy conversion in the air gap-coupled electromagnetic circuit. The electrical input power supplied at the stator terminals converts to magnetic power and crosses the air gap. The air gap power P_{ag} is converted to mechanical power delivered at the shaft after overcoming the losses in the rotor circuit.

Although the per-phase equivalent circuit is not enough for designing controllers with good dynamic performance like that required in an electric or hybrid vehicle, the circuit helps develop the basic understanding of induction machines. The vast majority of applications of induction motors are for adjustable speed drives where controllers designed for good steady-state performance are adequate. The circuit does allow the analysis of a number of steady-state performance features. The parameters of the circuit model are as follows:

E_{ms} = stator-induced emf per phase

V_s = stator terminal voltage per phase

I_s = stator terminal current

R_s = stator resistance per phase

X_{ls} = stator leakage reactance

X_m = magnetizing reactance

X_{lr}' = rotor leakage reactance referred to stator

R_r' = rotor resistance referred to stator

I_r' = rotor current per phase referred to stator

Note that the voltages and currents described here in relation to the per-phase equivalent circuit are phasors and not space vectors. The power and torque relations are

$$P_{ag} = \text{Air gap power} = 3\left|I_r'\right|^2 \frac{R_r'}{s}$$

$$\begin{aligned} P_{dev} &= \text{Developed mechanical power } = 3\left|I_r'\right|^2 \frac{(1-s)R_r'}{s} \\ &= (1-s)P_{ag} \\ &= T_e\omega_m \end{aligned}$$

$$P_R = \text{Rotor copper loss } = 3\left|I_r'\right|^2 R_r'$$

The electromagnetic torque is given by

$$\begin{aligned} T_e &= 3\left|I_r'\right|^2 \frac{(1-s)R_r'}{s\omega_m} \\ &= \frac{3R_r'}{s\omega_s} \frac{V_s^2}{(R_s + R_r's)^2 + (X_s + X_r')^2} \end{aligned} \quad (6.42)$$

The steady-state speed–torque characteristics of the machine are as shown in Figure 6.25. The torque produced by the motor depends on the slip and the stator currents among other variables. The induction motor starting torque, while depending on the design, is lower than the peak torque achievable from the motor. The motor is always operated in the linear region of the speed–torque curve to avoid the higher losses associated with the high

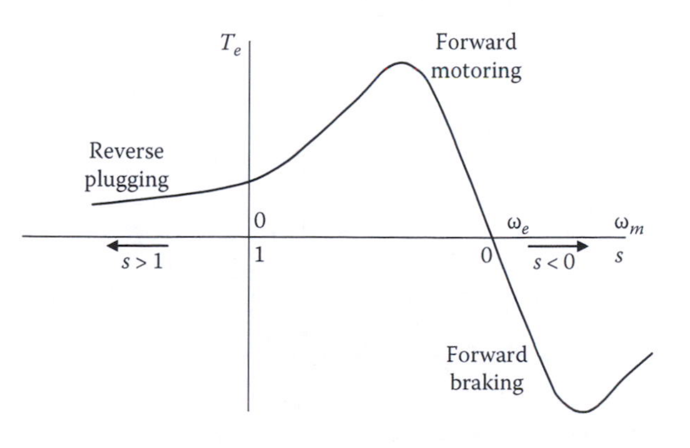

FIGURE 6.25
Steady-state torque–speed characteristics of an induction motor.

slip operation. In other words, operating the machine at small slip values improves the efficiency.

The value of the rotor circuit resistance determines the speed at which the maximum torque will occur. In general, the starting torque is low and the maximum torque occurs close to the synchronous speed when the slip is small. The motor draws a large current during line starting from a fixed AC source, which gradually subsides as the motor reaches the steady-state speed. If the load requires a high starting torque, the motor will accelerate slowly. This will make a large current flow for a longer time, thereby creating a heating problem.

The nonlinearity at speeds below the rated condition is due to the effects of the leakage reactances. At higher slip values, the frequencies of the rotor variables are large resulting in dominating impedance effects from the rotor leakage inductance. The air gap flux cannot be maintained at the rated value under this condition. Also, large values of rotor current (which flows at high slip values) cause a significant voltage drop across the stator winding leakage impedance ($R_s + j\omega L_{ls}$), which reduces the induced voltage and in turn the stator mmf flux density $\hat{B}_{ms}$.

6.4.2 Simplified Torque Expression

A simplified linear torque expression is sufficient to analyze the motor–load interaction, since the induction motor is invariably always operated in the linear region at maximum flux density $\hat{B}_{ms}$ with the help of a power electronic feed circuit. The segment of interest in the torque speed characteristic curve of the induction motor is shown by the solid line in Figure 6.26. The synchronous speed is set by the applied voltage frequency and the slope of the linear region is set by the design parameters and material properties. Hence, assuming that the stator flux density is kept constant

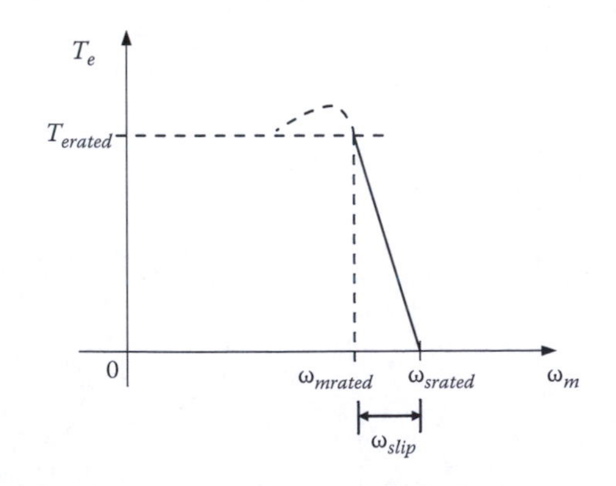

FIGURE 6.26

Torque–speed characteristics of induction motor for rated flux condition.

at its rated value, the steady-state torque can be expressed linearly as a function of slip:

$$T_e = K_{IM}\omega_{slip}$$

Here K_{IM} is a constant; the constant originates from the dependence of torque production with the geometry and the number of turns of the induction machine. The electromagnetic torque is produced by the tendency of the stator and rotor mmfs trying to align with each other. The rotor mmf is due to the rotor current. The principle of torque production essentially lies in the Lorentz force law ($F = Bil$). Therefore, the electromagnetic torque produced in an induction machine at steady state can be expressed as

$$T_e = k_M \hat{B}_{ms} \hat{I}_r' \tag{6.43}$$

where

k_M is a machine constant

$\hat{B}_{ms}$ is the equivalent peak stator mmf flux density for the three-phase machine

$\hat{I}_r'$ peak equivalent rotor current

Note that these are different from the single-phase equivalent circuit per-phase quantities. Mohan in [5] showed that this machine constant is $k_M = \pi r l (N_S/2)$ where r is the radius to the air gap, l is axial length of the machine, and N_S is the equivalent number of turns.

In order to find a relation between $\hat{B}_{ms}$ and rotor current $\hat{I}_r'$, let us denote the rotor mmf by the space vector $\vec{F}_r(t)$. The stator windings must carry currents in addition to the magnetizing current $\vec{i}_M(t)$ to support the currents induced in the rotor by transformer action to create $\vec{F}_r(t)$. These rotor currents referred to stator, or in other words, the additional stator current is represented by $\vec{i}_r'(t)$ (with magnitude-$\hat{I}_r'$) and is related to $\vec{F}_r(t)$ by

$$\vec{i}_r'(t) = \frac{\vec{F}_r}{N_S/2}$$

The total stator current is the sum of magnetizing current and referred rotor current

$$\vec{i}_S(t) = \vec{i}_M(t) + \vec{i}_r'(t) \tag{6.44}$$

These space vectors are shown in Figure 6.27. The rotor leakage inductance L_{lr}' has been neglected in this diagram for simplification. Although this is

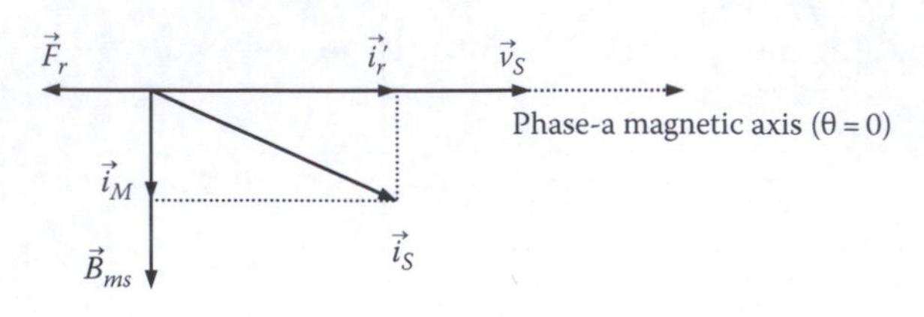

FIGURE 6.27
Space vector diagram of stator and rotor variables at $t=0$.

an ideal situation, it is a very important assumption that helps one grasp the basic concepts of torque production in induction machines. Neglecting the rotor leakage reactance is equivalent to assuming that all of the flux created by the rotor bar currents crosses the air gap and links the stator windings, and that there are no leakage fluxes in the rotor. The rated speed on induction machines is close to the synchronous speed and the machine usually operates near the rated condition with a very small slip. At small slip values, the slip speed ω_{slip} is very small and it is justified to neglect the effect of rotor leakage inductance, which is small anyway. Mohan [5] showed that under this simplifying assumption, the rotor bar currents induced by Faraday's law are proportional to the stator flux density and slip speed, the peak of which can be given by

$$\hat{I}_r' = k_r \hat{B}_{ms} \omega_{slip} \tag{6.45}$$

where k_r is a machine design constant. Substituting $\hat{I}_r'$ from Equation 6.45 in the torque equation

$$T_e = k_m k_r \hat{B}_{ms}^2 \omega_{slip}$$

The electromagnetic torque for fixed stator flux density is

$$T_e = K_{IM} \omega_{slip} \tag{6.46}$$

where $K_{IM} = k_m k_r \hat{B}_{ms}^2$.

The simple torque expression presents a convenient method of defining the torque–speed relationship of an induction machine linearly near the rated operating point very much similar to the DC machine relationship. The expression can be used to find the steady-state operating point of an induction machine-driven electric and hybrid vehicles by finding the point of intersection of the machine torque–speed characteristics and the road load force–speed characteristics. When the motor rotates at synchronous speed, the slip speed is zero and the motor does not produce any torque. In practice, the machine never reaches the synchronous speed even in an unloaded condition, since a small electromagnetic torque is needed to overcome the no-load losses that include the friction and windage losses. The slip speed is small up to the rated torque of the machine, and hence, it is reasonable to neglect the rotor leakage inductance, which gives the linear torque–speed relationship. The machine runs close to the synchronous speed under no-load condition with a very small slip. As the machine is loaded from the no-load condition, the slip starts to increase and the speed approaches the rated speed condition. Beyond the rated condition, the machine operates with a higher slip and the assumption of neglecting the leakage inductance starts to fall apart. This portion of the torque–speed characteristics is shown by a dotted curve in Figure 6.26.

The induction motors for electric and hybrid vehicles and other high-performance applications are supplied from a variable voltage, variable frequency AC source. Varying the frequency changes the rated flux and synchronous speed of the machine, which essentially causes the linear torque–speed curve of Figure 6.26 to move horizontally along the speed axis toward the origin.

Example 6.4

The vehicle road load characteristics on a level road is $T_{TR} = 24.7 + 0.0051\omega_{wh}^2$. The induction motor torque–speed relationship in the linear region is given by $T_e = K_{IM}(40 - \omega_m)$, including the gear ratio of the transmission system. The rated torque of 40 N m is available at a speed of 35 rad/s. Find the steady-state operating point of the vehicle.

Solution

The induction motor torque constant is

$$K_{IM} = \frac{40}{(40 - 35)} = 8\ \text{N m/rad/s}$$

The steady-state operating point is obtained by solving the vehicle road load characteristic and the motor torque–speed characteristic, which gives

$$\omega^* = 36.08\ \text{rad/s} \quad \text{and} \quad T^* = 31.34\ \text{N m}$$

6.4.3 Speed Control Methods

The speed of an induction motor can be controlled in two ways by varying the stator terminal voltage and by varying the stator frequency. Changing the terminal voltage changes the torque output of the machine, as is evident from Equation 6.42. Note that changing the applied voltage does not change the slip for maximum torque. The speed control through changing the

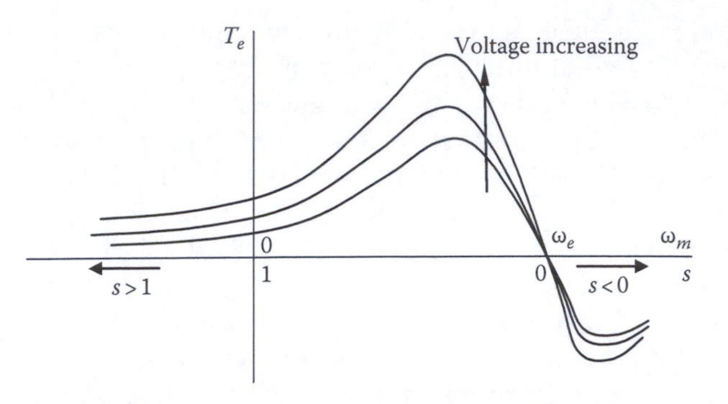

FIGURE 6.28
Torque–speed profile at different voltages with fixed supply frequency.

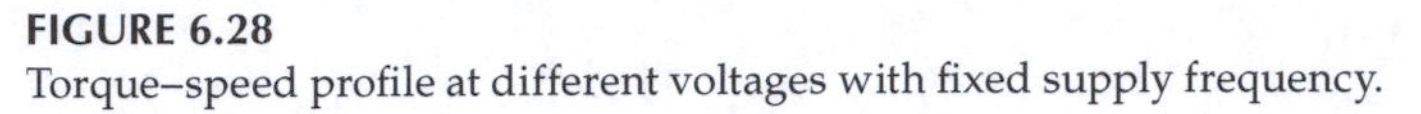

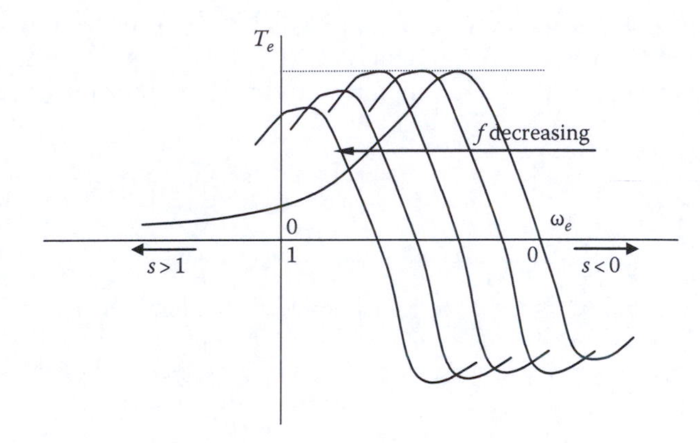

FIGURE 6.29
Torque–speed profile with variable frequency, but constant V/f ratio.

applied frequency is based on the frequency and synchronous speed relation $\omega_e = 4\pi f/p$; changing f changes ω_e. Figures 6.28 and 6.29 show the variations in torque–speed characteristics with changes in voltage and frequency, respectively. What is needed to drive the induction motor is a power electronics converter that will convert the available constant voltage into a variable voltage, variable frequency output according to the command torque and speed. The top-level block diagram of such a drive system is shown in Figure 6.30. The first generation controllers of induction motor drives used in electric vehicles employed slip control (constant V/Hz control) using a table of slip versus torque. The performance of such a drive for vehicle applications is very poor, since the concept of V/Hz control is based on steady-state equivalent circuit of the machine. The dynamic performance of the machine improves significantly using vector control. The *dq*-axes transformation theory for induction motors pertaining to vector control theory will be covered in Chapter 9, which discusses high-performance AC motor control methods.

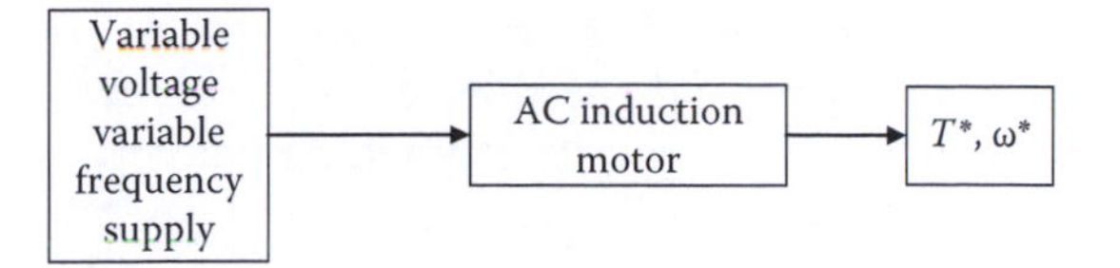

FIGURE 6.30
Induction motor drive.

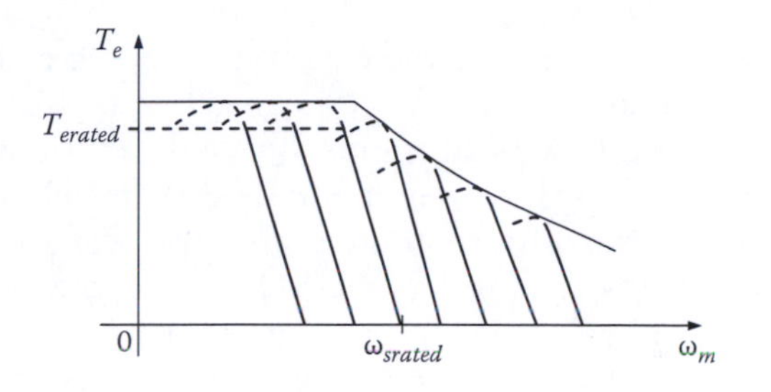

FIGURE 6.31
Torque speed operating envelop for the induction motor

Figure 6.31 shows the envelope of the torque–speed characteristics of an induction motor. Using a power electronics controlled drive, it is possible to achieve constant power characteristics from an induction motor at higher speeds, a feature that is so important for electric and hybrid vehicle motor drives.

6.4.4 Regenerative Braking

One of the advantages of using electric motors for vehicle propulsion is to save energy during vehicle braking through regeneration. The regenerated energy can be used to recharge the batteries of an electric or hybrid vehicle. It is important to note that it will not be possible to capture all of the energy that is available during vehicle braking, especially when sudden stops are commanded. The energy that is available during braking is the kinetic energy that was acquired by the vehicle during acceleration. The energy is typically too high to be processed by the electric motor that is used for propulsion. Processing high energy at a relatively short time would require a large motor, or, in other words, a motor with very high power ratings, which is impractical. Hence, electric and hybrid vehicles must be equipped with the mechanical brake system even though the electric motor drive is designed with regeneration capability. However, regeneration can recapture a significant portion of kinetic energy extending the vehicle range. The vehicle supervisory controller decides the amount of braking that is needed from the mechanical system based on the driver braking command, the amount of regeneration possible, and the vehicle velocity.

In the regenerative braking mode, the vehicle kinetic energy drives the shaft of the electric machine and the flow of energy is from the wheels to the energy storage system. The electric machine converts the mechanical power available from the vehicle kinetic energy and converts it to electrical energy. From the machine perspective, this is no different than operating the machine in the generator mode. Regenerative braking can increase the range of electric vehicle by a small percentage (about 10%–15%).

The induction machine works as a generator when it is operated with a negative slip, i.e., the synchronous speed is less than the motor speed ($\omega_m > \omega_e$). The negative slip makes the electromagnetic torque negative during regeneration or generating mode. In the negative slip mode of operation, the voltages and currents induced in the rotor bars are of opposite polarity compared to those in the positive slip mode. The electromagnetic torque acts on the rotor to oppose the rotor rotation, thereby decelerating the vehicle.

The motor drives for electric and hybrid vehicles are always four-quadrant drives meaning that the electric motor is controlled by the drive to deliver positive or negative torque at positive or negative speed. The transition from forward motoring to regeneration can be explained with the help of Figure 6.32 for four-quadrant induction motor drives. The linear segments of the induction motor toque–speed curves for several operating frequencies are shown in the figure. Consider the frequencies f_1 and f_2. The curves are extended in the negative torque region to show the characteristics during regeneration. Suppose initially, the vehicle is moving forward being driven by the positive torque delivered by the induction motor, and the steady-state operating point in this condition is at point "1." Now, the vehicle driver presses the brake medal to slow down the vehicle. The vehicle system controller immediately changes the motor drive frequency to f_2 such that $\omega_{s2} < \omega_{m1}$. The operating point shifts to point "2" immediately, since the motor speed cannot change instantaneously due to the inertia of the system. At point

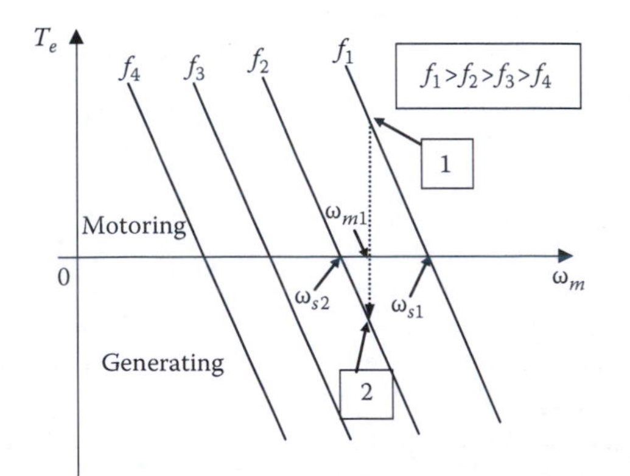

FIGURE 6.32
Transition from motoring to generating using a four-quadrant drive.

"2," the slip and the electromagnetic torque are negative and the motor is regenerating. The vehicle will decelerate from this condition onward. As the motor speed decreases and falls below the synchronous speed, the operating frequency needs to be changed to a lower value so that generating mode operation can be maintained. The power electronics drive is responsible for establishing the shifted linear torque–speed curves with different synchronous speeds for the induction machine at different frequencies, as shown. The drive circuit does so by changing the frequency of the supply voltage. The regenerative braking mode continues as long as there is kinetic energy available and the driver wishes to slow down the vehicle. Similar to starting, the regeneration also has to be achieved in a controlled way so that the power rating of the machine is not exceeded. The amount of kinetic energy to be converted within the desired stopping time determines the power that is to be handled by the machine.

6.5 Permanent Magnet Machines

The machines that use magnets to produce air-gap magnetic flux instead of field coils as in DC commutator machines or magnetizing component of stator current as in induction machines are called *PM machines*. This configuration eliminates the rotor copper loss as well as the need for maintenance of the field exciting circuit. The PM machines can be broadly classified into two categories:

- PM synchronous machines (PMSM): These machines have uniformly rotating stator field as in induction machines. The induced waveforms are sinusoidal and hence *dq* transformation and vector control are possible.
- PM trapezoidal or brushless DC machines (PM BLDC): The induced voltages in these machines are trapezoidal in nature; the phase currents are rectangular or square wave in nature. These PM machines are also known as *square wave* or *electronically commutated machines*. The stator field is switched in discrete steps with square wave pulses.

There are several advantages of using PMs for the field excitation in AC machines. The PMs provide a loss-free excitation in a compact way without complications of connections to the external stationary electric circuits. This is especially true for smaller machines, since there is always an excitation penalty associated with providing the rotor field through electrical circuits. The large synchronous machines use rotor conductors to provide the excitation, since the losses in the exciter circuit, referred to as excitation penalty, are small, especially when compared to the high costs of magnets. For smaller machines, the mmf required is small and the resistive effects often become comparable and dominating, resulting in lower efficiency. The smaller cross-sectional area of the windings for small power machines further deteriorates the resistive loss effect. Moreover, the cross-sectional area available for winding decreases as the motor size gets smaller. The loss-free excitation of PM in smaller machines with a compact arrangement is a definite plus with the only drawback being the high costs of the PMs. Nevertheless, PM machines are a strong contender for electric and hybrid vehicle drives despite the larger size of these machines. All the production passenger hybrid vehicles use PM machines for the traction motor. The factors guiding the trend are the excellent performance and the high power density achievable from the PM machine drives.

6.5.1 Permanent Magnets

The PMs are a source of mmf, much like a constant current source with relative permeability μ_r just greater than air, i.e., $\mu_r \approx 1.05$–1.07. The PM characteristics are displayed in the second quadrant of the *B*–*H* plot, as shown in Figure 6.33, conforming with the fact that these are sources of mmf. The magnets remain permanent as long as the operating point is within the linear region of its *B*–*H* characteristics. However, if the flux density is reduced beyond the knee-point of the characteristics (B_d), some magnetism will be lost permanently. On removal of the demagnetizing field greater than the limit, the new characteristics will be another straight line parallel to but lower than the original. The common types of magnets used in PM machines are the ferrites, samarium cobalt (SmCo), and neodymium-iron-boron (NdFeB). The features and properties of these three magnets are discussed below [6].

6.5.1.1 Ferrites

- They have been available for decades.
- Their cost is low.

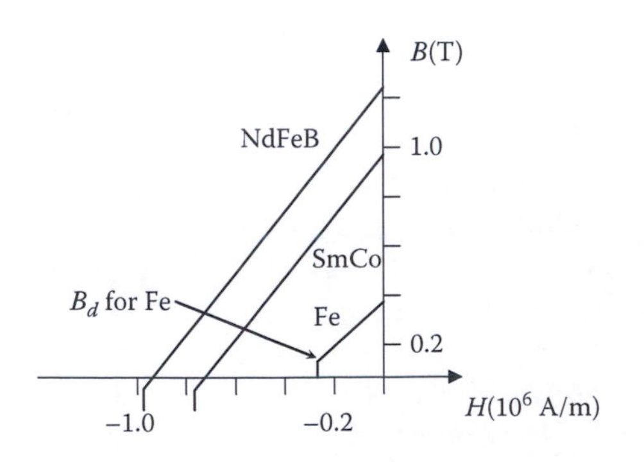

FIGURE 6.33
Characteristics of commonly used PMs.

- The residual flux density B_r at 0.3–0.4T is much lower than the desired range of gap flux density for high power density.
- B_d is higher for those ferrites for which B_r is higher.
- Ferrites have high resistivity and low core losses.
- They can be operated up to 100°C.
- An increase in temperature increases B_r and decreases B_d.

6.5.1.2 Samarium Cobalt

- This material has a higher value of B_r, 0.8–1.1 T.
- B_d is well into the third quadrant.
- B_r decreases somewhat with temperature, while B_d increases. This leads to increased sensitivity to demagnetization as temperature increases.
- The resistivity is 50 times that of Cu.
- Their cost is relatively high, reflecting the cost of rare earth element and an expensive metal.

6.5.1.3 Neodymium–Iron–Boron

- Sintered NdFeB developed in Japan in 1983 provided the major impetus to PM motors.
- B_r is in the range of 1.1–1.25 T at room temperature. This is adequate to produce a flux-density of 0.8–0.9 T across a relatively large air gap.
- B_r decreases by about 0.1% for each degree rise in temperature.
- The knee-point of flux density (B_d) increases rapidly with temperature. This imposes a limit on maximum temperature for NdFeB in the range of 100°C–140°C, depending on the detailed composition.
- The cost of these sintered NdFeB materials is still high mainly because of the manufacturing complexity of the sintering process.
- The cost may reduce in the future with increase in volume use. Fe(77%), B(8%) cost relatively little and Nd is one of the more prevalent rare earth elements.
- Bonded NdFeB magnets can be produced at a lower cost, but B_r is lower at about 0.6–0.7 T.

PM machines are designed with adequate considerations for magnet protection. Demagnetization may occur if flux density is reduced below the knee-point of flux density B_d. Most PM motors are designed to withstand considerable overload currents (2–4 times the rated) without danger to the magnets.

6.5.2 PM Synchronous Motors

The PMSM is a synchronous motor with sinusoidal mmf, voltage, and current waveforms where the field mmf is provided by PMs. The use of rare earth magnet materials increases the flux density in the air gap and accordingly increases the motor power density and torque-to-inertia ratio. In high-performance motion control systems that require servo-type operation, the PMSM can provide fast response, high power density, and high efficiency. In certain applications like robotics and aerospace actuators, it is preferable to have the weight as low as possible for a given output power. The PMSM, similar to the induction and DC machines, is fed from a power electronic inverter for its operation. The smooth torque output is maintained in these machines by shaping the motor currents, which requires a high-resolution position sensor and current sensors. The control algorithm is implemented in a digital processor using feedback from the sensors. A flux weakening mode that enables a higher speed operation in the constant power region is possible in PMSM by applying a stator flux in opposition to the rotor magnet flux. The motor high speed limit depends on the motor parameters, its current rating, the back-emf waveform, and the maximum output voltage of the inverter.

PMSM and induction motors have good performance in terms of torque response and have rugged motor structures, although broken magnet chips in PM machines is a concern. The slip speed calculation makes the induction motor control more complicated than that of the PMSM. Without a rotor cage, the PMSM has a lower inertia that helps the electrical response time, although the induction motor electrical response characteristics will be the fastest because of the smaller time constant. The electrical time constant of magnetic circuits is determined by the L/R ratio. The load current transient in induction machines is limited only by the small leakage inductance, where the time constant inductance in PM machines is the much higher self-inductance. With a higher power density, the PMSM is smaller in size compared to an induction motor with the same power rating. The PMSM is more efficient and easier to cool due to the absence of rotor copper loss compared to the induction machines. The induction motor has lower cost and zero cogging torque because of the absence of PMs. Also, it is less sensitive to higher operating temperatures. The induction motor can sustain a higher peak stator current at several times the rated current without the danger of demagnetizing the magnets. Both the induction motors and the PMSM suffer from limited field weakening speed range.

The PMs in PMSMs are not only expensive, but also sensitive to temperature and load conditions, which constitute the major drawbacks of PM machines. Most of the PMSMs are found in small-to-medium power applications, although there are some high-power applications for which PMSMs are being used.

PMSM has a stator with a set of three-phase sinusoidally distributed copper windings similar to the windings described in Section 6.2 on AC machines. A balanced set of applied three-phase voltages forces a balanced set of sinusoidal currents in the three-phase stator windings, which in turn establishes the constant amplitude rotating mmf in the air gap. The stator currents are regulated using rotor position feedback so that the applied current frequency is always in synchronism with the rotor. The PMs in the rotor are appropriately shaped and their magnetization directions are controlled such that the rotor flux linkage created is sinusoidal. The electromagnetic torque is produced at the shaft by the interaction of these two stator and rotor magnetic fields.

The PMSMs are classified according to the position and shape of the PMs in the rotors. The three common arrangements of PMs in the rotors are surface mounted, inset, and interior or buried, which are shown in Figure 6.34. The difference between surface-mounted and inset magnets is that the magnets in the latter are inside the rotor surface, but still exposed to the air gap. The surface-mounted and inset rotor PMSMs are often collectively called

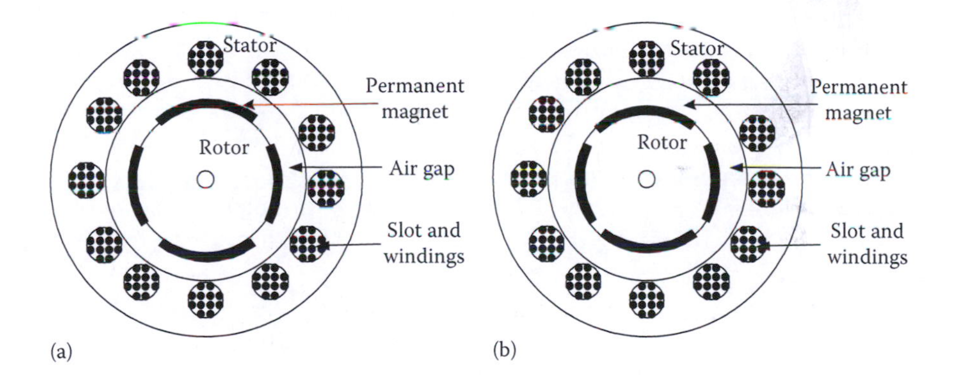

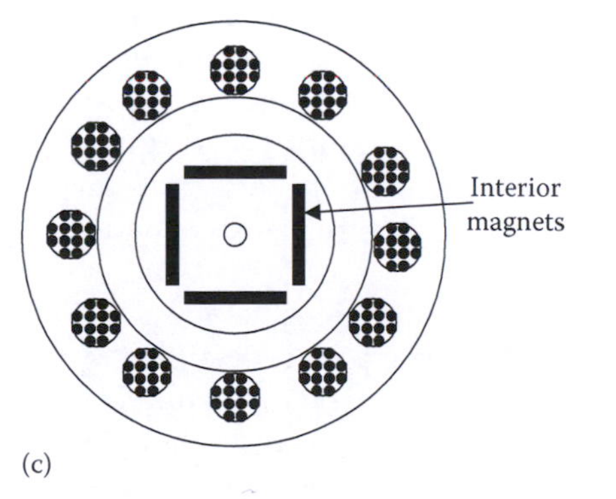

FIGURE 6.34
PM machines: (a) surface-mounted, (b) inset, and (c) interior.

the *surface-mount PMSMs*. In the surface-mounted PMSM, the magnets are epoxy-glued or wedge-fixed to the cylindrical rotor. Nonmagnetic stainless steel or carbon fiber sleeves are also used to contain the magnets. The manufacturing of this kind of rotor is simple, although the mechanical strength of the rotor is only as good as that of the epoxy glue. In the inset PMSM, the magnets are put into the rotor surface slots, which make the magnet more secured. The other type of PMSM is the interior PMSM, so named because the magnets are buried inside the rotor. The manufacturing process is complicated and expensive for the interior PMSM.

6.5.3 PMSM Models

The modeling and control of PMSMs can be done using either scalar or vector methods. In the scalar control method, the per-phase equivalent circuit is derived from the three-phase *abc* system. With vector control method, the three-phase variables are transformed into an equivalent two-phase system, and both amplitude and angle of the current are controlled. The three-phase *abc* reference directions, and the fictitious two-phase $\alpha\beta$ and *dq* reference directions are shown in Figure 6.35. For the *dq* systems, *d* stands for direct axis and *q* stands for quadrature axis. The *abc* and $\alpha\beta$ reference frames are fixed in the stator with the direction of α-axis chosen in the same direction as the *a*-axis. The β-axis lags the α-axis by 90° space angle. The *abc* to $\alpha\beta$ transformation essentially transforms the three-phase stationary variables to a set of two-phase stationary variables. In the *dq* reference, the *d*-axis is aligned with the magnet flux direction, while the *q*-axis lags the *d*-axis by 90° space angle.

The direct and quadrature-axes inductances of a PMSM play an important role in the control of the machine. For the surface-mounted PMSM, the two inductances are approximately equal, since the permeability of the flux path between the stator and the rotor is equal all around the stator circumference. The uniformity in the magnetic path despite the presence of magnets is because the permeability of magnets is approximately equal to that of the air. The space needed to mount the magnets increases the radial distance of the effective air gap, making the self-inductance relatively smaller in PMSMs.

The direct and quadrature-axes reluctances are unequal in inset PMSMs, since space is occupied by magnet in the direct axis and by iron in the quadrature axis. The *q*-axis inductance L_q is larger than the *d*-axis inductance L_d, since the *d*-axis flux path has larger effective air gap, and hence, higher reluctance, although the length of the air gap between the stator and rotor is the same. These inductances

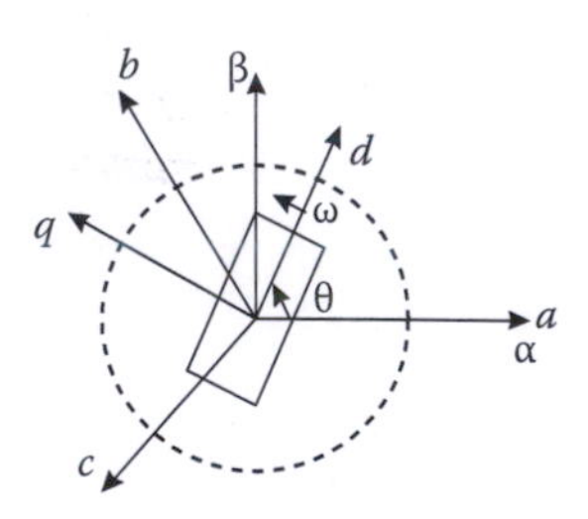

FIGURE 6.35
The stationary and synchronous frames in PMSM.

consist of the magnetizing inductances and the leakage inductances, and are given as

$$L_q = L_{ls} + L_{mq} \tag{6.47}$$

and

$$L_q = L_{ls} + L_{md} \tag{6.48}$$

The interior PMSM has its magnets buried inside the rotor. The q-axis inductance L_q in the interior PMSM can be much larger than the d-axis inductance L_d. The larger difference in the d and q-axes inductances make the interior PM more suitable for flux weakening operation, delivering a wider constant power region compared to the surface-mount or inset PMSMs. The extended constant power range capability is extremely important for an electric and hybrid vehicle application to eliminate the use of multiple gear ratios. Because of the unequal reluctance paths in the direct and quadrature axes, a reluctance torque exists in buried and inset PMSMs.

The per-phase equivalent circuit model and the scalar control of PMSM are discussed in the following section. The vector control methods are presented in Chapter 9 after the theory of reference frame transformation is addressed in further detail.

6.5.3.1 Voltage Equations

The stator circuit of a PMSM is similar to that of an induction motor or a wound rotor synchronous motor with the applied voltage being balanced by the stator winding resistance drop and the induced voltage in the winding. The PMSM model is derived below, assuming that the stator windings in the three phases are symmetrical and sinusoidally distributed. The eddy current and hysteresis losses are neglected in the model and the damper or cage windings are also not considered. The damper windings are not necessary in PMSMs, since PM is a poor electrical conductor and eddy currents are negligible. Large armature currents can be tolerated in these machines without significant demagnetization. The stator phase voltage equation in the stationary abc reference frame is

$$\vec{v}_{abcs} = \bar{R}_s \vec{i}_{abcs} + \frac{d}{dt}\vec{\lambda}_{abcs} \tag{6.49}$$

where
$(\vec{f}_{abcs})^T = [f_{as}\, f_{bs}\, f_{cs}]$ (f represents v, i or λ)
$\bar{R}_s = \text{diag}[R_s\ R_s\ R_s]$

The flux linkages are

$$\vec{\lambda}_{abcs} = \bar{L}_s \vec{i}_{abcs} + \bar{\lambda}_f \tag{6.50}$$

where $\bar{L}_s$ is the inductance matrix which is the same for all synchronous machines. $\bar{\lambda}_f$ is due to the PM and is given by

$$\bar{\lambda}_f = \lambda_f \begin{bmatrix} \sin\theta_r \\ \sin\left(\theta_r - \frac{2\pi}{3}\right) \\ \sin\left(\theta_r + \frac{2\pi}{3}\right) \end{bmatrix} \tag{6.51}$$

where λ_f is the amplitude of the flux linkage established by the PM, as viewed from the stator phase windings.

6.5.3.2 Per-Phase Equivalent Circuit

A simplified per-phase equivalent circuit can be derived for the special case of surface-mount PM machines and also when the magnets are only slightly inset in the rotor. The air-gap inductances along the *d* and *q*-axes are equal in this case, and can be denoted by

$$L_m = L_{md} = L_{mq} \tag{6.52}$$

In the per-phase equivalent circuit, the fundamental frequency components of voltages and currents of the inverter-driven machine are regarded as balanced three-phase sets for steady-state and slowly varying transient conditions. The PM machine can then be represented by the per-phase equivalent circuit shown in Figure 6.36a. The current source I_F represents the root mean square (rms) value of the equivalent magnet current i_{fd} that would create

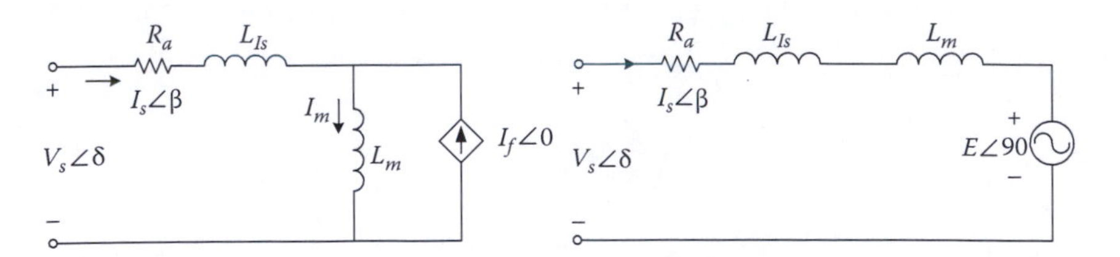

FIGURE 6.36
PM synchronous motor equivalent circuits showing (a) equivalent magnet current and (b) induced voltage.

the PM flux λ_f. The PM flux, in terms of the equivalent current, is given by $\lambda_f = L_m i_{fd}$. Assuming sinusoidal distribution of stator windings, only the fundamental component of magnet flux density link with the stator windings. The magnetizing inductance L_m for PM machines is much smaller than that of induction machines because of much larger effective air gap.

From the stator terminals, the effect of the PM flux is seen as a back-emf voltage given by $E = j\omega L_m I_m$. This back-emf or induced voltage is shown in the alternative representation of the per-phase equivalent circuit in Figure 6.36b. The stator per-phase rms current is denoted by I_s, which leads the reference axis (*d*-axis in this case) by an angle γ. The PM machine provides continuous torque only when rotor speed is synchronized with the stator frequency. For the per-phase model, the torque is given by

$$T_e = 3 \times \frac{P}{2} L_m I_F I_S \sin\gamma \tag{6.53}$$

The torque is maximum when $\gamma = 90°$. The vector diagram for this maximum torque condition is shown in Figure 6.37a. The input power factor in this condition is lagging. Operation in this mode is practical up to the limit of the inverter supply voltage. At higher speeds, the appropriate operating condition is that for unity power factor, which maximizes the inverter volt-amp rating utilization. The vector diagram for this condition is given in Figure 6.37b. The motor can also be operated with a leading power factor by increasing the voltage angle δ, as shown in Figure 6.37c. This condition is desired for some large synchronous drives that allows the use of a load-commutated inverter.

The per-phase equivalent circuit is useful for developing an understanding of the basic operation of synchronous machines. The per-phase equivalent circuit and the corresponding torque equations signify that torque is maximized at $\gamma = 90°$ where the stator current is aligned in quadrature with the magnet field axis. The magnet field axis is along the direct or *d*-axis

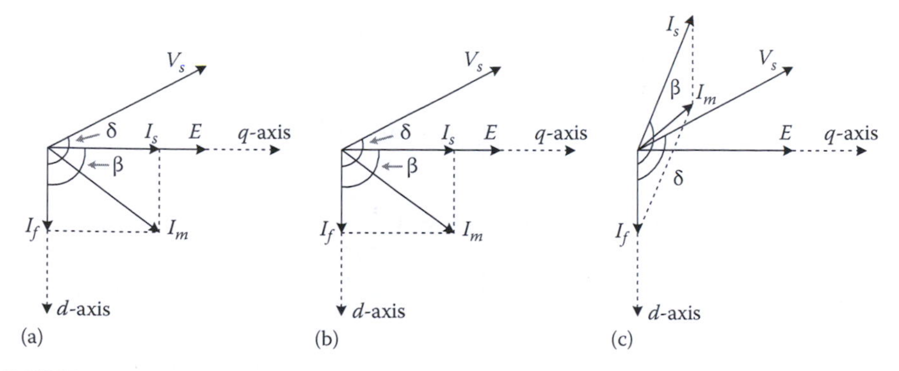

FIGURE 6.37
Vector diagrams with three power factors for the per-phase equivalent circuit: (a) lagging, (b) unity, and (c) leading.

and the stator current is along the quadrature or q-axis in this condition. Maintaining such orthogonality consistently and continuously through both magnitude and angle controls of the stator current form the basis of vector controls, which is used for traction motor drives. Motor operation for $\gamma > 90°$ is known as field weakening operation, which is also used in vector control methods. It can also be noted that for regeneration, the rotor angle γ is made negative through current control.

6.5.4 PM Brushless DC Motors

The PM AC machines with trapezoidal back-emf waveforms are the PM brushless DC (PM BLDC) machines. The trapezoidal-shaped back-emf waveforms in these machines are due to the concentrated windings of the machine used instead of the sinusoidally distributed windings used in the PMSMs. The PM BLDC motors are used in a wide variety of applications ranging from computer drives to sophisticated medical equipment. The reason behind the popularity of these machines is the simplicity of control. Only six discrete rotor positions per electrical revolution are needed in a three-phase machine to synchronize the phase currents with the phase back-emfs for effective torque production. A set of three Hall sensors mounted on the stator facing a magnet wheel fixed to the rotor and placed 120° apart can easily give this position information. This eliminates the need for a high-resolution encoder or position sensor required in PMSMs, but the penalty paid for position sensor simplification is in the performance. Vector control is not possible in PM BLDC machines because of the trapezoidal shape of the back-emfs.

The three-phase back-emf waveforms and the ideal phase currents of a PM BLDC motor are shown in Figure 6.38. The back-emf waveforms are fixed with respect to the rotor position. Square wave phase currents are supplied such that they are synchronized with the peak back-emf segment of the respective phase. The controller achieves this objective using rotor position feedback information. The motor basically operates like a DC motor with its electronic controller; hence, the motor is called the brushless DC motor.

6.5.4.1 PM BLDC Machine Modeling

The PM in the rotor can be regarded as a constant current source, giving rise to the back-emfs in the stator windings. The three stator windings for the three phases are assumed to be identical with 120° (electrical) phase displacement among them. Therefore, the stator winding resistances and the self-inductance of each of the three phases can be assumed to be identical. Let

R_s = stator phase winding resistance
$L_{aa} = L_{bb} = L_{cc} = L$ = stator phase self-inductance
$L_{ab} = L_{ac} = L_{bc} = M$ = stator mutual inductance

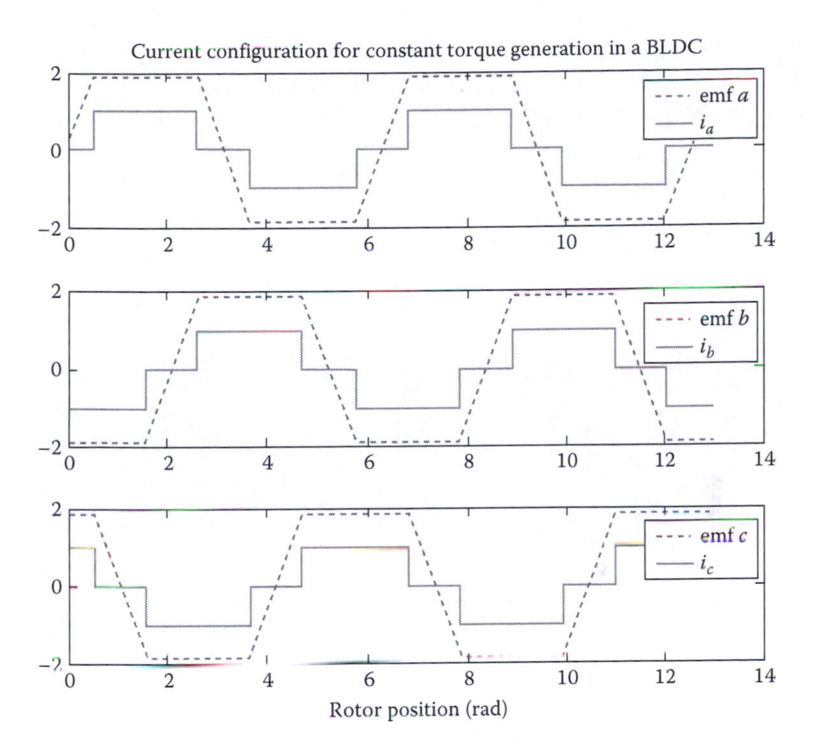

FIGURE 6.38
Back-emf and ideal phase currents in the three phases of a PM brushless DC motor.

The voltage balance equations for the three phases are

$$\begin{bmatrix} v_a \\ v_b \\ v_c \end{bmatrix} = R \cdot \begin{bmatrix} i_a \\ i_b \\ i_c \end{bmatrix} + \begin{bmatrix} L & M & M \\ M & L & M \\ M & M & L \end{bmatrix} \cdot p \cdot \begin{bmatrix} i_a \\ i_b \\ i_c \end{bmatrix} + \begin{bmatrix} e_a \\ e_b \\ e_c \end{bmatrix} \tag{6.54}$$

where

p is the operator d/dt

e_a, e_b, and e_c are the back-emfs in the three phases, respectively

The back-emf is related to the phase flux linkage as

$$e = \frac{d\lambda}{dt} = \frac{d\lambda}{d\theta} \cdot \frac{d\theta}{dt}$$

However, $d\theta/dt = \omega_r$, which is the rotor speed. Then,

$$e = \omega_r \cdot \frac{d\lambda}{d\theta} \tag{6.55}$$

Similar to the back-emfs, the currents are also shifted by 120°, and they satisfy the condition $i_a+i_b+i_c=0$. Therefore, we have $M\cdot i_b+M\cdot i_c=-M\cdot i_a$; similar expressions exist for the two other phases. Equation 6.9 can then be simplified as

$$\begin{bmatrix} v_a \\ v_b \\ v_c \end{bmatrix} = R\cdot\begin{bmatrix} i_a \\ i_b \\ i_c \end{bmatrix} + \begin{bmatrix} L-M & 0 & 0 \\ 0 & L-M & 0 \\ 0 & 0 & L-M \end{bmatrix}\cdot p\cdot\begin{bmatrix} i_a \\ i_b \\ i_c \end{bmatrix} + \begin{bmatrix} e_a \\ e_b \\ e_c \end{bmatrix}$$

The rate of change of currents with the applied voltages can be expressed as

$$p\cdot\begin{bmatrix} i_a \\ i_b \\ i_c \end{bmatrix} = \frac{1}{L-M}\cdot\left[\begin{bmatrix} v_a \\ v_b \\ v_c \end{bmatrix} - R\cdot\begin{bmatrix} i_a \\ i_b \\ i_c \end{bmatrix} - \begin{bmatrix} e_a \\ e_b \\ e_c \end{bmatrix}\right] \tag{6.56}$$

The electrical power transferred to the rotor is equal to the mechanical power $T_e\omega_r$ available at the shaft. Using this equality, the electromagnetic torque for the PM BLDC motor is

$$T_e = \frac{e_a\cdot i_a + e_b\cdot i_b + e_c\cdot i_c}{\omega_r} \tag{6.57}$$

For the control strategy described previously where only two-phase currents are active at a time, the torque expression for equal currents in two phases simplifies to

$$T_e = \frac{2\cdot e_{\max}\cdot I}{\omega_r} \tag{6.58}$$

Since the currents are controlled to synchronize with the maximum back-emf only, e_{max} has been used in Equation 6.58 instead of e as a function of time or rotor position. Assuming magnetic linearity, Equation 6.55 can be written as

$$e = K\cdot\omega_r\cdot\frac{dL}{d\theta}$$

Hence, the maximum back-emf is

$$e_{max} = K\cdot\left[\frac{dL}{d\theta}\right]_{max}\cdot\omega_r \quad \text{or} \quad e_{max} = K'\cdot\omega_r \tag{6.59}$$

Equations 6.58 and 6.59 are very similar to $E = K \cdot \phi \cdot \omega$ and $T = K \cdot \phi \cdot I$ equations associated with regular DC machines. Therefore, with the described control strategy, a PM BLDC machine can be considered to behave like a DC machine.

6.6 Switched Reluctance Machines

The switched reluctance motor (SRM) is a doubly salient, singly excited reluctance machine with independent phase windings on the stator. The stator and the rotor are made of magnetic steel laminations, with the latter having no windings or magnets. The SRMs can be of various stator–rotor pole combinations related to different phase configurations. The cross-sectional diagrams of a four-phase, 8–6 SRM and a three-phase, 12–8 SRM are shown in Figure 6.39. The three-phase, 12–8 machine is a two-repetition version of the basic 6–4 structure within the single stator geometry. The two-repetition machine can alternately be labeled as a four-poles/phase machine, compared to the 6–4 structure that can be called a two-poles/phase machine. The stator windings on diametrically opposite poles are connected either in series or in parallel to form one phase of the motor. When a stator phase is energized, the most adjacent rotor pole-pair is attracted toward the energized stator in order to minimize the reluctance of the magnetic path. Therefore, it is possible to develop constant torque in either direction of rotation by energizing consecutive phases in succession.

The aligned position of a phase is defined to be the orientation when the stator and rotor poles of the phase are perfectly aligned with each other,

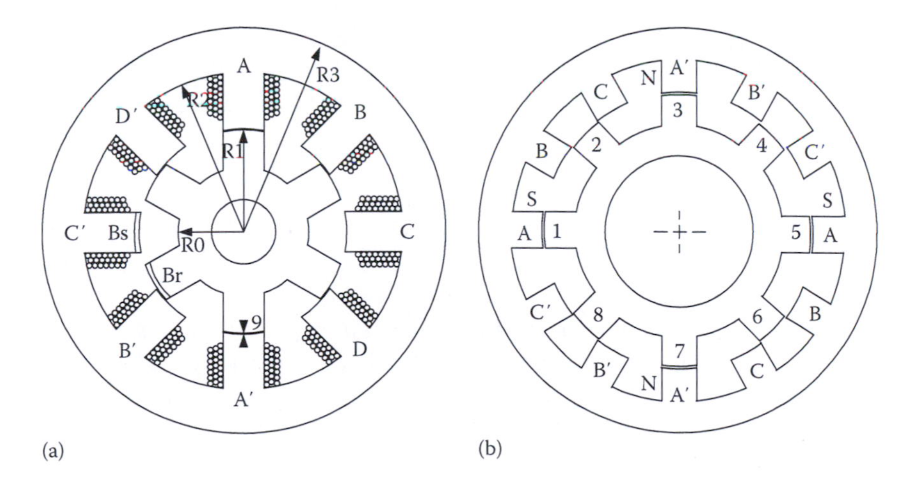

FIGURE 6.39
Cross-sections of three-phase SR machines: (a) four-phase 8/6 structure; (b) 12/8, two-repetition (two-channel) structure.

attaining the minimum reluctance position. The unsaturated phase inductance is maximum (L_a) in this position. The phase inductance decreases gradually as the rotor poles move away from the aligned position in either direction. When the rotor poles are symmetrically misaligned with the stator poles of a phase, the position is said to be the unaligned position. The phase has the minimum inductance (L_u) in this position. Although, the concept of inductance is not valid for a highly saturating machine like SRM, the unsaturated aligned and unaligned inductances are two key reference positions for the controller.

Several other combinations of the number of stator and rotor poles exist, such as 10–4, 12–8, etc. A 4–2 or a 2–2 configuration is also possible, but these have the disadvantage that if the stator and rotor poles are aligned exactly, then it would be impossible to develop a starting torque. The configurations with higher number of stator/rotor pole combinations have less torque ripple and do not have the problem of starting torque.

6.6.1 Advantages and Disadvantages

The switched reluctance machines or motors possess few unique features that make them strong competitors to existing AC and DC motors in various adjustable speed drive and servo applications. The machine construction is simpler due to the absence of rotor winding and PMs. The bulk of the losses in the machine occur in the stator, which is relatively easier to cool. The starting torque of the SRM can be very high without the problem of excessive inrush current due to its higher self-inductance. Another great advantage of SRMs is that the maximum permissible rotor temperature is higher. The SRMs can be designed with a wide constant power region, which is a feature particularly attractive for traction applications. The SRM also has some unique fault tolerance features that do not exist in other AC machines. Since each phase winding is connected in series with converter switching elements, there is no possibility of shoot-through faults between the DC buses in the SRM drive converter. The independent stator phases enable drive operation in spite of loss of one or more phases, and the drive can be brought to a safe shutdown instead of a sudden stop.

The SRM also comes with a few disadvantages among which torque ripple and acoustic noise are the most critical. The double saliency construction and the discrete nature of torque production by the independent phases lead to higher torque ripple compared to other machines. The higher torque ripple also causes the ripple current in the DC supply to be quite large, necessitating a large filter capacitor. The doubly salient structure of the SRM also causes higher acoustic noise compared to other machines. The main source of acoustic noise is the radial magnetic force-induced resonance with the circumferential mode-shapes of the stator. The drawbacks have been addressed through research, and solutions for torque ripple minimization and acoustics noise reduction do exist. Besides, all applications are not highly sensitive to torque ripple and acoustic noise.

The absence of PMs imposes the excitation burden on the stator windings and converter, which increases the converter kVA requirement. Compared to PM brushless machines, the per-unit stator copper losses will be higher, reducing the efficiency and torque per ampere. However, the maximum speed at constant power is not limited by the fixed magnet flux as in the PM machine, and hence, an extended constant power region of operation is possible in SRMs. The control can be simpler than the field-oriented control of induction machines, although for torque ripple minimization significant computations may be required for an SRM drive.

6.6.2 SRM Design/Basics

The fundamental design rules governing the choice of phase numbers, pole numbers, and pole arcs are discussed in detail by Lawrenson et al. [7] and also by Miller [8]. From a designer's point of view, the objectives are to minimize the core losses, to have good starting capability, to minimize the unwanted effects due to varying flux distributions and saturation, and to eliminate mutual coupling. The choice of the number of phases and poles is open, but a number of factors need to be evaluated prior to making a selection. A comprehensive design methodology of SRM appears in [9].

The fundamental switching frequency is given by

$$f = \frac{n}{60} N_r \text{ Hz} \tag{6.60}$$

where
n is the motor speed in rpm
N_r is the number of rotor poles

The "step angle" or "stroke" of an SRM is given by

$$\text{Step Angle } \varepsilon = \frac{2\pi\delta}{N_{ph} \cdot N_{rep} \cdot N_r} \tag{6.61}$$

The step angle is an important design parameter related to the frequency of control per rotor revolution. N_{rep} represents the multiplicity of the basic SRM configuration, which can also be stated as the number of pole pairs per phase. N_{ph} is the number of phases. N_{ph} and N_{rep} together set the number of stator poles.

The regular choice of the number of rotor poles in an SRM is

$$N_r = N_s \pm k_m \tag{6.62}$$

where
k_m is an integer such that $k \bmod q \neq 0$
N_s is the number of stator poles

Some combinations of parameters allowed by Equation 6.62 are not feasible, since sufficient space must exist between the poles for the windings. The most common choice of Equation 6.62 for the selection of stator and rotor pole numbers is $k_m = 2$ with the negative sign.

The SRM is always driven into saturation to maximize the utilization of the magnetic circuit, and hence, the flux-linkage λ_{ph} is a nonlinear function of stator current and rotor position given by

$$\lambda_{ph} = \lambda_{ph}(i_{ph}, \theta)$$

The electromagnetic profile of an SRM is defined by the $\lambda - i - \theta$ characteristics shown in Figure 6.40. The highest possible saliency ratio (the ratio between the maximum and minimum unsaturated inductance levels) is desired to achieve the highest possible torque per ampere, but as the rotor and stator pole arcs are decreased, the torque ripple tends to increase. The torque ripple adversely affects the dynamic performance of an SRM drive. For many applications, it is desirable to minimize the torque ripple, which can be partially achieved through appropriate design. The torque dip observed in the T–i–θ characteristics of an SRM (see Figure 6.41) is an indirect measure of the torque ripple expected in the drive system. The torque dip is the difference between

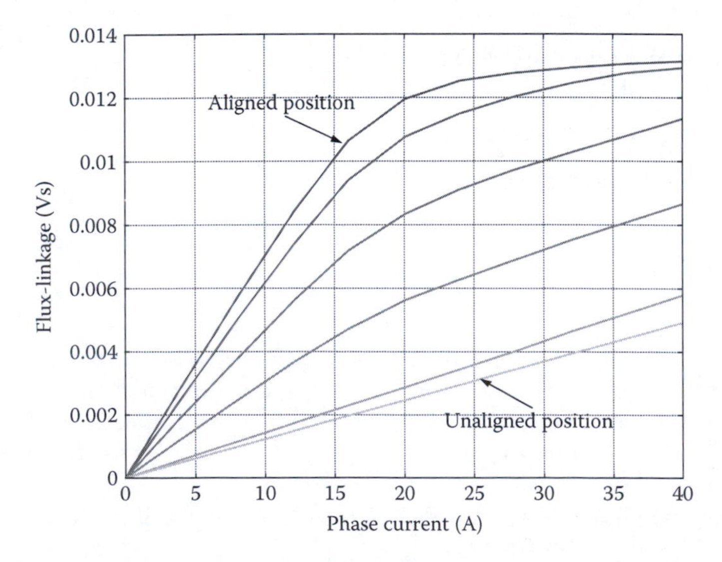

FIGURE 6.40
Flux–angle–current characteristics of a four-phase SRM.

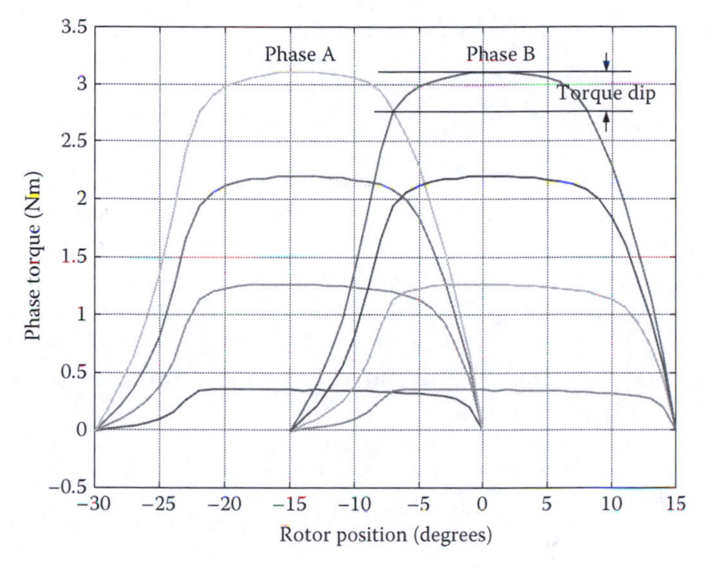

FIGURE 6.41
Torque–angle–current characteristics of a four-phase SRM for four constant current levels.

the peak torque of a phase and the torque at an angle where two overlapping phases produce equal torque at equal levels of current. The smaller the torque dip, the less will be the torque ripple. The T–i–θ characteristics of the SRM depend on the stator–rotor pole overlap angle, pole geometry, material properties, number of poles, and number of phases. A design tradeoff needs to be considered to achieve the desired goals. The T–i–θ characteristics must be studied through finite element analysis during the design stage to evaluate both the peak torque and torque dip values.

6.6.3 Principle of Operation

6.6.3.1 Voltage-Balance Equation

The general equation governing the flow of stator current in one phase of an SRM can be written as

$$V_{ph} = i_{ph}R_s + \frac{d\lambda_{ph}}{dt} \tag{6.63}$$

where

- V_{ph} is the DC bus voltage
- i_{ph} is the instantaneous phase current
- R_s is the winding resistance
- λ_{ph} is the flux linking the coil

The stator phase voltage can be expressed as

$$V_{ph} = i_{ph}R_s + \frac{\partial \lambda_{ph}}{\partial i_{ph}}\frac{di_{ph}}{dt} + \frac{\partial \lambda_{ph}}{\partial \theta}\frac{d\theta}{dt} = i_{ph}R_s + L_{inc}\frac{di_{ph}}{dt} + k_v\omega \tag{6.64}$$

where
L_{inc} is the incremental inductance
k_v is the current-dependent back-emf coefficient
$\omega = d\theta/dt$ is the rotor angular speed

Assuming magnetic linearity (where $\lambda_{ph} = L_{ph}(\theta)i_{ph}$), the voltage expression can be simplified as

$$V_{ph} = i_{ph}R_s + L_{ph}(\theta)\frac{di_{ph}}{dt} + i_{ph}\frac{dL_{ph}(\theta)}{dt}\omega \tag{6.65}$$

The last term in Equation 6.65 is the "back-emf" or "motional-emf" and has the same effect on SRM as the back-emf has on DC motors or PM BLDC motors. However, the back-emf in SRM is generated in a different way from the DC machines or PM BLDC motors where it is caused by a rotating magnetic field. In an SRM, there is no rotor field and back-emf depends on the instantaneous rate of change of phase flux linkage and the phase current.

In the linear case, which is always valid for lower levels of phase current, the per-phase equivalent circuit of an SRM consists of a resistance, an inductance, and a back-emf component. The back-emf vanishes when there is no phase current or the phase inductance is constant relative to the rotor position. Depending on the magnitude of current and rotor angular position, the equivalent circuit changes its structure from being primarily an *R-L* circuit to primarily a back-emf-dependent circuit.

6.6.3.2 Energy Conversion

The energy conversion process in an SRM can be evaluated using the power balance relationship. Multiplying Equation 6.65 by i_{ph} on both sides, the instantaneous input power can be expressed as

$$P_{in} = V_{ph}i_{ph} = i_{ph}^2R_s + \left(L_{ph}i_{ph}\frac{di_{ph}}{dt} + \frac{1}{2}i_{ph}^2\frac{dL_{ph}}{d\theta}\omega\right) + \frac{1}{2}i_{ph}^2\frac{dL_{ph}}{d\theta}\omega$$

$$= i_{ph}^2R + \frac{d}{dt}\left(\frac{1}{2}L_{ph}i_{ph}^2\right) + \frac{1}{2}i_{ph}^2\frac{dL_{ph}}{d\theta}\omega \tag{6.66}$$

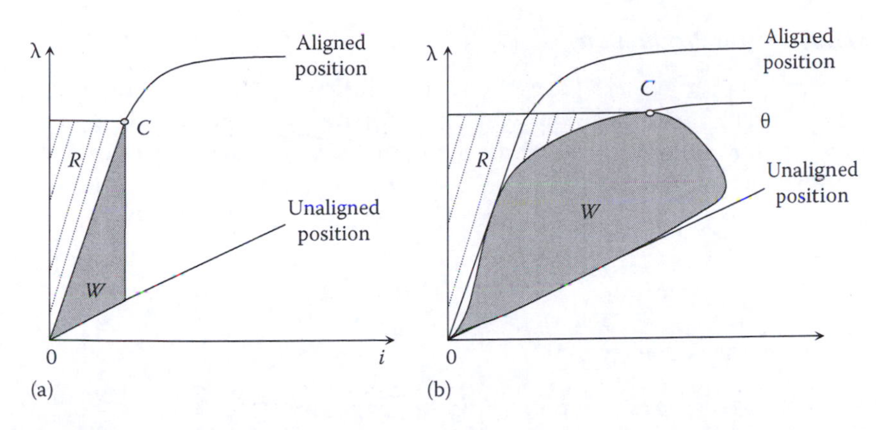

FIGURE 6.42
Energy partitioning during one complete working stroke. (a) Linear assumption. (b) Practical case. W = energy converted into mechanical work. R = energy returned to the DC supply.

The first term represents the stator winding loss, the second term denotes the rate of change of magnetic stored energy, while the third term is the mechanical output power. The rate of change of magnetic stored energy always exceeds the electromechanical energy conversion term. The most effective use of the energy supplied is when the current is maintained constant during the positive $dL_{ph}/d\theta$ slope. The magnetic stored energy is not necessarily lost, but can be retrieved by the electrical source if an appropriate converter topology is used. In the case of a linear SRM, the energy conversion effectiveness can be at most 50%, as shown in the energy division diagram of Figure 6.42a. The drawback of lower effectiveness is the increase in converter volt-amp rating for a given power conversion of the SRM. The division of input energy increases in favor of energy conversion if the motor operates under magnetic saturation. The energy division under saturation is shown in Figure 6.42b, where the effectiveness is clearly much more than 50%. This is the primary reason for operating the SRM always under saturation. The term energy ratio instead of efficiency is often used for SRM, because of the unique situation of the energy conversion process. The energy ratio is defined as

$$ER = \frac{W}{W + R} \tag{6.67}$$

where

- W is the energy converted into mechanical work
- R is the energy returned to the source using a regenerative converter

The term *energy ratio* is analogous to the term *power factor* used for AC machines.

6.6.3.3 Torque Production

The torque is produced in the SRM by the tendency of the rotor to attain the minimum reluctance position when a stator phase is excited. The general expression for instantaneous torque for such a device that operates under the reluctance principle is

$$T_{ph}(\theta,i_{ph}) = \left.\frac{\partial W'(\theta,i_{ph})}{\partial\theta}\right|_{i=\text{constant}} \tag{6.68}$$

where W' is the coenergy defined as

$$W' = \int_0^i \lambda_{ph}(\theta,i_{ph})di$$

Obviously, the instantaneous torque is not constant. The total instantaneous torque of the machine is given by the sum of the individual phase torques

$$T_{inst}(\theta,i) = \sum_{phases} T_{ph}(\theta,i_{ph}) \tag{6.69}$$

The SRM electromechanical properties are defined by the static $T-i-\theta$ characteristics of a phase shown in Figure 6.39. The average torque is a more important parameter from the user's perspective and can be derived mathematically by integrating Equation 6.69.

$$T_{avg} = \frac{1}{T}\int_0^T T_{inst}dt \tag{6.70}$$

The average torque is also an important parameter during the design process.

When magnetic saturation can be neglected, the instantaneous torque expression becomes

$$T_{ph}(\theta,i) = \frac{1}{2}i_{ph}^2\frac{dL_{ph}(\theta)}{d\theta} \tag{6.71}$$

The linear torque expression also follows from the energy conversion term (last term) in Equation 6.66. The phase current needs to be synchronized with the rotor position for effective torque production. For positive or motoring torque, the phase current is switched such that rotor is moving from the unaligned position toward the aligned position. The linear SRM model is very insightful in understanding these situations. Equation 6.71 clearly shows that for motoring torque, the phase current must coincide with the

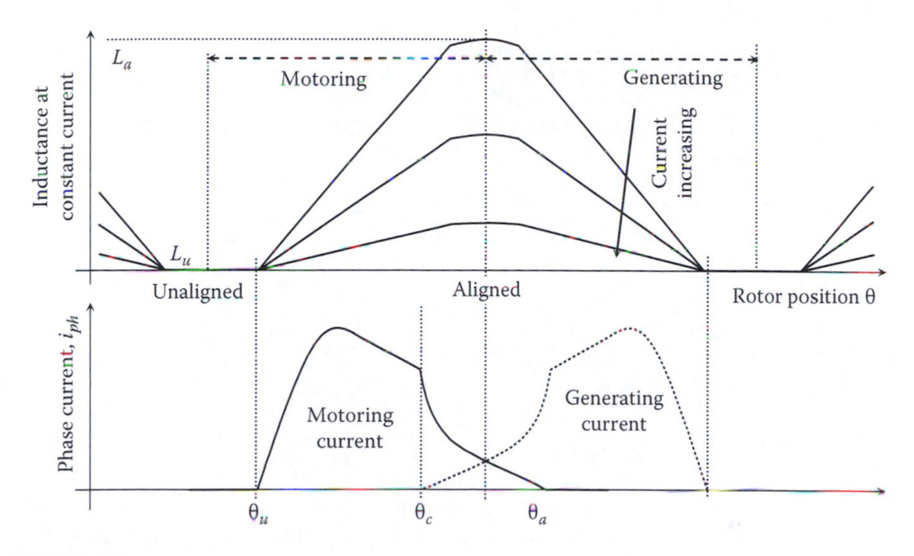

FIGURE 6.43
Phase currents for motoring and generating modes with respect to rotor position and idealized inductance profiles.

rising inductance region. On the other hand, the phase current must coincide with the decreasing inductance region for braking or generating torque. The phase currents for motoring and generating modes of operation are shown in Figure 6.43 with respect to the phase inductance profiles. The torque expression also shows that the direction of current is immaterial in torque production. The optimum performance of the drive system depends on the appropriate positioning of phase currents relative to the rotor angular position. Therefore, a rotor position transducer is essential to provide the position feedback signal to the controller.

6.6.3.4 Torque–Speed Characteristics

The torque–speed plane of an SRM drive can be divided into three regions, as shown in Figure 6.44. The constant torque region is the region below the base speed ω_b, which is defined as the highest speed when maximum rated current can be applied to the motor at rated voltage with fixed firing angles. In other words, ω_b is the lowest possible speed for the motor to operate at its rated power.

Region 1: In the low-speed region of operation, the current rises almost instantaneously after turn-on, since the back-emf is small. The current can be set at any desired level by means of regulators, such as hysteresis controller or voltage pulse-width modulation (PWM) controller.

As the motor speed increases, the back-emf soon becomes comparable to the DC bus voltage, and it is necessary to phase advance the turn-on angle

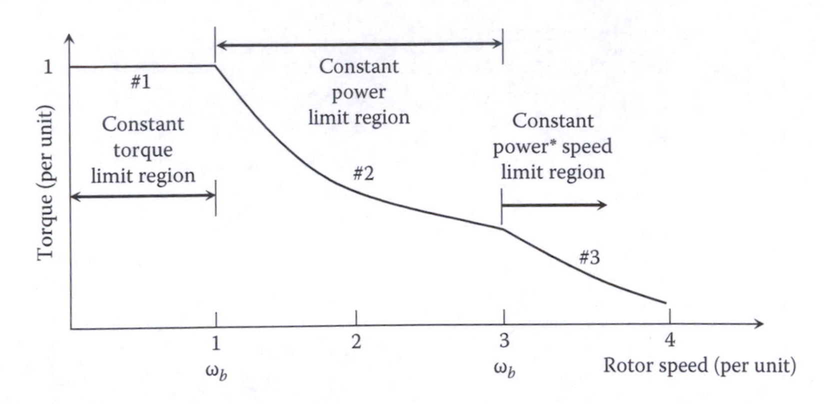

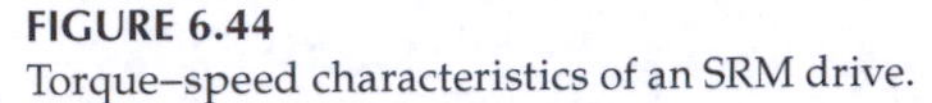

FIGURE 6.44
Torque–speed characteristics of an SRM drive.

so that the current can rise up to the desired level against a lower back-emf. Maximum current can still be forced into the motor by PWM or chopping control to maintain the maximum torque production. The phase excitation pulses also need to be turned off at a certain time before the rotor passes alignment to allow the freewheeling current to decay so that no braking torque is produced.

Region 2: When the back-emf exceeds the DC bus voltage in high-speed operation, the current starts to decrease once pole overlap begins and PWM or chopping control is no longer possible. The natural characteristic of the SRM, when operated with fixed supply voltage and fixed conduction angle θ_{dwell} (also known as the dwell angle), begins when the phase excitation time falls off inversely with speed and so does the current. Since the torque is roughly proportional to the square of the current, the rapid fall in torque with speed is countered by adjusting the conduction angle. Increasing the conduction angle increases the effective amps delivered to the phase. The torque production is maintained at a level high enough in this region by adjusting the conduction angle θ_{dwell} with the single-pulse mode of operation. The controller maintains the torque inversely proportional to the speed; hence, this region is called the constant power region. The conduction angle is also increased by advancing the turn-on angle until the θ_{dwell} reaches its upper limit at speed ω_p.

The medium speed range through which constant power operation can be maintained is quite wide and very high maximum speeds can be achieved.

Region 3: The θ_{dwell} upper limit is reached when it occupies half the rotor pole pitch, i.e., half the electrical cycle. θ_{dwell} cannot be increased further because otherwise the flux would not return to zero and the current conduction would become continuous. The torque in this region is governed by the natural characteristics, falling off as $1/\omega^2$.

The torque–speed characteristics of the SRM are similar to that of a DC series motor, which is not surprising considering that the back-emf is proportional to current, while the torque is proportional to the square of the current.

Problems

6.1 A 460 V, 60 Hz, 6 pole, 1176 rpm, Y-connected induction motor has the following parameters referred to the stator at rated condition:

$$R_s = 0.19\ \Omega,\ R_r = 0.07\ \Omega,\ X_s = 0.75\ \Omega,\ X_r = 0.67\ \Omega \text{ and } X_m = \infty$$

Find the speed of the motor for a braking torque of 350 N m and the inverter frequency of 40 Hz when the motor is supplied at rated voltage.

6.2 A three-phase induction machine is operated from a variable voltage, fixed frequency source. (a) Derive an expression for machine efficiency in terms of slip (not in terms of torque and speed). Include only stator and rotor copper losses and core loss in P_{loss}. Model core loss by a constant resistance in the equivalent circuit. To simplify the analysis, assume that core loss resistance and magnetizing reactance are large compared to the other parameters. Under this assumption, you can use an approximate equivalent circuit where the core loss resistance and magnetizing reactance are directly across the stator terminals.
(b) Does motor efficiency depend on terminal voltage? Calculate the slip that maximizes motor efficiency.

6.3 Find the condition of operation that minimizes the losses in a separately excited DC machine. (Start by writing an equation for P_{loss} in terms of the field currents and armature currents. Assuming linearity for all the nonlinear functions, establish the relation between armature current and field current and then find the condition for minimum P_{loss}.)

6.4 Present an argument for why it is impossible to achieve maximum efficiency at every operating point (T^*,ω^*) for a PM DC machine. (Start by writing an equation for P_{loss} in terms of T, ω, and machine flux ϕ.)

6.5 Proceeding as in Problem 6.2, explain why it is impossible to minimize losses at any operating point (T^*,ω^*) for a series DC motor.

6.6 (a) A PM brushless DC has a torque constant of 0.12 N m/A referred to the DC supply. Estimate its no-load speed in rpm when connected to a 48 V DC supply.
(b) If the armature resistance is 0.15 Ω/phase and the total voltage drop in the controller transistors is 2 V, determine the stall current and the stall torque.

6.7 Consider a three-phase 6/8 SRM. The stator phases are excited sequentially with a total time of 25 ms required to excite all the three phases. Find the angular velocity of the rotor. Express your answer both in rad/s and rpm.

6.8 The following flux equation describes the nonlinear characteristics of a three-phase, 6/4 SRM:

$$\lambda_j(i,\theta) = \lambda_s(1 - \exp(-i_j f_j(\theta)), \quad i_j \geq 0$$

where λ_S = saturation flux = 0.2 V s and $f(\theta) = a + b^* \cos(N_r\theta - (j-1)2\pi/m)$.

Here, j = 1, 2, 3 denotes the phase number and m = 3. Also, given a = .024 and b = .019

(a) Derive the expression for the phase torque $T_j(i,\theta)$.

(b) Plot the λ–i–θ characteristics for six angles between and including the unaligned and aligned positions. Take the maximum current as 100 A.

(c) Plot the T–i–θ characteristics between the unaligned and aligned rotor positions. Take the maximum current as 100 and 10 A current steps for the torque characteristics.

References

1. G. Dubey, *Power Semiconductor Controlled Drives*, Prentice Hall, Englewood Cliffs, NJ, 1989.
2. R.H. Park, Two-reaction theory of synchronous machines—Generalized method of analysis—Part I, *AIEE Transactions*, 48, 716–727, July 1929.
3. P. Vas, *Electric Machines and Drives: A Space-Vector Theory Approach*, Oxford University Press, Oxford, U.K., 1992.
4. D.W. Novotny and T.A. Lipo, *Vector Control and Dynamics of AC Drives*, Oxford University Press, Inc., New York, 1996.
5. N. Mohan, *Electric Drives—An Integrated Approach*, MNPERE, Minneapolis, MN, 2001.
6. T.J.E. Miller, *Brushless Permanent Magnet and Switched Reluctance Motor Drives*, Oxford University Press, Oxford, U.K., 1989.
7. P.J. Lawrenson, J.M. Stephenson, P.T. Blenkinsop, J. Corda, and N.N. Fulton, Variable-speed switched reluctance motors, *IEE Proceedings*, Pt. B, 127(4), 253–265, July 1980.
8. T.J.E. Miller, *Switched Reluctance Motors and Their Control*, Magna Physics Publishing, Hillsboro, OH and Oxford Science Publications, Oxford, U.K., 1993.
9. M.N. Anwar, I. Husain, and A.V. Radun, A comprehensive design methodology for switched reluctance machines, *IEEE Transactions on Industry Applications*, 37(6), 1684–1692, November–December 2001.

7

Power Electronic Converters

Power electronic circuits convert electrical energy from one voltage level and frequency to another for an electrical load. The bases of the power electronics circuits are the power semiconductor devices. Power devices act as an electronic switch in a power circuit to change circuit configuration. Power devices are available commercially since the 1970s and have seen tremendous advancements over the past 40 years. These devices are now available with ratings up to thousands of volts and thousands of amperes in a single package. Two or six switch configurations, suitable for certain power converter topologies, are available in a single package.

The power devices are connected to form a circuit or topology that can assume two or more configurations depending on the device-switching states. Commands generated in a controller circuit bring about the changes in the switching states. These power electronic circuit topologies are known as power converter circuits. Additional energy storage circuit elements of inductors and capacitors are often used in the power converter circuits. Resistors are not used in power converters to avoid losses, but parasitic resistances of power devices, storage elements, wires, and connectors invariably exist. The converters process power based on signals from a control unit. Four basic power converter circuits are

- DC/DC converter: Converts DC voltages from one level to another.
- DC/AC inverter: Converts DC voltage to AC voltages. The AC voltages can be square wave or sinusoidal voltages composed of pulse width modulated (PWM) signals.
- AC/DC converter or rectifier: Rectifies AC voltages to DC voltages.
- AC/AC converter: Converts AC voltages at one level and frequency directly into another voltage level and frequency. These are known as cycloconverters.

In this book, the power converter circuits used in the electric and hybrid vehicle powertrain and high-voltage segments are discussed. The DC/DC converters are used both in the electric power transmission path as well as for power supply to the 12 V electronics. The DC/AC inverters are used for the electric drive for AC electric machines. There are many different topologies available for DC/DC power conversion. The converter topologies commonly useful in electric and hybrid vehicles are discussed in detail in this book. The DC/DC converters used for static power conversion are presented in this chapter; the DC/DC converters for DC electric motor drives and DC/AC inverters for AC electric motor drives are covered in Chapter 8.

7.1 Power Electronic Switches

The power electronic devices or switches are of either two or three terminals; the third terminal, when it exists, is for the small signal control input. Device operation switches from "on" to "off" state or vice versa based on either signal changes to its control terminal or changes in circuit conditions. Two-terminal devices change state when circuit conditions dictate, whereas three-terminal device change state with changes in its control input.

Let us consider an ideal controlled switch shown in Figure 7.1a prior to discussing the practical power devices. The switch has three terminals for power connections and a control input. Switch operation is defined in terms of quadrants established by the voltage and current axes. The quadrants are labeled in Figure 7.1b in the counterclockwise direction of the i–v plane of switch conduction. The ideal switch shown is a four-quadrant switch, which means that it can handle bidirectional current as well as bidirectional voltage. In the ideal switch, there is no voltage drop across the device when it conducts. Ideally, the operating point will be along the axes with the ideal switches either carrying current with zero voltage drop across it or blocking a voltage with zero current flowing through it. The ideal switch also turns on and off instantaneously without any delay; consequently, ideal switch has no power loss. The practical semiconductor switch differs considerably from the ideal switch. Practical switches have conduction voltage drops and require

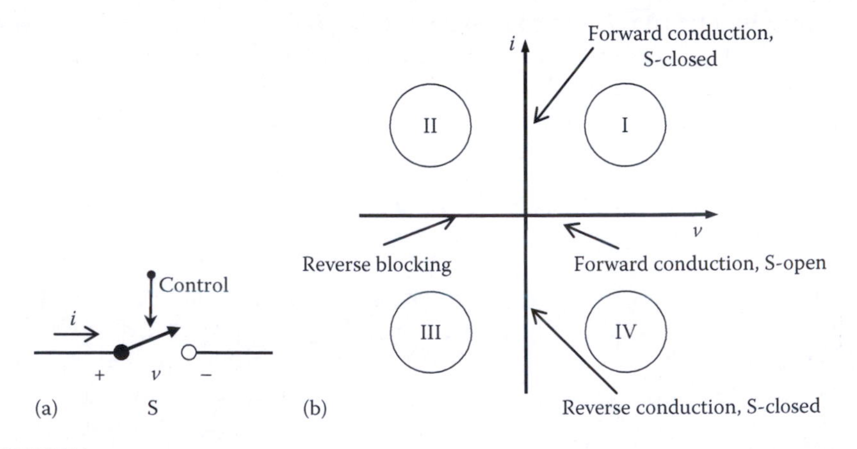

FIGURE 7.1
Ideal switch and its conduction characteristics. (a) Switch symbol; (b) switch characteristics in four quadrants.

finite time to turn-on and turn-off; furthermore, practical devices do not have four-quadrant capability unless it is combined with one other device.

Each of the practical devices used in power converters have their own specific characteristics [1,2]. The particular application and its power requirements determine the type of device to be used in a converter topology. The bipolar junction transistors (BJTs) have higher power ratings and excellent conduction characteristics, but the base drive circuit is very complicated, since these are current-driven devices. On the other hand, the metal oxide semiconductor field effect transistors (MOSFETs) are voltage-driven devices, and hence the gate drive characteristics are much simpler. The switching frequency of a MOSFET is much higher compared with BJT, but the maximum available device power ratings would be much smaller for the former. The insulated gate bipolar transistor (IGBT) device, invented in the early 1980s, combines the attractive features of both the MOSFET and the BJT. The IGBT is the device of choice today in most medium- to high-power applications due to their availability in high-power ratings. Previously, when DC motors were the primary machine choice, the high-power converters were typically made of silicon controller rectifiers (SCRs), which are available in very high-power ratings. However, unlike the other devices, SCRs cannot be turned off through a gate signal and a commutation circuit is required to turn them off. The gate turn-off thyristor (GTO) device is similar to SCR except that it can be turned off through a gate signal, although the current required in the gate signal to turn them off is typically four to five times the current required to turn them on. Attempts to combine the gating characteristics of a MOSFET and the conduction characteristics of an SCR resulted in the device called MOS-controlled thyristor (MCT) in the late 1980s. However, their failure under certain conditions did not make these devices popular. In addition to the switches mentioned above, there is an additional two-terminal device called diode, which is universally used in all converters. The diodes are used in conjunction with other controlled devices in the power converter to provide current paths for inductive circuits or for blocking reverse voltages. The important features of the devices discussed above are summarized in Table 7.1. Further details of the operating characteristics of the devices that have a significance in electric and hybrid vehicle applications are discussed in the following text.

7.1.1 Diode

The diode is a two-terminal, uncontrollable switch, which is turned on and off by the circuit. A positive voltage across the anode and the cathode of the diode turns the device on allowing current conduction up to its rated value. There will be a small forward voltage drop during diode conduction, as shown in Figure 7.2. The diode conducts current in one direction only and blocks voltage in the negative direction, which makes it a quadrant II switch. The diode can block a reverse voltage up to its breakdown level.

TABLE 7.1
Summary of Power Devices

Name	Symbol	Turn-On[a]	Turn-Off[a]	Comments
Diode	Cathode, Anode	Positive anode to cathode voltage	Reverse anode current Recovery time before turning off	Turn off and on depend on circuit conditions High-power capabilities
SCR	Cathode, Gate, Anode	Small gate pulse (current) Slow to medium turn-on time (~5 μs)	Anode current goes below holding Delay time before forward voltage can be applied (10–200 μs)	Very high power Needs additional circuit to turn-off On voltage ≈ 2.5 V
GTO	Cathode, Gate, Anode	Small gate pulse (current) Slow to medium turn-on time (~10 μs)	Remove charge from gate (medium current) Medium speed (~20 μs)	High power Easier to turn off than SCR On voltage ≈ 2.5 V
BJT	Collector, Base, Emitter	Medium current to base to turn-on Medium speed (~1 μs)	Remove current from base Medium speed (~5 μs)	Medium power Easy to control Medium drive requirements On voltage ≈ 1.5 V
MOSFET	Drain, Gate, Source	Voltage to gate (v_{GS}) Very high speed (~0.6 μs)	Remove voltage from gate Very high speed (~0.5 μs)	Low power Very easy to control Simple gate drive requirement High on losses ≈ 0.1 Ω on resistance
IGBT	Collector, Gate, Emitter	Voltage to gate (v_{GE}) High speed (~0.2 μs)	Remove voltage from gate High speed (~0.5 μs)	Medium power Very easy to control On voltage ≈ 3.0 V Combines MOS and BJT technologies

[a] The turn-on and turn-off times, and the on-state voltage drops are nominal values. The actual values depend on the particular device and its ratings.

7.1.2 Power BJT

The power BJT is a three-terminal controlled switch. The circuit symbol and the i–v characteristics of a BJT are shown in Figure 7.3. When sufficient positive base current i_B flows through the base of an npn-BJT, the transistor action allows large positive collector current i_C to flow through the junctions

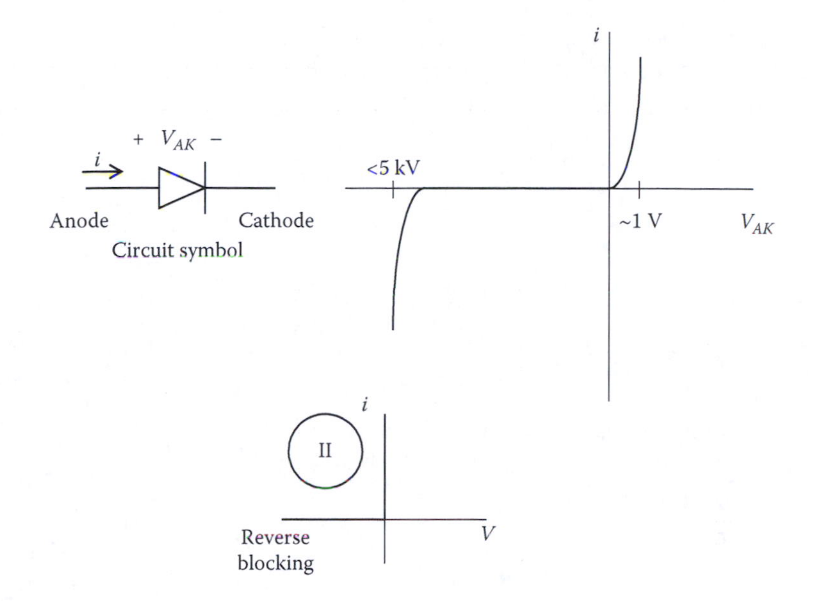

FIGURE 7.2
Diode symbol, characteristics, and operating quadrant.

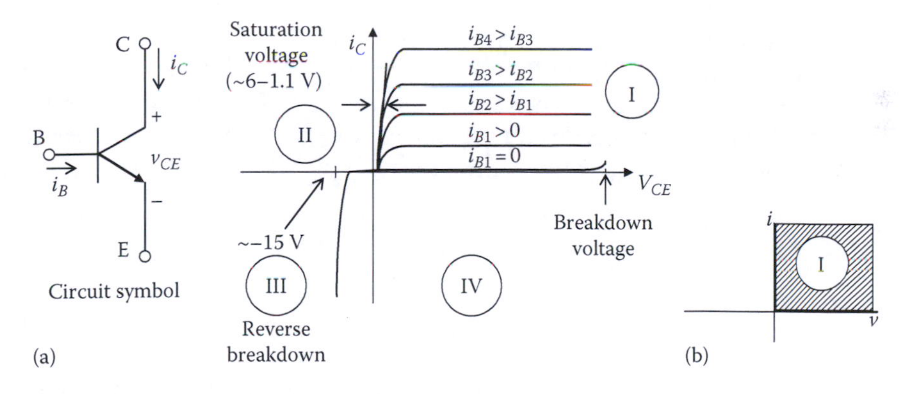

FIGURE 7.3
An npn power BJT: (a) device symbol and i–v characteristics and (b) operating quadrant.

of the device with a small positive collector to emitter voltage v_{CE} (knows as saturation voltage) drop. The amplitude of the base current determines the amplification in the collector current. The power transistors are always operated as a switch either at saturation with high enough base current or at cutoff with zero base current. The power BJT is a controllable switch that can be turned on or off with the help of the base current. The device allows forward current or blocking voltage, and hence is a quadrant I switch. Steady state operations in quadrants II and IV are not possible.

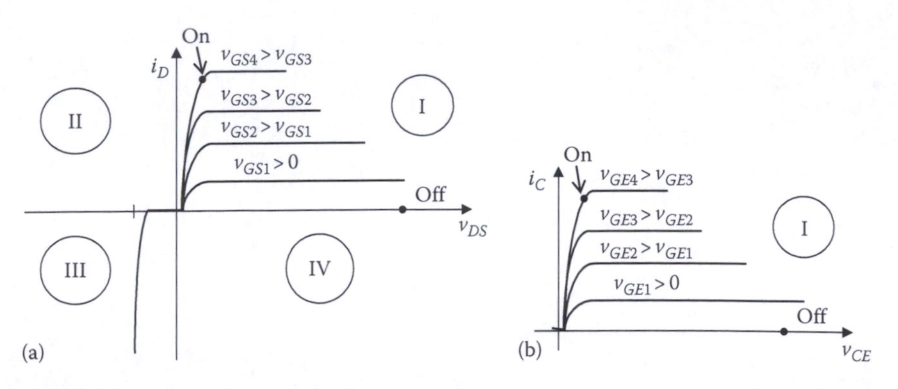

FIGURE 7.4
Current–voltage characteristics of (a) MOSFET and (b) IGBT.

7.1.3 Power MOSFET

The power MOSFET is a controlled switch similar to the power BJT. The three terminals of the MOSFET are gate, drain, and source, which are analogous to the base, collector, and emitter, respectively, of the BJT. Similar to the power BJT, the MOSFET is a quadrant one switch allowing forward flow of current and blocking forward voltage between drain and source. The device works as a controlled switch through the control of the voltage between gate and drain. The current–voltage characteristics of a MOSFET are illustrated in Figure 7.4a. Unlike the BJT, MOSFET is a voltage-driven device with much faster switching capabilities. This allows power MOSFET operation at much higher frequencies compared with the BJT, although the available maximum device voltage and current ratings are much smaller than those of the BJT.

7.1.4 IGBT

The excellent conduction characteristics of the power BJT has been combined with the excellent gate characteristics of the power MOSFET in the IGBT. The IGBT has high input impedance like that of a MOSFET and low conduction losses like that of a BJT. There is no second breakdown problem in the IGBT. By chip design and structure, the equivalent drain-to-source resistance R_{DS} is controlled to behave like that of a BJT. The conduction characteristics of an IGBT are shown in Figure 7.4b.

7.1.5 Bidirectional Switch

It was mentioned in the earlier chapters that electric and hybrid vehicles can recapture energy during vehicle braking or when it slows down through regenerative mode of operation of the electric machine. The energy recovered is used to recharge the batteries. For the regenerative mode of operation, power converters need to allow bidirectional power flow, which requires the

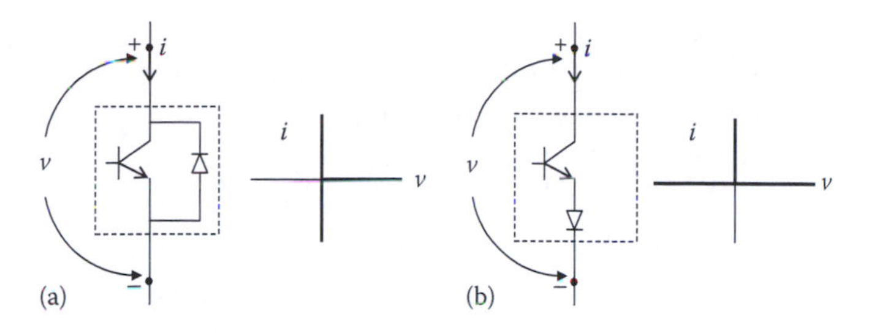

FIGURE 7.5
(a) Bidirectional current switch; (b) bidirectional voltage switch.

power devices or switches of the DC/DC or DC/AC converter topologies to be bidirectional. Figure 7.5 gives an example of how bidirectional switches can be made by combining a BJT and a diode.

7.2 DC/DC Converters

The DC/DC converters change the system voltages from one level to another. The input to the converter is a filtered DC voltage, although it may be unregulated. The output is a regulated DC voltage; multiple outputs are also designed for many applications. The converters have inductive and capacitive storage elements in addition to the switching devices. The switching frequencies of the power devices are very high, often reaching as high as few hundred kilohertz. The high switching frequency helps minimize the size of the storage elements. The efficiencies of these switching converters are also very high in the upper 90 percentages.

Depending on whether the transformer isolation is used, the switching converters can be broadly classified into isolated and non-isolated DC/DC converters. The basic non-isolated DC/DC converter has a single switch and a single diode; these types of converters may also have one inductor and one capacitor as the energy storage elements. There are also other types of non-isolated DC/DC converters that use two switches and two diodes in the circuit topology and additional energy storage elements. A lot of applications, including certain ones in electric and hybrid vehicles, require electrical isolation between the input and the output. The isolated DC/DC converters are derived from the basic topologies, but have high-frequency transformer isolation.

The high-power DC/DC converters are primarily used in three places of electric and electric vehicles: voltage boost for electric powertrain, high- to low-voltage interface for 12 V electronics, and converters for energy storage cell balancing. A non-isolated DC/DC converter is often used in the powertrain of electric and hybrid vehicles to boost the battery voltage to higher levels for the propulsion electric motor drive. The 12 V electronics in the electric and hybrid vehicles are supplied with a high- to low-voltage DC/DC converter, which needs to be of isolated type. These non-isolated and isolated converter topologies are discussed below. The DC/DC converters used in energy storage cell balancing are discussed in Section 7.3.

7.2.1 Non-Isolated DC/DC Converters

DC/DC converters can be designed to either step-up or step-down the input DC voltage. Depending on the location of the two switches and the inductor and capacitor, the three basic second-order DC/DC converter topologies are buck, boost, and buck–boost converters. A buck converter is a step-down converter, a boost converter is a step-up converter, and a buck–boost converter can be operated in either mode. Improved topologies are available with two switching devices and two diodes, and additional storage elements. Consequently, the order of the converter also increases with the additional components. Examples of higher order converters are Cuk and SEPIC converters. All of these DC/DC converters are used to convert an unregulated DC voltage into a regulated DC output voltage; hence, the converters are known as switching regulators.

The DC/DC converters can be operated either in the *continuous conduction mode* (CCM) or in the *discontinuous conduction mode* (DCM). The conduction relates to the inductor current in the DC/DC converters. If the inductor current remains continuous and is above zero at all times, then the converter is in CCM. If the lowest value of the inductor current is zero, then the converter is in DCM. The converter operations are explained below with the help of CCM waveforms.

7.2.1.1 Buck Converter

The circuit topology of a buck converter with a MOSFET implementation is shown in Figure 7.6a. A BJT could also be used for the switch S. The voltage and current waveforms in the circuit are shown in Figure 7.6b. The average output voltage $\langle v_o \rangle$ is smaller than the input DC voltage V_{in}; hence, the circuit is known as a buck converter. The operation of the converter primarily involves the switching of the power device according to the controller command. The transistor switch is turned on for a duty cycle D defined as

$$D = \frac{\text{Switch-on time}}{\text{Time period } (T)} \tag{7.1}$$

The time period T is fixed and sets the switching frequency of the converter. The circuit operation has two modes: (1) when the switch S is on and (2) when the switch is off. The switch is on for a period of DT s and off for a period of $(1-D)T$ s. The inductor current increases when the switch is on, thereby charging the inductor. When the switch is off, the diode is forced to

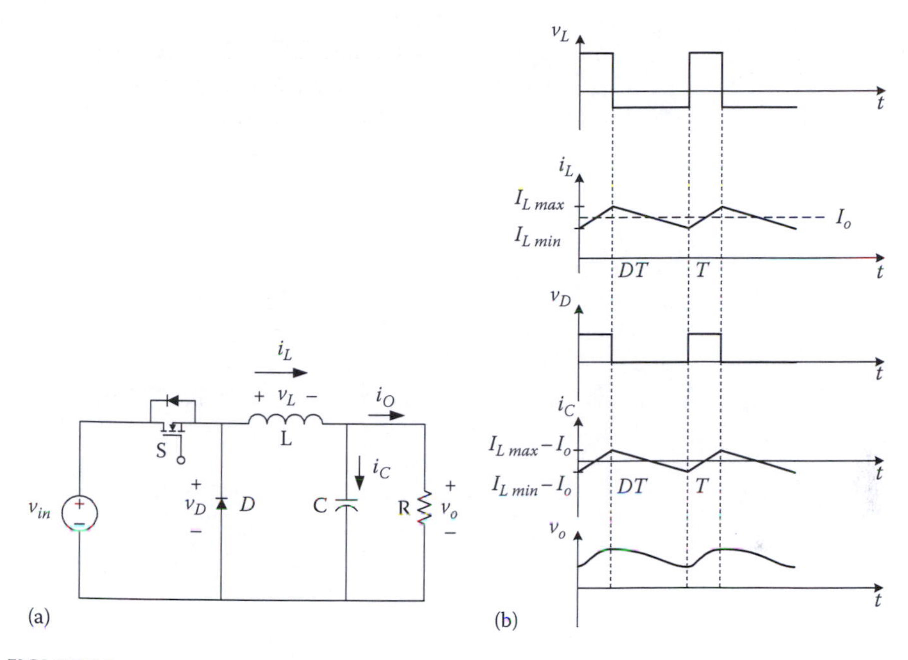

FIGURE 7.6
(a) Buck converter with MOSFET-diode implementation; (b) voltage and current waveforms.

turn-on to maintain the continuity of the inductor current; the inductor gets discharged and its current decreases. Through the charging and discharging of the inductor, energy gets transferred from the input to the output. The filter capacitor C along with the output resistor R provides a very large time constant compared with the switching frequency. Hence, the output voltage ripple will be very low; typically, the ripple is less than 1%.

The input and output average voltages and currents are related to the duty cycle as

$$\frac{\langle v_o \rangle}{V_{in}} = \frac{\langle i_{in} \rangle}{I_o} = D \tag{7.2}$$

where

$\langle v_o \rangle$ and $\langle i_{in} \rangle$ are the average output voltage and the average input current, respectively

$I_o = \langle i_o \rangle$ is the average output current

7.2.1.2 Boost Converter

A boost converter steps up a DC voltage from one level to another. The circuit topology of a boost converter with a MOSFET implementation is shown

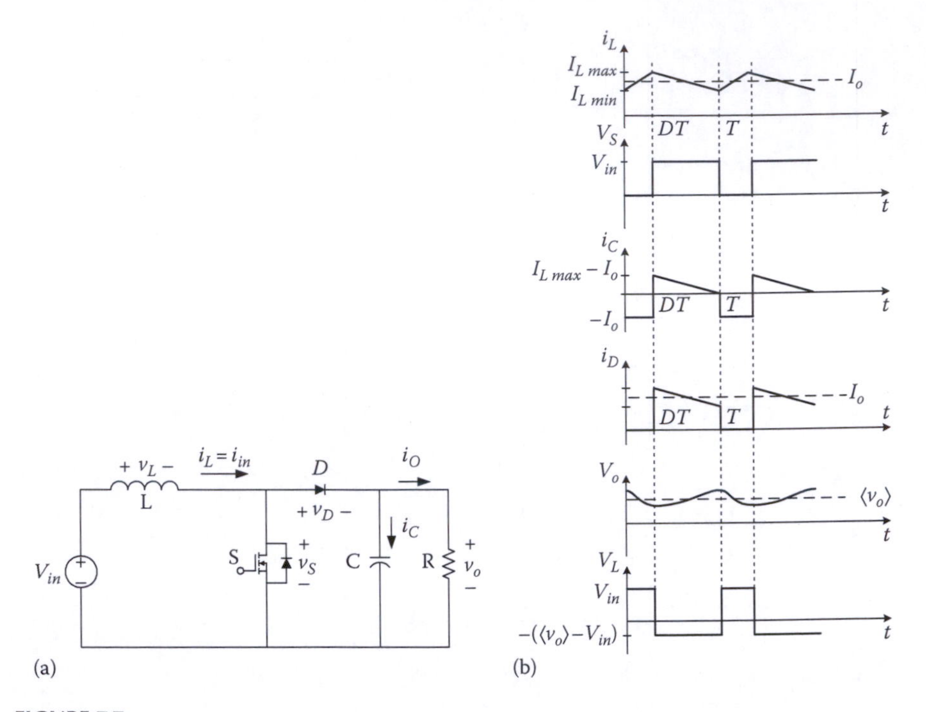

FIGURE 7.7
(a) Boost converter with MOSFET-diode implementation; (b) voltage and current waveforms.

in Figure 7.7a. A BJT could also be used for the switch S. The current and voltage waveforms are shown in Figure 7.7b. The circuit has two modes of operation in a switching cycle, depending on whether the switch is on or off. When the switch is on for a period of DTs, the input current i_{in} increases, thereby storing energy in the inductor. Input current is the same as the inductor current i_L. When the switch is turned off, the diode becomes forward biased and starts conducting. Energy from the inductor flows to the load. The inductor current decreases until the end of the period when the switch is turned on again. A large filter capacitor C is required compared with the buck converter since high rms current flows through it.

The input and output average voltages and currents are related to the duty cycle as

$$\frac{\langle v_o \rangle}{V_{in}} = \frac{\langle i_{in} \rangle}{I_o} = \frac{1}{1-D} \tag{7.3}$$

where

$\langle v_o \rangle$ and $\langle i_{in} \rangle$ are the average output voltage and the average input current, respectively

$I_o = \langle i_o \rangle$ is the average output current

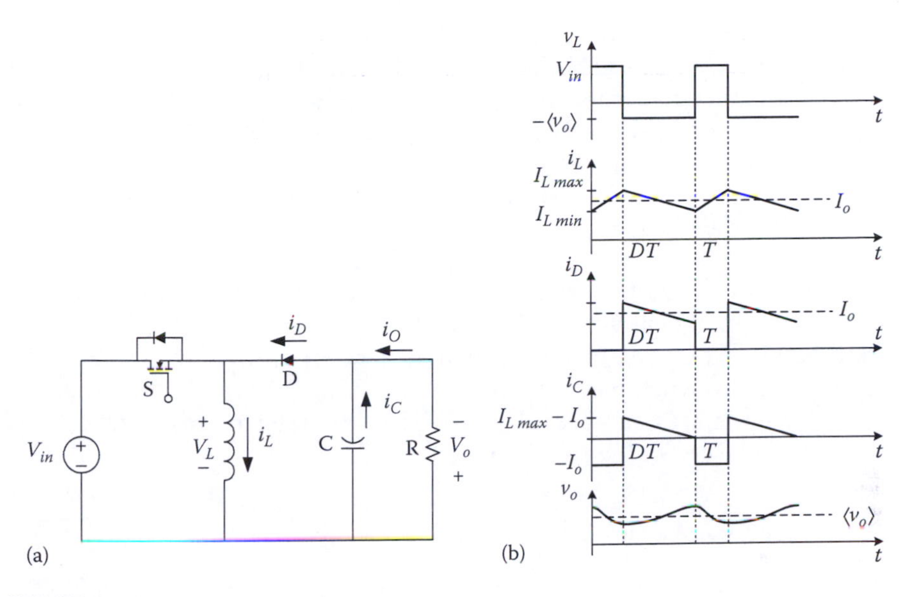

FIGURE 7.8
(a) Buck–boost converter with MOSFET-diode implementation; (b) voltage and current waveforms.

7.2.1.3 Buck–Boost Converter

The buck–boost converter topology can either step up or step down the input voltage; hence, the name of the converter. The circuit topology of a buck–boost converter with a MOSFET implementation is shown in Figure 7.8a. The current and voltage waveforms are shown in Figure 7.8b. Similar to the buck and boost converters, the buck–boost converter also has two modes of operation in one switching cycle. When the switch S is on for a period of *DT*s, the input current flows through the inductor storing energy in it. The diode is reverse biased in this mode, and the filter capacitor C maintains the load current. When the switch is off for the period $(1-D)T$ s, the diode becomes forward biased and the stored inductor energy is transferred to the load. The inductor current decreases until the switch is turned on again in the CCM. The output voltage polarity of a buck–boost converter is inverted.

The input and output average voltages and currents are related to the duty cycle as

$$\frac{\langle v_o\rangle}{V_{in}}=\frac{\langle i_{in}\rangle}{I_o}=\frac{D}{1-D} \tag{7.4}$$

where

$\langle v_o\rangle$ and $\langle i_{in}\rangle$ are the average output voltage and the average input current, respectively

$I_o=\langle i_o\rangle$ is the average output current

The CCM of the basic converters have been discussed so far. The boundary between the CCM and DCM depends on the size of the inductor relative to the other parameters of the circuit. The converters enter the DCM when the inductor size is reduced below the critical size L_{crit}, provided other parameters remain the same. The smaller the L_{crit} value, the wider is the choice for inductor design. For the same switching frequency and load resistance, the buck converter has the highest L_{crit}, while the boost converter has the smallest L_{crit}.

In the DCM, the minimum inductor current in each of the three basic topologies is zero. For a short interval, the inductor current remains zero. The DC/DC converters frequently enter DCM, since they often operate under no load conditions. In the DCM, the voltage gain is not only a function of *D*, but also of switching frequency, load, and circuit parameters; this makes the analysis in DCM much more complicated. The readers are referred to the references [2,3] for the analysis of the DCM operation.

7.2.1.4 Fourth-Order DC/DC Converters

The DC/DC converters presented so far have only one switch and a diode, and are of second order. The minimal switch number ensures high efficiency of the converters, although the stress on the switches will also be high. However, the basic converters are not suitable under certain input–output conditions, especially when the input voltage has a wide range. Cascade and cascode combinations of the basic converters result in other attractive DC/DC converter topologies such as the SEPIC and Cuk converters [3]. These converters have two inductors and two capacitors; hence, the order of the converter is fourth.

7.2.1.5 Powertrain Boost Converter

A boost-type non-isolated DC/DC converter is used in the electric power transmission path in many electric and hybrid vehicles. The DC/DC converter boosts the battery-pack output voltage to a higher level for the efficient operation of the traction electric drive in a battery-equipped electric or hybrid electric vehicle. The converter needs to be bidirectional to recover and recharge the battery pack during vehicle regenerative braking; however, it does not have to be isolated. The higher DC-link voltage enables high-speed operation of the electric machine, which helps minimize its size and increases its power density. The other advantages of adding the boost converter are smaller battery-pack size, reduction in filter choke and DC-link capacitor sizes of the motor drive, and reduction in wiring harness size due to lower current requirements. Regulated output voltage is possible to the extent desired in these converters. The drawbacks of adding the boost segment in the electric powertrain are increase in cost, requirement of an input side filter, and increase in the number of failure points.

The boost DC/DC converter is essential in the powertrain of a fuel cell electric vehicle since the output voltage of a fuel cell is typically low and unregulated. The DC/DC converter increases the voltage level and regulates the output voltage. This converter topology needs to be of unidirectional only since regenerative fuel cells are yet to be developed.

The DC/DC converters can also be used to interface between two energy storage devices in an electric or a hybrid vehicle. For example, when a battery–ultracapacitor combination is used for the energy storage system, one of the components is connected directly to the DC bus, while the other is buffered through a DC/DC converter. The ultracapacitor can be used to capture as much of regenerative braking energy as possible, since its power density is high; the battery pack can be designed to provide the zero emission range of the vehicle.

The half-bridge bidirectional non-isolated DC/DC converter, derived from the basic topologies, is one common converter topology used for boost operation of the main DC bus. The converter topology is shown in Figure 7.9a. Power flows from the low-voltage side (which would be the battery-pack side) to the high-voltage DC-link side in the boost mode with lower switch S2 and the diode D1 in operation. In buck mode, upper switch S1 and the diode D2 come into operation; controlling S1 allows power to flow from the high-voltage DC-link side to the battery pack.

The two other converter topologies suitable for the mains DC-boost are shown in Figure 7.9b and c [4]. Figure 7.9b shows the Cuk converter that boosts voltage when power flows from V_{in} to DC bus V_{DC}. In this boost mode, S1 and D2 are active, and S2 and D1 are inactive. In the buck mode, voltage is stepped down from V_{DC} to V_{in} when S2 and D1 are active, and S1 and D2 are inactive. C_t and C2 are transfer capacitor and DC-link capacitor, respectively. Figure 7.9c shows the combined SEPIC/Luo converter, which operates as a SEPIC converter when voltage boost operation is required and as a Luo converter when buck operation is required. In the SEPIC operation mode, power flows from low-voltage battery-pack side to high-voltage DC-link side boosting the input battery voltage V_{in}. In this mode, S1 and D2 are active, and S2 and D1 are inactive. In the Luo operation mode, power flows from the high-voltage DC-link side to the low-voltage battery-pack side stepping down the input DC-link voltage. In this mode, S2 and D1 are active, and S1 and D2 are inactive. Table 7.2 shows the stress on active and passive components for the three-converter topologies [4]. For comparison, the advantages of half-bridge converter over Cuk and SEPIC/Luo converters are a smaller size of inductor, reduced voltage and current ratings of active components, less number of components, less point of failure, and higher efficiency. The main drawback of half-bridge is that the output current might become discontinuous due to small inductor size, which impacts the size of the DC-link capacitor.

7.2.2 Isolated DC/DC Converters

The DC/DC converters can provide electrical isolation between the input and the output by incorporating a transformer in the power stage. These

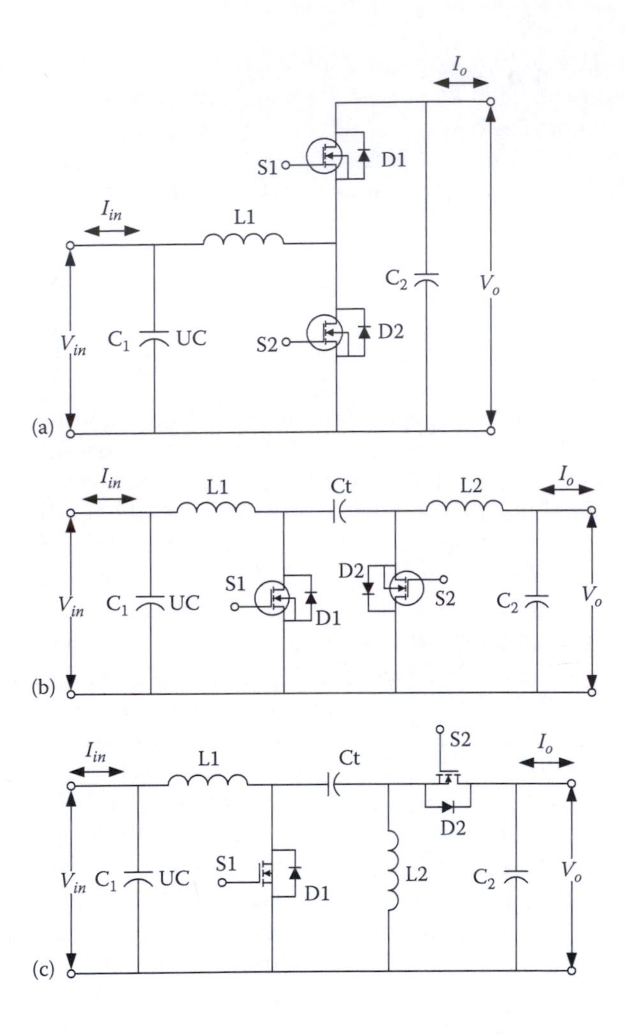

FIGURE 7.9
Non-isolated DC/DC converters: (a) half-bridge, (b) Cuk, and (c) SEPIC/Luo converters.

transformers operate at very high frequencies and are much smaller in size and weight compared with line frequency transformers. A transformer stage in DC/DC converters is also highly suitable in designs where the input–output voltage levels differ by a large ratio. The transformer in DC/DC converters also assist in reducing device-switching stress. The output voltage polarity can be easily set with the transformer isolation. The main disadvantage of adding the transformer isolation is the increased volume and mass. The converter efficiency also reduces slightly, and the controller has to be carefully designed to avoid core saturation with transformers.

The isolated DC/DC converters can be derived from either the buck topology or the boost topology. In addition, depending on the number of switches used, the DC/DC converters can be single-switch type or multiple-switch

TABLE 7.2

Stresses on Active and Passive Components in Non-Isolated DC/DC Converters

	Half-Bridge	Cuk	SEPIC/Luo
Transfer capacitor voltage rating	—	$V_{in} + V_o$	V_{in}
Switch voltage rating	V_o	$V_{in} + V_o$	$V_{in} + V_o$
Diode voltage rating	V_o	$V_{in} + V_o$	$V_{in} + V_o$
Inductor average current	$\frac{I_o}{1-d_1}$, Boost	$I_{L1} = I_o \frac{d_1}{1-d_1}$	$I_{L1} = I_o \frac{d_1}{1-d_1}$
	$\frac{I_o}{d_2}$, Buck	$I_{L2} = I_o$	$I_{L2} = I_o$, Boost
			$I_{L1} = I_o \frac{(1-d_2)}{d_2}$
			$I_{L2} = I_o$, Buck

Note: d_1 and d_2 are duty cycles for boost and buck operations, respectively.

type. The single-switch converters are similar to the buck or boost converter, but with a transformer isolation. The two single-switch isolated DC/DC converters are the forward and flyback converters. The forward converter is derived from the buck topology, whereas the flyback converter is derived from the boost topology. These are the simplest isolated DC/DC converters with the lowest component count. However, the magnetic components are used in a unipolar, single-quadrant mode. The switching stress on the power device is $2V_{in}$. These topologies are used in low power applications, about 100 W or less.

The multiswitch DC/DC converters fully utilize the transformer magnetics and are used in higher power applications. Depending on the number of switches and the type of transformer used, there are three types of isolated multiswitch DC/DC converters: push–pull, half-bridge, and full-bridge. The input side of push–pull converters uses a center-tapped transformer, while half-bridge and full-bridge converters use noncenter-tapped transformers. The secondary side of the transformers can use a half-bridge diode configuration with a center tap or a full-bridge diode configuration without a center-tap. The multiswitch converters can be either buck-derived or boost-derived. The full-converter topology is the most suitable for applications exceeding 500 W. The common high-power DC/DC converter in electric and hybrid vehicle applications for the high- to low-voltage interface is of the full-bridge converter type.

The circuit topology of a buck-derived push–pull converter is shown in Figure 7.10. The circuit uses two switches on the input side and an inductor for energy storage on the output side. The center-tapped transformer is used for voltage scaling and electrical isolation. Switches S1 and S2 are turned on and off alternately with equal duty ratios within one switching period.

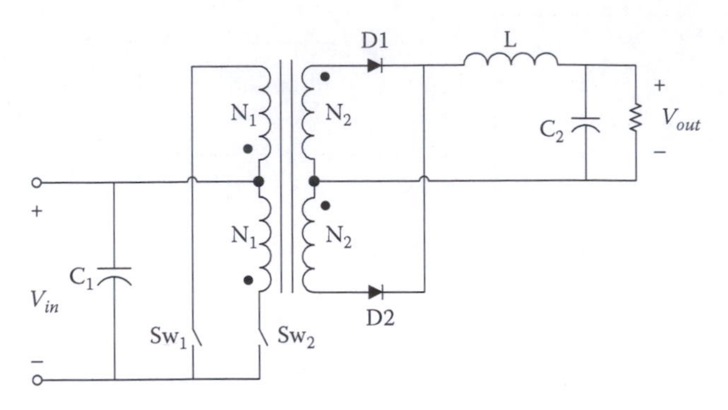

FIGURE 7.10
Push–pull isolated DC/DC converter.

The duty ratio for each switch can be varied between 0 and 0.5. A minimum dead time when both switches are off is guaranteed by the controller between the switching transitions of the two devices to avoid the possibility of shoot-through across the supply voltage. The voltage gain of the push–pull converter is

$$\frac{V_o}{V_{in}} = 2\frac{N_2}{N_1}D, \quad 0 < D < 0.5$$

The converter topology is simpler than the half-bridge and full-bridge topologies, but the main disadvantage is that the diode and power device-switching stresses are in excess of $2V_{in}$. In addition, the transformer core is prone to saturation due to any slight non-ideality in the switching pattern. The push–pull isolated converter is used in medium-power applications up to about 500 W.

The half-bridge converter use active switches to generate symmetrical AC waveforms at the primary side of the transformer. The transformer core flux is excited with bidirectional waveforms, which gives better core utilization and makes it less prone to flux saturation. The circuit topology of a half-bridge converter with a center-tapped secondary is shown in Figure 7.11. The switches operate the same way as in the push–pull converter, and the maximum duty cycle for each switch is 0.5. The voltage gain of the half-bridge converter is

$$\frac{V_o}{V_{in}} = \frac{N_2}{N_1}D, \quad 0 < D < 0.5$$

The maximum blocking voltage for each switch in the half-bridge converter is V_{in} rather than $2\,V_{in}$ like that in the push–pull converter. The half-bridge converter is the most commonly used isolated DC/DC converter in the

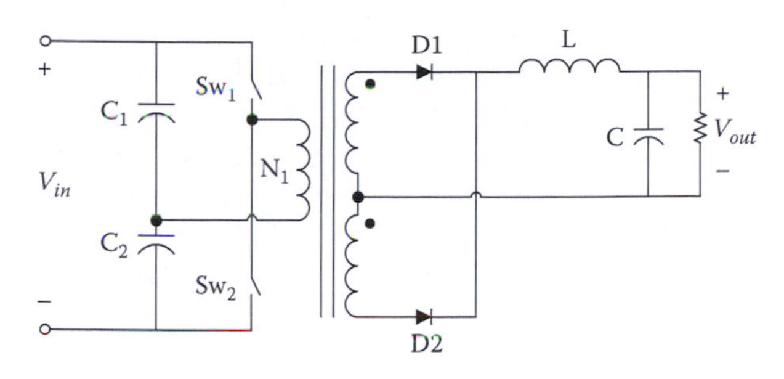

FIGURE 7.11
Half-bridge isolated DC/DC converter.

medium-power range. The switching device stress is V_{in}, and the switch count is less than the full-bridge converter.

The full-bridge converter topology also generates symmetrical AC waveforms at the primary side of the transformer, but with four active switches. The circuit topology of a full-bridge converter with a full diode bridge at the secondary side is shown in Figure 7.12a. The switches are operated in pairs; switches S1–S4 and S2–S3 are switched alternately with the maximum duty ratio for each switch being 0.5. A deadtime is also inserted between the transitions from one pair to another to avoid shoot-through across the supply voltage. The voltage gain of the full-bridge converter is

$$\frac{V_o}{V_{in}} = 2\frac{N_2}{N_1}D, \quad 0 < D < 0.5$$

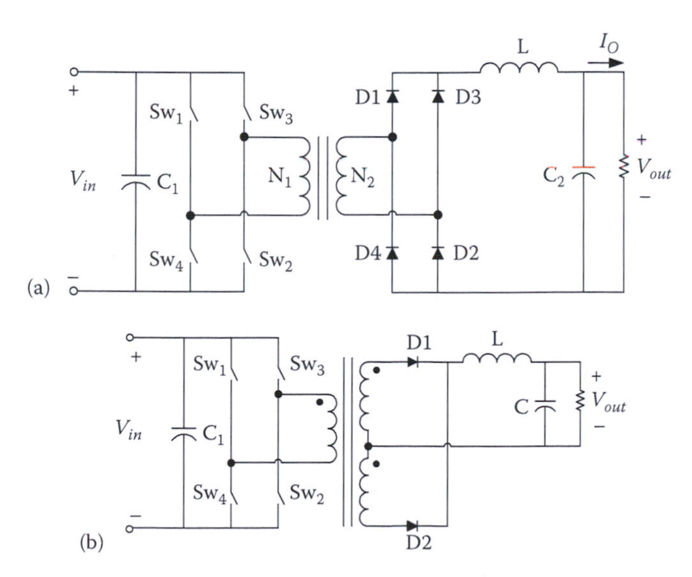

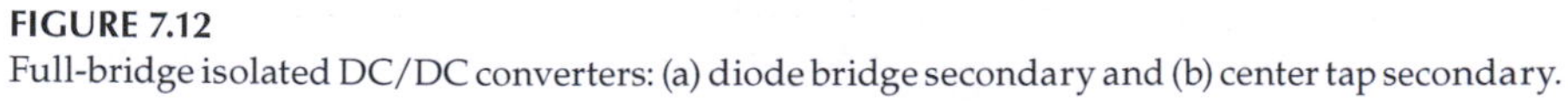
FIGURE 7.12
Full-bridge isolated DC/DC converters: (a) diode bridge secondary and (b) center tap secondary.

An alternative to the diode bridge used on the low-voltage side of the full-bridge converter is to use a center-tap transformer with two diodes. The topology is shown in Figure 7.12b. The alternative is often used for cost minimization since the diodes carry large currents and are expensive. The topology shown in Figure 7.12b is unidirectional, but can easily be converted into a bidirectional topology by replacing the two diodes with two controlled switches. The appropriate control of the secondary switches allows the topology to be operated in the boost mode with power flowing from the low- to the high-voltage side. The use of switches is also more efficient than using diodes.

The typical power rating of the high- to low-voltage DC/DC converter in mid-size electric or hybrid vehicles is around 2 kW. The low-voltage output supplies the 12 V electronics in the vehicle. The 12 V ground is always connected to the vehicle chassis; hence, the requirement for this converter topology to be of isolated type. The converter topology commonly used for this application is one of the full-bridge types. The switching frequency of this converter used in vehicles is around 100 kHz. In most electric and hybrid vehicles, this converter is unidirectional, although there are advantages of using a bidirectional converter. The boost mode operation, possible with the topologies shown in Figure 7.12 but with switches in the secondary, helps eliminate the 12 V starter for IC engines in hybrid electric vehicles. The hybrid vehicles have a high-power electric machine that can start the engine more efficiently and much faster than the conventional 12 V starter. The boost mode operation also helps in extreme weather starting conditions. In extremely cold cranking situations, the high-voltage battery may not be able to provide all of the required engine cranking power; the 12 V battery can supply additional starting power to the high-power electric machine to help start the IC engine.

7.3 Cell Balancing Converters

The individual battery cell voltages as well as the charge will vary in a series-connected string as the entire battery pack goes through charge–discharge cycles. The charge imbalance in a series-connected string of cells reduces the overall capacity as well as the average life of the pack. Battery cell charge equalization is used to avoid long-term imbalance. Individualized cell balancing methods monitor and balance the charge or the voltage during charging and discharging of an electrochemical pack using passive or active components. The cell balancing electronics can also monitor *SoC* and/or voltage during discharge, so that the capacity of the pack is not limited by the weakest cell in the string. The overall cell management methods were discussed earlier in Chapter 4. In this section, the power electronics included in the balancing electronics are presented. Some of the cell balancing circuits are dissipative in nature, while others use DC/DC converter circuits to transfer charges from one cell to another or between a cell and the supply. Based on the operational feature, the balancing methods can be classified broadly into two types: passive or dissipative balancing methods, and active or non-dissipative balancing methods. The two types of cell balancing circuits are discussed below.

7.3.1 Passive Balancing Methods

In passive balancing methods, the individual cell voltages are equalized by either dissipating the excess energy or providing a shunt path to bypass the charging current. In the most common approaches, a dissipative component such as a resistor or a zener diode is connected across a cell to prevent overcharging of the cell. The dissipative resistors connected directly across each cell will always draw currents from the cells as shown in Figure 7.13a, but a large resistor value is used so that the current bleeding is minimal. The value of the resistor is chosen to ensure that charges are equalized during both cell charging and discharge. The magnitude of the current drawn by the resistor is larger for the higher voltage cells; the cells with higher voltages discharge faster and the current gradually decreases as the voltage drops. All the cells in the series string achieve balance through this power dissipation in the resistors.

An improvement over the continuous dissipation in the resistors can be achieved by adding a switch in series with the resistor as shown in

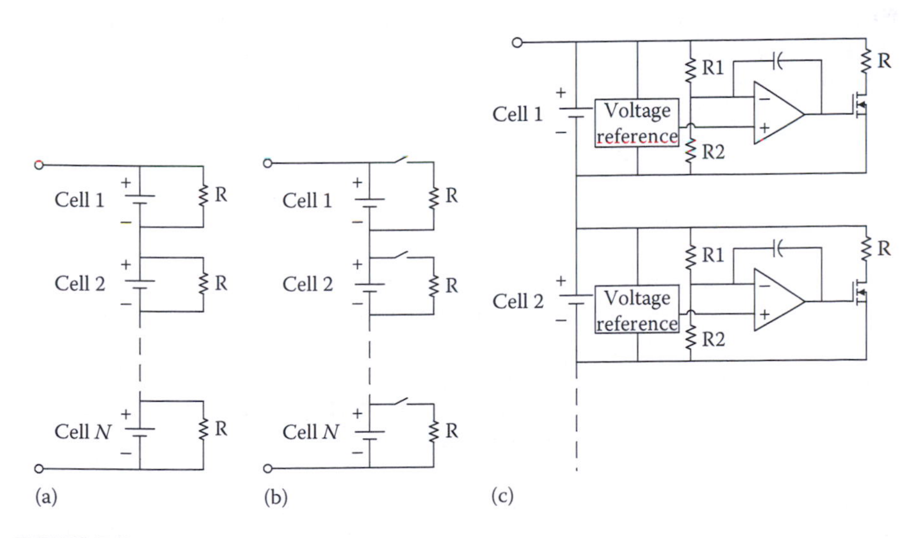

FIGURE 7.13
Passive cell voltage equalization circuits: (a) resistive shunt, (b) resistor with switch, and (c) analog shunt equalization.

Figure 7.13b [5]. The switch can be realized using a MOSFET or a BJT. Each cell voltage is measured and monitored to control the turn-on and turn-off of the switches. Cell voltage monitoring and control of switches can be done centrally by a microprocessor. During cell charging, all the switches will remain open initially. Once a cell voltage reaches its limit, the switch for that cell is turned on and the charging current is bypassed through the resistor. The switched resistor method minimizes the loss in the resistors, but adds cost and complexity of the switches, voltage sensors, and the central processor. The use of a central processor can be avoided by using a local analog shunt electronic circuit as shown in Figure 7.13c [6]. An electronic circuit monitors and compares each cell voltage against a reference value, which is set to the limit of the cell voltage. When the cell reaches the reference voltage, the op-amp comparator turns the Darlington switch connecting the resistor R in parallel to the cell. The current is proportionally shunted through the resistor, and the cell is charged at a constant voltage thereafter. The battery-pack charging continues until all the cells are completely charged. The analog shunt equalization circuit can balance the voltage of even highly unmatched cells, and is completely local to each cell.

The cell balancing circuits using zener diodes are shown in Figure 7.14. In one method, a sharp-knee zener clamping diode is connected across each cell or a group of cells to provide a current shunt path for cell voltage equalization. The balancing circuit is shown in Figure 7.14a. The minimum available zener voltage is higher than the voltage of most electrochemical cells; hence, a zener has to be connected across at least a pair of cells. The zeners break down when the cell voltages reach the zener breakdown voltage, clamping the voltage across two or more cells to the zener breakdown voltage. The drawback of the method is the large power dissipation when the zener diode clamps the cell voltage, especially when the cell is being recharged with high charging currents.

An analog equalization method can be designed with a shunt transistor switch, and zener diode is shown in Figure 7.14b. The zener clamping diode breaks down when the cell voltage reaches its upper limit. The zener diode and resistor circuit then provide the turn-on signal for the transistor switch, which then bypasses the charging current. The entire equalization circuitry is local and there is no need for a central processor.

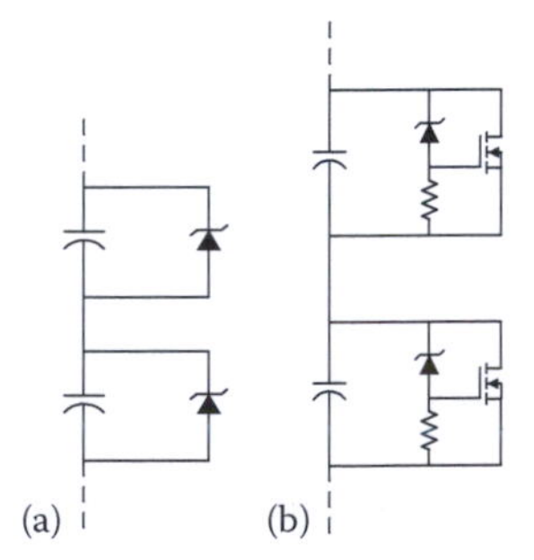

FIGURE 7.14
Passive cell voltage equalization circuits: (a) Zener diode clamping and (b) Zener with switch.

The main drawback of the passive methods is the wastage of energy in the dissipative components. Additionally, relatively high currents may need to be bypassed through the equalizing circuit for cell balancing. The components in the analog balancing circuitry also have strong dependence on temperature, which may not be tolerable. These methods are otherwise less expensive and easier to implement.

7.3.2 Active Balancing Methods

In active cell balancing methods, non-dissipative elements (capacitors and inductors) are used in conjunction with switches to selectively transfer energy from the overcharging cells to the undercharged cells. The electronic switches and associated circuitry actively control the turn-on and turn-off of the switches to transfer energy among cells. A central microprocessor monitors the pack and cell conditions to control the switches for cell voltage equalization. The current charge/discharge rates in energy storage systems of electric and hybrid vehicles are fairly high with short durations. Consequently, the charge equalization currents could be of the same order of magnitude. The charge transfer among the cells using non-dissipative elements for cell equalization minimizes the energy losses in electric and hybrid vehicles. The active balancing circuits are more complex and difficult to implement, but ensure the highest possible charge/discharge efficiency. In some active balancing methods, charges are locally redistributed among neighboring cells for cell voltage equalization, while in more complex methods, charges are globally redistributed relative to a constant reference voltage.

The concept of cell balancing using transfer of energy from the strongest cell to the weakest cell can be explained using a switched capacitor network. The circuit schematic of a switched capacitor network for *N* number of cells with a pair of generic switches and a flying capacitor across each battery cell is shown in Figure 7.15 [7]. Each neighboring cell has a similar circuit. The equalizer needs single-pole, double-throw (SPDT) switches. The switches can be implemented using two transistor switches—zener diodes and transformer; alternative transformerless SPDT implementations are also possible, which are given in [7]. The transformer-coupled switches are desirable for equalizers that have high isolation requirements. The charge from one cell can be transferred to the neighboring cell of lower voltage using the flying capacitors. For example, when cell 1 in the figure reaches its limit, flying capacitor C_1 across Sw_1 and Sw_2 are connected to the top rails until C_1 is charged to the level of cell 1 voltage. The voltage of cell 1 drops by a small amount in the process. The switch network then connects C_1 to cell 2 to bring up its voltage to the same level. Using this charge pump technique, charges can be transferred across neighboring cells, and subsequently to distant cells to balance the entire pack. The disadvantage

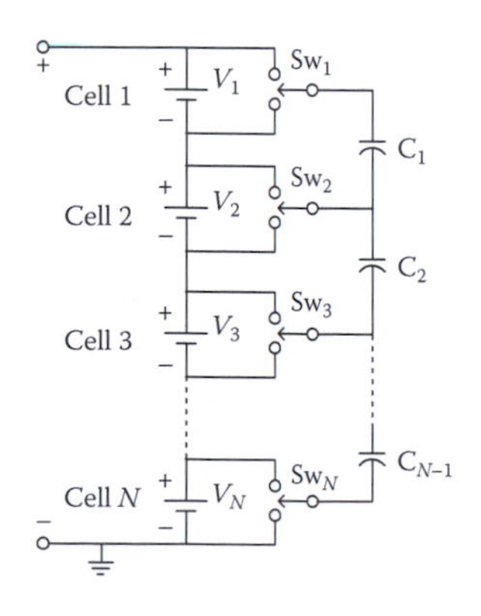

FIGURE 7.15
Switched capacitor circuit.

of the method is that the voltages are equalized with neighboring cells rather than with respect to a reference voltage. This charge transfer technique is inherently slow for a long series string of cells.

Charge can be transferred among the battery cells using DC/DC converters instead of routing them through flying capacitors. The degree of cell voltage equalization and efficiency of charge transfer depend on the type of DC/DC converter circuit used [8,9]. Also, charges can be transferred to distant or neighboring cells depending on the circuit topology used. Three categories of DC/DC converter-based cell balancing circuits are discussed: (1) individual DC/DC converter, (2) centralized DC/DC converter, and (3) current diverter DC/DC converter.

7.3.2.1 Individual DC/DC Converter

The charge equalization method using isolated DC/DC converters is shown in Figure 7.16a. The DC/DC converters can be unidirectional transferring energy only in one direction from the overcharged cell to the battery DC-bus during charging or can be bidirectional transferring energy back and forth from the unbalanced cells to the battery-pack DC-bus or to other cells. The method is highly efficient and precisely regulates the voltage across each cell. However, the high component count with individual DC/DC converters for each cell increases the cost of cell balancing circuitry. The DC/DC converter modules can be connected across a group of cells instead of individual cells for large battery systems, such as that in electric and hybrid vehicles.

A unidirectional flyback DC/DC converter module for a group of cells is shown in Figure 7.16b. When an overvoltage is detected in the group of cells, the excess energy is transferred to the DC-bus by PWM control of the transistor switch, allowing precise regulation of the cell voltage. The control of the switch can be either from a central microprocessor or from a simple

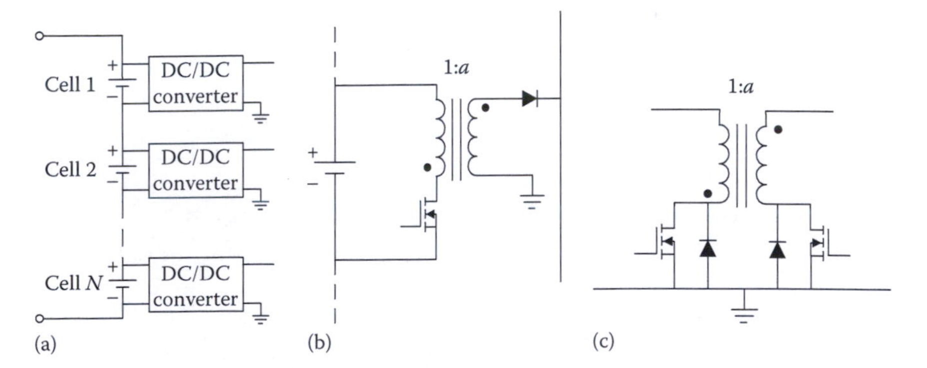

FIGURE 7.16
Active cell voltage equalization with: (a) isolated DC/DC converters, (b) unidirectional flyback converter, and (c) bidirectional flyback converter.

comparator with a preset voltage. The isolation transformer turns ratio must satisfy the condition $a \leq N$, where N is the number of cell groups. A bidirectional flyback DC/DC converter circuit for cell voltage equalization is shown in Figure 7.16c. The bidirectional cell equalization circuit transfers energy during both charging and discharging, thereby increasing pack utilization and enhancing cell life. Energy can be transferred in either direction by controlling the two switches of each module. Cell equalization during charging takes place the same way as with the flyback DC/DC converter with the antiparallel diode of the switch conducting during energy transfer from the magnetization inductance to the DC-bus. The transistor switch on the DC-bus side is turned off during this mode. During pack discharging, energy can be transferred from the DC-bus to the weaker cells by PWM control of the converter switch on the DC-bus side, thus maintaining cell voltage equalization.

7.3.2.2 Centralized DC/DC Converter

The control and hardware complexity with individual DC/DC converters can be simplified by using a centralized flyback or forward DC/DC converter built with a multiwinding transformer [8,9]. A flyback converter configuration with centralized DC/DC converter is shown in Figure 7.17a. The primary of the transformer is connected to the DC-bus, while the secondaries are distributed among N group of cells. The isolation transformer turns ratio must satisfy the condition $a \leq N$, where N is the number of cell groups. The cell equalizing circuit is designed to transfer energy from the

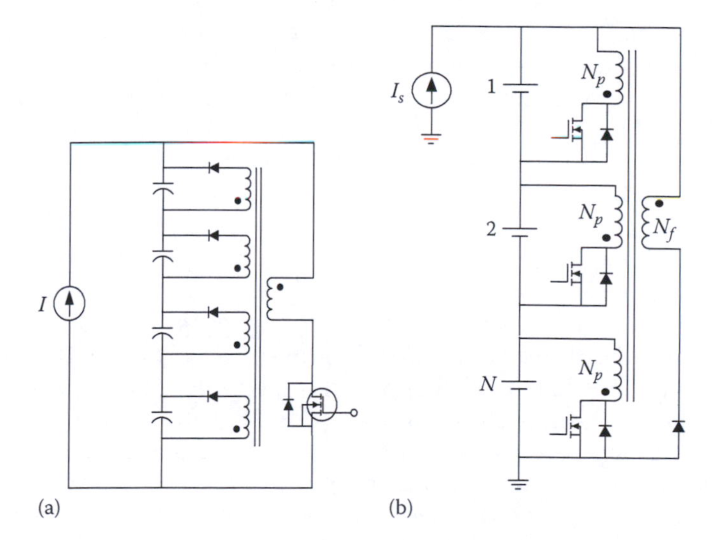

FIGURE 7.17
Charge equalization with (a) centralized flyback converter and (b) centralized forward converter.

battery pack to the weaker cells that have the undervoltages. The centralized DC/DC converter uses only one switch that is kept off until a weak cell is detected. When undervoltage is detected in one of the cells or cell groups during charging, the switch is turned on to transfer pack energy to the magnetizing inductance of the transformer. Once the magnetizing inductance is charged, the switch is turned off and the charging currents are automatically steered to the cell equalizer circuit segment that has the lowest voltage at the cathode of its diode. In the ideal case, the largest portion of the stored energy will be transferred to the cells with the lowest voltage without any additional control. Thus, the flyback converter preferentially transfers the energy to the cell with the lowest voltage. The charge transfer continues until all the cell voltages are equalized. In practice, transformer parasitics, especially the transformer's leakage inductance, dictates the distribution of the magnetizing inductance stored energy among the cells. The transformer leakage inductance L_{ls} is the main component for controlling the charging current. The charging current is given by

$$I_{ch}(t) = \frac{V_{bat} - aV_{bx}}{L_{ls}} t \tag{7.5}$$

where

V_{bat} is the total battery voltage
V_{bx} is the cell-group module voltage

The practical difficulty is the matching of all the leakage inductances among the secondary windings. The coaxial winding transformers can be used for these converters to tightly control the parasitics [9].

Forward converters and a multiwinding transformer, as shown in Figure 7.17b, can also be used to transfer energy from the overcharged cells to the weak cell-group modules. When a cell group becomes overcharged, the corresponding DC/DC converter module switch is turned on to transfer the energy from that cell to the others via the magnetizing inductance of the transformer. Energy is transferred from the overcharged module to the weaker modules with the most energy diverted to the lowest voltage module. The charge transfer current is directly proportional to the difference between the terminal voltages of the overcharged module and the weaker module, and is inversely proportional to the leakage inductance in the secondary windings similar to Equation 7.5. Consequently, the lowest voltage cell group accepts the most charge. The leakage inductance plays a critical role in determining the charge transfer currents. The winding N_f and the diode D_f provides a guaranteed reset path when all the active switches are turned off and there is still stored energy left in the magnetizing inductance. The diodes of the modules also turn off when all the switches are off. The residual energy in that case is returned to the pack

DC-bus through diode D_f. The turn ratio between the reset winding and the module windings is given by

$$a = \frac{N_f}{N_m} \leq N$$

where N is the number of batter cell-group modules.

7.3.2.3 Current Diverter DC/DC Converter

The current diverter cell equalization circuits transfer energy to the neighboring cells using DC/DC converters [8]. A current diverter circuit that can only transfer energy downward through a stack of series connected cells is shown in Figure 7.18a. Each diverter module consists of a MOSFET switch, a diode, and an energy storage element. A flyback converter module is necessary for the last cell group to transfer excess energy back to the battery pack when required.

A bidirectional current diverter circuit topology is shown in Figure 7.18b with two MOSFET switches and an inductor per module. Charge can be transferred to the cell either above or below an overcharged cell. There is also no need for a separate flyback module for the last cell. The diverter modules along with two corresponding pack cell groups form a half-bridge converter feeding an inductive load. During normal operation, the diverter modules are disabled. When a cell group is overcharged, the corresponding MOSFET switch is turned on to transfer energy from the cell group to the inductor. When the switch is turned off, the stored energy in the inductor

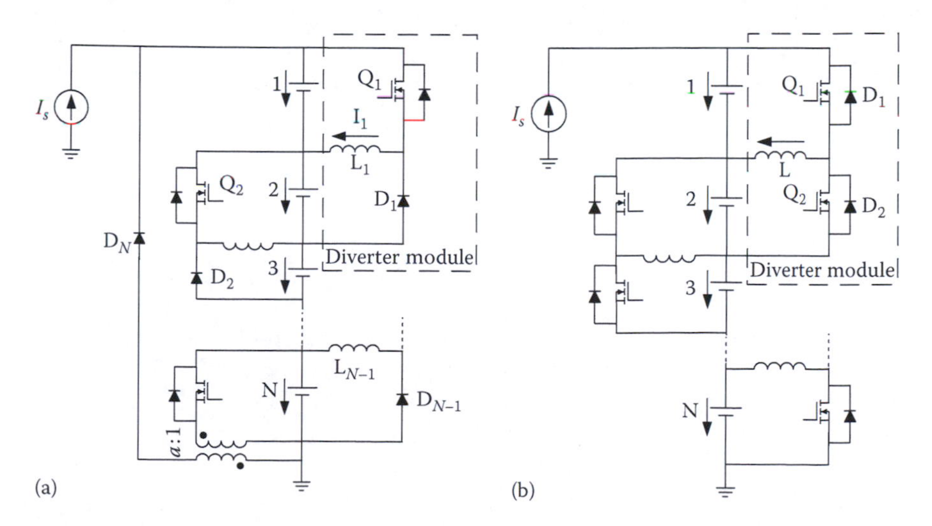

FIGURE 7.18
Current diverter circuits: (a) unidirectional and (b) bidirectional.

transfers to the cell group up or down the stack depending on the location of the inductor.

References

1. B.J. Baliga, *Power Semiconductor Devices*, PWS Publishing Company, Boston, MA, 1995.
2. J.G. Kassakian, M.F. Schlecht, and G.C. Verghese, *Principles of Power Electronics*, Addison Wesley Publishing Company, Reading, MA, 1991.
3. I. Batarseh, *Power Electronic Circuits*, John Wiley & Sons Inc., Hoboken, NJ, 2004.
4. R.M. Schupbach and J.C. Balda, Comparing DC-DC converters for power management in hybrid electric vehicles, *IEEE International Electric Machines and Drives Conference*, 3, 1369–1374, June 2003.
5. M.J. Isaacson, R.P. Hollandsworth, P.J. Giampaoli, F.A. Linkowsky, A. Salim, and V.L. Teofilo, Advanced lithium ion battery charger, in *The Fifteenth Annual IEEE Battery Conference on Applications and Advances*, Long Beach, CA, January 2000, pp. 193–198.
6. Y. Eguchi, H. Okada, K. Murano, and A. Sanpei (inventors), Sony Corporation (assignee), Battery protection circuit, U.S. Patent 5,530,336, 1993.
7. C. Pascual and P.T. Krein, Switched capacitor system for automatic series battery equalization, in *IEEE Applied Power Electronics Conference, Twelfth Annual*, February 1997, Vol. 2, pp. 848–854.
8. N.H. Kutkut and M.D. Divan, Dynamic equalization techniques for series battery stacks, in *IEEE Telecommunications Energy Conference*, Boston, MA, October 1996, pp. 514–521.
9. N.H. Kutkut, H.L.N. Wiegman, D.M. Divan, and D.W. Novotny, Charge equalization for an electric vehicle battery system, *IEEE Transactions on Aerospace and Electronic Systems*, 34, 235–246, January 1998.

8

Electric Motor Drives

The electric motor drive converts the stiff DC battery voltage to either a DC voltage for DC motors or an AC voltage for AC motors. The drive processes the power flow from the energy storage system to the electric machine and vice versa. The controller generates the gating signals for the power electronic circuit devices based on the driver input command, the sensor feedback signals, and the control algorithm for the type of electric machine used. The driver input is translated into a torque command for the motor drive. The torque command in conjunction with the feedback signals set up the operating point parameters for the electric motor and accordingly controls the turn-on and turn-off of the power switches inside the drive system. During vehicle propulsion, the motor drive delivers power at the desired voltage and frequency to the motor, which in turn provides the desired torque at the wheels. During regeneration, the motor drive processes power flow from the wheels to the energy storage system following the driver commands.

8.1 Electric Drive Components

A motor drive consists of a power electronic converter and the associated controller. The power electronic converter is made of solid state devices and handles the flow of bulk power from the source to the motor input terminals. The controller is made of microcontroller or digital signal processor and associated small signal electronics. The function of the controller is to process information and generate the switching signals for the power converter semiconductor switches. The interaction between the components of the motor drive with the source and the electric motor is shown schematically in Figure 8.1.

The power converter can be either a DC drive supplying a DC motor or an AC drive supplying an AC motor. The converter functions of the two types of drives are shown in Figure 8.2. A power converter is made of high-power fast-acting semiconductor devices, such as those discussed in Chapter 7. The solid state devices in the power electronic circuit of the motor drive function as an on–off electronic switches to convert the fixed supply voltage into a variable voltage and variable frequency supply. All these devices have a control input gate or base through which the devices are turned on and off

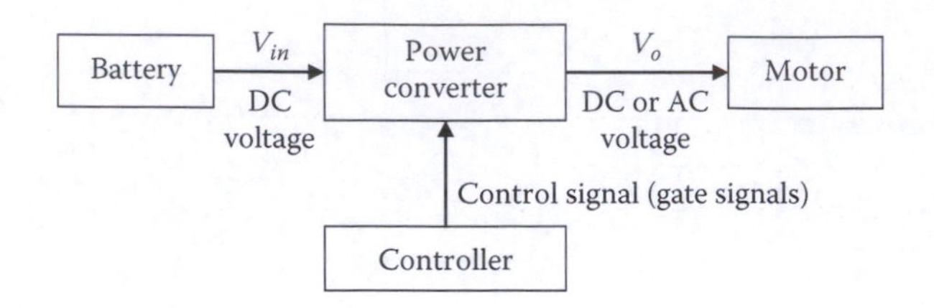

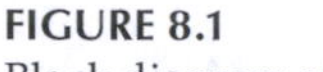
FIGURE 8.1
Block diagram of a motor drive.

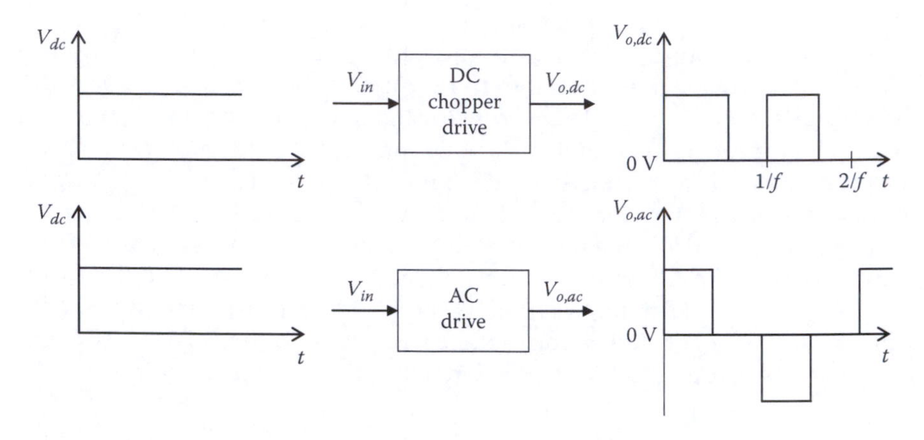

FIGURE 8.2
DC/DC and DC/AC converter functions.

according to the command generated by the controller. The tremendous advances in the power semiconductor technology over the past two decades enabled the development of compact, efficient, and reliable DC and AC electric motor drives. The commonly used power device for the electric motor drive in electric and hybrid vehicles is the IGBT. The voltage and current ratings of the IGBT and their switching frequency is well within the range required for electric propulsion drive systems.

The drive controller manages and processes the system information to control the flow of power in the electric drivetrain. The controller acts on the command input from the user while following a motor control algorithm. Numerous motor control algorithms have been developed over the past decades for the various types of electric machines. Some of these control algorithms are for high-performance drives, while others are for less-demanding adjustable speed drives. The motor drive for the electric powertrain requires fast response characteristics with high efficiency; hence, these fall under the category of high-performance drives. These motor drive control algorithms are computationally intensive requiring faster processors and interfacing with a relatively higher number of feedback signals. The modern controllers are digital instead of analog, which helps minimize drift and error and improves performance through their capability of processing complex algorithms in a short time. The controllers are essentially an embedded system where microprocessors and digital signal processors are used for signal processing along with peripheral and interface electronic modules. The interface circuits consisting of A/D and D/A converters are required for communication between the processor and the other components of the system. The digital I/O circuitry brings in the digital input signals and carries out the gate driver signals for the power semiconductor devices.

This chapter discusses the DC and AC drives used for controlling the power flow and energy conversion in electric machines with special emphasis to those used in electric and hybrid vehicle propulsion. Although DC machines are complex in construction, the DC drives are much simpler than their AC counterparts. The simpler DC drive models serve as a useful tool in understanding the interaction between the propulsion system and the vehicle road load. On the other hand, the complexity of AC motor drives is in the drive system, although the AC electric machines are less complicated than the DC machines. The AC drives use a standard six-switch inverter topology and a PWM strategy to generate the three-phase sinusoidal waveforms in the machines. The inverter topology and PWM strategies are covered in this section. The AC machine control algorithms are covered in Chapter 9. The switched reluctance motor (SRM) drives require a power electronics topology different from the six-switch inverter; these SRM drives are discussed toward the end of the chapter.

8.2 DC Drives

The DC drives for electric and hybrid vehicle applications are the DC/DC converters, such as DC-choppers, resonant converters, or isolated full-bridge converters. DC-choppers are more commonly used for the propulsion DC electric motor drives. A two-quadrant chopper is analyzed in this chapter as representative DC drives. The simplicity of the two-quadrant chopper and the torque–speed characteristics of the separately excited DC motor will be used to present the interaction of a power electronic motor drive and the vehicle road load. The details on various other types of DC drives can be found in several textbooks on electric motor drives [1–4].

8.2.1 Two-Quadrant Chopper

The two-quadrant DC chopper allows bidirectional current and power flow with unidirectional voltage supply. The schematic of a two-quadrant chopper is shown in Figure 8.3. The motor current i_0 is inductive current, and therefore cannot change instantaneously. The transistor Q_1 and diode D_1

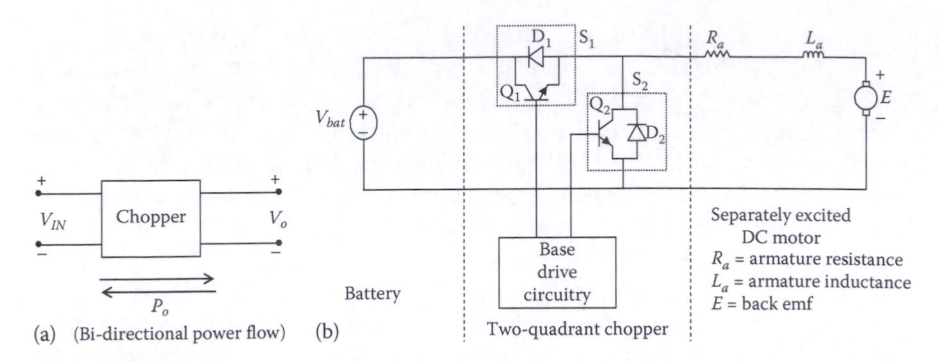

FIGURE 8.3
DC electric motor drive.

combined makes the bidirectional current switch S_1. Similarly, switch S_2 is made of transistor Q_2 and diode D_2. The on and off conditions of the two switches make four switching states (SWS), two of which are allowed and two are restricted as shown in Table 8.1. In the allowed switching states SWS1 and SWS2, the switches S_1 and S_2 has to withstand positive voltage when they are off, and both positive and negative currents when they are on. Therefore, bidirectional current switches have been used.

In quadrant 1 operation, turning on Q_1 allows current and power to flow from the battery to the motor. Both the motor terminal voltage v_o and current i_0 are greater than or equal to zero. Q_2 is required to remain continuously off in quadrant 1 operation, and hence, $i_{b2}=0$. When Q_1 turns off, D_2 turns on since i_0 is continuous. Quadrant 1 operation takes place during the acceleration and constant velocity cruising of a vehicle. The chopper operating modes toggle between SWS1 and SWS2 in this quadrant as shown in Figure 8.4.

The transistor Q_1 switches at fixed chopper frequency to maintain the desired current and torque output of the DC motor. The motor current i_0 is been shown in Figure 8.5 with exaggerated ripple, where in practice the ripple magnitude is much smaller compared with the average value of i_0.

TABLE 8.1
Switching States of Two-Quadrant Chopper

Switching State	S_1	S_2	Comments
SWS0	0 (Off)	0 (Off)	Not applicable in CCM, since i_0 is inductive
SWS1	0	1 (On)	$v_0=0$; $v_{S1}=v_{IN}$ (allowed) $i_{IN}=0$; $i_{S2}=-i_0$
SWS2	1	0	$V_0=v_{IN}$; $v_{S2}=v_{IN}$ (allowed) $i_{IN}=i_0$; $i_{S1}=i_0$
SWS3	1	1	Not allowed since v_{IN} will get shorted

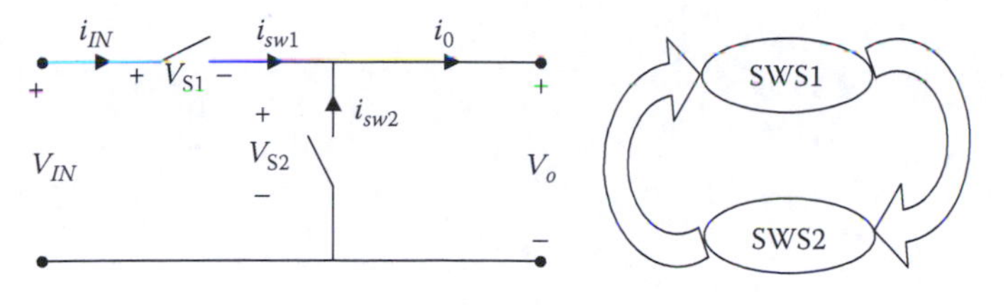

FIGURE 8.4
Quadrant I operation.

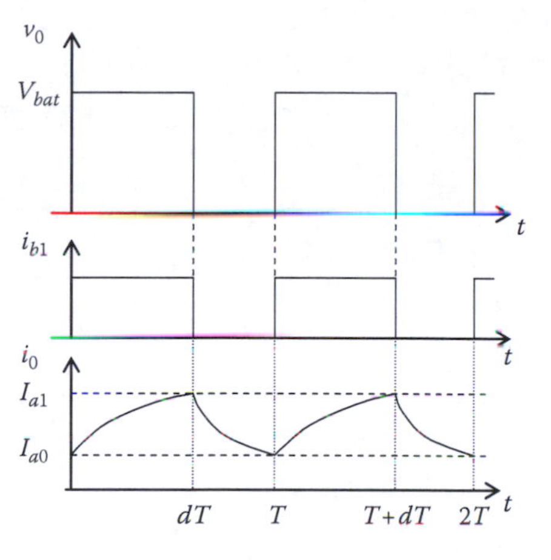

FIGURE 8.5
Output voltage, gate drive for Q_1, and motor current.

The filtering requirements set the time period of switching such that a smooth current and torque output are available. The output of the outer-loop vehicle controller desiring a specific torque output of the motor is the duty ratio d, which sets the on-time of the transistor Q_1. d is a number between 0 and 1, which when multiplied by the time period T gives the on-time of the transistor. The gate drive signal for Q_1 is a function of d, and consequently, assuming ideal switching conditions, the input voltage to the motor is also dependent on d. The circuit configurations in the allowed SWS are shown in Figure 8.6 to aid the steady-state analysis of the drive system. The steady-state analysis is carried out assuming the ideal conditions that there are no switching losses and no delay in the turn-on and turn-off of the devices.

8.2.2 Open Loop Drive

From a systems perspective, the two-quadrant chopper drives the DC motor that delivers power to the transmission and wheels for vehicle propulsion as shown in Figure 8.7. The input to the system comes from the driver of the vehicle through the acceleration pedal and the brake pedal. The acceleration

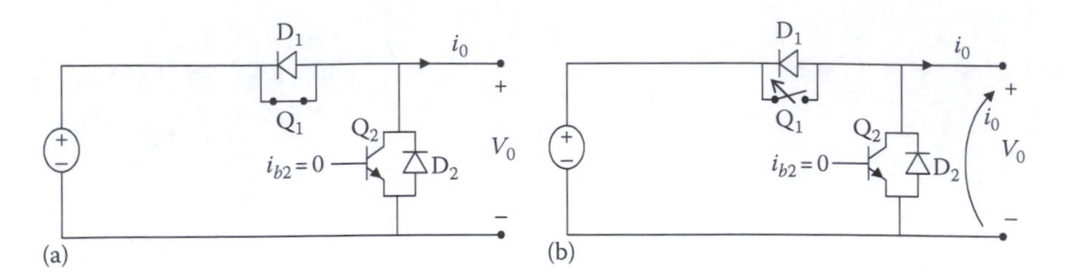

FIGURE 8.6
Circuit condition for (a) switch Q_1 on and (b) for switch Q_2 off.

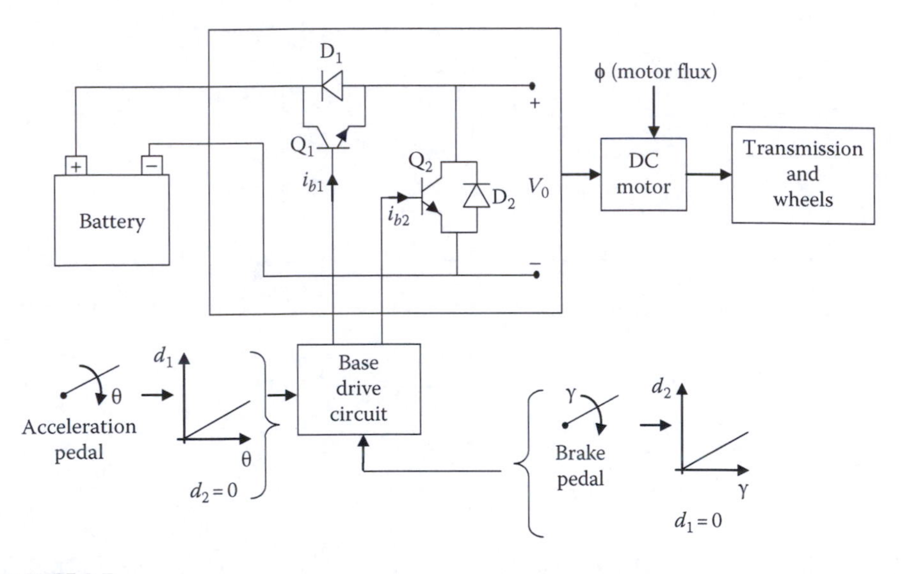

FIGURE 8.7
Open loop drive for bidirectional power flow.

and constant speed cruising is controlled by Q_1 in quadrant 1 operation, while braking is controlled by Q_2 in quadrant 2 operation. In a simplified vehicle control strategy, the slope of the acceleration pedal dictates the desired vehicle response and the angle of the pedal is used to set the duty ratio d_1 for Q_1. Similarly, the slope of the brake pedal relates to the amount of braking desired, and the angle of the brake pedal is used to set the duty ratio d_2 for Q_2 (Figure 8.8). The two pedals must not be depressed simultaneously.

As mentioned in Chapter 6, one of the advantages of using electric motors for vehicle propulsion is to save energy during vehicle braking through regeneration. The energy from the wheels is processed by the power converter and delivered to the battery or any other type of energy storage device during regenerative braking. For the two-quadrant chopper, the amount of regeneration per cycle is a function of the duty ratio as will be shown later. Therefore, the real duty ratio commands of an electric or hybrid vehicle will

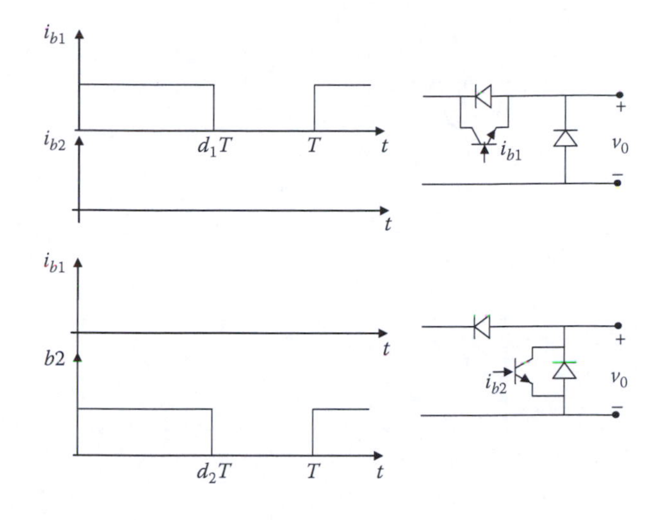

FIGURE 8.8
Base drive signals for the two transistors during acceleration and regeneration.

be nonlinearly related to the pedal angle than is assumed in the simplified analysis to follow. However, the simplistic assumption will give a good insight into the system control.

8.2.2.1 *Steady-State Analysis of Quadrant 1*

During acceleration of the vehicle with the two-quadrant chopper, the power and the current flow into the motor from the source. The current can be continuous or discontinuous depending on the required torque, although the average value of the current is a nonzero positive value. These two scenarios give the two modes of operation of the two-quadrant chopper, namely, the continuous conduction mode (CCM) and the discontinuous conduction mode (DCM). Let us first analyze the CCM by assuming that the required torque and the chopping frequency to propel the vehicle forward are high enough to maintain a positive current into the motor continuously as shown in Figure 8.5. The current i_0 is the armature current. Therefore, $i_0 = i_a$. Neglecting ripple in the back-emf E due to ripple in ω, we can assume that E is a constant.

During $0 \le t \le dT$, the equivalent circuit of the motor is as shown in Figure 8.9. Applying Kirchhoff's voltage law (KVL) around the circuit loop,

$$V_{bat} = R_a i_a + L_a \frac{di_a}{dt} + E$$

The initial condition is $i_a(0) = I_{a0} > 0$. This is a first-order linear differential equation, solving which one gets

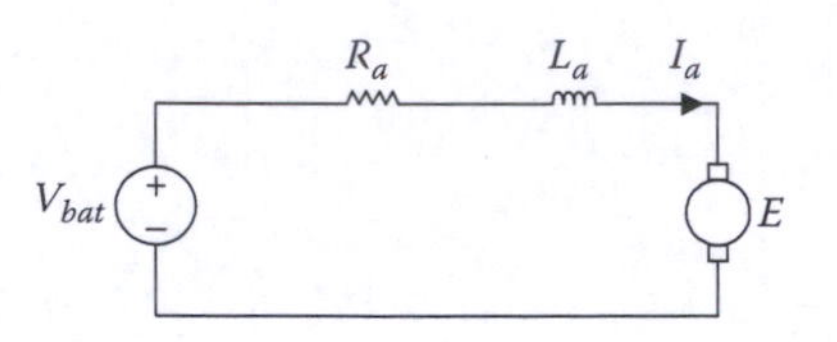

FIGURE 8.9
Equivalent circuit with Q_1 on.

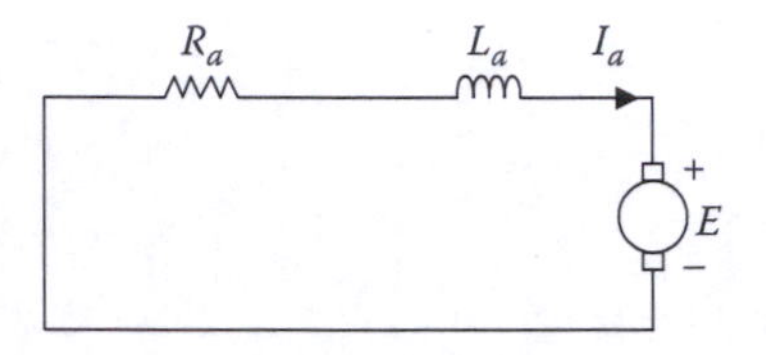

FIGURE 8.10
Equivalent circuit when Q_2 is off.

$$i_a(t) = \frac{V_{bat} - E}{R_a}\left(1 - e^{-t/\tau}\right) + I_{a0}e^{-t/\tau}$$

The final condition of SWS2 is

$$i_a(dT) = \frac{V_{bat} - E}{R_a}\left(1 - e^{-dT/\tau}\right) + I_{a0}e^{-dT/\tau} = I_{a1} \tag{8.1}$$

During $dT < t \leq T$, the equivalent circuit of the motor is as shown in Figure 8.10.
Let $t' = t - dT$. Applying KVL,

$$0 = R_a I_a + L_a \frac{di_a}{dt} + E$$

Solving the linear differential equation,

$$i_a(t') = -\frac{E}{R_a} + \left(I_{a1} + \frac{E}{R_a}\right)e^{-t'/\tau} = \frac{E}{R_a}\left(-1 + e^{-t'/\tau}\right) + I_{a1}e^{-t'/\tau}$$

In steady state, $i_a\,(t' = T - dT) = I_{a0}$. Therefore,

$$I_{a0} = \frac{E}{R_a}\left(-1 + e^{-T(1-d)/\tau}\right) + I_{a1}e^{-T(1-d)/\tau} \tag{8.2}$$

Using Equations 8.1 and 8.2 to solve for I_{a0} and I_{a1}, we get

$$I_{a0} = \frac{V_{bat}}{R_a}\left(\frac{e^{dT/\tau}-1}{e^{T/\tau}-1}\right) - \frac{E}{R_a}$$

$$I_{a1} = \frac{V_{bat}}{R_a}\left(\frac{1-e^{-dT/\tau}}{1-e^{-T/\tau}}\right) - \frac{E}{R_a}$$

The armature current ripple is

$$\Delta i_a = I_{a1} - I_{a0}$$

$$= \frac{V_{bat}}{R_a}\left[\frac{1+e^{T/\tau}-e^{dT/\tau}-e^{(1-d)T/\tau}}{e^{T/\tau}-1}\right] \tag{8.3}$$

If the armature current i_a has ripple, motor torque T_e will have ripple, since the motor torque is proportional to the armature current ($T_e = K\ \Phi i_a$). The speed is also proportional to the electromagnetic torque (see Equation 6.16). Therefore, for electric and hybrid vehicle applications, significant ripple in T_e is undesirable, since ripple in torque causes ripple in speed ω, resulting in a bumpy ride. For a smooth ride, ripple in T_e needs to be reduced.

8.2.2.2 Ripple Reduction in i_a

The armature current ripple can be reduced in one of two ways: (1) adding a series armature resistance or (2) increasing the chopper switching frequency.

Adding a series inductance in the armature increases the electric time constant τ. The new time constant is

$$\tau = \frac{L_f + L_a}{R_a}$$

where L_f is the added series inductance (Figure 8.11). As τ increases, Δi_a decreases for fixed switching period T. The trade-off is the increase in i^2R losses due to L_f, since practical inductances have series resistance. Also, the electrical response time will increase due to the increase in the time constant.

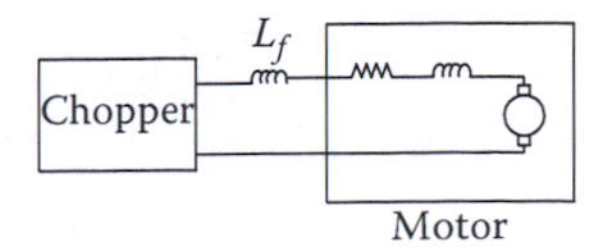

FIGURE 8.11
Series inductance L_f added in series with the chopper.

Increasing the switching frequency of the chopper, that is, decreasing T, will also reduce the armature current ripple. The upper limit on switching frequency depends on the type of switch used. The switching frequency must also be smaller than the controller computational cycle time, which depends on the speed of the processor used and the complexity of the control algorithm. The trade-off of using a higher switching frequency is the higher switching losses in the power devices.

8.2.2.3 Acceleration (CCM)

In the acceleration mode, the current and the power flow into the motor from the battery or energy source. The condition for CCM mode is

$$I_{a0} > 0 \Rightarrow I_{a0} = \frac{V_{bat}}{R_a}\left(\frac{e^{d_1T/\tau}-1}{e^{T/\tau}-1}\right) - \frac{E}{R_a} > 0$$

$$\Rightarrow V_{bat}\left(\frac{e^{d_1T/\tau}-1}{e^{T/\tau}-1}\right) > E.$$

Note that

$$0 \le \frac{e^{d_1T/\tau}-1}{e^{T/\tau}-1} \le 1$$

since $0 \le d_1 \le 1$. It follows that

$$V_{bat} > V_{bat}\left(\frac{e^{d_1T/\tau}-1}{e^{T/\tau}-1}\right)$$

Therefore, the condition for CCM is

$$V_{bat} \ge V_{bat}\left(\frac{e^{d_1T/\tau}-1}{e^{T/\tau}-1}\right) > E \tag{8.4}$$

The electrical time constant of the power converter is much faster than the mechanical time constant of the motor and the vehicle. The analysis of the interaction between the motor torque–speed characteristics and the vehicle force–velocity characteristics is best conducted on an average basis over one time period. The KVL applied around the motor armature circuit loop gives

$$v_a(t) = R_a i_a(t) + L_a\frac{di_a}{dt} + K\phi\omega(t)$$

Averaging both sides yields

$$\langle v_a \rangle = R_a\langle i_a \rangle + k\phi\langle \omega \rangle \tag{8.5}$$

The average armature circuit can be represented by circuit in Figure 8.12. The average armature voltage in the CCM is

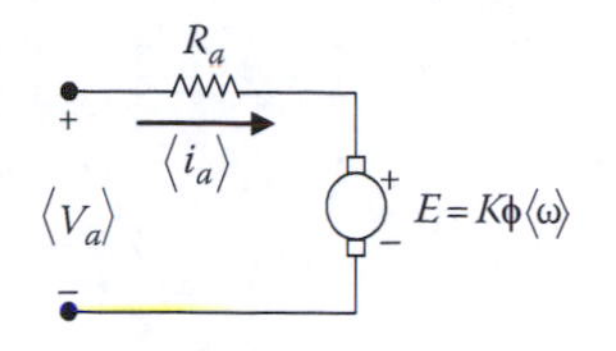

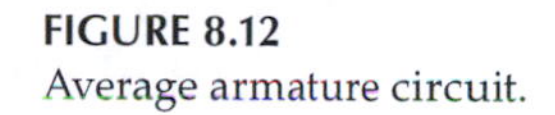

FIGURE 8.12
Average armature circuit.

$$\langle v_a \rangle = \frac{1}{T}\int_0^T v_a(\tau)\,d\tau = \frac{V_{bat} d_1 T}{T} = d_1 V_{bat} \tag{8.6}$$

The average torque equation is

$$T_e(t) = K\phi i_a(t)$$

$$\Rightarrow \langle i_a \rangle = \frac{\langle T_e \rangle}{K\phi}$$

Substituting the average current in Equation 8.6

$$d_1 V_{bat} = R_a \frac{\langle T_e \rangle}{K\phi} + K\phi\langle\omega\rangle$$

$$\Rightarrow \langle\omega\rangle = \frac{d_1 V_{bat}}{K\phi} - \frac{R_a}{(K\phi)^2}\langle T_e \rangle \tag{8.7}$$

The average speed–torque characteristic of a separately excited DC motor driven by a two-quadrant chopper in the CCM given by Equation 8.7 is shown qualitatively below in Figure 8.13. The effect of increasing the duty ratio d_1 in the acceleration mode is to shift the no-load speed and rest of the characteristics vertically upward in the first quadrant.

8.2.2.4 Acceleration (DCM)

When the torque required from the motor in the acceleration mode is not high enough, the chopper may enter the DCM where the armature current becomes discontinuous as shown in Figure 8.14.

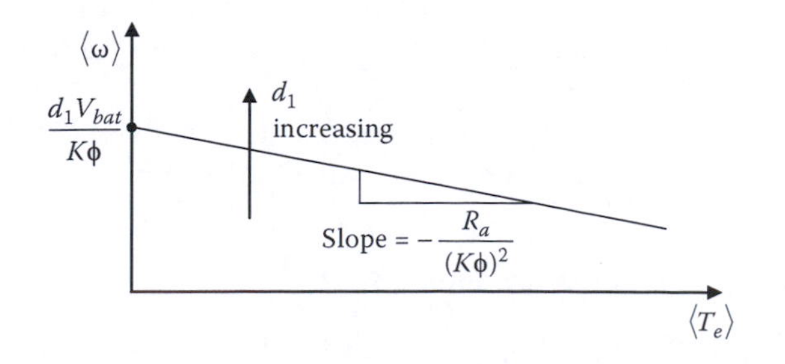

FIGURE 8.13
Torque–speed characteristics of the chopper-fed DC drive.

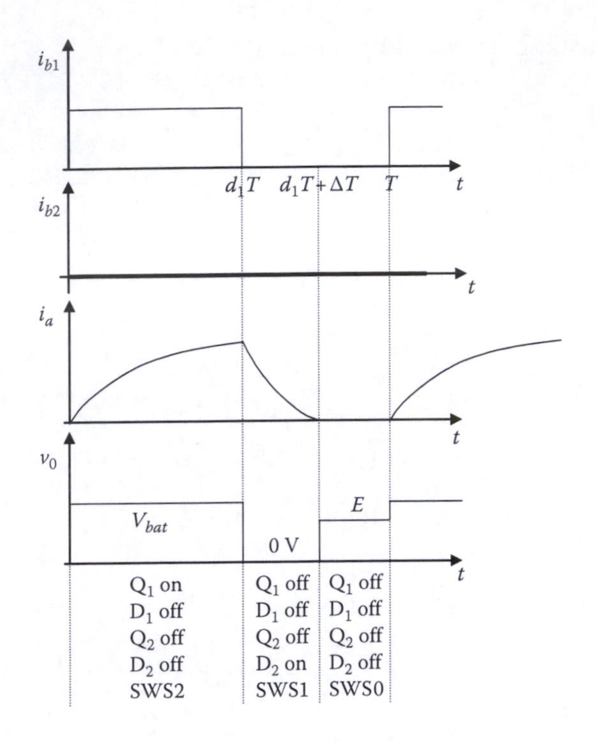

FIGURE 8.14
Voltage and current waveforms during acceleration in the DCM.

In the DCM, $I_{a0} \leq 0$ and $V_{bat} > E$ (since power is still flowing from the energy source into the motor). The operation is still in quadrant 1. The condition for DCM of operation is

$$V_{bat}\left(\frac{e^{d_1 T/\tau} - 1}{e^{T/\tau} - 1}\right) \leq E < V_{bat} \tag{8.8}$$

The motor armature current cannot go negative without the use of switch Q_2 in the DCM. Q_1 is used to control the flow of power from the source to the motor only. In the interval between $d_1T + \Delta T$ and T, for i_a to become negative, D_1 must turn on. However, during this interval, $v_0 = V_{bat} > E$; hence, i_a cannot go negative. Also, i_a cannot become positive since that would require $v_0 = V_{bat}$, which would require Q_1 to be on, but $i_{b1} = 0$. Hence, $i_a = 0$ over this interval.

8.2.2.5 Acceleration (Uncontrollable Mode)

When a vehicle is rolling down a steep slope, it is possible for the propulsion motor to attain a large value of back-emf. In such a case, if $E > V_{bat}$, current cannot be forced into the motor and the use of Q_1 becomes meaningless.

The supply voltage saturation limit prevents the driver from supplying more power into the motor to move faster than the velocity attained due to gravity. Therefore, the driver cannot control the vehicle using the acceleration pedal; he or she can only slow down the vehicle by using the brake pedal. If the brake pedal is not used in this situation, then the vehicle enters the uncontrollable mode (UNCM). When $E > V_{bat}$, i_a starts to decrease and once it reaches zero, diode D_1 becomes forward biased and turns on. i_a continues to increase in the negative direction until it reaches its steady-state value of

$$i_a = \text{Constant} = \frac{-E + V_{bat}}{R_a} \tag{8.9}$$

The mode of operation in this stage is in quadrant II. The current and switch conditions are shown in Figure 8.15. Depressing the acceleration pedal to increase d_1 does not control the vehicle in any way, and the vehicle is in fact rolling downward, while regenerating into the source in an uncontrolled way. The controller fault protection algorithm must kick in at this stage to prevent any overcharging of the batteries. Of course, the driver can regain control by switching to the brake pedal from the acceleration pedal and forcing controlled regeneration through the use of Q_2. This will help slow down the vehicle on a downhill slope.

8.2.2.6 Braking Operation (CCM in Steady State)

The most efficient way of recovering energy during vehicle braking is through regeneration in the motor drive system. Let us assume that $E < V_{bat}$ and the brake pedal is depressed. Q_1 is kept off during this period, while braking is controlled through the gate signal i_{b2}. For regeneration, the power flow must be from the motor to the energy source storage requiring armature current i_a to be negative. Turning Q_2 on helps i_a become negative (from a previously positive value), and an average negative current can be established in a relatively short time for vehicle braking and regeneration. The voltage and current waveforms during braking operation in CCM are shown in Figure 8.16.

An analysis similar to the CCM during acceleration will yield the I_{a1} and I_{a2} values in the steady-state CCM during braking as

$$\begin{aligned} I_{a1} &= \frac{1}{R_a}\left\{V_{bat}\left(\frac{1-e^{-d_2'T/\tau}}{1-e^{-T/\tau}}\right)-E\right\}, \quad d_2' = 1 - d_2 \\ I_{a2} &= \frac{1}{R_a}\left\{V_{bat}\left(\frac{e^{d_2'T/\tau}-1}{e^{T/\tau}-1}\right)-E\right\} < 0 \end{aligned} \tag{8.10}$$

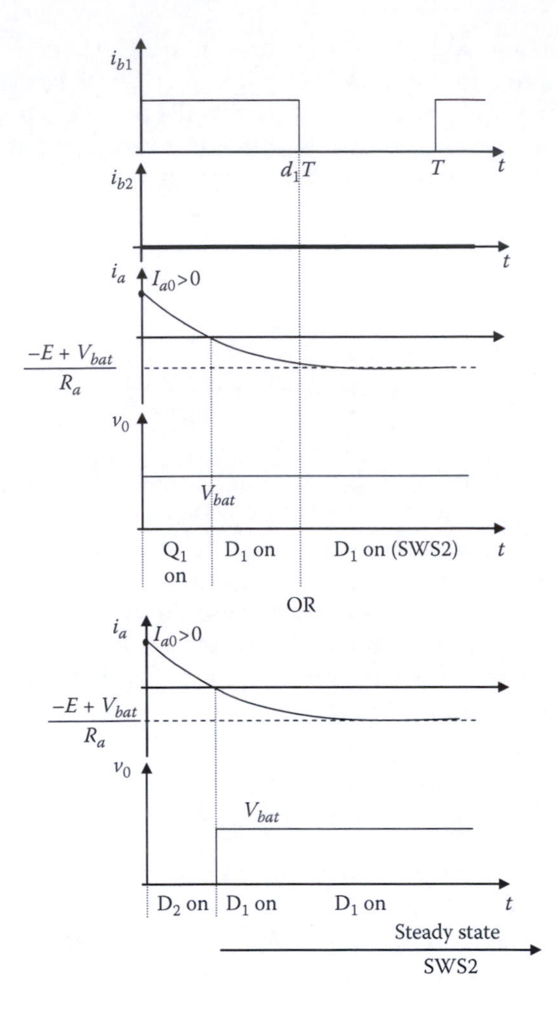

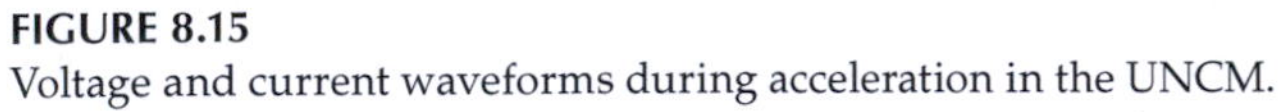

FIGURE 8.15
Voltage and current waveforms during acceleration in the UNCM.

The current ripple during braking is

$$\Delta i_a = \frac{V_{bat}}{R_a}\left[\frac{-e^{-d_2'T/\tau}+1+e^{T/\tau}-e^{d_2T/\tau}}{e^{T/\tau}-1}\right] \tag{8.11}$$

During braking CCM, $I_{a1}<0$ and $E<V_{bat}$. Also, note that

$$0<\left(\frac{1-e^{-d_2'T/\tau}}{1-e^{-T/\tau}}\right)=\frac{e^{T/\tau}-e^{d_2T/\tau}}{e^{T/\tau}-1}<1$$

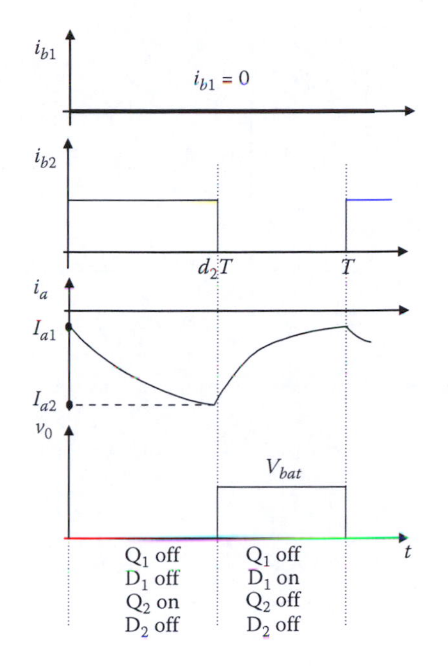

FIGURE 8.16
Voltage and current waveforms during braking operation in CCM.

Therefore, the condition for continuous conduction during braking is

$$0 < V_{bat}\left(\frac{1-e^{-d_2'T/\tau}}{1-e^{-T/\tau}}\right) < E < V_{bat} \tag{8.12}$$

The average voltage equation in the braking CCM is

$$\langle v_a \rangle + R_a \langle i_a \rangle + 0 = E = K\phi \langle \omega \rangle \tag{8.13}$$

The average motor torque is

$$\langle T_e \rangle = -K\phi \langle i_a \rangle \tag{8.14}$$

The average motor terminal voltage is

$$\langle v_a \rangle = \frac{1}{T}\int_0^T v_a(\tau)d\tau = \frac{1}{T}V_{bat}(T - d_2T) = (1-d_2)V_{bat} \tag{8.15}$$

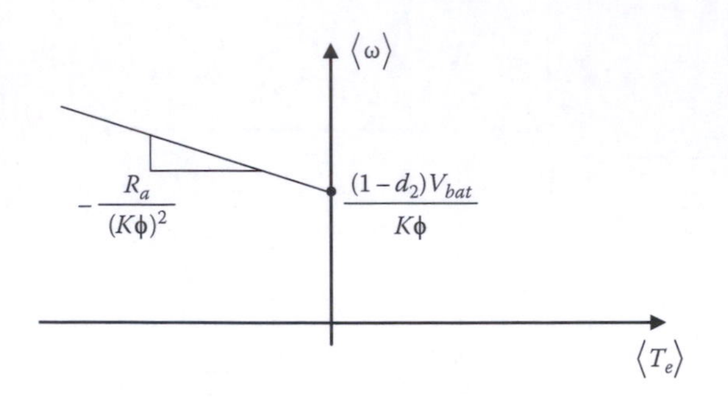

FIGURE 8.17
DC motor drive speed–torque characteristics during braking.

Substituting in $\langle v_a \rangle$ from Equation 8.15 and $\langle i_a \rangle$ from Equation 8.14 into Equation 8.13 gives

$$(1-d_2)V_{bat} - \frac{R_a}{K\phi}\langle T_e \rangle = K\phi\langle \omega \rangle$$

The average speed–torque characteristics is (Figure 8.17)

$$\langle \omega \rangle = \frac{(1-d_2)V_{bat}}{K\phi} - \frac{R_a}{(K\phi)^2}\langle T_e \rangle, \quad \text{for } \langle T_e \rangle < 0,\ \langle \omega \rangle > 0 \tag{8.16}$$

8.2.2.7 Regenerative Power

The power is regenerated into the energy source only during the part of the cycle when current flows into the battery from the motor as shown in Figure 8.18. When transistor Q_2 is on, power is only being dissipated in the switch and contact resistances. When Q_2 is off and diode D_1 is conducting, the current termed $i_{bat}(t)$ is flowing into the battery. Therefore, the instantaneous regenerative power is $P_{reg}(t) = V_{bat} \times i_{bat}(t)$. The average regenerative power is

$$\langle P_{reg} \rangle = \frac{1}{T}\int_{d_2T}^{T} P_{reg}(\gamma)d\gamma \tag{8.17}$$

Using an analysis similar to that done for the CCM during acceleration and the results of Equation 8.10, the battery current can be derived as

$$i_{bat}(t) = \begin{cases} \left[I_{a2} - \frac{E - V_{bat}}{R_a}\right]e^{-(t-d_2T)/\tau} + \frac{E - V_{bat}}{R_a} & \text{for } d_2T < t \le T \\ 0 & \text{otherwise} \end{cases} \tag{8.18}$$

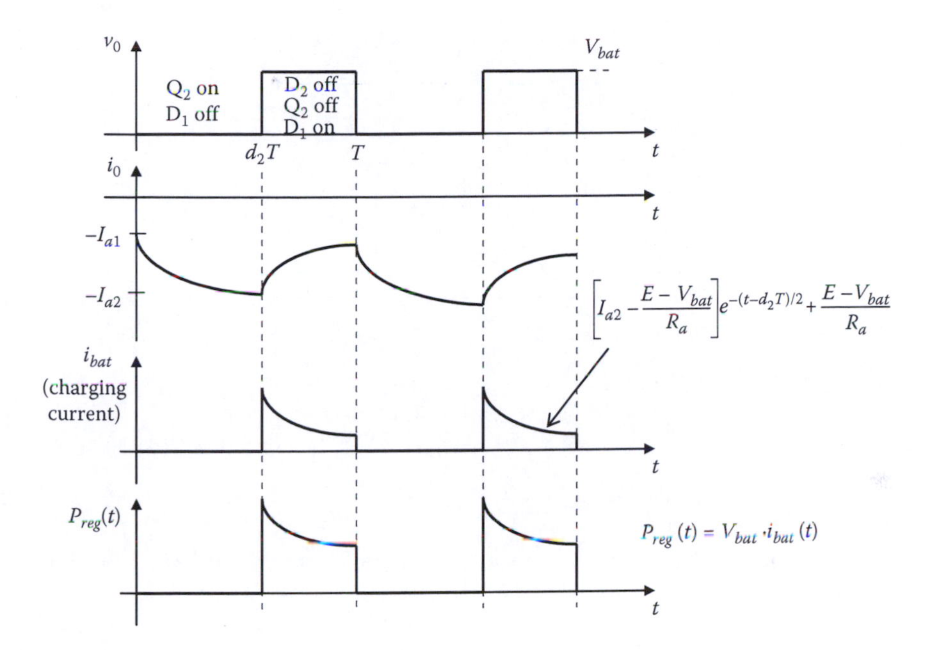

FIGURE 8.18
Voltage, current and power waveforms during regenerative braking.

The average regenerative power obtained by inserting Equation 8.18 into Equation 8.17 and integrating is given by

$$\langle P_{reg} \rangle = \frac{V_{bat}^2}{R_a}\left[\left(\frac{E}{V_{bat}} - 1\right)(1 - d_2) + \frac{\tau}{T}\left\{\frac{e^{(1-d_2)T/\tau} + e^{d_2T/\tau} - e^{T/\tau} - 1}{1 - e^{T/\tau}}\right\}\right] \quad (8.19)$$

The regenerative energy per cycle is

$$\int_0^T P_{reg}(\gamma)d\gamma = \int_{d_2T}^T P_{reg}(\gamma)d\gamma = \langle P_{reg} \rangle T$$

8.3 Operating Point Analysis

The following discussion presents the steady-state operating point analysis of the vehicle system at the intersection of the motor torque–speed characteristics with the road load characteristics. Four operating points are chosen

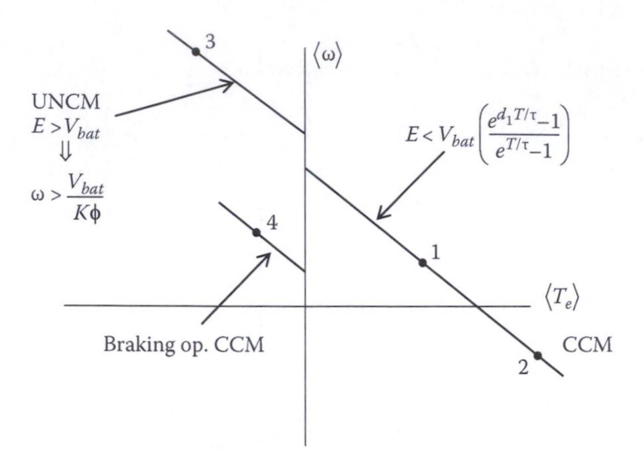

FIGURE 8.19
Motor speed–torque characteristics for four chopper modes.

for analysis in three different quadrants of the motor speed–torque plane, representing four chopper operating modes discussed in Sections 8.2.2.3 through 8.2.2.6. These modes are acceleration CCM in quadrant 1 (scenario 1), acceleration CCM in quadrant IV (scenario 2), acceleration UNCM in quadrant 2 (scenario 3), and braking CCM in quadrant 2 (scenario 4). The four scenarios are shown in Figure 8.19.

Now, recall the tractive force versus vehicle speed characteristics of Chapter 2, which essentially defines the speed–torque characteristics of the electric or hybrid vehicle load. The tractive force versus velocity characteristics of Figure 2.11 can be converted into an equivalent vehicle load speed–torque characteristics, knowing the transmission gear ratio and the vehicle wheel radius. The steady-state operating point of the vehicle for certain conditions of the motor drive system and road load characteristics can be obtained by overlaying the two speed–torque characteristics on the same plot. The steady-state operating points at the intersection of the motor and road load characteristics are shown in Figure 8.20.

Scenario 1
In this scenario, the vehicle is moving forward on a level roadway with a constant velocity. The chopper is in the acceleration CCM of operation.

Scenario 2
The chopper is operating in the acceleration CCM, yet the vehicle is moving backward on a steep uphill road. If the duty ratio has not yet reached 100%, d_1 can be increased all the way up to 1 by depressing the acceleration pedal further and increasing the torque output of the motor. Increasing d_1 will raise the motor speed–torque characteristics vertically upward enabling a possible steady-state operating point in the first quadrant. If the motor rating has reached its limit, then the torque cannot be increased further to overcome the road load resistance. The slope is too steep for the capacity of

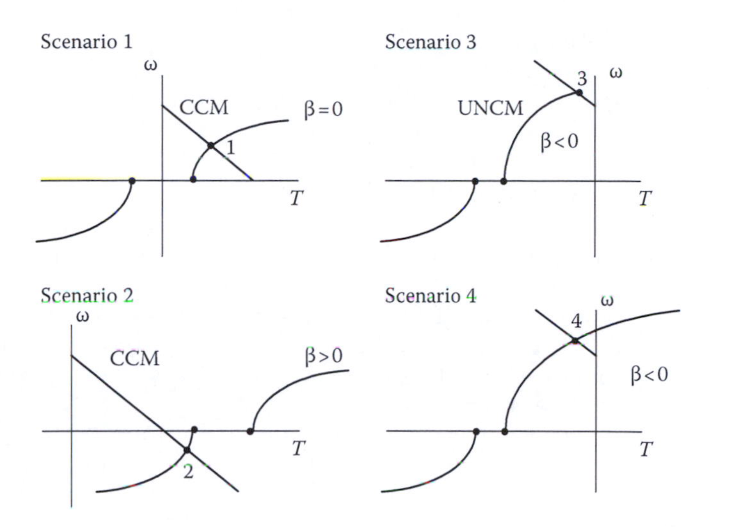

FIGURE 8.20
Operating point analysis for the scenarios.

the motor and the vehicle rolls backward. The wheel back drives the motor in this case.

Scenario 3
The vehicle is going downhill with the chopper operating in the acceleration mode. The angle of the acceleration pedal will have no bearing on the steady-state operating point. There will be uncontrolled regeneration into the energy source.

Scenario 4
The vehicle is going downhill in a controlled fashion using the brake pedal. The speed of the vehicle going down the slope is in control of the driver.

Example 8.1

An electric vehicle drivetrain with a 72 V battery pack is powered by a two-quadrant chopper-driven DC motor as shown in Figure 8.7. d_1 is the duty ratio for the acceleration operation, while d_2 is the duty ratio for braking operation.

The various parameters are given below:

The EV parameters are

$$m = 1000\text{ kg},\ C_D = 0.2,\ A_F = 2\text{ m}^2,\ C_0 = 0.009,\ C_1 = 0$$

$$\rho = 1.1614\text{ kg/m}^3 \text{ and } g = 9.81\text{ m/s}^2$$

$$r_{wh} = \text{radius of wheel} = 11\text{ in.} = 0.28\text{ m}$$

Motor and controller parameters are

Rated armature voltage $V_{arated} = 72\,V$,
Rated armature current $I_{arated} = 400\,A$,
$R_a = 0.5\,\Omega$, $L_a = 8\,mH$, $k\phi = 0.7\,V\,s$,
f_s = chopper switching frequency = 400 Hz

(a) Find the series filter inductance L_f, so that worst case motor armature current ripple is 1% of I_{arated}.
(b) The vehicle road load characteristics on a level road is $T_{TR} = 24.7 + 0.0051\omega_{wh}^2$. What is the EV steady-state speed on a level road if $d_1 = 0.7$? Assume an overall gear ratio of 1 for the transmission system. Find the chopper mode of conduction [Scenario 1].
(c) What is the percent ripple in the armature current for the operating point in (b)?
(d) The vehicle road load characteristics on a grade of 5.24% ($\beta = 3°$) is $T_{TR} = 119.1 + .0051\omega_{wh}^2$. What is the EV steady-state speed for $d_1 = 0.7$? [Scenario 2].
(e) What is the EV speed on grade of −5.24% using brake pedal with $d_2 = 0.5$? [Scenario 4].

Solution

(a) From Equation 8.3 with $x = e^{T_P/\tau}$, $x > 1$

$$\Delta i_a = I_{a1} - I_{a0}$$

$$= \frac{V_{bat}}{R_a}\left[\frac{1 + x - x^{d_1} - x^{(1-d_1)}}{x - 1}\right]$$

The worst case condition is when $d_1 = 0.5$. Therefore,

$$0.01 \times 400 = \frac{72}{0.5}\left[\frac{1 + x - x^{1/2} - x^{1/2}}{x - 1}\right] \Rightarrow 1.0278 + 0.972x - 2\sqrt{x} = 0$$

Solving for x we get, $x = 1, 1.118$. Since $x > 1$, take $x = 1.118$
Therefore,

$$1.118 = e^{T_P/\tau} \Rightarrow \tau = 8.96T_P = 22.4\text{ ms}$$

Now,

$$\tau = \frac{L_a + L_f}{R_a} = 22.4 \times 10^{-3}$$

$$\Rightarrow L_f = 3.2\text{ mH}$$

(b) The motor steady-state torque–speed characteristics using Equation 8.7 is

$$\langle\omega_m\rangle = \frac{0.7\times 72}{0.7} - \frac{0.5}{0.7^2}\langle T_e\rangle \quad \Rightarrow \langle\omega_m\rangle = 72 - 1.02\langle T_e\rangle$$

The steady-state operating point is the point of intersection of the motor torque–speed characteristic and the vehicle road load characteristic $T_{TR} = 24.7 + 0.0051\omega_{wh}^2$. Solving the two equations, the operating point is

$$T^* = 32.4\ \text{N} \quad \text{and} \quad \omega^* = 38.9\ \text{rad/s}$$

Now,

$$V_{bat}\left(\frac{e^{d_1T_P/\tau} - 1}{e^{T_P/\tau} - 1}\right) = 72\times\frac{e^{0.7\times 2.5/22.4} - 1}{e^{2.5/22.4} - 1} = 49.6$$

$$\text{And } E = k\phi\langle\omega\rangle = 0.7\times(38.9) = 27.2$$

Therefore,

$$V_{bat}\left(\frac{e^{d_1T/\tau} - 1}{e^{T/\tau} - 1}\right) > E$$

and hence, the chopper is operating in CCM. It can also be shown that $I_{a0} > 0$, verifying that the operation is indeed in CCM.

(c) The ripple in the armature current for the operating point in part (b) is

$$\Delta i_a = \frac{72}{.5}\left[\frac{1 + 1.118 - 1.118^{.7} - 1.118^{(1-.7)}}{1.118 - 1}\right] = 3.37\ \text{A}$$

The ripple current magnitude is still less than 1% of rated current.

$$\langle i_a\rangle = \frac{\langle T^*\rangle}{k\phi} = \frac{32.4}{0.7} = 46.3\ \text{A}$$

Therefore, the percentage ripple is

$$\%\ \text{ripple} = \frac{\Delta i_a}{\langle i_a\rangle}\times 100\% = \frac{3.37}{46.7}\times 100\% = 7.28\%$$

(d) The steady-state operating point for a grade of 5.24% is at the intersection of the motor torque–speed characteristic $\langle\omega_m\rangle = 72 - 1.02\langle T_e\rangle$ and the vehicle road load characteristic $T_{TR} = 119.1 + 0.0051\omega_{wh}^2$. The two solutions are $\omega^* = -40.8$ and 233. The correct solution is $\omega^* = -40.8$ since the other speed is too high. The vehicle in this case is actually rolling backward due to insufficient power from the propulsion unit. It can be verified that the maximum gradability of the vehicle is 2.57%.

(e) The vehicle road load characteristic for a slope of −5.24% is $T_{TR} = -119.1 + 0.0051\omega_{wh}^2$. The motor torque–speed characteristic in the braking CCM from Equation 8.14 is

$$\langle\omega_m\rangle = \frac{(1-d_2)V_{bat}}{K\phi} - \frac{R_a}{(K\phi)^2}\langle T_e\rangle \Rightarrow \langle\omega_m\rangle = 51.4 - 1.02\langle T_e\rangle$$

Solving for the operating point from the two torque–speed characteristics, $T^* = -57.4, 347$. The negative value is the solution since the chopper is in braking mode. This gives the steady-state speed as $\omega^* = 51.4 - 1.02(-57.4) = 109.9$ rad/s. The vehicle speed is $V^* = 0.28 \times 109.9 = 30.77$ m/s $= 68.8$ mi/h.

Now,

$$E = 0.7 \times 109.9 = 76.9 \geq 72\frac{1-e^{-0.5\times 0.1116}}{1-e^{-0.1116}} = 37$$

which satisfies the condition of CCM for braking given by Equation 8.12.

8.4 AC Drives

The synchronous speed of AC machines is proportional to the supply frequency ($\omega_s = 120\ f/P$), which means that the speed can be controlled by varying the frequency of the AC input voltage. The power electronic converter supplying the variable voltage—variable frequency output to an AC machine (induction or synchronous) from the available energy source (typically in the form of a fixed DC voltage)—is known as the inverter. The inverter consists of six switches and, through appropriate control, shapes the available DC voltage into balanced three-phase AC voltage of the desired magnitude and frequency. The inverter can be broadly classified as a voltage source inverter or a current source inverter. The voltage source inverter, shown in Figure 8.21, is common for electric and hybrid vehicle applications, where the source typically delivers a stiff voltage. The six-switch voltage source inverter can operate in the six-step mode or in the pulse width modulation (PWM) mode. The inverter output invariably has a number of harmonic components in addition to the desired fundamental

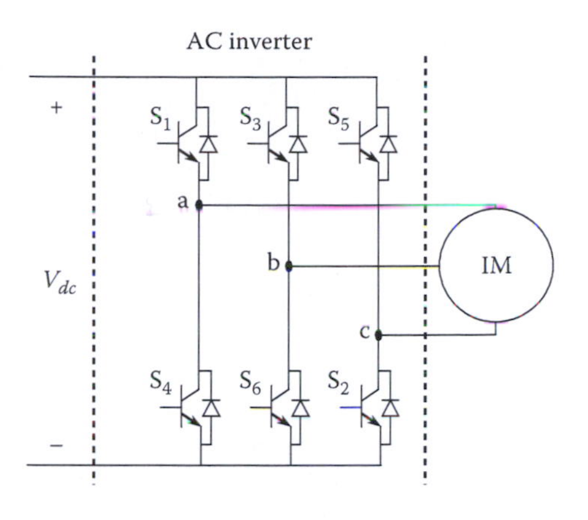

FIGURE 8.21
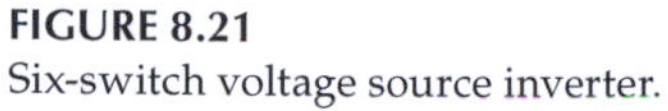
Six-switch voltage source inverter.

voltage component. The PWM is used to minimize the harmonic contents of the output voltage signal. The electronic controller generates the gate switch signals for the inverter power devices using a PWM method and control commands to process the power flow and deliver the output voltage at desired voltage and frequency.

The six-switch inverter topology is used for induction machines as well as for PM or any other synchronous machines. The most common method of speed control used in industrial applications of induction motors is the volts/hertz scalar control method. The performance requirements of an electric and hybrid electric vehicle motor drive are much more rigorous than what can be achieved from a scalar control method; hence, vector control methods based on the reference frame transformation theory are used. The power electronics circuit used for the induction motor drive is the six-switch topology irrespective of the type of control method used.

8.4.1 Six-Step Operation

The six-step operation is the simplest approach of generating AC voltage using the six-switch inverter. For the sake of analysis, let us replace the transistor and diodes with ideal switches, which results in a simplified equivalent inverter circuit shown in Figure 8.22. The DC voltage is represented here as V_{dc}. The characteristics of the ideal inverter are as follows: (1) switches can carry current in both directions and (2) total number of possible SWS is $2^6 = 64$. Some of the SWS are not allowed. For example, S_1 and S_4 cannot be on at the same time. The operation of the inverter can be divided into six intervals between 0 and 2π radians with each interval being of $\pi/3$ radians duration. In each interval, three of the switches are on and three are off. This operation is known as six-step operation. The six

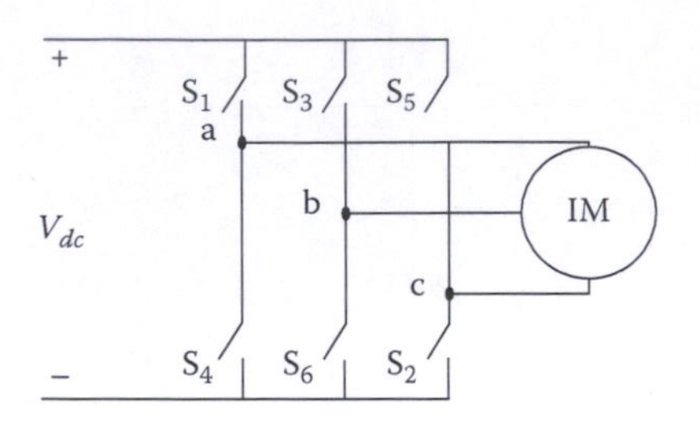

FIGURE 8.22
Ideal six-switch voltage source inverter.

interval periods identified with the operating switches during that period are as follows:

$$\begin{array}{ll} 1 \rightarrow 0 < wt < \pi/3: & S_1S_5S_6 \\ 2 \rightarrow \pi/3 < wt < 2\pi/3: & S_1S_2S_6 \\ 3 \rightarrow 2\pi/3 < wt < \pi: & S_1S_2S_3 \\ 4 \rightarrow \pi < wt < 4\pi/3: & S_4S_2S_3 \\ 5 \rightarrow 4\pi/3 < wt < 5\pi/3: & S_4S_5S_3 \\ 6 \rightarrow 5\pi/3 < wt < 2\pi: & S_4S_5S_6 \end{array}$$

The gating signals for the electronic switches and the resulting output AC voltage for six-step operation are shown in Figure 8.23. The line-to-line voltage and the line-to-neutral voltage, that is, voltage of one phase, are shown in the figure.

In three-phase machines, three-wire systems are used, where the line terminals *a*, *b*, and *c* are connected to the motor with neutral terminal *n* kept hidden. The relationships between line-to-line voltages (i.e., line voltages) and line-to-neutral voltages (i.e., phase voltages) are

$$\begin{aligned} v_{ab} &= v_{an} - v_{bn} \\ v_{bc} &= v_{bn} - v_{cn} \\ v_{ca} &= v_{cn} - v_{an} \end{aligned} \tag{8.20}$$

There are three unknowns and two linearly independent equations (v_{an}, v_{bn}, v_{cn}). Therefore, we need another equation to find the inverse solution. For a balanced three-phase electrical system, we know that $v_{an}+v_{bn}+v_{cn}=0$. Therefore, the line and phase voltage relationships can be written in matrix format as

$$\begin{bmatrix} 1 & -1 & 0 \\ 0 & 1 & -1 \\ -1 & 0 & 1 \\ 1 & 1 & 1 \end{bmatrix} \begin{bmatrix} v_{an} \\ v_{bn} \\ v_{cn} \end{bmatrix} = \begin{bmatrix} v_{ab} \\ v_{bc} \\ v_{ca} \\ 0 \end{bmatrix}$$

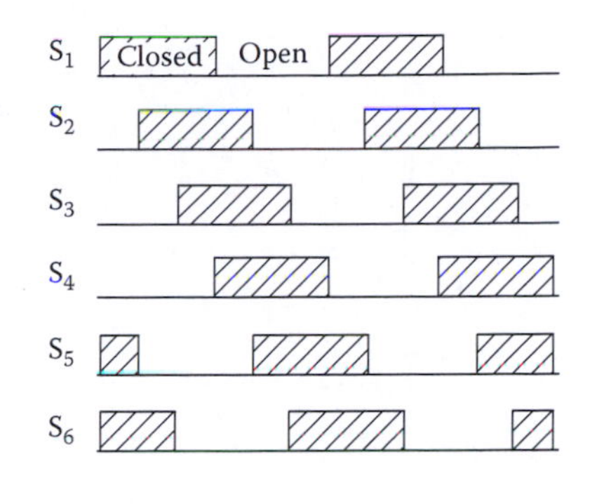

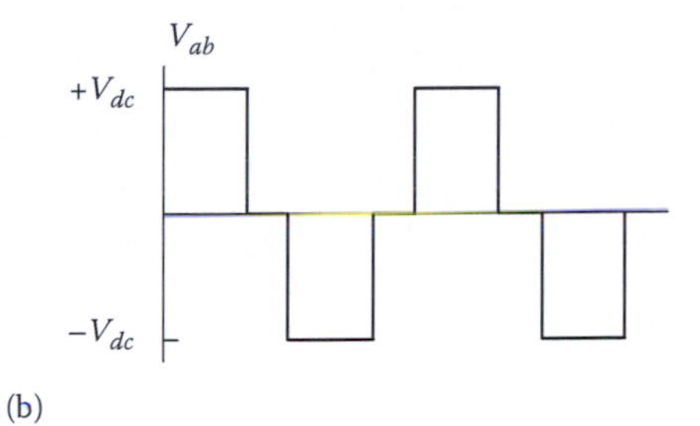

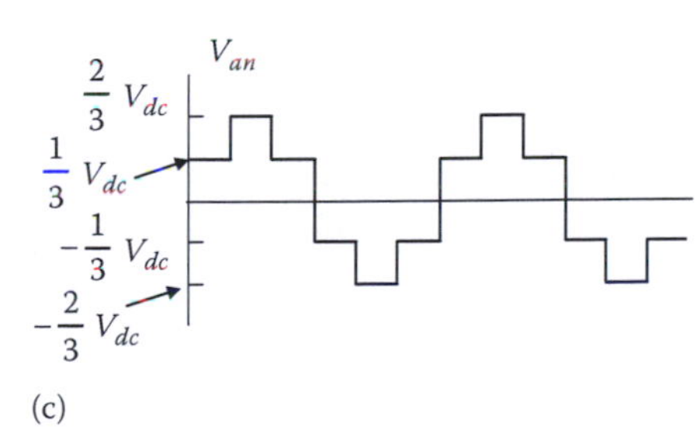

FIGURE 8.23
Six-step inverter gate signals and output voltages.

The solutions for the phase voltages are

$$
\begin{aligned}
v_{an} &= -\frac{1}{3}\left[v_{bc} + 2v_{ca}\right] \\
v_{bn} &= \frac{1}{3}\left[v_{bc} - v_{ac}\right] \\
v_{cn} &= -\frac{1}{3}\left[v_{bc} + v_{ac}\right]
\end{aligned}
\tag{8.21}
$$

The phase voltages are useful for per phase analysis of the three-phase systems. The phase voltage v_{an} of the six-step inverter output can be derived using Equation 8.21 as shown in Figure 8.24. Now, the question

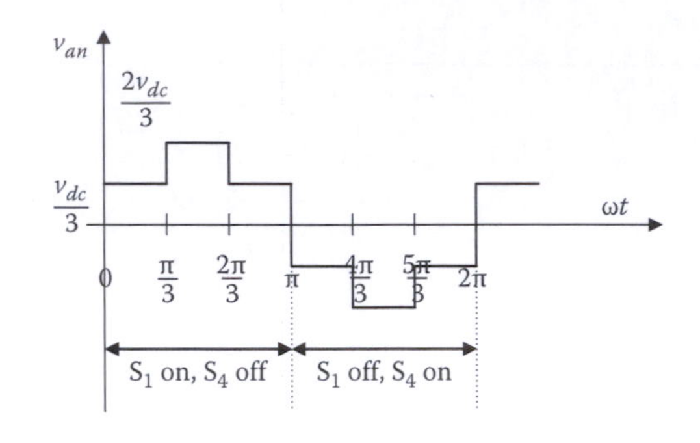

FIGURE 8.24
Phase voltage during switch S_1 operation.

is what type of switch is required for the inverter. Let us first consider switch S_1. When S_1 is off, $v_{S1} = V_{dc}$. When S_1 is on, S_4 is off. The phase voltage of the six-step inverter is shown in Figure 8.24. The output voltage can be filtered to make the supply more sinusoidal. The supply is naturally filtered when supplying an inductive load, such as the electric motor. It is only the fundamental component of the supply that contributes to electromagnetic torque production, while the higher harmonic components are responsible for part of the losses in the system. The interaction of the induction motor with the higher harmonic components is analyzed in Section 8.4.1.1. The fundamental component of the phase voltage and the current through the switch into phase a winding is shown in Figure 8.25. Since the motor is inductive, line current lags the phase voltage. Therefore, we must use bidirectional current switches since i_a is both positive and negative.

8.4.1.1 Harmonic Analysis

Let us consider the switching interval number 3 where $2\pi/3 < \omega t < \pi$. The switches that are on during this period are S_1, S_2, and S_3. The inverter

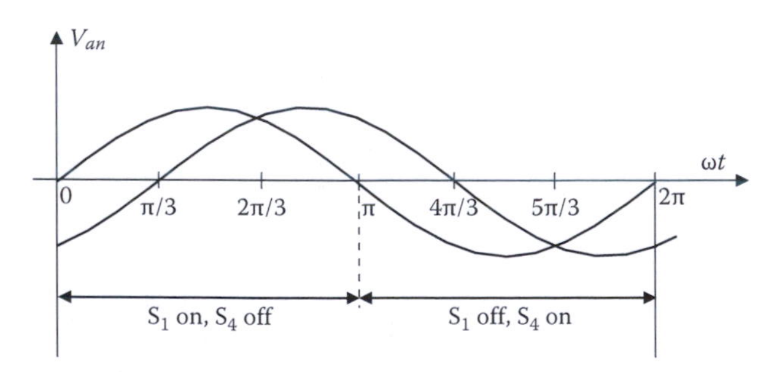

FIGURE 8.25
Phase voltage and current during S_1 operation.

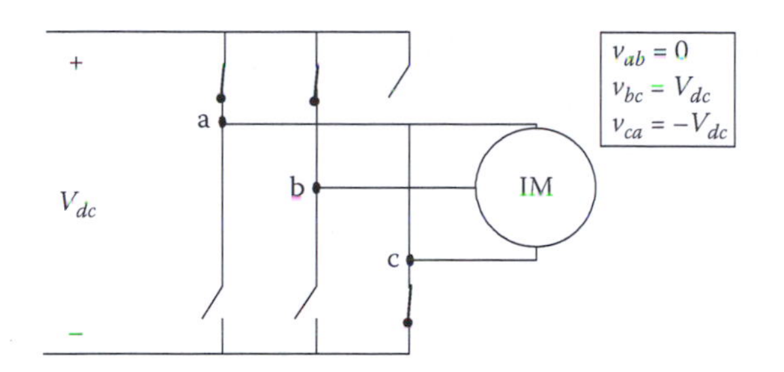

FIGURE 8.26
Inverter switch connection in switching interval state number 3.

configuration in this interval is shown in Figure 8.26. The line-to-line square wave voltage can be written in terms of the fundamental and harmonic components using Fourier series analysis as

$$v_{ab} = \frac{2\sqrt{3}}{\pi} V_{dc} \left\{ \sin\left(\omega t + \frac{\pi}{6}\right) + \frac{1}{5}\sin\left(5\omega t - \frac{\pi}{6}\right) + \frac{1}{7}\sin\left(7\omega t + \frac{\pi}{6}\right) + \cdots \right\} \tag{8.22}$$

Note that the harmonics present are the $6n \pm 1$ (n is an integer) components and that the triplen harmonics are absent. The harmonic phase voltages that are 30 degrees phase shifted from the line voltages are

$$v_{an} = \frac{2}{\pi} V_{dc} \left\{ \sin(\omega t) + \frac{1}{5}\sin(5\omega t) + \frac{1}{7}\sin(7\omega t) + \cdots \right\} \tag{8.23}$$

The dominant harmonic components are shown in Figure 8.27.

The harmonics do not contribute to the output power, but they certainly increase the power losses that reduce the efficiency and increase the thermal loading of the machine. The harmonic losses do not vary significantly with the load. The interactions of fundamental air gap mmf with harmonic air gap mmfs produce torque pulsations, which may be significant at low speeds.

8.4.2 Pulse Width Modulation

The PWM techniques are used to mitigate the adverse effects of harmonics in the inverter [3,5]. The harmonics in the output PWM voltage are not eliminated but shifted to a much higher frequency, which makes the filtering a lot easier. The controllability of the amplitude of the fundamental output voltage is another advantage of PWM. The PWM techniques developed over the years are sinusoidal PWM, uniform sampling PWM, selective harmonic elimination PWM, space

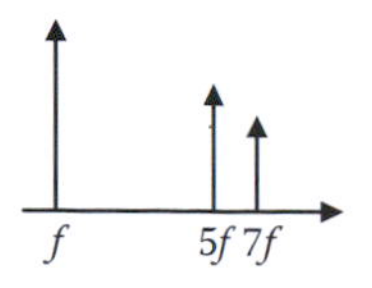

FIGURE 8.27
Dominant harmonic components in inverter output voltages.

vector (SV) PWM, and random PWM. The two commonly used PWM techniques, the sinusoidal PWM and the space vector PWM, are discussed in the following sections.

8.4.2.1 Sinusoidal PWM

In the sinusoidal PWM method, a sinusoidal control signal V_a is compared with a high-frequency triangular waveform V_T to generate the inverter switch gating signals. The frequency of V_T establishes the inverter switching frequency f_c. The magnitude and frequency of the sinusoidal signal are controllable, but the triangular signal magnitude and frequency are kept constant. The sinusoidal control signal V_a modulates the switch duty ratio, and its frequency f is the desired fundamental frequency of the inverter. For the three-phase voltage generation, the same V_T is compared with three sinusoidal control voltages v_A, v_B, and v_C, which are 120° out of phase with respect to each other to produce a balanced output. The switches are controlled in pairs (S_1,S_4), (S_2,S_5), and (S_3,S_6). When one switch in a pair is closed, the other switch is open. In practice, there has to be a blanking pulse between the changes of control signals for the switches in a pair to ensure that there is no short circuit in the inverter. This is necessary since practical switches take finite time to turn-on and turn-off. The three-phase sinusoidal PWM signals are shown in Figure 8.28. The switch control signals follow the logic given below:

S_1 is on when $v_A > V_T$
S_2 is on when $v_C > V_T$
S_3 is on when $v_B > V_T$
S_4 is on when $v_A < V_T$
S_5 is on when $v_C < V_T$
S_6 is on when $v_B < V_T$

The amplitude modulation index is

$$m = \frac{V_{a,peak}}{V_{T,peak}} = \frac{A}{A_m} \tag{8.24}$$

where
A is the amplitude of the reference sine wave
A_m is the amplitude of the triangular carrier wave

Let us define the ratio of carrier frequency and fundamental frequency as

$$p = \frac{f_C}{f}$$

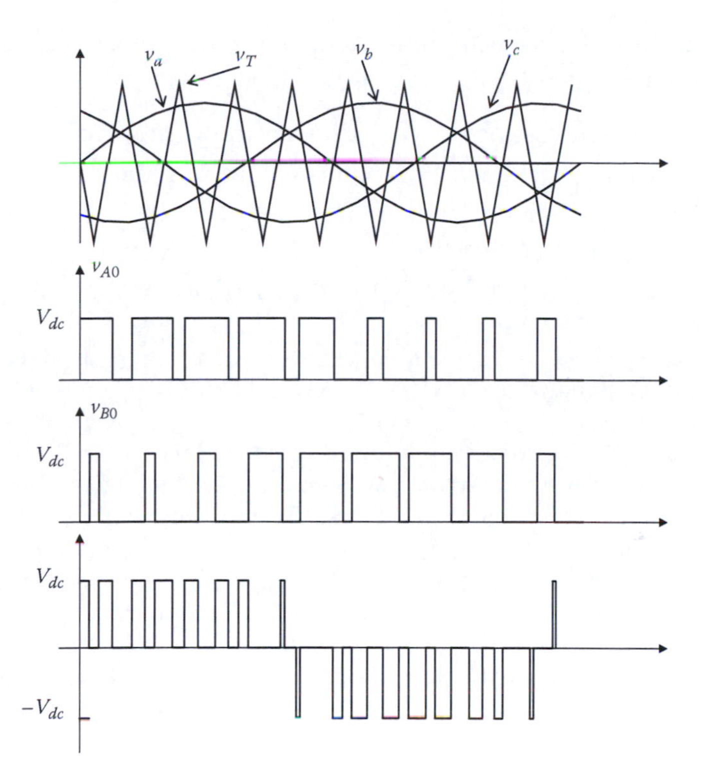

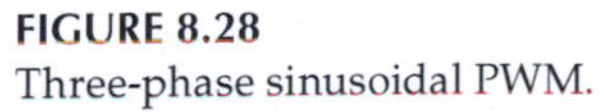
FIGURE 8.28
Three-phase sinusoidal PWM.

The rms value of the fundamental inverter output phase voltage V_{ph} is

$$V_{ph,1} = m\frac{V_{dc}}{2\sqrt{2}}, \quad m \le 1 \tag{8.25}$$

where V_{dc} is the DC input voltage to the inverter. Note that $V_{ph,1}$ increases with m until $m=1$. For $m>1$, modulation ceases to be sinusoidal PWM wherein it is referred to as over-modulation. The rms value of the fundamental in the line-to-line voltage is

$$V_{LL,1} = m\frac{\sqrt{3}V_{dc}}{2\sqrt{2}}, \quad m \le 1 \tag{8.26}$$

There are two types of modulations depending on the value of p, called synchronous modulation and asynchronous modulation. In synchronous modulation, $p=3n$, $n=1, 2, \ldots$, making the carrier wave symmetrical with respect to the three-phase reference voltages v_a, v_b, and v_c. Otherwise, the modulation is called asynchronous modulation. The characteristics of asynchronous modulation are

1. Pulse pattern does not repeat itself identically from cycle to cycle.
2. Subharmonics of f and a DC component are introduced.
3. Subharmonics cause low-frequency torque and speed pulsations known as frequency beats.
4. For large p, frequency beats are negligible. For small p, frequency beats may be significant.
5. For small p, synchronous modulation is used. Preferably, p is an odd multiple of three.

The boundary of sinusoidal modulation is reached when $m=1$. The relationship between the fundamental component of phase voltage and m ceases to be linear for values greater than 1. For sufficiently large values of m, the output phase voltage becomes a square wave with maximum amplitude of the fundamental equal to $2V_{dc}/\pi$ (relate to Equation 8.23). This is essentially the six-step operation of the inverter. Note that the amplitude of the fundamental on the boundary of linear sinusoidal PWM is only 78.5% of the maximum value. The modulation with $m>1$ is known as overmodulation; lower-order harmonics are introduced in this range of modulation.

8.4.2.2 Harmonics in Sinusoidal PWM

The inverter output phase voltage contains harmonics that are of odd multiples of the carrier frequency f_C (i.e., f_C, $3f_C$, $5f_C$, …). The waveform also contains side bands centered around multiples of f_C and given by

$$f_h = k_1 f_C \pm k_2 f = \left(k_1 p \pm k_2\right) f$$

where k_1+k_2 is an odd integer. The harmonic frequency centers are at $k_1 f$ for odd integers of k_1, while the side bands are symmetrically located without a center for even integers of k_1. Note that the magnitudes of band frequency harmonics decrease rapidly with increasing distance from band center. The harmonic frequency components for $k_1=1$ and 2 along with the bandwidth are shown in Figure 8.29. The bandwidth of harmonics is the range of frequency over which the harmonics are considered dominant. The frequency bandwidth increases with m. The dominant harmonic among the side bands are at frequencies $(p\pm2)f$ and $(2p\pm1)f$.

8.4.2.3 Space Vector PWM

The voltage space vectors embedded in the dq models and vector controllers of AC machines present a highly compatible method for the control of

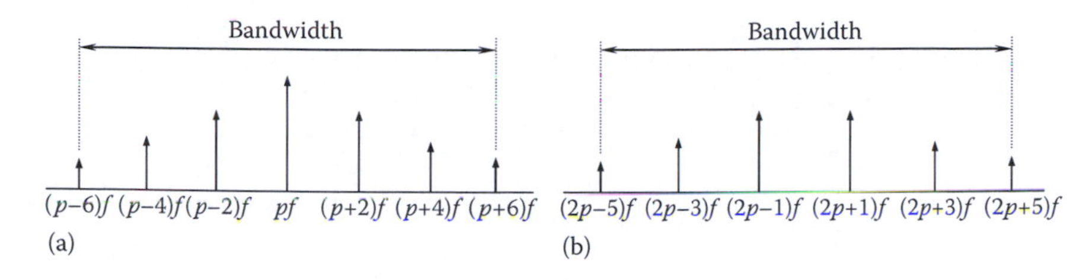

FIGURE 8.29
Harmonic frequency components for $k_1 = 1$ (a) and $k_1 = 2$ (b).

three-phase PWM inverters. The *dq* voltage commands for the direct and quadrature axes voltages generated by the controller to drive a motor are converted into an equivalent PWM signal to be applied to the gates of the six inverter switches. For variable frequency drive applications, space vector PWM is very popular because of its superior performance compared with other voltage PWM techniques.

The three-phase voltage source inverter can assume one of eight ($2^3 = 8$) states only, since each switch in a phase leg can assume either an *on* or *off* position. Each of these eight states is associated with a specific set of values of the three-phase voltages. Based on the state, the machine phase winding terminals are connected to the upper or lower side of the DC link. For the terminal voltages, eight voltage vectors are defined using 3-bit digital states as 000, 100, 110, 010, 011, 001, 101, and 111. For example, the state 011 stands for Phase-a connected to the lower side of the DC bus, while Phases b and c connected to the upper side. Alternatively, a 0 indicates that the lower switch is turned on and a 1 indicates that the upper switch is on. This means that the inverter can generate eight stationary voltage vectors $\vec{V}_0$ through $\vec{V}_7$ with the subscript indicating the corresponding state of the inverter. These voltage vectors are

$$\vec{V}_1 = \begin{bmatrix} 0 \\ 0 \\ 0 \end{bmatrix}, \quad \vec{V}_1 = \begin{bmatrix} 1 \\ 0 \\ 0 \end{bmatrix}, \quad \vec{V}_2 = \begin{bmatrix} 1 \\ 1 \\ 0 \end{bmatrix}, \quad \vec{V}_3 = \begin{bmatrix} 0 \\ 1 \\ 0 \end{bmatrix}$$

$$\vec{V}_4 = \begin{bmatrix} 0 \\ 1 \\ 1 \end{bmatrix}, \quad \vec{V}_5 = \begin{bmatrix} 0 \\ 0 \\ 1 \end{bmatrix}, \quad \vec{V}_6 = \begin{bmatrix} 1 \\ 0 \\ 1 \end{bmatrix}, \quad \vec{V}_7 = \begin{bmatrix} 1 \\ 1 \\ 1 \end{bmatrix}$$

The SWS, voltage vectors and the associated switch conditions are given in Table 8.2. The eight voltage vectors represent two null vectors ($\vec{V}_0$ and $\vec{V}_7$) and six active-state vectors forming a hexagon (Figure 8.30). A simple circuit analysis will reveal that the phase voltage magnitudes vary between ±(2/3)

TABLE 8.2
Space Vector Switching States

Switching State	On devices	Voltage Vector
0	$S_4S_6S_2$	$\vec{V}_0$
1	$S_1S_6S_2$	$\vec{V}_1$
2	$S_1S_3S_2$	$\vec{V}_2$
3	$S_4S_3S_2$	$\vec{V}_3$
4	S_4S_3S	$\vec{V}_4$
5	$S_4S_6S_5$	$\vec{V}_5$
6	$S_1S_6S_5$	$\vec{V}_6$
7	$S_1S_3S_5$	$\vec{V}_7$

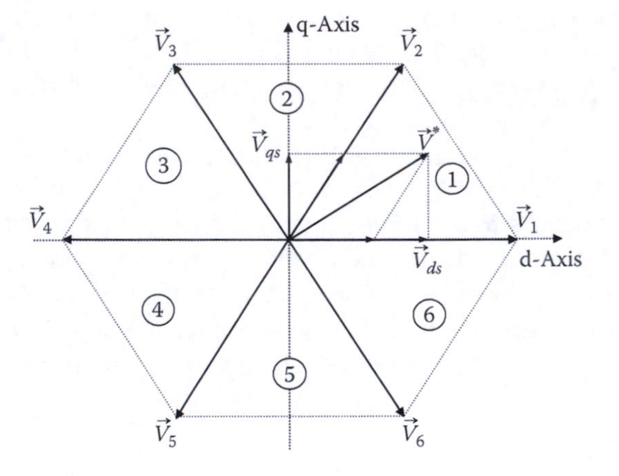

FIGURE 8.30
Inverter SWS and space vectors.

V_{dc} in steps of $(1/3)V_{dc}$ in the six active states. This is analogous to the six-step operation where the voltage magnitude variations are similar for the six steps. For example, in vector state $\vec{V}_1$, switches S_1, S_2, and S_6 are on with Phase-a connected to the upper side of the DC bus, and Phase-b and Phase-c connected to the lower side. In this state, the inverter phase voltages are $v_{an} = (2/3)V_{dc}$, $v_{bn} = -(1/3)V_{dc}$, and $v_{cn} = -(1/3)V_{dc}$. This is the step in the interval $\pi/3 < \omega t < 2\pi/3$ in Figures 8.23 and 8.24 for six-step operation. By transforming the three-phase *abc* variables into *dq* variables, the *d* and *q* voltages are $v_d = (2/3)V_{dc}$ and $v_q = 0$. Therefore, voltage vector $\vec{V}_1$ has a magnitude of $(2/3)V_{dc}$ and is along the *d*-axis in the *dq*-plane. It can be shown similarly that each of the six active-state vectors has a magnitude of (2/3)V_{dc} and is displaced by 60° with respect to each other in the *dq*-plane. The six active-state vectors form a hexagon as shown in Figure 8.30 with the null vectors remaining in the origin. The six active-state vectors can be represented in space vector form as

$$\vec{V}_k = \frac{2}{3} V_{dc} e^{j(k-1)\pi/3} \tag{8.27}$$

where $k=1, \ldots 6$. Note that the same notation $\vec{V}_k$ is used to denote the voltage vector in the *abc* frame in relation to the switch location and the space vector in mathematical terms. The space vector can be easily related to the *dq* variables in the two co-ordinate *dq* reference frame.

The SV PWM operates in a modulated fashion including the null vectors within a fixed time period in contrast to the six-step operation of the inverter. The six-step operation uses only the six active states moving in the sequence $\vec{V}_1 \to \vec{V}_2 \to \vec{V}_3 \to \vec{V}_4 \to \vec{V}_5 \to \vec{V}_6 \to \vec{V}_1$ with a duration of 60° in each one of the states. The objective in SV PWM is to generate the gating signals such that harmonically optimized PWM voltage is obtained at the output of the inverter.

8.4.2.4 Generation of SV PWM Switching Signals

The continuous space vector modulation technique is based on the fact that every reference voltage vector $\vec{V}^*$ inside the hexagon can be expressed as a combination of the two adjacent active-state vectors and the null-state vectors. Therefore, the desired reference vector imposed in each cycle is achieved by switching among four inverter states. The sector where the space vector $\vec{V}^*$ lies determines the two active-state vectors that will be used in generating the gate switching signals of a PWM period. The phase angle is evaluated from $\theta = \arctan(V_{qs}/V_{ds})$ and $\theta \in [0,2\pi]$. The phase angle is related to the relative magnitudes of V_{qs} and V_{ds}. For example, in sector 1, $0 \le \arctan(V_{qs}/V_{ds}) < \pi/3$; hence, $0 < V_{qs} < \sqrt{3}V_{ds}$. The following conditions are used to determine the sector where the space vector is located:

$$\text{Sector } 1: 0 < V_{qs} < \sqrt{3}V_{ds}$$

$$\text{Sector } 2: V_{qs} > \left|\sqrt{3}V_{ds}\right|$$

$$\text{Sector } 3: 0 < V_{qs} < -\sqrt{3}V_{ds}$$

$$\text{Sector } 4: 0 > V_{qs} > \sqrt{3}V_{ds}$$

$$\text{Sector 5}: V_{qs} < -\left|\sqrt{3}V_{ds}\right|$$

$$\text{Sector 6}: 0 > V_{qs} > -\sqrt{3}V_{ds}$$

Let us assume that $\vec{V}_k$ and $\vec{V}_{k+1}$ are the components of the reference vector $\vec{V}^*$ in sector k and in the adjacent active sectors $(k+1)$, respectively. In order to obtain optimum harmonic performance and minimum switching frequency for each of the power devices, the state sequence is arranged such that the transition from one state to the next is performed by switching only one inverter leg. This condition is met if the sequence begins with one zero state and the inverter poles are toggled until the other null state is reached. To complete the cycle, the sequence is reversed, ending with the first zero state. If, for instance, the reference vector sits in sector 1, the state sequence has to be…0127210…, whereas in sector 4, it is…0547450…. The central part of the space vector modulation strategy is the computation of both the active and the zero-state times for each modulation cycle. These are calculated by equating the applied average voltage to the desired value. Figure 8.31 demonstrates the on-times of vectors.

In the following equation, T_k denotes the required on-time of active-state vector $\vec{V}_k$, T_{k+1} denotes the required on-time of active-state vector $\vec{V}_{k+1}$, and (T_0+T_7) is the time for the null-state vector to cover the complete time period T_s [6]. The on-time is evaluated by the following equation:

$$\int_0^{T_s/2} \vec{V}dt = \int_0^{T_0/2} \vec{V}_0 dt + \int_{T_0/2}^{T_0/2+T_k} \vec{V}_k dt + \int_{T_0/2+T_k}^{T_0/2+T_k+T_{k+1}} \vec{V}_{k+1} dt + \int_{T_0/2+T_k+T_{k+1}}^{T_s/2} \vec{V}_7 dt \tag{8.28}$$

where $T_0+T_k+T_{k+1}+T_7=T_s$.

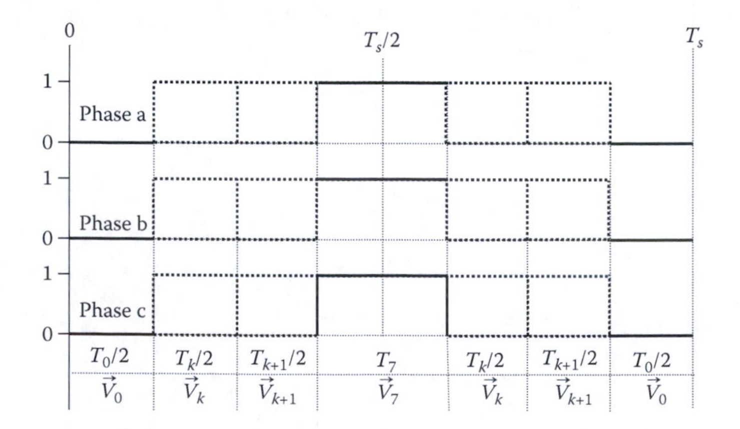

FIGURE 8.31
Diagram of on-times of vectors.

By taking $\vec{V}_0$ and $\vec{V}_7$ to be zero vectors, and $\vec{V}^*, \vec{V}_k$, and $\vec{V}_{k+1}$ to be constant vectors in a PWM period, Equation 8.28 can be simplified as

$$\frac{T_s}{2}\vec{V} = \frac{T_0}{2}\vec{V}_0 + \frac{T_k}{2}\vec{V}_k + \frac{T_{k+1}}{2}\vec{V}_{k+1} + \frac{T_7}{2}\vec{V}_7$$

$$\Rightarrow T_s\vec{V} = T_k\vec{V}_k + T_{k+1}\vec{V}_{k+1} \tag{8.29}$$

Using the space vector relation for $\vec{V}$ and Equations 8.27 and 8.29, it can be shown that

$$\begin{bmatrix} T_k \\ T_{k+1} \end{bmatrix} = \frac{\sqrt{3}T_s}{V_{dc}} \begin{bmatrix} \sin\frac{k\pi}{3} & -\cos\frac{k\pi}{3} \\ -\sin\frac{(k-1)\pi}{3} & \cos\frac{(k-1)\pi}{3} \end{bmatrix} \begin{bmatrix} V_{ds} \\ V_{qs} \end{bmatrix} \tag{8.30}$$

The duty ratios are related to the on-time as

$$\begin{bmatrix} T_k \\ T_{k+1} \end{bmatrix} = T_s \begin{bmatrix} D_k \\ D_{k+1} \end{bmatrix} \tag{8.31}$$

Comparing Equations 8.30 and 8.31, one can obtain the duty ratios as

$$\begin{bmatrix} D_k \\ D_{k+1} \end{bmatrix} = \sqrt{3} \begin{bmatrix} \sin\frac{k\pi}{3} & -\cos\frac{k\pi}{3} \\ -\sin\frac{(k-1)\pi}{3} & \cos\frac{(k-1)\pi}{3} \end{bmatrix} \begin{bmatrix} \frac{V_{ds}}{V_{dc}} \\ \frac{V_{qs}}{V_{dc}} \end{bmatrix} \tag{8.32}$$

The on-time for the null vectors are obtained as follows:

$$\begin{bmatrix} T_0 \\ T_7 \end{bmatrix} = \frac{1}{2} \begin{bmatrix} T_s - T_k - T_{k+1} \\ T_s - T_k - T_{k+1} \end{bmatrix} \tag{8.33}$$

Therefore, Equations 8.31 and 8.33 give the on-times T_k, T_{k+1}, T_0, and T_7 with the duty ratio given by Equation 8.32 within a PWM period. The on-times of the three-phase PWM generator for the controller are determined as follows:

$$\begin{bmatrix} T_A \\ T_B \\ T_C \end{bmatrix} = \begin{bmatrix} \vec{V}_k & \vec{V}_{k+1} & \vec{V}_7 \end{bmatrix} \begin{bmatrix} T_k \\ T_{k+1} \\ T_7 \end{bmatrix} \quad (k = 1, \ldots 6) \tag{8.34}$$

where

T_A, T_B, and T_C are the on-times of each phase
$\vec{V}_k$ is one of the six active operating states defined previously

For the instance shown in Figure 8.29, $\vec{V}^*$ is in sector 1. The on-times are obtained as

$$\begin{bmatrix} T_A \\ T_B \\ T_C \end{bmatrix} = \begin{bmatrix} \vec{V}_1 & \vec{V}_2 & \vec{V}_7 \end{bmatrix} \begin{bmatrix} T_k \\ T_{k+1} \\ T_7 \end{bmatrix} = \begin{bmatrix} 1 & 1 & 1 \\ 0 & 1 & 1 \\ 0 & 0 & 1 \end{bmatrix} \begin{bmatrix} T_k \\ T_{k+1} \\ T_7 \end{bmatrix}$$

The on-times of the switching gates are set up by the SV PWM controller by decomposing the rotating reference vector $\vec{V}^*$ into two components made of its neighboring space vectors. The duty cycles calculated from the two space vectors help establish the balanced three-phase voltages for the AC electric machine.

It was shown that only 78.5% of the inverter's capacity is used with sinusoidal PWM method. In SV PWM method, inverter's capability is improved by using a separate modulator for each of the inverter legs, generating three reference signals forming a balanced three-phase system. In this way, the maximum obtainable output voltage is increased with ordinary SV PWM up to 90.6% of the inverter capability. The SV PWM algorithm is fairly complex and computation intensive, but still well within the capabilities of the digital signal processors available today.

8.4.3 Current Control Methods

In systems where the output current depends not only on the input voltage but also on the load, a closed loop PWM method is necessary. The current PWM methods use current feedback information from sensors and generate PWM signals for the inverter gates in a closed loop algorithm. The measured three-phase currents are compared with the three reference current commands generated by the outer-loop controller. The error between the measurements and the reference signals are utilized in a PWM scheme to generate the gate switching signals. A number of current-controlled PWM methods have been developed ranging from fairly simple to rather complex ones. Some of these methods are hysteresis current controller, ramp-comparison controller, predictive current controller etc. The predictive current controller is one of the more complex ones where parameters of the load are used to predict the reference currents. Two of the simpler methods, the hysteresis current controller and the ramp-comparison controller, are described briefly in the following.

8.4.3.1 Hysteresis Current Controller

In the hysteresis current controller, the error between the measured current and the reference current is compared with a hysteresis band as shown in Figure 8.32a. If the current error is within the band, then PWM output remains unchanged. If the current error exceeds the band, then the PWM output is reversed forcing a change in the sign of *di/dt* slope. Mathematically stated, the PWM output is obtained from

$$\text{PWM} = \begin{cases} 0 & \text{if } \Delta i < -h/2 \\ 1 & \text{if } \Delta i > h/2 \end{cases}$$

A "0" PWM output signal cuts off the supply voltage in the controlled phase forcing the current to decay, while a "1" PWM output applies voltage to the

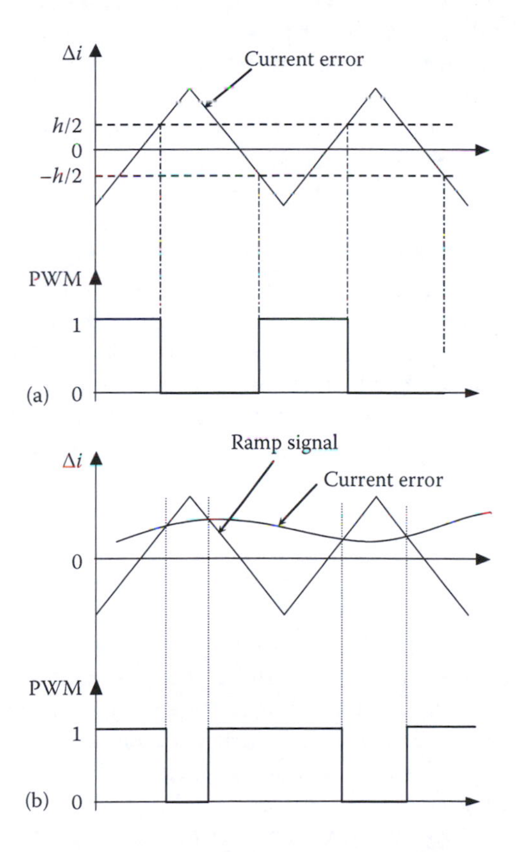

FIGURE 8.32
(a) Hysteresis current control and PWM output. (b) Ramp comparison control and PWM output.

phase resulting in an increase of the current. The voltage will then force the current to vary in such a way that it remains within the hysteresis band.

The advantage of such a controller is that the error remains within a certain band, which is known by the user. In the hysteresis controller, the switching frequency is unknown, which makes it difficult to design the filters. The switching frequency should be carefully monitored so that the inverter limits are not exceeded. In practical implementations, a frequency limit is used so as not to exceed the inverter maximum switching frequency. The hysteresis band is designed according to the limits of the device switching frequency. If the hysteresis band h is chosen to be very narrow, then the switching frequency will be very high, and not be compatible with the maximum switching frequency of the power devices. On the other hand, if the band is too wide then the current error will be too large.

The hysteresis current controller can be used in a three-phase PWM inverter with each phase having its own PWM controller. If the actual current is higher than the reference current by the amount of half of the hysteresis band, the lower leg switch of the bridge inverter is turned on to reduce the phase current. The difficulty in three-phase hysteresis control is that there may be conflicting requirements of switch conditions for the phases based on the output of the hysteresis controller. The difficulty arises from the interaction of the phases of the three-phase system and the independent hysteresis controllers for each phase. The consequence of this difficulty is that the current does not remain within the hysteresis band. For example, a current-increase command in Phase-a needs a return path through Phase-b or Phase-c lower legs. If Phase-b and Phase-c happen to have upper leg switches on during this instant, the current in Phase-a will not increase to follow the command, but will freewheel. In this case, it is possible that the current error of Phase-a will exceed the hysteresis current band. Using the *dq* transformation theory, it is possible to first transform the three-phase currents into two-phase *dq* currents and then impose the hysteresis control in the *dq* reference frame.

8.4.3.2 Ramp Comparison Controller

Another way of controlling the required stator current is to use a controller-based fixed-frequency ramp signal that stabilizes the switching frequency. The current error is first fed into a linear controller, which is typically of the proportional integral (PI) type. The output of the linear controller is compared with a high-frequency sawtooth-shaped triangular signal to generate the PWM switch signals. If the error signal is higher than the triangular signal, the PWM output signal will be "1" and in the other case, the output will be "0." The control actions are shown graphically in Figure 8.32b. The stator voltages will vary in order to minimize the current error signal. Three identical controllers are used in three-phase systems.

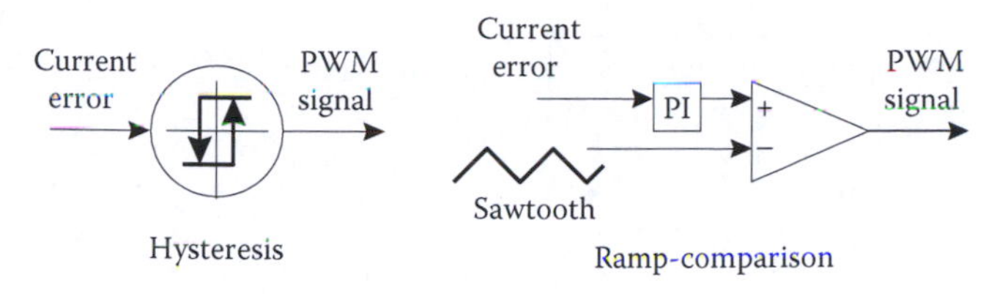

FIGURE 8.33
Hysteresis and ramp comparison techniques.

The ramp-comparison method has a fixed switching frequency set by the sawtooth wave frequency that makes it easier to ensure that the inverter switching frequency is not exceeded. There are more parameters to adjust in the ramp comparison controller, allowing greater flexibility of control compared with the hysteresis controller. The control parameters for the ramp comparison controller include gains of the linear controller and the magnitude and frequency of the sawtooth wave, whereas the only control parameter in a hysteresis controller is the width of the hysteresis band. The functional differences in the two methods are depicted in Figure 8.33.

The primary disadvantage of the ramp comparison controller is an increase in response time due to transport delay. The situation can be improved by using a high-gain proportional controller instead of a PI controller and by increasing the switching frequency of the sawtooth signal.

8.5 SRM Drives

The power electronic drive circuits for SRM drives are quite different from those of AC motor drives. The torque developed in an SRM is independent of the direction of current flow. Therefore, unipolar converters are sufficient to serve as the power converter circuit for the SRM unlike induction motors or synchronous motors, which require bidirectional currents. This unique feature of the SRM together with the fact that the stator phases are electrically isolated from one another generated a wide variety of power circuit configurations. The type of converter required for a particular SRM drive is intimately related to motor construction and the number of phases. The choice also depends on the specific application.

8.5.1 SRM Converters

The most flexible and the most versatile four-quadrant SRM converter is the bridge converter shown in Figure 8.34a, which requires two switches and

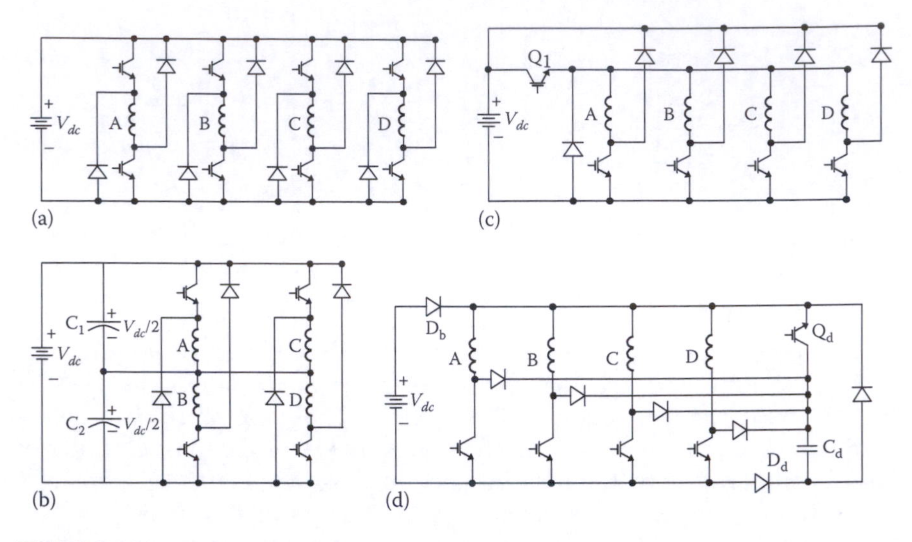

FIGURE 8.34
Converter topologies for SRM: (a) classic bridge power converter; (b) split-capacitor converter; (c) Miller converter, and (d) energy-efficient converter.

two diodes per phase [7,8]. The switches and the diodes must be rated to withstand the supply voltage plus any transient overload. During the magnetization period, both the switches are turned on and the energy is transferred from the source to the motor. Chopping or PWM, if necessary, can be accomplished by switching either one or both the switches during the conduction period according to the control strategy. At commutation, both switches are turned off and the motor phase is quickly defluxed through the freewheeling diodes. The main advantage of this converter is the independent control of each phase, which is particularly important when phase overlap is desired. The only disadvantage is the requirement of two switches and two diodes per phase. This converter is especially suitable for high-voltage, high-power drives.

The split-capacitor converter shown in Figure 8.34b has only one switch per phase but requires a split DC supply [7]. The phases are energized through either the upper or the lower DC bus rail and the midpoint of the two capacitors. Therefore, only one-half of the DC bus voltage can be applied for torque production. In order to maintain a power flow balance between the two supply capacitors, the switching device and the freewheeling diode are transposed for each phase winding, which means that the motor must have an even number of phases. Also the power devices must be rated to withstand the full-DC supply voltage.

In low-speed applications, where PWM current control is desirable over the entire range of operation, the bridge converter can be reduced to the circuit shown in Figure 8.34c, developed by Miller [8]. In this converter, chopping is performed by one switch common to all phases. The circuit requires $(n+1)$ switches for an n-phase motor. The main limitation of this circuit is that at higher speeds, the off-going phase cannot be de-energized fast enough because the control switch Q_1 keeps turning on intermittently, disabling forced demagnetization. A class of power converter circuits with less than two switches per phase for SRMs having four or more phases has been developed by Pollock [9].

The energy-efficient C-dump converter shown in Figure 8.34d is a regenerative converter topology with reduced number of switches [10]. The topologies were derived from the C-dump converter proposed earlier by Miller [8]. The energy-efficient converter topologies eliminate all the disadvantages of the C-dump converter without sacrificing its attractive features, and also provide some additional advantages. The attractive features of the converters are lower number of power devices, full regenerative capability, freewheeling in chopping or PWM mode, simple control strategy, and faster demagnetization during commutation. The energy-efficient C-dump converter has one-switch plus one-diode forward voltage drop in the phase magnetization paths.

Converters with reduced number of switches are typically less fault-tolerant compared with the bridge converter. The ability to survive component or motor phase failure is a highly attractive feature for high-reliability applications. On the other hand, in low-voltage applications, the voltage drop in two switches can be a significant percentage of the total bus voltage, which may not be affordable. Among other factors to be considered in selecting a drive circuit are cost, complexity in control, number of passive components, number of floating drivers required, and so on. The drive converter must be chosen to serve the particular needs of an application.

8.5.2 SRM Controls

Appropriate positioning of the phase excitation pulses relative to the rotor position is the key in obtaining effective performance out of an SR motor drive system. The turn-on time, the total conduction period, and the magnitude of the phase current determine torque, efficiency, and other performance parameters. The type of control to be employed depends on the operating speed of the SRM.

8.5.2.1 Control Parameters

The control parameters for an SRM drive are the turn-on angle (θ_{on}), the turn-off angle (θ_{off}), and the phase current. The conduction angle is defined as $\theta_{dwell} = \theta_{off} - \theta_{on}$. The complexity of the determination of the control parameters depends on the chosen control method for a particular application. The current command can be generated for one or more phases depending on the controller. In voltage-controlled drives, the current is indirectly regulated by controlling the phase voltage.

At low speeds, the current rises almost instantaneously after turn-on due to the negligible back-emf, and the current must be limited by either controlling the average voltage or regulating the current level. The type of control used has a marked effect on the performance of the drive. As the speed increases, the back-emf increases and opposes the applied bus voltage. Phase advancing is necessary to establish the phase current at the onset of rotor and stator pole overlap region. The voltage PWM or chopping control is used to force maximum current into the motor in order to maintain the desired torque level. Also, the phase excitation is turned off early enough so that the phase current decays completely to zero before the negative torque-producing region is reached.

At higher speeds, the SRM enters the single-pulse mode of operation, where the motor is controlled by advancing the turn-on angle and adjusting the conduction angle. At very high speeds, the back-emf will exceed the applied bus voltage once the current magnitude is high and the rotor position is appropriate, which causes the current to decrease after reaching a peak, even though a positive bus voltage is applied during the positive $dL/d\theta$ region. The control algorithm outputs θ_{dwell} and θ_{on} according to speed. At the end of θ_{dwell}, the phase switches are turned off so that negative voltage is applied across the phase to commutate the phase as quickly as possible. The back-emf reverses polarity beyond the aligned position and may cause the current to increase in this region if the current does not decay to insignificant levels. Therefore, the phase commutation must precede the aligned position by several degrees so that the current decays before the negative $dL/d\theta$ region is reached.

In the high-speed range of operation, when the back-emf exceeds the DC bus voltage, the conduction window becomes too limited for current or voltage control and all the chopping or PWM has to be disabled. In this range, θ_{dwell} and θ_{adv} are the only control parameters and control is accomplished based on the assumption that approximately θ_{dwell} regulates torque and θ_{adv} determines efficiency.

8.5.2.2 Advance Angle Calculation

Ideally, the turn-on angle is advanced such that the reference current level i^* is reached just at the onset of pole overlap. In the unaligned position, phase inductance is almost constant, and hence, during turn-on, back-emf can be neglected. Also, assuming that the resistive drop is small, Equation 6.65 can be written as

$$V_{ph} = L(\theta)\frac{\Delta i}{\Delta\theta}\omega \tag{8.35}$$

Now, $\Delta i = i^*$ and $\Delta\theta = \theta_{overlap} - \theta_{on} = \theta_{adv}$, where $\theta_{overlap}$ is the position where pole overlap begins, θ_{on} is the turn-on angle, and θ_{adv} is the required phase turn-on advance angle. Therefore, we have

$$\theta_{adv} = L_u\omega\frac{i^*}{V_{dc}} \tag{8.36}$$

The above simple advance angle θ_{adv} calculation approach is sufficient for most applications, although it does not account for the errors due to neglecting the back-emf and the resistive drop in the calculation.

8.5.2.3 Voltage-Controlled Drive

In low-performance drives, where precise torque control is not a critical issue, fixed-frequency PWM voltage control with variable duty cycle provides the simplest means of control of the SRM drive. A highly efficient variable speed drive having a wide speed range can be achieved with this motor by optimum use of the simple voltage feeding mode with closed loop position control only. The block diagram of the voltage-controlled drive is shown in Figure 8.35. The angle controller generates the turn-on and turn-off angles for a phase depending on the rotor speed, which simultaneously determines the conduction period, θ_{dwell}. The duty cycle is adjusted according to the voltage command signal. The electronic commutator generates the gating signals based on the control inputs and the instantaneous rotor position. A speed feedback loop can be added on the outside as shown when precision speed control is desired. The drive usually incorporates a current sensor typically placed on the lower leg of the DC-link for overcurrent protection. A current feedback loop can also be added that will further modulate the duty cycle and compound the torque–speed characteristics just like the armature voltage control of a DC motor.

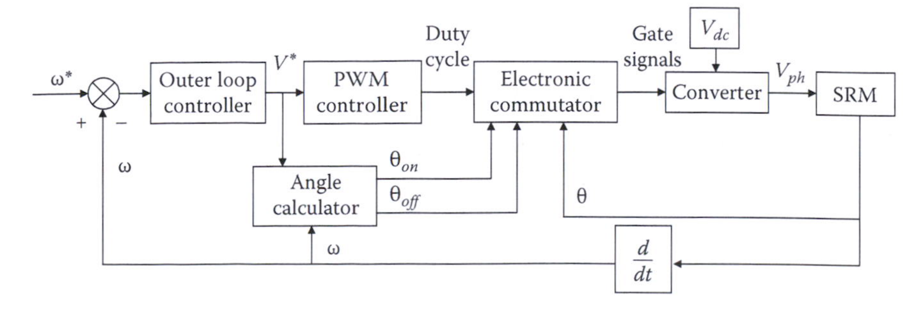

FIGURE 8.35
Voltage-controlled drive.

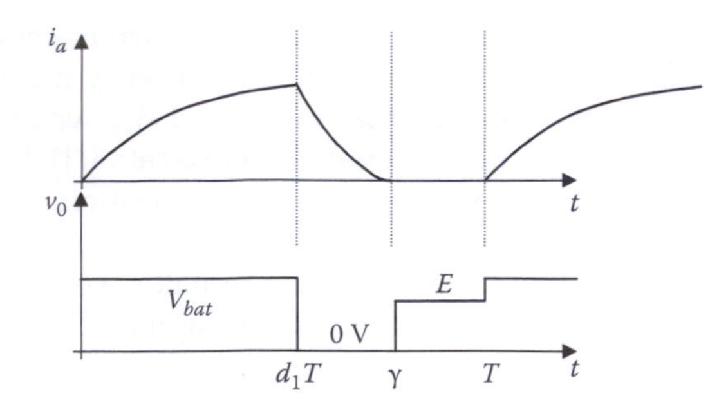

FIGURE P8.1

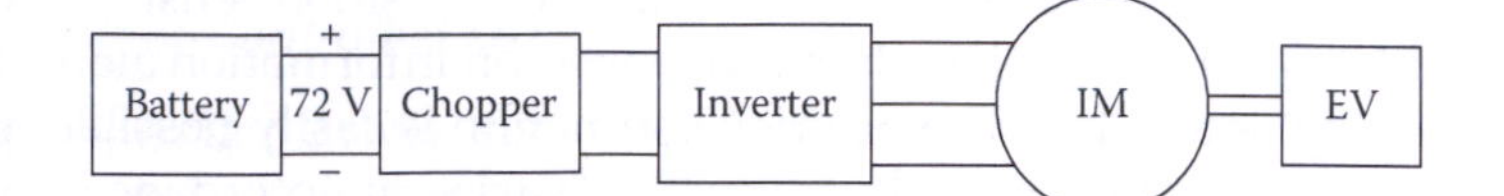

FIGURE P8.2

than 10 A, what is the value of the filter inductance, or what value should the switching frequency be changed to?

8.3 Find the regions in the T–ω plane for DCM, CCM, and UNCM acceleration operation of a two-quadrant chopper-fed DC motor. That is, find the restrictions on T and ω for each mode.

Hint: Start with the condition on E. Solve the inequality for d_1. Then use the ω–T characteristics to eliminate d_1. Also, remember that $0 \le d_1 \le 1$.

Plot these regions for the given parameters and also plot the safe operating area given:

$$-100\,\text{N m} \le T \le 100\,\text{N m}$$
$$-300\,\text{rad/s} \le \omega \le 300\,\text{rad/s}$$
$$-30\,\text{hp} \le P \le 30\,\text{hp}$$

8.4 Describe the UNCM of braking operation. Draw waveforms of armature current and terminal voltage. Calculate the speed–torque characteristics for this mode. What quadrant in the ω–T plane is this mode?

8.5 Consider the electric vehicle drivetrain driven by a two-quadrant chopper shown in Figure P8.5. d_1 is the duty ratio for the acceleration operation, while d_2 is the duty ratio for braking operation. The various parameters are given below:

The EV parameters are:

$$m = 1050\text{ kg},\ M_B = 150\text{ kg},\ C_D = 0.25,\ A_F = 2\text{ m}^2,\ C_0 = 0.01,\ C_1 = 0$$

$\rho = 1.1614\,\text{kg/m}^3$ and $g = 9.81\,\text{m/s}^2$

r_{wh} = radius of wheel = 0.28 m

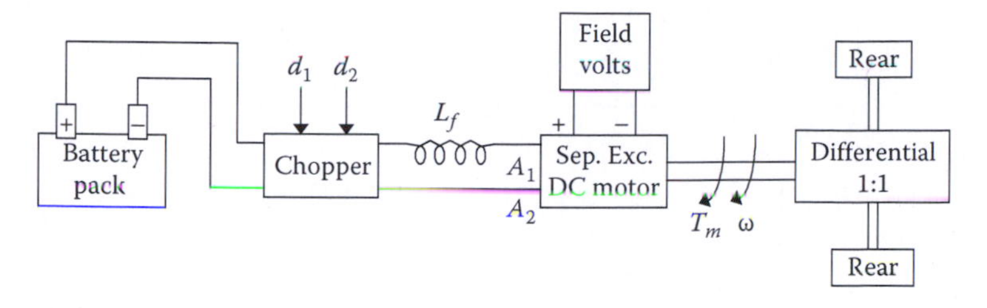

FIGURE P8.5

Motor and controller parameters are:

$R_a = 0.1\ \Omega$, $L_a = 2\,\text{mH}$, $k\phi = 0.6\ \text{V s}$, $I_{arated} = 200\,\text{A}$
f_s = chopper switching frequency = 500 Hz
L_f = series filter inductance = 1.6 mH

In each of the following cases, determine whether steady-state operation is in CCM, DCM, or UNCM.

(a) $d_1 = 0.4$, $d_2 = 0$, $V = 25\,\text{m/s}$
(b) $d_1 = 0.8$, $d_2 = 0$, $V = 45\,\text{m/s}$
(c) $d_2 = 0$, $V = 25\,\text{m/s}$, $T = 40\,\text{N m}$

Note: V is the vehicle steady-state velocity and T is the motor torque. Also neglect friction and windage loss and assume zero power loss between the motor shaft and the vehicle wheels.

8.6 An AC inverter is operated in a sinusoidal pulse mode. The transistor base current waveforms are shown in Figure P8.6. Sketch line-to-line voltages v_{AB}, v_{BC}, and v_{CA}, and line to neutral voltage v_{AN}. Briefly comment on the voltages. Are they balanced? (i_{ci} is for $i = 1$–6 are the base currents for transistors 1 to 6, respectively).

8.7 A 460 V, 60 Hz, 6 pole, 1176 rpm, Y-connected induction motor has the following parameters referred to the stator at rated condition:

$R_s = 0.19\,\Omega$, $R_r = 0.07\,\Omega$, $X_s = 0.75\,\Omega$, $X_r = 0.67\,\Omega$, and $X_m = \infty$.

The motor is fed by a six-step inverter. The inverter is fed from a battery pack through a DC/DC converter.

The battery pack voltage is 72 V. Neglecting all the losses:

(i) Determine the output of the DC/DC converter.
(ii) Mention the type of the converter and its conversion ratio.

8.8 The motor in problem 8.7 is employed to drive an electric vehicle that requires 300 N m to propel the vehicle on a level road at constant velocity. The configuration is shown below. Determine its operating speed and slip while the frequency and voltage are kept constant at rated value.

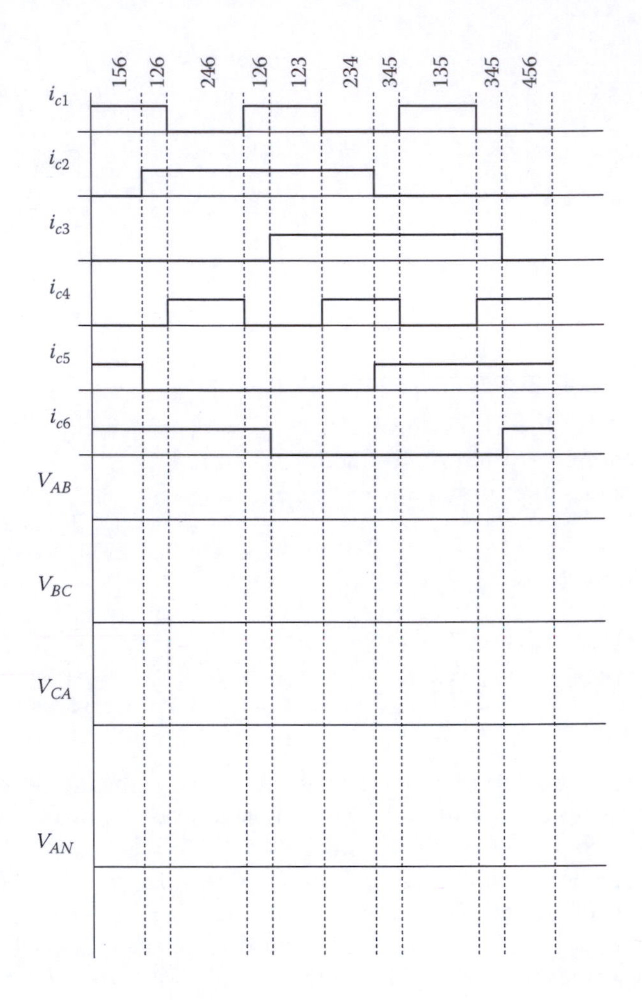

FIGURE P8.6

8.9 The vehicle in problem 8.7 is moving downward so that it requires 250 N m.

(i) What will be the input voltage for the motor from the inverter? Hence, determine the conversion ratio of the converter. Frequency is kept constant at rated value and the motor is running at the rated speed.

(ii) What should be the operating frequency of the inverter if the input voltage to the motor is kept constant at rated value and the motor is running at rated speed?

8.10 The DC-link voltage of a PWM inverter feeding a vector-controlled induction motor drive is 300 VDC. Space Vector PWM is used to supply the command voltages to the motor through the inverter. The voltage command in the stationary reference frame is $\vec{V}^S = 210e^{j0.44}$. What is the sector for this voltage command? Which two inverter voltage vectors are to be used? Calculate the *x*, *y*, and *z* percentages to apply the command voltage.

8.11 The following flux equation describes the nonlinear characteristics of a 3-phase, 6/4 SRM:

$$\lambda_j(i,\theta) = \lambda_s\left(1 - \exp(-i_j f_j(\theta)\right), \quad i_j \geq 0$$

where λ_S = saturation flux = 0.2 Vs and

$$f(\theta) = a + b * \cos\left(N_r\theta - (j-1)\frac{2\pi}{m}\right)$$

Here, j = 1, 2, and 3 denotes the phase number and m = 3. Given that a = 0.024 and b = 0.019. Also given, phase resistance R_{ph} = 0.3.

Write a computer program and plot the phase A currents when the motor is energized in the single pulse mode between the mechanical angles of 45°–75° for speeds of 100, 1,000, and 10,000 rpm. The applied voltage is 24 V.

References

1. J.G. Kassakian, M.F. Schlecht, and G.C. Verghese, *Principles of Power Electronics*, Addison Wesley Publishing Company, Reading, MA, 1991.
2. M.H. Rashid, *Power Electronics: Circuits, Devices and Applications*, Prentice Hall, Upper Saddle River, NJ, 2003.
3. N. Mohan, T.M. Undeland, and W.P. Robins, *Power Electronics: Converters, Applications and Design*, John Wiley & Sons Inc., Chichester, U.K., 1995.
4. G.K. Dubey, *Power Semiconductor Controlled Drives*, Prentice Hall, Englewood Cliffs, NJ, 1989.
5. A.M. Trzynadlowski, *Introduction to Modern Power Electronics*, John Wiley & Sons Inc., New York, 1998.
6. C. Hou, DSP implementation of sensorless vector control for induction motors, MS thesis, University of Akron, Akron, OH, 2001.
7. R.M. Davis, W.F. Ray, and R.J. Blake, Inverter drive for switched reluctance motor: Circuits and component ratings, *IEE Proceedings*, 128(pt. B, no. 2), 126–136, March 1981.
8. T.J.E. Miller, *Switched Reluctance Motors and their Control*, Magna Physics Publishing, Hillsboro, OH, 1993.
9. C. Pollock, Power converter circuits for switched reluctance motors with the minimum number of switches, *IEE Proceedings*, 137(pt. B, no. 6), 373–384, November 1990.

10. S. Mir, I. Husain, and M. Elbuluk, Energy-efficient C-dump converters for switched reluctance motors, *IEEE Transactions on Power Electronics*, 12(5), 912–921, September 1997.
11. S. Mir, I. Husain, and M. Elbuluk, Switched reluctance motor modeling with on-line parameter adaptation, *IEEE Transactions on Industry Applications*, 34(4), 776–783, July–August 1998.
12. M.S. Islam, M.N. Anwar, and I. Husain, Design and control of switched reluctance motors for wide-speed-range operation, *IEE Proceedings: Electric Power Applications*, 150(4), 425–430, July 2003.

9

Control of AC Machines

The robustness of the AC machines compared with DC machines makes them attractive in various motor drive applications including those for electric and hybrid vehicles. The robustness advantages are of benefit only when AC motor controllers can deliver performance comparable to or better than that of DC machines. The control simplicity in DC machines is due to the mechanical arrangements of commutators and brushes that maintain the orthogonality between the stator and rotor fluxes without any controller involvement. Fast response is a critical requirement for traction applications; motor drives must respond instantaneously with torque changes following a command change. In DC machines, the torque is directly proportional to the armature current without any dynamics involved, which makes the response instantaneous. The AC motor drives can be designed with fast response characteristics, but at the expense of controller complexity.

A brief overview of induction machine controls was presented in Chapter 6. However, those were scalar control methods that do not have the fast response characteristics desirable for traction applications. The desired control method for AC induction machines is the *vector control* where both the magnitude and the phase angle of the excitation current are controlled [1,2]. Vector control of the stator current is used in AC machines to overcome the effects of stator resistance, inductance, and induced voltage in the same manner as in DC machines. The stator current is transformed into a reference frame where the two parameters can be controlled independently, one for the rotor flux and the other for the electromagnetic torque. A 90° spatial orientation of the rotor flux is maintained with respect to the stator torque-producing current component using sensor feedback signals and model-based computations. Section 9.1 addresses the *dq* modeling essential for vector controls, and then the control algorithms are presented.

The vector control methods based on *dq* modeling are also used for permanent magnet (PM) synchronous machines. Any machine with sinusoidal excitations can employ the vector control method. However, vector controls cannot be applied to machines where it is not possible to transform the machine electrical variables into a suitable reference frame. The PM brushless DC (BLDC) and switched reluctance (SR) machines use square wave–type excitations for torque production, and the controller algorithms are different as has already been described in Chapter 8.

9.1 Vector Control of AC Motors

The electric and hybrid electric vehicle propulsion drives require accurate speed control with fast response characteristics. The induction motor drive is capable of delivering high performance similar to that of DC motors using the vector control approach. Although vector control complicates the controller implementation, the lower cost and rugged construction of the induction machines are an advantage over the DC machines.

The key variable for control in speed- and position-controlled applications is the torque. Although torque is never measured directly, torque estimators from machine models are frequently used to generate the current commands. A current controller in the innermost control loop regulates the motor current by comparing the command currents with feedback current measurements coming from the sensors. Speed control, if necessary, is achieved in the outer loop by comparing command speed signal with the feedback speed signal. With the two loops arranged in a cascade, the speed controller output of the outer loop is the current command for the inner loop. In certain high-performance position-controlled applications, such as in the actuator drives for accessories in electric and conventional vehicles, the position is controlled in the outermost loop putting the speed controller in an intermediate loop. The ability to produce a step change in torque with a step change in command generated from the outer loop represents the degree of control over the motor drive system for high-performance applications. The vector control in induction motors enables the machine to produce step changes in torque with instantaneous transition from one steady state to another steady state, which dramatically improves the dynamic performance of the drive system.

The objective of vector control or field orientation is to make the induction motor emulate the separately excited DC motor or the PM BLDC motor (i.e., PM trapezoidal motors). To understand vector control, let us revisit the torque production mechanism in DC machines. A simplified diagram of a DC motor with the field produced by separate excitation is shown in Figure 9.1. The field flux linkage space vector $\vec{\lambda}_f$ is stationary and is along the d-axis of the motor. The armature current space vector $\vec{i}_a$ is always along the q-axis due to the commutator and brush actions even though the rotor is revolving. The orthogonality between the field and the armature current ensures the optimal condition for torque production providing the highest torque per ampere ratio. The electromagnetic torque of the DC machine is given by

$$T_e = k_T \lambda_f i_a \tag{9.1}$$

where k_T is a machine constant depending on geometry and design parameters of the machine. The vector notations are dropped, since these variables

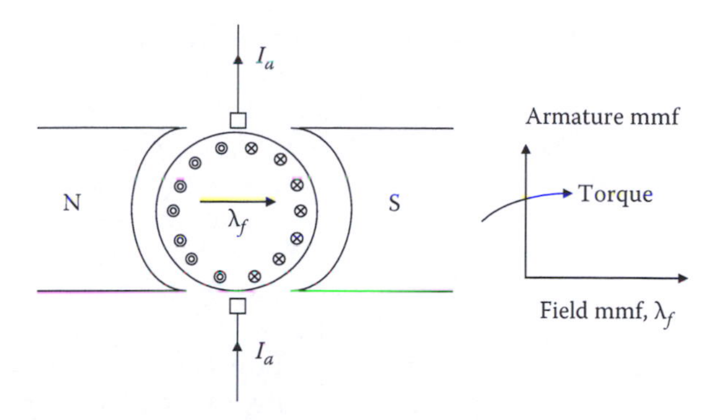

FIGURE 9.1
Torque in a separately excited DC machine.

are constants in DC machines. The armature and field circuits in a separately excited DC machine are completely independent or decoupled, allowing independent control over torque and magnetic field. The independent flux control is especially desirable for electric vehicle type applications, where flux weakening is used at higher speeds above rated torque conditions in the constant power region of torque–speed characteristics. The constant power range helps minimize the transmission gear requirements for propulsion drives.

In the case of PM BLDC motors, the rotor position sensor and power electronic converter replaces the mechanical commutators and brushes of DC motors; these components work in synchronism to maintain the orthogonality between the stator current space vector $\vec{i}_s(t)$ and the rotor flux vector $\vec{\lambda}_r(t)$ on an average basis. The back-emfs in these machines are trapezoidal, and not sinusoidal; hence, vector control is not possible. Torque ripple is a problem in PM BLDC motors, since square wave currents are used for torque production in synchronism with the trapezoidal back-emfs. In contrast, the orthogonality between armature and field mmfs is continuously maintained in DC commutator machines that help deliver a smoother torque. However, in the case of PM synchronous machines, the back-emfs are sinusoidal, and using vector control, smooth torque control can be achieved like that in an induction machine.

In light of the discussion presented above, the primary requirements of instantaneous torque control are controllability of the armature current, a controlled or constant field flux, and an orthogonal spatial angle between stator mmf axis and rotor mmf axis. In the case of DC and PM BLDC motors, the last two requirements are automatically met with the help of commutators and brushes, and position sensors and inverter switching, respectively. However, that the orthogonality is maintained on an average basis only for PM BLDC motors, the effect of which shows up in performance. In the case of induction machines and PM synchronous machines, these requirements are met with the help of *dq* models and reference frame transformations. Instantaneous torque control is achieved when the three requirements are met at every instant of time. Note that the armature of a machine is the component that carries the bulk of the current delivered by the source. In DC machines, the armature is in the rotor, whereas for AC machines, the armature is in the stator. The control of armature currents is achieved with the help of current regulators, such as the hysteresis current regulator or a PI current regulator. Armature current control is necessary to overcome the effects of armature winding resistance, leakage inductance, and induced voltage in the stator windings. For field-weakening operation, the rotor flux needs to be reduced, which is achieved through field current control. The task is simple in separately excited DC machines. In induction and PM sinusoidal machines, *dq* modeling that decouples the torque- and flux-producing components of currents and subsequent control of these components help achieve the objective.

9.2 *dq* Modeling

The *dq* modeling relates to the transformation of three-phase variables in the *abc* coordinate system into an equivalent two-phase coordinate system that has an arbitrary speed in a given reference frame [3,4]. In the *dq* coordinate system, the *d*-axis is along the direct magnetic axis of the resultant mmf, while the *q*-axis is in quadrature to the direct axis. The *dq* modeling of AC machines enables the development of electric machine controller, which operates in the inner loop with respect to the outer loop of the system level controller. The *dq* modeling analysis provides all the necessary transformation equations required to implement the inner loop controller. Furthermore, the electromagnetic torque T_e expression in terms of machine variables (current, flux linkage etc.) in the *dq* model is often used to estimate the torque for closed loop control. These machine variables are easily measurable using sensors, which is much simpler than torque measurement. The simplified torque expression of Equation 6.46 or the steady state torque expression of Equation 6.42 are inadequate for a dynamic controller implementation, which will be evident through later discussions.

A significant coupling exists between the stator and rotor variables of a three-phase AC machine. The objective of *dq* modeling is to transform the three-phase *abc* variables into *dq* variables in a suitable reference frame such that the coupling disappears. The space vector approach is retained in the *dq* reference frame, since these vectors help express the complex three-phase equations of AC machines in a compact form and also provides a simple relation for transformation between *abc* and *dq* reference frames. We have seen earlier in Chapter 6 that a balanced set of three-phase currents

$i_a(t)$, $i_b(t)$, and $i_c(t)$ flowing through 120° space-displaced balanced set of *a*, *b*, and *c* stator windings, respectively, establishes a rotating mmf $\vec{F}_S(t)$; this resultant rotating mmf has a constant peak amplitude that rotates around the stator circumference at synchronous speed. The *dq* transformation is based on the concept that *d*- and *q*-axes equivalent stator currents, designated as $i_{ds}(t)$ and $i_{qs}(t)$, flowing through a fictitious set of orthogonal windings along the *d*- and *q*-axes establish the same stator rotating mmf $\vec{F}_S(t)$. In space vector form, the rotating mmf is given by (repeated here from Chapter 6 for convenience)

$$\begin{aligned}\vec{F}_s(t) &= \vec{F}_a(t) + \vec{F}_b(t) + \vec{F}_c(t)\\ &= \hat{F}\angle\theta_F\\ &= \frac{N_S}{2}\vec{i}_S(t)\end{aligned}$$

Here, the space vector representation of stator current is

$$\begin{aligned}\vec{i}_S(t) &= i_a(t) + i_b(t)\angle 120 + i_c(t)\angle 240\\ &= \hat{I}_S\ \angle\theta_I\end{aligned}$$

The *dq* transformation is applicable to not just currents, but to all the electrical and magnetic variables in the three-phase system. The general variable symbol *f* will be used as in Chapter 6 to represent the three-phase AC machine variables of voltage, current, and mmf or flux. Several choices exist to define the relationship between the *abc*-axes variables and the *dq*-axes variables, since the transformation is from a three-phase set of variables to a two-phase set of variables. One choice is to take the *dq* variables to be 2/3 times the projection of $\vec{f}_{abc}$ on the *d*- and *q*-axes; $(2/3)\cdot\vec{f}_{abc}$ gives the same space vector magnitude as the peak value of the individual phase time-phasor variables as we have seen earlier. Another possible choice is to take the *dq* variables to be $\sqrt{2/3}$ times the projection of $\vec{f}_{abc}$ on the *d*- and *q*-axes. The $\sqrt{2/3}$ ratio between the *dq* variables and the *abc* variables conserves power without any multiplying factor in the *dq* and *abc* reference frames, and hence is known as the power invariant transformation. In order for the equivalent *dq* windings to establish the same stator mmf $\vec{F}_S(t)$ as is done by the *abc* windings, the number of turns in the equivalent sinusoidally distributed orthogonal windings must be $2/3N_S$ or $\sqrt{2/3}N_S$, depending on the ratio chosen for transformation. We shall arbitrarily choose the multiplying factor of 2/3 to define the transformation. In this case, the *q*- and *d*-axes variables are the projections of the *a*, *b*, and *c* variables on the *q*- and *d*-axes, respectively, multiplied by 2/3. Mathematically stated, for a general variable

f (representing voltage, current, or flux linkage), the q- and d-axes variables in terms of the a, b, and c variables are

$$
\begin{aligned}
f_q &= \frac{2}{3}\left[f_a \cos\theta + f_b \cos\left(\theta - 120^0\right) + f_c \cos\left(\theta + 120^0\right)\right] \\
f_d &= \frac{2}{3}\left[f_a \sin\theta + f_b \sin\left(\theta - 120^0\right) + f_c \sin\left(\theta + 120^0\right)\right]
\end{aligned}
\tag{9.2}
$$

The transformation relation holds good for stator and rotor variables. A third variable is required to obtain a unique transformation, which comes from the neutral terminal. Representing the neutral terminal variable as a zero-sequence component, we have

$$
f_0 = \frac{1}{3}\left[f_a + f_b + f_c \right] \tag{9.3}
$$

All the three-phase systems considered here are assumed to be balanced. Furthermore, the AC machine windings are assumed to be connected either in Δ or Y without a neutral connection. Therefore, for all practical purposes,

$$
f_a + f_b + f_c = 0 \quad \Rightarrow \quad f_0 = 0
$$

Henceforth, we will concentrate only on the d and q variables, ignoring the zero-sequence component.

The variables in the dq reference frame can be expressed in the space vector form as

$$
\begin{aligned}
\vec{f}_{qd}(t) = f_q(t) - jf_d(t) &= \frac{2}{3} e^{-j\theta}\left[f_a(t) + e^{j\frac{2\pi}{3}} f_b(t) + e^{-j\frac{2\pi}{3}} f_b(t) \right] \\
&= \frac{2}{3} e^{-j\theta} \vec{f}_{abc}(t)
\end{aligned}
\tag{9.4}
$$

The result shows that the dq space vector $\vec{f}_{qd}(t)$ is two-third of the space vector $\vec{f}_{abc}(t)$ along the d- and q-axes. Equating the real and imaginary parts of Equation 9.4, the transformation matrices between the abc and dq variables are

$$
\begin{bmatrix} f_q(t) \\ f_d(t) \\ 0 \end{bmatrix} = T_{abc->qd} \begin{bmatrix} f_a(t) \\ f_b(t) \\ f_c(t) \end{bmatrix}, \quad \text{where } T_{abc \to qd} = \frac{2}{3} \begin{bmatrix} \cos(\theta) & \cos(\theta - 2\pi/3) & \cos(\theta + 2\pi/3) \\ \sin(\theta) & \sin(\theta - 2\pi/3) & \sin(\theta + 2\pi/3) \\ 0.5 & 0.5 & 0.5 \end{bmatrix}
\tag{9.5}
$$

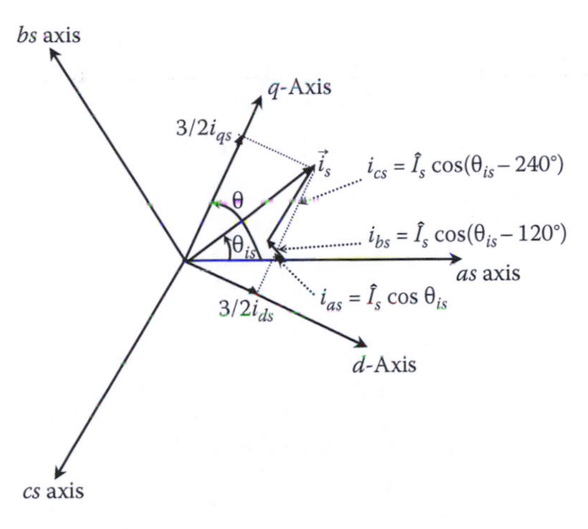

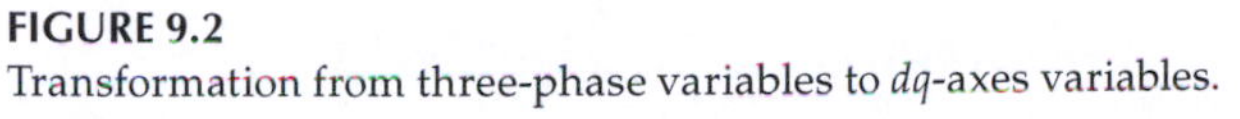

FIGURE 9.2
Transformation from three-phase variables to *dq*-axes variables.

This transformation is known as Park's transformation. The *abc* variables are obtained from the *dq* variables through the inverse of the Park transform:

$$T_{qd->abc} = \begin{bmatrix} \cos(\theta) & \sin(\theta) & 1 \\ \cos(\theta - 2\pi/3) & \sin(\theta - 2\pi/3) & 1 \\ \cos(\theta + 2\pi/3) & \sin(\theta + 2\pi/3) & 1 \end{bmatrix} \quad (9.6)$$

The projections from both the *abc* frame and the *dq* frame to form the same space vector for a three-phase AC machine current variable is shown in Figure 9.2. The orientation of the *dq*-axes with respect to the *abc*-axes is at an arbitrary angle θ. The direct-axis and quadrature-axis stator currents i_{ds} and i_{qs} are a set of fictitious two-phase current components, and the same is true for other *dq* variables. The *dq* reference frame can be stationary with respect to the stator or rotating at an arbitrary speed, such as at rotor speed or at synchronous speed. Again, when a stationary *dq* reference frame is used, the *dq*-axes may be at any arbitrary angle with respect to our chosen reference Phase-*a* axis.

9.2.1 Rotating Reference Frame

Although the speed of the *dq* windings can be arbitrary, three of those are the most suitable for machine analysis. These three speeds of *dq* windings are 0, ω_m, and ω_e. The zero speed is known as the stationary reference frame, where more commonly the stationary *d*-axis is aligned with the Phase-*a* axis of the stator. The angle of transformation θ in Equations 9.4 through 9.6 is 0 in this case. The *d*- and *q*-axes variables oscillate at the synchronous frequency in the balanced sinusoidal steady state. When ω_e is chosen as the speed of the reference dq frame, all the associated variables in the stator and in the rotor dq windings appear as DC variables in the balanced sinusoidal steady state. For an arbitrary speed of the reference dq windings, the angle of transformation is

$$\theta = \int_0^t \omega(\xi)d\xi + \theta_0 \tag{9.7}$$

Example 9.1

The three-phase currents in an AC machine are

$$i_a(t) = 10\cos 377t$$

$$i_b(t) = 10\cos\left(377t - \frac{2\pi}{3}\right)$$

$$i_c(t) = 10\cos\left(377t + \frac{2\pi}{3}\right)$$

Calculate the currents in the dq reference frame in (a) the stationary reference frame and (b) the synchronous reference frame.

Solution

(a) Using Equation 9.5 and $\theta = 0$, the dq current variables in the stationary reference frame are

$$i_d^s(t) = -10\sin(377t)$$

$$i_q^s(t) = 10\cos(377t)$$

(b) Using Equation 9.5 and $\theta = \omega t = 377t$ with $\theta_0 = 0$, the dq current variables in the synchronously rotating reference frame are

$$i_d^e(t) = 0$$

$$i_q^e(t) = 10$$

9.2.2 Induction Machine *dq* Model

Let us assume that the stator and rotor voltages and currents in the three-phase model vary arbitrarily in time. The voltage balance equations in the stator and rotor circuits (refer to Figure 6.23) of the three-phase induction machine in space vector form are

$$\vec{v}_{abcs} = R_s \vec{i}_{abcs} + p\vec{\lambda}_{abcs}$$

$$\vec{v}_{abcr} = R_r \vec{i}_{abcr} + p\vec{\lambda}_{abcr}$$

The rotor winding voltage $\vec{v}_{abcr}$ is zero for squirrel cage induction machine with short-circuited rotor windings, but will be represented as such for generality. The rotor variables are represented here without the rotor-to-stator referral symbol for the sake of simplicity, but are implicitly incorporated in the equations. The stator and rotor flux linkages include coupling effects between the windings of stator and rotor circuits as well as between stator and rotor windings. Accounting for all the magnetic coupling and assuming magnetic linearity ($\lambda = Li$), the phase-variable form of the voltage equations in the *abc* frame can be derived as

$$\vec{v}_{abcs} = R_s \vec{i}_{abcs} + L_s(p\vec{i}_{abcs}) + L_m(p\vec{i}_{abcr})e^{j\theta_r} + j\omega_r L_m \vec{i}_{abcr} e^{j\theta_r} \tag{9.8}$$

$$\vec{v}_{abcr} = R_r i_{abcr} + L_r(p\vec{i}_{abcr}) + L_m(p\vec{i}_{abcs})e^{-j\theta_r} - j\omega_r L_m \vec{i}_{abcs} e^{-j\theta_r} \tag{9.9}$$

where
$\omega_r = p\theta_r = d\theta_r/dt$ is the rotor speed
$L_s = L_{ls} + L_m$
$L_r = L_{lr} + L_m$

L_{ls} and L_{lr} are the stator and rotor leakage inductances, respectively. The derivation can be carried out using either the space vector approach [1] or reference frame transformation [4].

Multiplying by $e^{j\theta}$ and applying *dq* transformation, Equations 9.8 and 9.9 can be transformed into a general reference frame rotating at a speed ω as

$$\vec{v}_{qds} = R_s i_{qds} + L_s(p\vec{i}_{qds}) + L_m(p\vec{i}_{qdr}) + j\omega(L_s \vec{i}_{qds} + L_m \vec{i}_{qdr}) \tag{9.10}$$

$$\vec{v}_{qdr} = R_r i_{qdr} + L_r(p\vec{i}_{qdr}) + L_m(p\vec{i}_{qds}) + j(\omega - \omega_r)(L_r \vec{i}_{qdr} + L_m \vec{i}_{qds}) \tag{9.11}$$

The matrix form of the induction motor model in the arbitrary *dq* reference frame is

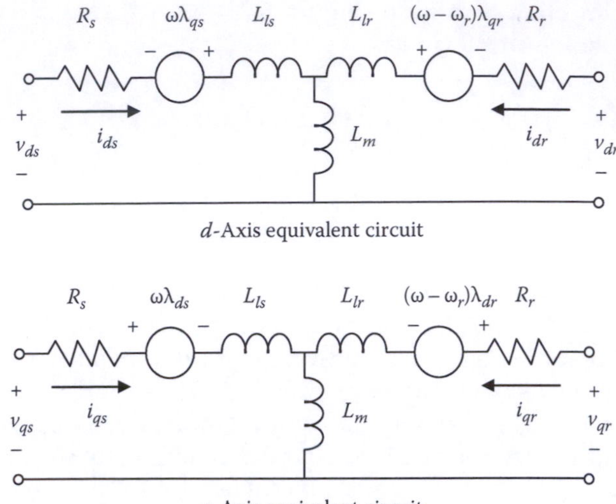

FIGURE 9.3
d- and *q*-axes circuits of the induction machine.

$$\begin{bmatrix} v_{ds} \\ v_{qs} \\ v_{dr} \\ v_{qr} \end{bmatrix} = \begin{bmatrix} R_s & -\omega L_s & 0 & -\omega L_m \\ \omega L_s & R_s & \omega L_m & 0 \\ 0 & -(\omega - \omega_r)L_m & R_r & -(\omega - \omega_r)L_r \\ (\omega - \omega_r)L_m & 0 & (\omega - \omega_r)L_r & R_r \end{bmatrix} \begin{bmatrix} i_{ds} \\ i_{qs} \\ i_{dr} \\ i_{qr} \end{bmatrix} + \begin{bmatrix} L_s & 0 & L_m & 0 \\ 0 & L_s & 0 & L_m \\ L_m & 0 & L_r & 0 \\ 0 & L_m & 0 & L_r \end{bmatrix} \begin{bmatrix} pi_{ds} \\ pi_{qs} \\ pi_{dr} \\ pi_{qr} \end{bmatrix} \tag{9.12}$$

The *dq* equivalent circuit model for the induction machine in circuit schematic form is shown in Figure 9.3.

9.2.3 Power and Electromagnetic Torque

Power into the three-phase machine needs to be analyzed in terms of *dq* variables to arrive at the electromagnetic torque expression, which is used in the motor control loop. The power into the induction machine is the product of the phase voltage and phase currents given as

$$P_{in} = (v_{as}i_{as} + v_{bs}i_{bs} + v_{cs}i_{cs}) + (v_{ar}i_{ar} + v_{br}i_{br} + v_{cr}i_{cr}) \tag{9.13}$$

The input power is embedded in the real part of the product of the voltage space vector and the current space vector conjugate of the stator and rotor variables. The real part can be calculated as follows:

$$\mathrm{Re}\left(\vec{v}_{abc}\vec{i}^{*}_{abc}\right) = \mathrm{Re}[(v_a + v_b\angle 120 + v_c\angle 240)\cdot(i_a + i_b\angle -120 + i_c\angle -240)]$$

$$= \frac{3}{2}(v_a i_a + v_b i_b + v_c i_c)$$

Using the above relation in the power input equation for the induction machine, we get

$$P_{in} = \frac{2}{3}\left[\mathrm{Re}(\vec{v}_{abcs}\vec{i}^{*}_{abcs}) + \mathrm{Re}(\vec{v}_{abcr}\vec{i}^{*}_{abcr})\right]$$

Now, using Equation 9.4

$$P_{in} = \frac{2}{3}\left[\mathrm{Re}\left[\left(\frac{3}{2}e^{j\theta}\vec{v}_{qds}\right)\left(\frac{3}{2}e^{-j\theta}\vec{i}^{*}_{qds}\right)\right] + \mathrm{Re}\left[\left(\frac{3}{2}e^{j\theta}\vec{v}_{qdr}\right)\left(\frac{3}{2}e^{-j\theta}\vec{i}^{*}_{qdr}\right)\right]\right]$$

$$= \frac{3}{2}\left[\mathrm{Re}(\vec{v}_{qds}\vec{i}^{*}_{qds}) + \mathrm{Re}(\vec{v}_{qdr}\vec{i}^{*}_{qdr})\right] \tag{9.14}$$

The input power expression in scalar form is

$$P_{in} = \frac{3}{2}(v_{ds}i_{ds} + v_{qs}i_{qs} + v_{dr}i_{dr} + v_{qr}i_{qr}) \tag{9.15}$$

The multiplying factor 3/2 is due to the choice of the factor 2/3 for the ratio of *dq* and *abc* variables. The input power expression of Equation 9.15 would not have this 3/2 factor had we chosen $\sqrt{2/3}$ as the proportionality constant between the *dq* and *abc* variables.

Upon expanding the right-hand side of Equation 9.15 and extracting the output electromechanical power P_e [1], we get

$$P_e = \frac{3}{2}\mathrm{Im}\left[\omega_r L_m \vec{i}_{qds}\vec{i}^{*}_{qdr}\right] \tag{9.16}$$

where ω_r is the rotor angular velocity of a two-pole machine. The electromagnetic torque for a *P*-pole machine is, therefore,

$$T_e = \frac{3}{2}\frac{P}{2}L_m \operatorname{Im}\left[\vec{i}_{qds}\vec{i}_{qdr}^{\,*}\right] = \frac{3}{2}\frac{P}{2}L_m(i_{qs}i_{dr} - i_{ds}i_{qr}) \tag{9.17}$$

Several alternative forms of the electromagnetic torque can be derived using the stator and rotor flux linkage expressions. A couple of these torque expressions are

$$T_e = \frac{3}{2}\frac{P}{2}\operatorname{Im}\left[\vec{i}_{qds}\vec{\lambda}_{qds}^{\,*}\right] = \frac{3}{2}\frac{P}{2}(\lambda_{ds}i_{qs} - \lambda_{qs}i_{ds}) \tag{9.18}$$

$$T_e = \frac{3}{2}\frac{P}{2}\frac{L_m}{L_r}\operatorname{Im}\left[\vec{i}_{qds}\vec{\lambda}_{qdr}^{\,*}\right] = \frac{3}{2}\frac{P}{2}\frac{L_m}{L_r}(\lambda_{dr}i_{qs} - \lambda_{qr}i_{ds}) \tag{9.19}$$

The torque expressions in terms of *dq* variables are used in the vector control of induction motor drives. The vector control implementations are accomplished in one of the several available choices of reference frames, such as rotor flux-oriented reference frame, stator flux-oriented reference frame, or air gap flux-oriented reference frame. The *abc* variables at the input of the controller are converted to *dq* variables in the chosen reference frame. The control computations take place in terms of *dq* variables and the generated command outputs are again converted back to *abc* variables. The inverter controller executes the commands to establish the desired currents or voltages in the drive system.

The torque expression derived in this section is useful for motor controller implementation, while the simplified torque expression of Equation 6.46 is useful for system level analysis.

9.3 Induction Machine Vector Control

Vector control refers to the control of both magnitude and phase angles of the stator and rotor critical variables in a chosen reference frame. More specifically, the control methods used in high-performance induction motor drives are known as field-oriented controls. The term *field orientation* is used as a special case of vector control where a 90° spatial orientation between the stator and rotor mmfs is continuously maintained by the motor controller. The term "vector control" is general and is also used for controls where the angle between the rotor and stator critical variables is different from 90°.

Transformation of the variables into a rotating reference frame facilitates the instantaneous torque control of an induction machine similar to that of a DC machine. Consider the electromagnetic torque expression in Equation 9.19 in the *dq* reference frame

$$T_e = \frac{3}{2}\frac{P}{2}\frac{L_m}{L_r}(\lambda_{dr}i_{qs} - \lambda_{qr}i_{ds})$$

If we choose a reference frame that rotates at synchronous speed with the rotor flux linkage vector $\vec{\lambda}_{qdr}(t)$ continuously locked along the *d*-axis of the *dq* reference frame, then $\lambda_{qr} = 0$. The resulting torque equation is then

$$T_e = \frac{3}{2}\frac{P}{2}\frac{L_m}{L_r}\lambda_{dr}i_{qs} \tag{9.20}$$

Now, note the similarity between the above torque expression and the DC motor torque expression in Equation 9.1. An instantaneous change in i_{qs} current with constant λ_{dr} will result in an instantaneous change in torque similar to the situation in DC motors. We can conclude that an AC machine can be made to appear like a DC machine with appropriate reference frame transformations, and hence controlled similarly. There are several ways of implementing the vector control on induction motors. The implementation can be carried out through transformation into any one of the several synchronously rotating reference frames, such as the rotor flux reference frame, air gap flux reference frame, and stator flux reference frame [1,5]. The task is the simplest in the rotor flux-oriented reference frame, which is described in Section 9.3.1. In addition, according to the method of rotor flux angle measurement, the vector control is labeled as either direct vector control or indirect vector control.

9.3.1 Rotor Flux-Oriented Vector Control

The rotor flux linkage vector direction is chosen as the *d*-axis of the reference frame in the rotor flux-oriented vector control methods. The reference frame is also assumed rotating at the speed of the rotor flux vector. In the rotor flux-oriented reference frame, *q*-axis rotor flux $\lambda_{qr}^{rf} = 0$ and the torque is

$$T_e = \frac{3}{2}\frac{P}{2}\frac{L_m}{L_r}\lambda_{dr}^{rf}i_{qs}^{rf} \tag{9.21}$$

The instantaneous control of the *q*-axis stator current in the rotor flux reference frame results in an instantaneous response of the motor torque provided that the rotor flux is held constant. The current controller can be

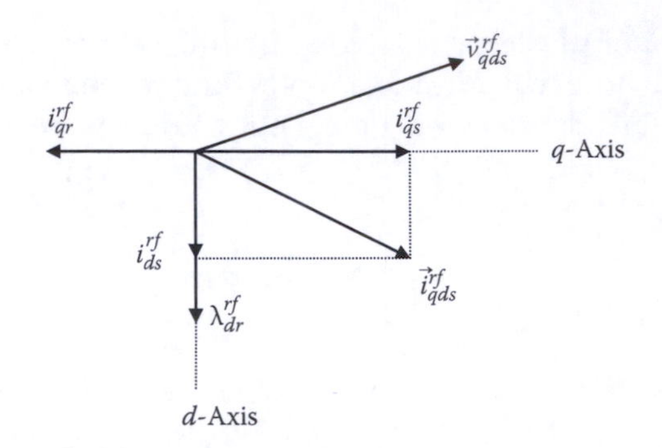

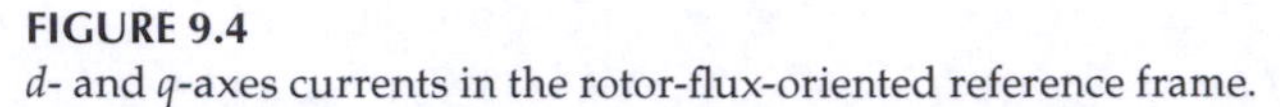

FIGURE 9.4
d- and *q*-axes currents in the rotor-flux-oriented reference frame.

assumed to have a high bandwidth, so that it establishes the stator command currents instantaneously. Therefore, the dynamics of the controller are primarily associated with the rotor circuit. The rotor flux and current and the stator current in the rotor flux-oriented reference frame are shown in Figure 9.4. The rotor voltage and current equations in the rotor flux reference frame, derived from Equation 9.12, are

$$v_{qr}^{rf} = R_r i_{qr}^{rf} + p\lambda_{qr}^{rf} + (\omega_{rf} - \omega_r)\lambda_{dr}^{rf}$$

$$v_{dr}^{rf} = R_r i_{dr}^{rf} + p\lambda_{dr}^{rf} - (\omega_{rf} - \omega_r)\lambda_{qr}^{rf}$$

The flux linkages from Figure 9.3 are

$$\lambda_{qr}^{rf} = L_{lr} i_{qr}^{rf} + L_m(i_{qs}^{rf} + i_{qr}^{rf}) = L_m i_{qs}^{rf} + L_r i_{qr}^{rf}$$

$$\lambda_{dr}^{rf} = L_{lr} i_{dr}^{rf} + L_m(i_{ds}^{rf} + i_{dr}^{rf}) = L_m i_{ds}^{rf} + L_r i_{dr}^{rf}$$

The rotor voltages are identically zero in squirrel cage induction machines. Moreover, the *q*-axis flux linkage is zero in the rotor flux-oriented reference frame ($\lambda_{qr}^{rf} = 0$). Therefore, the rotor *dq* circuit voltage balance equations becomes

$$0 = R_r i_{qr}^{rf} + (\omega_{rf} - \omega_r)\lambda_{dr}^{rf} \tag{9.22}$$

$$0 = R_r i_{dr}^{rf} + p\lambda_{dr}^{rf} \tag{9.23}$$

An important relation required to implement one form of vector control (indirect method) is the slip relation, which follows from Equation 9.22 as

$$\omega_{rf} - \omega_r = s\omega_{rf} = -\frac{R_r i_{qr}^{rf}}{\lambda_{dr}^{rf}} = \frac{R_r}{L_r}\frac{L_m i_{qs}^{rf}}{\lambda_{dr}^{rf}} \tag{9.24}$$

From the q-circuit flux linkage expression, we get

$$\lambda_{qr}^{rf} = L_m i_{qs}^{rf} + L_r i_{qr}^{rf} = 0 \quad \Rightarrow \quad i_{qr}^{rf} = -\frac{L_m}{L_r} i_{qs}^{rf} \tag{9.25}$$

Equation 9.25 describes that there will be an instantaneous response without any delay in the rotor current following a change in the torque command stator current i_{qs}^{rf}. The major dynamics in the rotor flux-oriented vector control method is in the behavior of d-axis rotor flux λ_{dr}^{rf}. λ_{dr}^{rf} is related to the d-axis rotor and stator currents according to

$$\lambda_{dr}^{rf} = L_r i_{dr}^{rf} + L_m i_{ds}^{rf} \tag{9.26}$$

Now, using Equations 9.23 and 9.26, the following dynamic relations can be easily established:

$$i_{dr}^{rf} = -\frac{L_m p}{R_r + L_r p} i_{ds}^{rf} \tag{9.27}$$

$$\lambda_{dr}^{rf} = \frac{R_r L_m}{R_r + L_r p} i_{ds}^{rf} \tag{9.28}$$

Equation 9.27 shows that i_{dr}^{rf} exists only when i_{ds}^{rf} changes, and is zero in the steady state. Equation 9.28 gives the dynamics associated in bringing a change in the rotor flux λ_{dr}^{rf}. In steady state, the rotor flux is

$$\lambda_{dr}^{rf} = L_m i_{ds}^{rf}$$

Therefore, in order to change the rotor flux command, i_{ds}^{rf} must be changed that will cause a transient occurrence of d-axis rotor current i_{dr}^{rf}. The time constant associated with these dynamics is $\tau_r = L_r/R_r$, which is commonly known as the rotor time constant.

9.3.2 Direct and Indirect Vector Control

The key to the implementation of the vector control is to find the instantaneous position of the rotor flux with respect to a stationary reference axis. Let this angle be defined as θ_{rf} with respect to the Phase-*a* reference axis as shown in Figure 9.5. The vector control methods can be implemented in one of two ways defined as direct and indirect methods according to the nature of measuring or calculating the rotor flux angle.

9.3.2.1 Direct Vector Control

In direct vector control methods, the rotor flux angle is calculated from direct measurements of machine electrical quantities. Using the *dq* model of the machine, the measurements are used to calculate the rotor flux vector, which directly gives the rotor flux angle θ_{rf}. The measurements of electrical variables can be carried out in one of several different ways. For example, flux sensing coils or Hall sensors placed in the air gap can be used to measure the air gap flux $\vec{\lambda}_{qdm}$. The subscript "*m*" stands for mutual flux in the air gap between the stator and the rotor. The *dq* model of the induction machine gives the following mathematical relations:

$$\vec{\lambda}_{qdm}^{s} = L_m(\vec{i}_{qds}^{s} + \vec{i}_{qdr}^{s}) \tag{9.29}$$

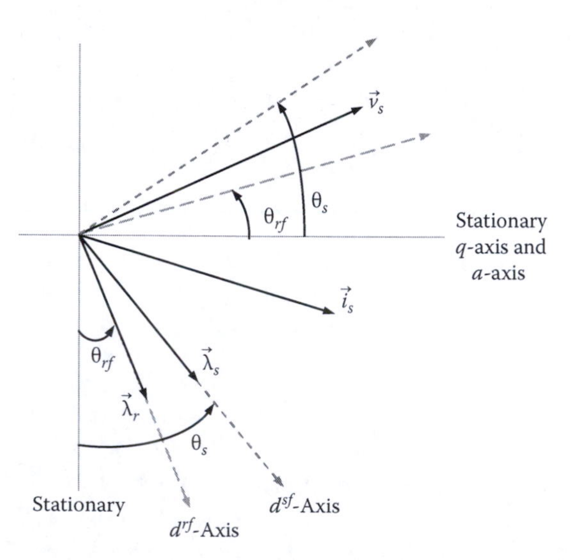

FIGURE 9.5
Relationship among stationary, rotor flux (*rf*)-oriented, and stator flux (*sf*)-oriented reference frames.

and

$$\vec{\lambda}^s_{qdr} = L_m \vec{i}^s_{qds} + L_r \vec{i}^s_{qdr} \tag{9.30}$$

The superscript "s" is used for the reference frame, since the air gap flux measurements are with respect to the stator in the stationary reference frame. Now, the objective is to write $\vec{\lambda}_{qdr}$ in terms of measurable quantities. From Equation (9.29), the rotor current can be written as

$$\vec{i}^s_{qdr} = \frac{\vec{\lambda}^s_{qdm}}{L_m} - \vec{i}^s_{qds}$$

Substituting the rotor current in Equation 9.30

$$\vec{\lambda}^s_{qdr} = \frac{L_r}{L_m}\vec{\lambda}^s_{qdm} - (L_r - L_m)\vec{i}^s_{qds}$$

$$\Rightarrow \lambda_r = \left|\vec{\lambda}^s_{qdr}\right|, \quad \theta_{rf} = \angle\vec{\lambda}^s_{qdr}$$

The approach is very straightforward requiring the knowledge of only two motor parameters, rotor leakage inductance, $L_r - L_m = L_{lr}$, and the ratio L_r/L_m. The rotor leakage inductance is a fairly constant value, while the L_r/L_m ratio varies only a little by magnetic flux path saturation. However, the big disadvantage is the need for flux sensors in the air gap.

The flux sensors can be avoided by using voltage and current sensors to measure the stator applied voltages and currents. The measurements are used to calculate the stator flux by direct integration of the phase voltage:

$$\vec{\lambda}^s_{qds} = \int (\vec{v}^s_{qds} - R_s \vec{i}^r_{qds})dt$$

The rotor flux vector can be obtained from the stator flux vector using the following mathematical relationship:

$$\vec{\lambda}^s_{qdr} = \frac{L_r}{L_m}\left(\vec{\lambda}^s_{qds} - L'_s \vec{i}_{qds}\right)$$

where $L'_s = L_s - (L_m^2/L_r)$. The method requires the knowledge of three motor parameters R_s, L'_s, and L_r/L_m. The major difficulty of the method is the significant variation of the stator resistance R_s due to temperature dependence and the integration of the phase voltage to obtain stator flux at low speeds. The integration is especially inaccurate at low speeds due to the dominating resistive voltage-drop term.

9.3.2.2 *Indirect Vector Control*

In the indirect vector control scheme, which is more commonly used, a rotor position sensor is used to derive the speed and rotor position information, and the slip relation of Equation 9.24 is used to derive the rotor flux angle. The implementation of rotor flux angle calculation in the indirect vector control method is shown in Figure 9.6.

9.3.2.3 *Vector Control Implementation*

The vector-controlled drive has three major components like any other motor drive system: electric machine, power converter, and controller. An implementation block diagram of a speed regulated vector-controlled drive is shown in Figure 9.7. The controller processes the input command signals and the feedback signals from the converter and the motor, and generates the gating signals for the PWM inverter or converter.

In the closed loop speed-controlled system, the input is a speed reference, which is compared with the measured speed feedback signal to generate

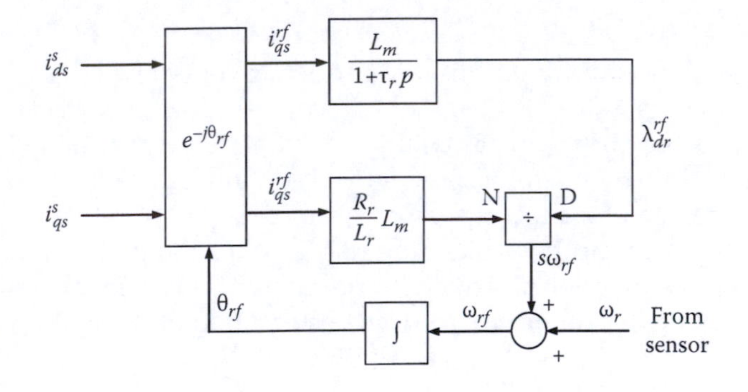

FIGURE 9.6
Rotor flux angle calculation in indirect vector control method.

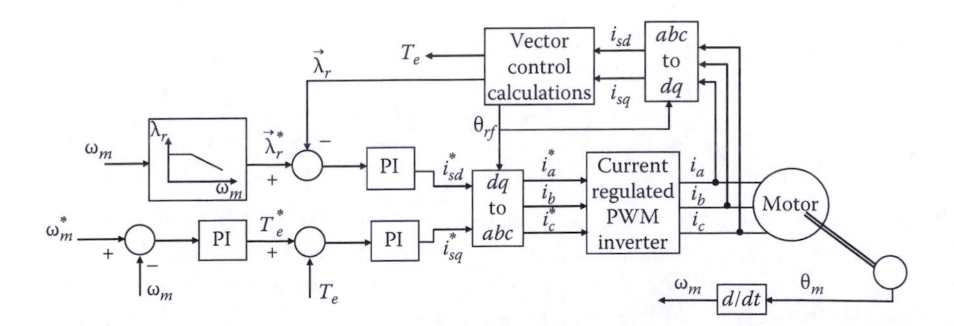

FIGURE 9.7
Implementation of vector control.

the control signals for maintaining the desired speed. The vector controller requires reference frame transformations and several computations that are typically implemented in a digital signal processor. The sensors provide current feedback information to the controller for vector calculations. The first step in the vector controller is the transformation calculations of the feedback current signals to a suitable *dq* reference frame. The following steps are the torque and reference frame angle calculations. The vector controller outputs the three-phase reference currents $i_a^*(t)$, $i_b^*(t)$, and $i_c^*(t)$ for the current-regulated PWM inverter. The current controller in the subsequent stage generates the PWM gating signals for the power electronics converter switches to establish the desired currents in the electric machine. The torque and angle calculations can be in the rotor flux reference frame where torque Equation 9.21 and the block diagram of Figure 9.7 are used. For speeds up to the rated speed of the machine, rated rotor flux $\vec{\lambda}_r$ is used. For higher speeds, the flux command is reduced to operate the machine in the constant power mode. This mode of operation is known as the flux weakening mode. In the indirect control method, position and speed sensors are used to provide the rotor position and motor speed feedback information (the implementation shown in Figure 9.7).

9.4 PM Machine Vector Control

9.4.1 Voltage and Torque in Reference Frame

For vector control of PM synchronous machines, the *abc* variables are transformed into the rotor reference frame, which is also the synchronous reference frame for synchronous machines in steady state. The stator phase voltage equation in the stationary *abc* reference frame were presented earlier in Chapter 6 and is repeated here for convenience:

$$\vec{v}_{abcs} = \overline{R}_s \vec{i}_{abcs} + \frac{d}{dt}\vec{\lambda}_{abcs}$$

where

$(\vec{f}_{abcs})^T = [f_{as}\, f_{bs}\, f_{cs}]$ (f represents voltage v, current i or flux linkage)

$\overline{R}_s = diag\,[R_s\ R_s\ R_s]$

The flux linkages are

$$\vec{\lambda}_{abcs} = \overline{L}_s \vec{i}_{abcs} + \overline{\lambda}_f$$

When transformed to the stator rotor reference frame, the *dq* equations of the PMSM become

$$v_q = R_s i_q + \frac{d}{dt}\lambda_q + \omega_r \lambda_d$$
$$v_d = R_s i_d + \frac{d}{dt}\lambda_d - \omega_r \lambda_q \tag{9.31}$$

where

$$\lambda_q = L_q i_q$$
$$\lambda_d = L_d i_d + \lambda_f$$

Here i_d and i_q are the dq axis stator currents, v_d and v_q are the dq axis stator voltages, R_s is the stator phase resistance, L_d and L_q are the dq axis phase inductances, λ_d and λ_q are dq axis flux linkages, and ω_r is the rotor speed in electrical radians per second. The subscript "s" used to refer to stator quantities earlier in the chapter for induction machines has been dropped here for simplicity. λ_f is the amplitude of the flux linkage established by the permanent magnet as viewed from the stator phase windings. The *d*- and *q*-axes inductances are

$$L_d = L_{ls} + L_{md} \quad \text{and} \quad L_q = L_{ls} + L_{mq}$$

L_{ls} is the leakage inductance. Note that the *d*- and *q*-axes mutual inductances can be different in the case of PM machines compared with inductance machines. The electromagnetic torque is given by

$$T_e = \frac{3}{2}\frac{P}{2}\left[\lambda_f i_q + (L_d - L_q) i_d i_q\right] \tag{9.32}$$

where P is the number of poles.

The rotor position information gives the position of *d*- and *q*-axes. The rotor position is given by

$$\theta_r = \int_0^t \omega_r(\xi)d\xi + \theta_r(0)$$

The control objective is to regulate the voltages v_d and v_q, or the currents i_d and i_q by controlling the firing angles of the inverter switches.

9.4.2 Simulation Model

The state-space representation of the PMSM dq model is useful for the computer simulation of the motor. The representation in the commonly used dq synchronous reference frame is

$$
\begin{aligned}
\frac{di_d}{dt} &= -\frac{R_s}{L_d} i_d + \frac{L_q}{L_d}\frac{P}{2}\omega_r i_q + \frac{v_d}{L_d} \\
\frac{di_q}{dt} &= -\frac{R_s}{L_q} i_q - \frac{L_d}{L_q}\omega i_d - \frac{1}{L_q}\omega_r \lambda_f + \frac{v_q}{L_q} \\
\frac{d\omega}{dt} &= \frac{1}{J}\left(T_e - T_l - B\omega - F\frac{\omega}{|\omega|}\right) \\
\frac{d\theta_r}{dt} &= \frac{P}{2}\omega
\end{aligned}
\tag{9.33}
$$

where

J is the rotor and load inertia
F is the coulomb friction
B is the viscous load
T_l is the load torque
P is the number of poles
ω_r is the mechanical rotor speed
θ is the electrical rotor position

The electrical rotor position is related to the mechanical rotor position θ_r as

$$\theta = \frac{P}{2}\theta_r$$

The electrical rotor speed is related to the mechanical rotor speed as

$$\omega = \frac{P}{2}\omega_r$$

The stationary $\alpha\beta$ reference frame is also sometimes used for modeling the PMSM. The $\alpha\beta$-model is

$$
\begin{aligned}
\frac{di_\alpha}{dt} &= -\frac{R_s}{L} i_\alpha + \frac{k_e}{L}\frac{P}{2}\omega \sin(\theta_r) + \frac{v_\alpha}{L} \\
\frac{di_\beta}{dt} &= -\frac{R_s}{L} i_\beta - \frac{k_e}{L}\frac{P}{2}\omega \cos(\theta_r) + \frac{v_\beta}{L} \\
\frac{d\omega}{dt} &= -\frac{3}{2}\frac{k_e}{J} i_\alpha \sin(\theta_r) + \frac{3}{2}\frac{k_e}{J} i_\beta \cos(\theta_r) - \frac{T_l}{J} - \frac{B}{J}\omega - \frac{F}{J}\frac{\omega}{|\omega|} \\
\frac{d\theta_r}{dt} &= \frac{P}{2}\omega
\end{aligned}
\tag{9.34}
$$

where
v_α and v_β are the αβ-stator voltages
i_α and i_β are the αβ-stator currents
L is the phase inductance with $L \approx 0.5^*(L_d + L_q)$
k_e is the motor torque constant

9.4.3 Transformation Equations

The transformation between the reference frames is given by Park's equations as

$$\begin{bmatrix} i_d \\ i_q \\ 0 \end{bmatrix} = T_{abc\to dq} \begin{bmatrix} i_a \\ i_b \\ i_c \end{bmatrix}, \quad \begin{bmatrix} i_\alpha \\ i_\beta \\ 0 \end{bmatrix} = T_{abc\to\alpha\beta} \begin{bmatrix} i_a \\ i_b \\ i_c \end{bmatrix}, \quad \begin{bmatrix} i_d \\ i_q \end{bmatrix} = T_{\alpha\beta\to dq} \begin{bmatrix} i_\alpha \\ i_\beta \end{bmatrix} \tag{9.35}$$

where

$$T_{abc\to dq} = \frac{2}{3} \begin{bmatrix} \cos(\theta) & \cos(\theta - 2\pi/3) & \cos(\theta + 2\pi/3) \\ \sin(\theta) & \sin(\theta - 2\pi/3) & \sin(\theta + 2\pi/3) \\ 0.5 & 0.5 & 0.5 \end{bmatrix}$$

$$T_{abc\to\alpha\beta} = \frac{2}{3} \begin{bmatrix} 1 & -0.5 & -0.5 \\ 0 & -\sqrt{3}/2 & \sqrt{3}/2 \\ 0.5 & 0.5 & 0.5 \end{bmatrix}, \quad T_{\alpha\beta\to dq} = \begin{bmatrix} \cos(\theta) & \sin(\theta) \\ -\sin(\theta) & \cos(\theta) \end{bmatrix} \tag{9.36}$$

and $T_{dq\to abc} = T_{abc\to dq}^{-1}$, $T_{\alpha\beta\to abc} = T_{abc\to\alpha\beta}^{-1}$, $T_{dq\to\alpha\beta} = T_{\alpha\beta\to dq}^{-1}$.

The *abc* variables are obtained from the *dq* variables through the inverse of the Park transformation equation:

$$T_{dq\to abc} = \begin{bmatrix} \cos(\theta) & \sin(\theta) & 1 \\ \cos(\theta - 2\pi/3) & \sin(\theta - 2\pi/3) & 1 \\ \cos(\theta + 2\pi/3) & \sin(\theta + 2\pi/3) & 1 \end{bmatrix} \tag{9.37}$$

9.4.4 PM Synchronous Motor Drives

The advantage of a PM synchronous machine is that it can be driven in the vector-controlled mode delivering high performance, unlike the PM trapezoidal machine that has to be driven with squarewave currents. Of course, a high precision position information is needed to implement the

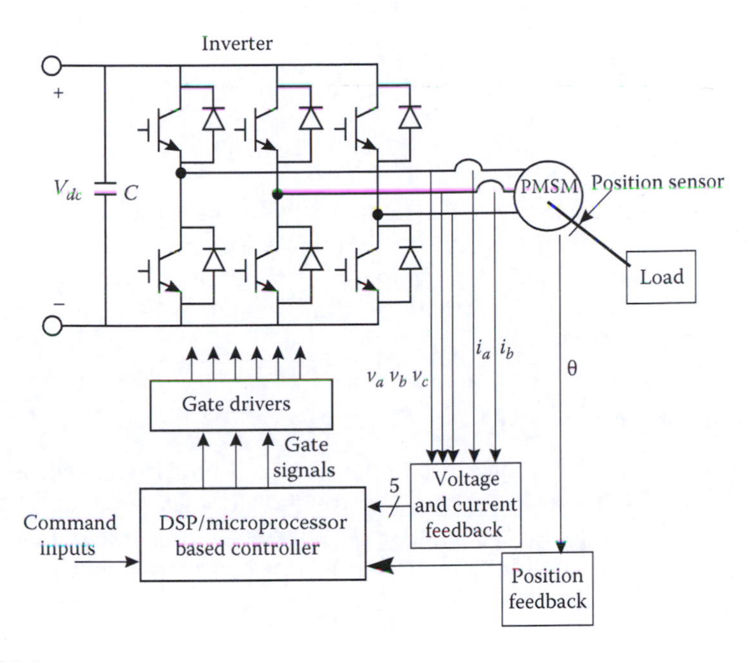

FIGURE 9.8
A typical PM synchronous motor drive structure.

vector control in PM synchronous machines. A typical PM synchronous motor drive consists of a PM synchronous motor, a three-phase bridge inverter, gate drivers, position sensor, current or voltage sensors, a microprocessor or digital signal processor, and interfacing circuits, as shown in Figure 9.8.

The vector control of a PM synchronous motor is simpler than that of an induction motor, since the motor always rotates at synchronous speed. In vector calculations, only the synchronously rotating reference frame is necessary. The system controller sets the reference or command signal, which can be position, speed, current, or torque. The variables needed for the controller are the feedback signals from the sensing circuits (position, speed, current, or voltage) or estimated values obtained within the signal processor. The error signals between the reference and actual variable signals are transformed to gate control signals for the inverter switches. The switches follow the gate commands to decrease the error signals by injecting desired stator currents into the three-phase stator windings.

Vector control is used for the PM synchronous motor drives in the electric and hybrid vehicle applications to deliver the required high performance. The torque expression in Equation 9.32 for the PM synchronous motor shows that if the d-axis current is maintained constant, the generated torque is proportional to the q-axis current. For the special case, when i_d is forced to be zero $\lambda_d = \lambda_f$ and

$$T_e = \frac{3}{2} \cdot \frac{P}{2} \cdot \lambda_f \cdot i_q$$
$$= k_e i_q \tag{9.38}$$

where $k_e = (3/2) \cdot (P/2) \cdot \lambda_f$ = motor constant. Since the magnetic flux linkage is a constant, the torque is directly proportional to the q-axis current. The torque equation is similar to that of a separately excited DC machine. Therefore, using reference frame transformations, the PM synchronous motor can be controlled like a DC machine.

9.4.4.1 Flux Weakening

The PM synchronous motor can be operated in the field-weakening mode similar to the DC motor to extend the constant power range and achieve higher speeds. An injection of negative i_d will weaken the air gap flux as seen from Equation 9.32. The implementation technique for the field weakening mode is shown in Figure 9.9.

9.4.4.2 Current and Voltage Controllers

The current and voltage control techniques described in Section 8.4 are all applicable to the PM synchronous motor drive. The current regulators in a PM synchronous motor drive can be designed in the dq reference frame where d- and q-axes current commands are generated by the outer loop controller. These dq current commands are then converted to abc-current commands using the rotor position feedback information. Current controllers in the abc-reference frame then generate the gate signals for the inverter. The simplified block diagram of a current controller is shown in Figure 9.10.

The current controllers in a PM synchronous motor drive can be avoided if the current feedback information is used to drive a linear regulator (such as a PI controller) to convert the dq current commands into dq voltage commands.

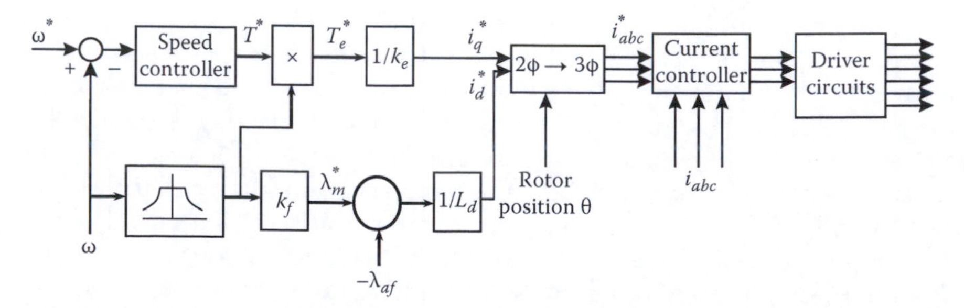

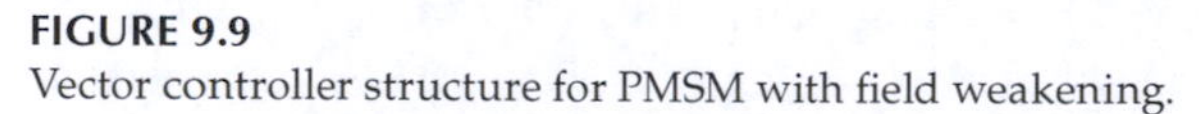
FIGURE 9.9
Vector controller structure for PMSM with field weakening.

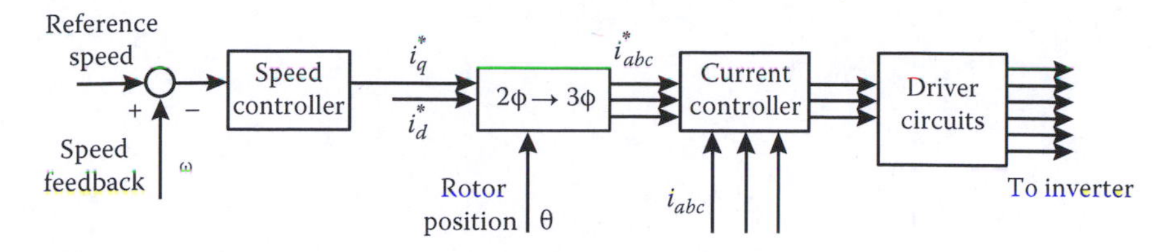

FIGURE 9.10
PM synchronous drive controls using a current controller.

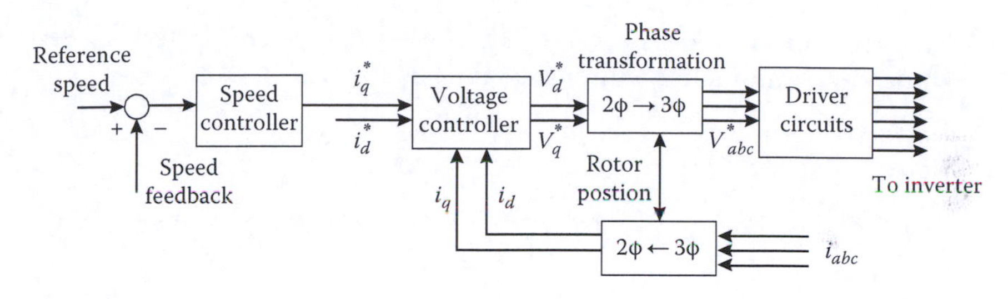

FIGURE 9.11
PM synchronous drive controls using a PWM voltage controller.

The voltage command signals can then be used for a voltage PWM scheme, such as the sinusoidal PWM or the space vector PWM. The block diagram of such a method is shown in Figure 9.11.

Problems

9.1 An induction motor has the following parameters:

5 hp, 200 V, three phase, 60 Hz, star connected
$R_s = 0.28\ \Omega$, $L_s = 0.055$ H, Turns ratio = 3
$R_r = 0.18\ \Omega$, $L_r = 0.056$ H, $L_m = 0.054$ H

The motor is supplied with its rated voltages. Find the *d*- and *q*-axes steady state voltages and currents in (a) synchronous reference frame and (b) rotor reference frame when the rotor is locked (i.e., $\omega_r = 0$).

9.2 The parameters of a 440 V, 10 hp, four-pole, 60 Hz induction machine with an indirect field-oriented control are given as

$$R_s = 0.768\ \Omega,\ X_{ls} = 2.56\ \Omega,\ X_m = 51.2\ \Omega, R_r = 0.768\ \Omega,\ X_{lr} = 2.56\ \Omega$$

Sketch the reference frame phasor diagrams for this machine for the following conditions:

(a) $\omega_r = \omega_{rated}$
   1. $T_e = 40\,\text{N m}$
   2. $T_e = 0$

(b) $\omega_r = 0$
   1. $T_e = 40\,\text{N m}$
   2. $T_e = 0$

The phasor diagrams should include the complex vector variables $\vec{\lambda}_r$, $\vec{I}_r$, $\vec{\lambda}_s$, $\vec{I}_s$, and $\vec{V}_s$ (e.g., $\vec{\lambda}_r = \lambda_{qr} - j\lambda_{dr}$).

9.3 The parameters of a 10 hp, 60 Hz, three-phase, wye-connected, six-pole, squirrel-cage induction motor are given as follows:

Rated voltage = 220 V
Rated speed = 1164 rpm

$$R_s = 0.294\,\Omega/\text{ph},\ X_{ls} = 0.524\,\Omega/\text{ph},\ R_r' = 0.156\,\Omega/\text{ph},\ X_{lr}' = 0.279\,\Omega/\text{ph}$$

$$X_m = 15.457\,\Omega/\text{ph},\quad J = 0.4\,\text{kg m}^2$$

(a) The motor operates in steady state under rated operating conditions. Calculate the stator and rotor currents in the stator reference frame and then calculate the developed torque.

(b) The motor operates under the same condition as in part (a). By using results obtained in (a), calculate the air gap flux and rotor flux vectors.

References

1. D.W. Novotny and T.A. Lipo, *Vector Control and Dynamics of AC Drives*, Oxford University Press, Inc., New York, 1996.
2. B.K. Bose, *Modern Power Electronics and AC Drives*, Prentice Hall, Upper Saddle River, NJ, 2001.
3. W.V. Lyon, *Transient Analysis of Alternating Current Machinery*, John Wiley & Sons, Inc., New York, 1954.
4. P.C. Krasue and O. Wasynchuk, *Analysis of Electric Machinery*, McGraw Hill, New York, 1986.
5. N. Mohan, *Electric Drives—An Integrated Approach*, MNPERE, Minneapolis, MN, 2001.

10

Internal Combustion Engines

Internal combustion (IC) engines are energy conversion devices that extract stored energy in a fuel through the combustion process and deliver mechanical power. The devices are also known as heat engines in the broader sense. Reciprocating-type IC engine is the primary choice of power plant in today's production hybrid vehicles. The IC engine is assisted by an electric motor of similar or lower power capability in these vehicles. IC engines for conventional vehicles are designed to operate over a wide range, and compromise is often necessary to deliver acceptable efficiency and performance throughout its operation regime. On the other hand, IC engines in hybrid vehicles require operation within a narrow band of torque–speed characteristics, especially in those with series-type hybrid architectures. However, IC engines are yet to be designed specifically for hybrid vehicles; significant opportunity exists to explore improved IC engine designs for hybrid vehicles.

The prime objectives of hybridizing road vehicles are to enhance fuel economy and reduce emissions. Fuel economy depends on the operating points of the IC engine, while emission control units target eliminating harmful pollutants from the exhaust stream. After developing an understanding of the working principles of IC engines, this chapter addresses fuel economy calculations and exhaust emission control methods. Discussions on IC engines include operation principles, air standard cycles, and combustion characteristics. For understanding engine operation efficiency, practical parameters such as brake mean effective pressure and brake-specific fuel consumption have been used.

10.1 Internal Combustion Engines

The devices that convert heat transfer to work are known as heat engines. Heat engines operate in cycles undergoing constant pressure, constant volume, constant entropy, etc., strokes within one cycle to convert thermal energy into useful work. The ideal cycle used for thermodynamic analysis is the Carnot cycle, although practical heat engines work with different cycles that have theoretical efficiencies lower than that of the Carnot cycle. The practical cycles have evolved due to the practical limitations of the Carnot cycles, and also because of the difference in the characteristics of the choices available for the energy source, working fluid, and hardware materials.

The heat engines of interest for the hybrid vehicle applications, primarily the IC engine and the gas turbine, will be discussed in this section. An IC engine is a heat engine that utilizes gas as a working fluid. The IC engines use heat cycles that gain their energy from the combustion of fuel within the engine. The IC engines can be reciprocating type, where the reciprocating motion of a piston is converted to linear motion through a crank mechanism. The IC engines used in automobiles, trucks, and buses are of the reciprocating type where the processes occur within reciprocating piston–cylinder arrangement. The gas turbines used in power plants are also IC engines where the processes occur in an interconnected series of different components. The Brayton cycle gas turbine engine has been adapted to automotive propulsion engine and has the advantage of burning fuel that requires little refining and the fuel burns completely. The gas turbines have fewer moving parts since there is no need to convert the rotary motion of the turbine. The disadvantages of gas turbines for automotive applications are complex construction, high noise levels, and relatively lower efficiency for smaller sized engines. Nevertheless, gas turbines have been considered for hybrid electric vehicles and prototype vehicles have been developed.

The performance of a heat engine is measured by the efficiency of the heat engine cycle defined as the ratio of the net work output per cycle W_{net} to the heat transfer into the engine per cycle. Another way of defining the performance of heat engines is to use the mean effective pressure MEP. The MEP is the theoretical constant pressure that, if exerted on the piston during the expansion stroke between the largest specific volume and the smallest volume, would produce the same net work as actually produced by the heat engine. Mathematically stated,

$$\text{MEP} = \frac{|W_{net}|}{\text{Displacement volume}} \tag{10.1}$$

The heat engine cycle performance analysis is carried out from the information available at certain convenient state points in the cycle. The parameters needed at the state points are pressure, temperature, volume, and entropy. If two parameters are known at two state points, then the unknown parameters are usually obtained from the process that the working fluid undergoes between the two state points (such as isobar, isentropic, etc.) and the laws of thermodynamics. The discussion of the laws of thermodynamics and efficiency analysis of heat engine cycles are beyond the scope of this book. Only a general introduction to the heat engine cycles of interest will be given. The concepts of entropy discussed earlier in Chapter 4 will be used here while discussing the thermodynamic cycles of the heat engines.

10.1.1 Reciprocating Engines

The two types of reciprocating IC engines are the spark-ignition (SI) engine and the compression-ignition (CI) engine. The two engines are commonly known as gasoline/petrol engine and diesel engine based on the type of fuel used for combustion. The difference in the two engines is in the method of initiating the combustion and in the processes of the cycle. In an SI engine, a mixture of air and fuel is drawn in and a spark plug ignites the charge. The intake of the engine is called the charge. Electronically controlled direct fuel injection is used in modern gasoline engine vehicles, which helps to measure out the right amount of fuel in response to driver demand. In a CI engine, air alone is drawn in and compressed to such a high pressure and temperature that combustion starts spontaneously when fuel is injected. The SI engines are relatively light and lower in cost and used for lower power engines as in the conventional automobiles. The CI engines are more suitable for power conversion in the higher power range, such as in trucks, buses, locomotives, ships, and in auxiliary power units. The fuel economy of CI engines is better than the SI engines, justifying their use in higher power applications [1].

Although gasoline engines have thus far been used in passenger hybrid vehicles, diesel engines have a lot of potential for replacing those in the future. Diesel engines are increasingly being used in several models of passenger IC engine vehicles that give high gas mileages because of their high fuel efficiency. The problem of NO_x emissions in diesel engines can be addressed through catalytic conversion and emission aftertreatment components. More on emissions and their controls appear in Section 10.4.

The sketch of a representative reciprocating IC engine including the terms standard for such engines is given in Figure 10.1. The engine consists of piston that undergoes a reciprocating motion within the engine cylinder. The position of the piston at the bottom of the cylinder when the volume inside is maximum is known as the *bottom dead center* (BDC). The position of the piston at the top of the cylinder when the volume inside is minimum is called the *top dead center* (TDC). This minimum cylinder volume when the piston is at the TDC is known as the *clearance volume*. A crank mechanism converts the linear motion of the piston into rotary motion and delivers the power to the crankshaft. The volume swept by the piston as it moves from the TDC to the BDC is known as the *displacement volume*, which is a parameter commonly used to specify the size of an engine. The *compression ratio* is defined as the ratio of the volume at BDC to the volume at TDC.

The diameter of the cylinder is called the *bore*. The bores in automotive SI engines are typically between 70 and 100 mm. Too small a bore leaves no room for valves while exceedingly large bores mean more mass and longer flame travel time. The smaller bores enable higher rpm of the engines. The vertical distance traversed by the piston from the BDC to the TDC is called the *stroke*. The stroke is typically between 70 and 100 mm. Too short a stroke means there will not be enough torque. The length of the stroke is limited

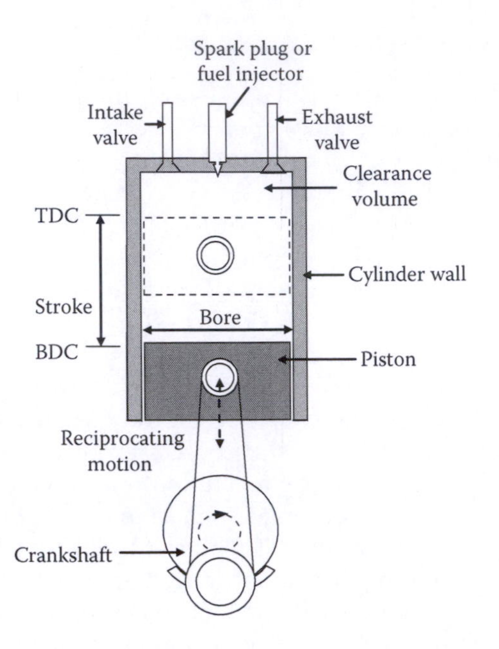

FIGURE 10.1
A reciprocating IC engine.

TABLE 10.1
Arrangement of Automotive Engine Cylinders

No. of Cylinders	Cylinder Arrangement
3	Inline
4	Inline, flat
5	Inline
6	Inline, flat, V (narrow, 60°, 90°)
8	V (90°)
10	V (90°)
12	V, flat (for exotics)

by the piston velocities. The minimum displacement of a cylinder can be 250cc, while the maximum can be up to 1000cc. The acceptable bore and stroke lengths lead to multiple cylinder engines. The multiple cylinders can be arranged inline, flat, or in a V-shaped configuration depending on the number of cylinders. The typical arrangements are given in Table 10.1.

For a good primary balance, the power strokes of the multiple cylinders are equally spaced. The engine arrangements that have good primary balance are: Inline 4 and 6 cylinders, 90° V 8 cylinders, and flat 4 and 6 cylinders. The arrangements that have poor primary balance are 90° V 6 cylinders and Inline 3 cylinders. In the arrangements with poor primary balance, counter-rotating balance shafts are used to cancel vibration.

The valve arrangement in the cylinder is known as the valve train. The valve train can be overhead valve (OHV), single overhead cam valve (SOHC), or dual overhead cam valve (DOHC). The OHV has camshaft, push rod, rocker, and valve, while the SOHC and DOHC have camshaft(s), rocker, and valve. The SOHC has one camshaft placed in the cylinder head, while DOHC has two camshafts in the cylinder head, one for the inlet valves and one for the exhaust valves. There can be 2, 3, 4, or 5 valves in a cylinder. The number of valve selection depends on the trade-off between flow and complexity.

10.1.2 Practical and Air-Standard Cycles

The automobile IC engines are typically four-stroke engines where the piston executes the four strokes of the cylinder for every two revolutions of the crankshaft. The four strokes are *intake, compression, expansion,* and *exhaust*. The operations within the four strokes are illustrated in the pressure–volume diagram of Figure 10.2. The numbers 1–5 in the diagram represent the distinct state points between the processes of the cycle. The *intake* is the process of drawing the charge into the cylinder with the intake valve open. The working fluid is compressed in the *compression* stage with the piston traveling from the BDC to the TDC. Work is done by the piston in the compression stage. In the next stage, heat is added during ignition of the compressed fluid by ignition for SI engines or by spontaneous combustion for CI engines. The next stage is the *expansion* process, which is also known as the power stroke. In this stroke, work is done by the cylinder on the piston. The *exhaust* process starts at the BDC with the opening of the exhaust valve. Unutilized heat is rejected from the engine during the exhaust process.

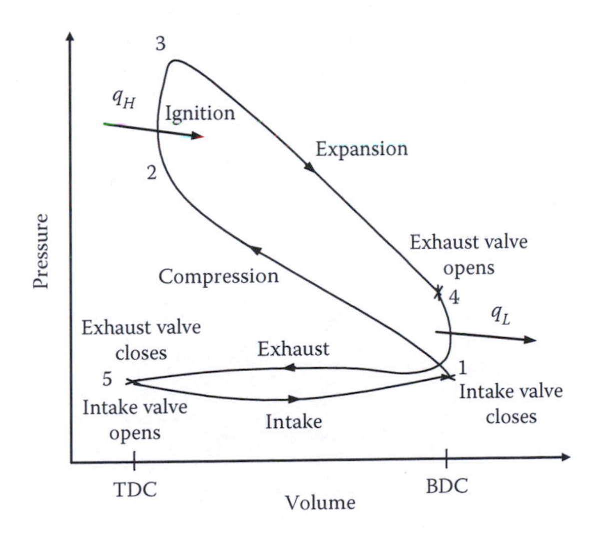

FIGURE 10.2
Pressure–volume diagram of a reciprocating-type IC engine.

The practical cycle entails significant complexity due to the irreversibilities associated with friction, pressure and temperature gradients, heat transfer between the gases and the cylinder walls, work required to compress the charge and to discharge the products of combustion. The complexity of the process typically calls for computer simulation for performance analysis. However, significant insight can be gained into the processes by making simplifying assumptions about the behavior of the processes that make up the cycle. Idealized processes can substitute the combustion and expansion processes within the cylinder. These idealized cycles are known as air-standard cycles. The air-standard analysis assumes that the working fluid is an ideal gas, the processes are all reversible, and the combustion process is replaced by a heat transfer process from an external source. A brief description of the two air-standard cycles, the Otto cycle and the Diesel cycle are given in the following.

10.1.2.1 Air-Standard Otto Cycle

The Otto cycle is the idealized air-standard version of the practical cycle used in SI engines. The air-standard Otto cycle assumes that the heat addition occurs instantaneously under constant volume when the piston is at the TDC. The cycle is illustrated on a p–v (pressure–volume) diagram in Figure 10.3. The *intake stroke* starts with the intake valve opening at the TDC to draw a fresh charge into the cylinder. The intake valve is open between 5 and 1 to take the fresh charge, which is a mixture of fuel and air. The volume of the cylinder increases as the piston moves down to allow more charge into the cylinder. The stroke ends with the piston reaching the BDC when the intake valve closes at that position. This state point at the BDC is labeled as 1. In the next cycle between 1 and 2, work is done on the charge by the piston to compress it, thereby increasing its temperature and pressure. This is the *compression stroke* when the piston moves up with both the valves closed.

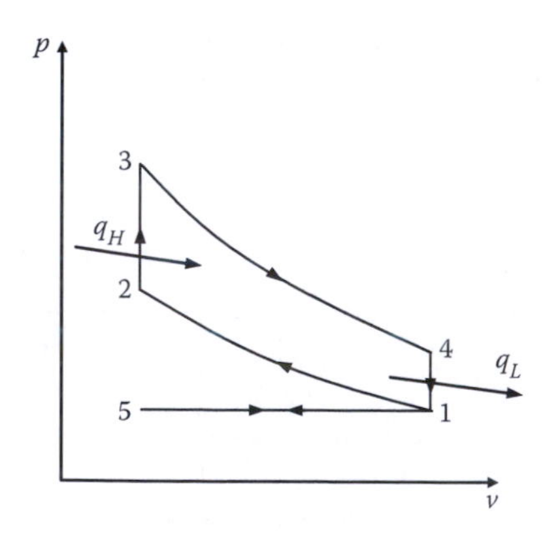

FIGURE 10.3
p–v diagram of an air-standard Otto cycle.

Process 1–2 is an isentropic (i.e., reversible as well as adiabatic) compression as the piston moves from the BDC to the TDC. The combustion starts near the end of the compression stroke in SI engines when the high-pressure, high-temperature fluid is ignited by the spark plug. The pressure thus rises at constant volume (i.e., almost instantaneously) to state point 3. Process 2–3 is the rapid combustion process when heat is transferred near the TDC to the air from the external source. The next stroke is the *expansion or power stroke*, when the gas mixture expands and work is done on the piston forcing it to return to the BDC. Process 3–4 represents the isentropic expansion when work is done on the piston. The final stroke is the *exhaust stroke*, which starts with the opening of the exhaust valve near 4. The piston begins to move upward purging the combustion products from the cylinder along the way. During process 4–1, the most of the heat is rejected while the piston is near BDC. Process 1–5 represents the exhaust of the burnt fuel at essentially constant pressure. At 5, the exhaust valve closes and the intake valve opens; the cylinder is now ready to draw in fresh charge for a repeat of the cycle.

The SI engines can be a four-stroke engine or a two-stroke engine. The two-stroke engines run on two-stroke Otto cycle where the intake, compression, expansion, and exhaust operations are all accomplished in one revolution of the crankshaft. The two-stroke cycles are used in smaller engines, such as those used in motorbikes.

Most SI engines, or gasoline engines as it is more commonly known, run on a modified Otto cycle. The air–fuel ratio used in these engines is between 10/1 and 13/1. The compression ratios are in the range of 9–12 for most production vehicles. The compression ratio of the engine is limited by the octane rating of the engine. A high compression ratio may lead to autoignition of the air–fuel mixture during compression, which is absolutely undesirable in an SI engine. The SI engines have been originally developed by limiting the amount of air allowed into the engine using a *carburetor* placed in the path of air intake. The function of the carburetor is to draw the fuel by creating a vacuum following "Bernoulli's principle." However, fuel injection, which is used for diesel engines, is now common for gasoline engines with SI instead of the carburetor. The control objective for fuel injection systems is to compute the mass flow rate of air into the engine at any instant of time and to mix the correct amount of gasoline with it, such that the air and fuel mixture is right for the engine running condition. In recent years, the requirements to meet the strict exhaust gas emission regulations have increased the demand for fuel injection systems.

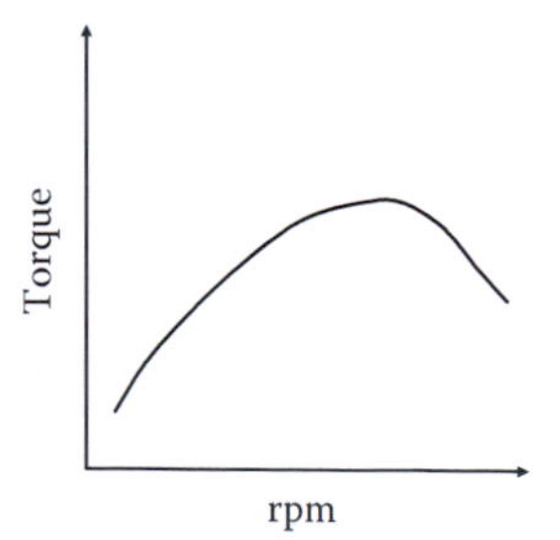

FIGURE 10.4
Torque–speed characteristics of a gasoline engine.

The torque–speed characteristics of an SI or gasoline engine are shown in Figure 10.4. The engine has a very narrow high torque range, which also requires high enough rpm of the engine. The narrow high torque region burdens the transmission gear requirements of SI engines.

The SI engines are widely used in automobiles and continuous development has resulted in engines that easily meet current emission and fuel economy standards. Currently, SI engines are the lowest cost engines, but the question remains whether it will be possible to meet future emission and fuel economy standards at a reasonable cost. The SI engine also has a few other drawbacks, which include the throttling plate used to restrict the air intake. The partial throttle operation is poor in SI engines due to throttle irreversibility, a problem that is nonexistent in diesel engines. In general, the throttling process leads to a reduction in the efficiency of the SI engine. The losses through bearing friction and sliding friction further reduce the efficiency of the engine.

10.1.2.2 Air-Standard Diesel Cycle

The practical cycles in diesel engines are based on the Diesel cycle. The Diesel cycle attempts to approach the Carnot efficiency as closely as possible. The air-standard Diesel cycle assumes that the heat addition takes place at constant pressure, while heat rejection occurs under constant volume. The cycle is shown on a p–v (pressure–volume) diagram in Figure 10.5.

The cycle begins with the *intake* of fresh air into the cylinder between 5 and 1. The intake valve is open between 5 and 1. The next process 1–2 is the same as in Otto cycle when isentropic (constant entropy) *compression* takes place as the piston moves from the BDC to the TDC. With a sufficiently high compression ratio, the temperature and pressure of air reaches such a level that the combustion starts spontaneously due to the injection of fuel near the end of the compression stroke. The heat is transferred to the working fluid under constant pressure during combustion in process 2–3, which also makes

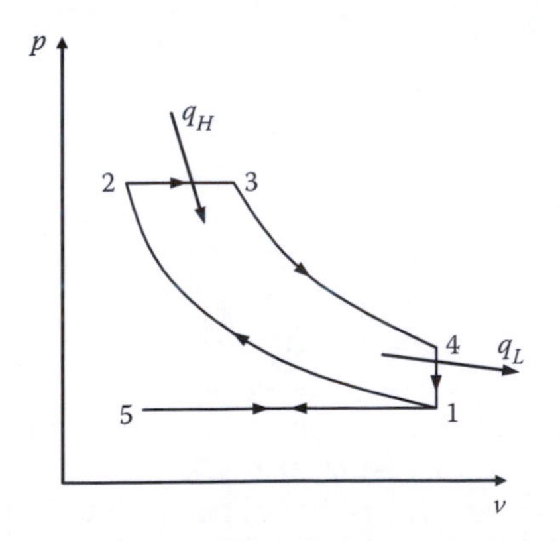

FIGURE 10.5
p–v diagram of an air-standard Diesel cycle.

up the first part of the *expansion* or power stroke. The isentropic expansion in process 3–4 makes up for the rest of the power stroke. The *exhaust* valve opens at state point 4 allowing the pressure to drop under constant volume during the process 4–1. Heat is rejected during this process while the piston is at BDC. The exhaust of the burnt fuel takes place during 1–5 at essentially constant pressure. The exhaust valve then closes and the intake valve opens and the cylinder is ready to draw in fresh air for a repeat of the cycle.

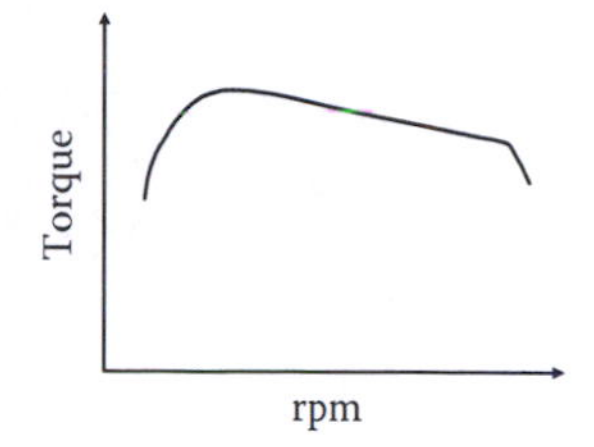

FIGURE 10.6
Torque–speed characteristics of a diesel engine.

The nominal range of compression ratios in CI engines is 13/1 to 17/1 and the air–fuel ratios used lie between 20/1 and 25/1. The higher compression ratio aided by the work produced during combustion results in higher efficiency in diesel engines compared to gasoline engines. The diesel engine has a lower specific power than the gasoline engine. The diesel engines also have a broad torque range as shown in Figure 10.6.

The major drawbacks of diesel engines include the requirement of stronger and heavier components that increase the mass of the engine and the speed limitation of the injection and flame propagation time. The improvements in diesel engine are directed toward reducing the nitrogen oxides in the exhaust; and the noise, vibration, and malodorous fumes from the engine. The recent automotive diesel engines developed addressing the aforementioned issues make them excellent candidates for hybrid electric vehicle applications.

10.1.3 Gas Turbine Engines

Gas turbines are used for stationary power generation as well as for transportation applications such as in aircraft propulsion and marine power plants. The gas turbines can be either an open type or a closed type. In the open type, the working fluid gains energy from the combustion of fuel within the engine, whereas in a closed type, the energy input by heat transfer is from an external source. The open type is used in vehicle propulsion systems and will be considered here. The use of gas turbines in transportation applications can be attributed to the favorable power output-to-weight ratio of gas turbines, which again makes it a viable candidate for hybrid electric vehicles.

The gas turbine engine runs on a Brayton cycle, which uses constant pressure heat transfer processes with isentropic compression and expansion processes in between. The major components of a gas turbine are shown in Figure 10.7a, along with the directions of energy flow. The corresponding p–v diagram for an air-standard cycle is illustrated in Figure 10.7b, which ignores the irreversibilities in air circulation through the components of the gas turbine. The constant entropy processes are shown as $s = c$.

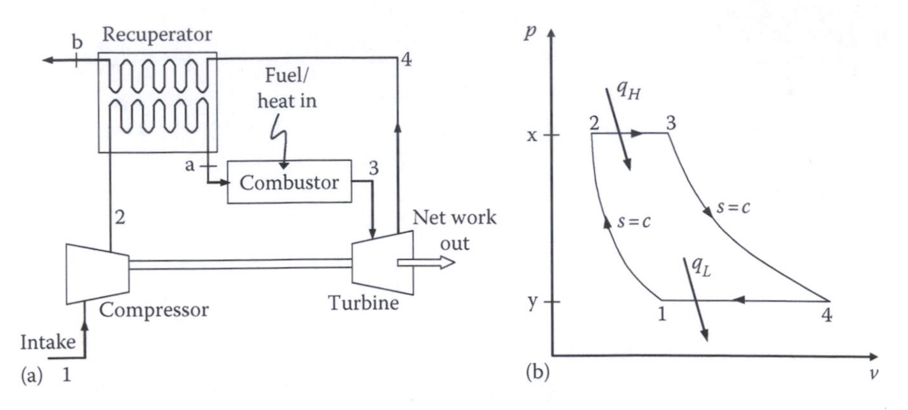

FIGURE 10.7
(a) Air-standard gas turbine cycle and (b) p–v diagram of ideal Brayton cycle.

The working fluid of air is considered to be an ideal gas in the air-standard analysis. At the beginning of the cycle at state 1, atmospheric air is continuously drawn in and compressed in the compressor to raise its pressure and temperature. The compressor is usually of the radial flow type for automotive applications. The air then moves into the combustion chamber or combustor, where combustion takes place with fuel injected to raise the temperature of the air. The high temperature, high pressure mixture is then expanded and cooled in the turbine, which produces power and delivers work. The hot gas turbine exhaust gas is utilized in a recuperator to preheat the air exiting the compressor before entering the combustor. This reduces the fuel needed in the combustor and increases the overall efficiency of the system. In the open-type gas turbine, fresh air is drawn in and exhaust gases are purged after going through the recuperator in each cycle. The turbine is designed to deliver work output that is greater than the required compressor work input. The excess work available at the shaft is used to propel a vehicle in automotive gas turbines or to generate electricity in stationary applications. The enclosed area in the figure represents the net work output. In particular, the area 12xy1 in the p–v diagram represents the compressor work input per unit mass and the area 34yx3 represents the turbine work output per unit mass.

The power output of a turbine is controlled through the amount of fuel injected into the combustor. Many turbines have adjustable vanes and/or gearing to decrease fuel consumption during partial load conditions and to improve acceleration. The major advantage of a gas turbine is that the only moving part is the rotor of the turbine. The turbine has no reciprocating motion, and consequently runs smoother than a reciprocating engine. Another big advantage of gas turbines is its multifuel capability. The turbine has the flexibility of burning any combustible fuel injected into the airstream, because the continuous combustion is not heavily reliant on the combustion characteristics of the fuel. The fuel in a gas turbine burns completely and cleanly, which keeps the emissions at a low level.

The gas turbine engine has a few drawbacks, which have prevented its widespread use in automotive applications. The complicated design of gas turbines increases the manufacturing costs. The response time of the gas turbine to changes in throttle request is slow relative to a reciprocating engine. The efficiency of the gas turbine decreases at partial throttle conditions making it less suitable for low-power applications. The turbine requires intercoolers, regenerators, and/or reheaters to reach efficiencies comparable to the SI or CI engines, which adds significant cost and complexity to a gas turbine engine [2].

10.2 BMEP and BSFC

Although torque is a measure of a particular engine's ability to do work, mean effective pressure (MEP) is a more useful parameter for standardized comparison of several different engines as was indicated earlier. For reciprocating engines, *brake mean effective pressure* (BMEP) is the more practical parameter, which measures the engine's work done per cycle per cylinder displaced volume. To understand BMEP, we first need to define indicated mean effective pressure (IMEP) and friction mean effective pressure (FMEP). The word "indicated" refers to the net work performed by the working mixture in the cylinder of the engine over the compression and expansion strokes. Part of this work goes into overcoming engine mechanical friction and losses to induct air–fuel mixture and expel excess charge. The remaining work or torque is what is available at the engine crankshaft. "Brake" refers to the work or torque available at the engine crankshaft as measured at the flywheel by a dynamometer. The dynamometer measurements help avoid mechanical losses associated with the powertrain of a vehicle. In terms of mean effective pressures, we can write

$$\text{IMEP} = \text{BMEP} + \text{FMEP}$$

The result of calculated BMEP gives the value for a theoretical constant pressure exerted in the combustion chamber as a result of combustion. While it does not describe the actual combustion pressure, it is a good standard for comparison.

Work done per cycle using the engine's output power is

$$\text{Work per cycle} = \frac{P_b n_R}{N} \tag{10.2}$$

where

P_b is engine brake output power
n_R is the number of crankshaft revolutions per power stroke
N is the engine speed in rps

n_R is the number of power strokes, which is 2 for a four-stroke engine. It must be noted that the engine power is the power output for the entire engine rather than just one cylinder. Therefore, the BMEP is obtained by dividing by the engine's entire displacement volume V_d as

$$\text{BMEP} = \frac{P_b n_R}{V_d N} \tag{10.3}$$

To simplify the computation, power can be expressed as the product of brake torque T_b, and engine speed N. Note that to be unit consistent with Equation 10.3, engine speed has to be expressed in terms of revolutions per unit time; thus multiplication by 2π is necessary. Using these values in Equation 10.3, the proper unit for BMEP with torque expressed in N m is

$$\frac{\left(\dfrac{\text{rev}}{\text{cycle}}\right)\left(2\pi\dfrac{\text{rad}}{\text{rev}}\right)(\text{N m})}{\left(\text{m}^3/\text{cycle}\right)} \Rightarrow \frac{\text{N}}{\text{m}^2} \text{ or Pa (Pascal)} \quad (1\text{ bar} = 100\text{ kPa})$$

BMEP in terms of torque is given by

$$\text{BMEP} = \frac{4\pi \times T_b}{V_d}\left(\text{N/m}^2\right) \tag{10.4}$$

The volume of one cylinder can be calculated from

$$V_d = \frac{\pi}{4} \times \text{bore}^2 \times \text{stroke}\left(\text{m}^3\right)$$

Fuel consumption in IC engines is measured in terms of fuel mass flow per unit time $\dot{m}_f$. The fuel flow rate per unit output power, called the *specific fuel consumption* (SFC), is a measurement of how effectively the engine is converting the stored chemical energy in the fuel into mechanical energy. For practical purposes, the more commonly used parameter is the ratio of the fuel's mass flow rate $\dot{m}_{fuel}$ and the measured output power P_b. Similar to BMEP, this parameter is called the *brake-specific fuel consumption* (BSFC) for reciprocating engines, indicating that power P_b or torque T_b is measured at the engine crankshaft using a dynamometer. The engine's output power at the flywheel

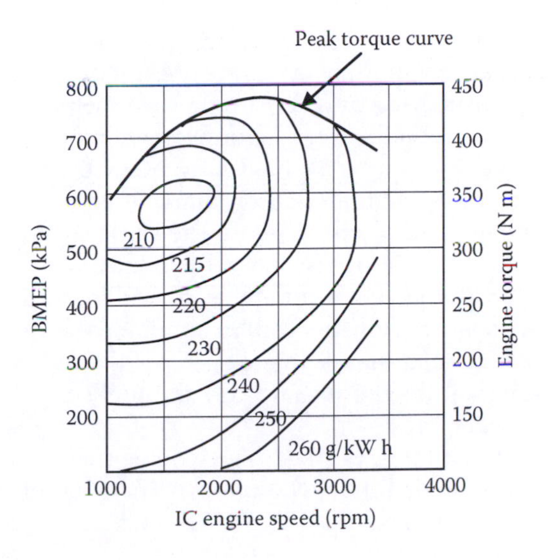

FIGURE 10.8
Example efficiency map of a diesel IC engine.

can be expressed as the product of the measured engine brake torque and speed. The resulting equation for calculating BSFC is

$$\text{BSFC} = \frac{\dot{m}_f}{P_b} = \frac{\dot{m}_f}{T_b 2\pi N} \tag{10.5}$$

Contour maps of BSFC under the engine's peak torque curve are obtained empirically through the laboratory testing of the engine at a variety of speed/load combination set point. The performance curves of engine power, torque, and BSFC versus speed is a useful tool for analyzing IC engines. An example efficiency map for a diesel engine is shown in Figure 10.8 with BSFC contours plotted in g/kW h [3]. Representative scales for BMEP in kPa and corresponding engine torque in N m have been used in the figure. Engine power available at the crankshaft for an operating point can be obtained from the product of the brake torque and the engine speed.

10.3 Vehicle Fuel Economy

The fuel economy is a key performance measure of a vehicle related to the efficiency of the powertrain components. The fuel economy is calculated based on fuel consumption over a standard drive cycle. The fuel economy of a production or already built vehicle can be calculated by actually driving the vehicle on the road or in a chassis dynamometer following a standard drive cycle such as a highway drive cycle or an urban drive cycle. Real world fuel economy is consumer specific; however, if representative highway or urban, standard drive cycles are used by a consumer, then the fuel economy should be close to the numbers obtained in a laboratory with a dynamometer.

The design of a vehicle can be evaluated by predicting the fuel economy through simulation. Subsystem models or characteristics obtained from test or simulation data are used to characterize the vehicle. Dynamic models of subsystems are avoided to simplify the vehicle level simulation. Standard drive cycles along with the vehicle model and road load characteristics can predict the fuel economy reasonably well in simulation with a fairly accurate characterization of the subsystems. In order to predict the fuel economy, the engine output power at the crankshaft and the fuel consumed to produce that power is necessary.

The resultant BSFC and the specific gravity of the fuel at prevailing conditions are necessary to predict the fuel economy. The BSFC of an engine is the mass of fuel consumed at the engine to produce a certain amount of energy. The specific gravity of the fuel is assumed constant for the drive cycle. This is an approximation since the specific gravity depends on temperature and other factors.

The fuel economy of a vehicle is given by

$$\text{FE}\,(\text{m/L}) = \frac{\text{Vehicle velocity, } v\,(\text{m/s})}{\text{Fuel flow rate, } Q_f\ (\text{L/h}) \times (\text{h/3600 s})} \tag{10.6}$$

The fuel flow rate is obtained from the mass flow rate of fuel as

$$Q_f(\text{L/h}) = \frac{\text{Mass flow rate, } \dot{m}_f\ (\text{g/h})}{\text{Specific gravity, } \rho_{fuel}\ (\text{g/L})} \tag{10.7}$$

The mass flow rate is obtained from the engine power at the crankshaft and the BSFC as

$$\dot{m}_f = P_b\ (\text{kW}) \times \text{BSFC}\ (\text{g/kW h}) \tag{10.8}$$

The fuel flow rate is then

$$Q_f(\text{L/h}) = \frac{P_b\ (\text{kW}) \times \text{BSFC}\ (\text{g/kW h})}{\rho_{fuel}\ (\text{g/L})}$$

Finally, the fuel economy is given by

$$\text{FE}\,(\text{m/L}) = \frac{3600 \times v\ (\text{m/s}) \times \rho_{fuel}\ (\text{g/L})}{P_b\ (\text{kW}) \times \text{BSFC}\ (\text{g/kW h})} \tag{10.9}$$

The fuel economy in mi/gal is given by

$$\text{FE (mi/gal)} = \frac{3600 \times v\,(\text{m/s}) \times \rho_{fuel}\,(\text{g/L}) \times (3.79\ \text{L/gal}) \times (\text{mi}/1609\ \text{m})}{P_b\,(\text{kW}) \times \text{BSFC}\,(\text{g/kW h})}$$

$$= 8.4797 \frac{v \times \rho_{fuel}}{P_b \times \text{BSFC}} \tag{10.10}$$

The engine power can be obtained from the traction power if the drivetrain efficiency is available. For a given traction power and drivetrain efficiency (η_{DT}), the engine power is

$$P_b = \frac{P_{TR}}{\eta_{DT}} \tag{10.11}$$

While BSFC can be useful in the standardized examination of fuel usage relative to power output, the evaluation of fuel efficiency tells what percentage of fuel being put into the engine is being used for mechanical work output. Efficiency for a combustion engine can be tabulated by dividing the measured brake power by the potential power of the injected fuel. The power potential of a particular fuel depends on both its energy content and the rate that it is injected. Standard fuel economy validation requires standard fuel formulation with well-established heat values. Energy content can be expressed in terms of a fuel's higher heating value (HHV), which is the amount of heat released per unit quantity when combusted and allowed to cool to its initial temperature (and the water in the exhaust is condensed). The product of HHV and the fuel mass flow rate gives the power potential of the fuel. The following expression is used to calculate efficiency for an IC engine.

$$\text{Fuel efficiency} = \frac{\text{Measured power}}{\text{Fuel power potential}} = \frac{P_b}{\dot{m}_f \times \text{HHV}} = \frac{T_b N}{\dot{m}_f \times \text{HHV}} \tag{10.12}$$

It must be emphasized that for efficiency calculations involving different chemical energy carriers, the HHV of fuels must be considered, and not the lower heating value (LHV) [4]. The definition of HHV is derived from the Heat of Formation, which includes all the energy changes of a chemical reaction between the initial state at 25°C and the final state, also at 25°C. The primary difference between HHV and LHV is that the latent heat of water vapor present in the combustion products is included in the former, i.e., water is assumed to be condensed at 1 atm and 25°C. The HHV is based on the true energy content representing the heat released by the oxidation of a fuel with air. As an example, the HHV of common diesel is 19,733 BTU/lb.

10.3.1 Fuel Economy in Hybrids

In charge-sustaining hybrids, all of the power is delivered by the engine. Even if the vehicle is driven with electric propulsion for a certain period of time, the energy coming from the energy storage system (battery or ultracapacitor or both) was originally supplied by the on-board fuel through the engine at some point in time. This is why state of charge (SoC) correction algorithms have to be used to make sure that all of the energy supplied by the fuel has been utilized to drive the vehicle. Otherwise, the fuel used to charge the energy storage system will be left unaccounted for, resulting in a worse fuel economy than what it actually is. On the other hand, if fuel used for charging the energy storage system is not accounted for in calculating the fuel economy, a better than actual result will be obtained. With the SoC corrections, the energy storage system is maintained at the same SoC at the beginning and end of the drive cycle for which fuel economy is being measured.

The fuel economy of plug-in hybrids must be calculated accounting for both the on-board fuel used and the off-board fuel supplied. The fuel economy in plug-in hybrids is measured according to the standard SAE J1711, which states that the total fuel supplied during the drive cycle is the sum of the on-board fuel consumed and the fuel equivalency of the off-board supplied electric charge. For the correct fuel economy calculation, the on-board energy storage system is recharged from the wall outlet until the SoC matches the initial value. The equivalent fuel for recharging the on-board energy storage system is obtained from

$$F_{elec} = \frac{E_{ob}}{\zeta_{fuel}} \tag{10.13}$$

where

$\zeta_{fuel} = 38.322$ kW h/gal is the fuel equivalency of the electric energy supplied
E_{ob} is the off-board electrical energy supplied by the utility

Example 10.1

A hybrid electric vehicle has a downsized engine, an electric motor/generator, and an ultracapacitor bank for electric propulsion assistance. The vehicle is driven for 30 min at a constant velocity of 25 m/s with engine operating at BSFC = 270 gal/kW h and electric motors. The traction power required for this constant velocity cruise is 15.2 kW. However, the ultracapacitor bank has an additional 960 kJ of energy captured during a regenerative braking. Calculate the fuel economy when all of the ultracapacitor energy is utilized for propulsion within the 30 min constant velocity cruising period.

Solution

Velocity $v = 25$ m/s; Energy captured by Ultracapacitor = 960 kJ

$$\text{Power supplied by ultracapacitor in 30 min} = \frac{960 \times 1000\ \text{J}}{30\ \text{min} \times 60\ \text{s/min}} = 533.33\ \text{W}$$

Given, $P_{TR} = 15.2$ kW

$$\text{Drivetrain efficiency } \eta_{DT} = 90\%$$

Power to the wheels is delivered both by ultracapacitor bank and engine over 30 min. Therefore,

$$P_{UC} + P_{Eng} = \frac{P_{TR}}{\eta_{DT}}$$

$$\text{Engine power } P_{Eng} = \frac{15.2}{0.9} - 0.533 = 16.36\ \text{kW}$$

$$\text{Fuel flow rate } Q_f = \frac{P_{TR}/\eta_{DT} \times \text{BSFC}}{\rho_{fuel}} = \frac{(16.36) \times 270}{720} = 6.135\ \text{L/h}$$

Fuel economy

$$\text{FE} = \frac{v}{Q_f} = \frac{25\ \text{m/s} \times 3600\ \text{s/h}}{6.135\ \text{L/h}} = 14{,}670\ \text{m/L}$$

$$= \frac{14{,}670\ \text{m/L} \times 3.79\ \text{L/gal}}{1609\ \text{m/mi}} = 34.55\ \text{mi/gal}$$

10.4 Emission Control System

The exhaust streams of both SI and CI engines are significant sources of environmental pollution [5]. Gasoline-fueled SI engine exhaust gases contain oxides of nitrogen (NO_x), carbon monoxide (CO), and hydrocarbons (HC). Diesel exhaust emissions contain lower levels of CO and HC, but similar levels of NO_x compared to gasoline engines. CO is a highly toxic substance and must be converted to CO_2 before being released to the atmosphere. CO_2 is not toxic, but does contribute to the greenhouse gases.

10.4.1 Generation of Pollutants

The origin of the pollutants in the exhaust stream of IC engines is in the combustion process itself. The complete combustion of hydrocarbons with excess oxygen would result in the formation of only water and carbon dioxide as follows:

$$[C_xH_y]+[O_2]\rightarrow[H_2O]+[CO_2]$$

However, due to practical limitations and constraints placed on IC engine operation, the complete combustion of hydrocarbons is not possible. Intermediate reaction products are formed since chemical reactions depend on pressure, temperature, concentration of species, and time. Exhaust streams contain much different types of hydrocarbons. While most of these hydrocarbons are intermediate reaction products, some unburned high molecular weight hydrocarbon fuel may also be present. CO is produced when there is insufficient oxygen. Retarded ignition in gasoline engines also aid in CO production.

Atmospheric nitrogen present in the intake stream reacts with excess oxygen at high combustion temperatures to produce the undesirable and highly toxic NO_x compounds. At high combustion temperatures, molecular nitrogen (N_2) and oxygen (O_2) in the combustion air disassociate and bond with each other; a process commonly referred to as thermal NO_x generation. Most of the NO_x generated in combustion is of the form nitric oxide (NO) and is governed by the following reactions:

$$O+N_2\rightarrow NO+N$$

$$N+O_2\rightarrow NO+O$$

$$N+OH\rightarrow NO+H$$

Although in much less quantities, a significant pollution source is the generation of nitrogen dioxide (NO_2), which is produced through the combination of NO and O_2 as follows:

$$2NO+O_2\rightarrow 2NO_2$$

NO released to the atmosphere will combine with O_2 to form NO_2, which is a toxic pollutant.

Both gasoline and diesel fuels contain some amount of sulfur with diesel fuel having a much greater content at 0.1%–0.3% weight as compared to less than 0.06% weight for gasoline. The combustion of sulfur in the engine compartment leads to the emission of sulfur oxides (SO_x).

The combustion of diesel fuel results in particulate material (PM) emissions in addition to the exhaust components discussed above. About 0.1%–0.5% of the fuel is emitted as small particulates, which consist primarily of soot with some additional adsorbed hydrocarbon material. Particulates smaller than 0.1 μm, which are present in both spark and compression engine exhaust streams, have the greatest deposition efficiency in lungs [6]. Soot emission is composed of small pieces of solid carbon matter emitted from the combustion of fuel in highly rich conditions, when there is insufficient amount of oxygen present to fully combust the fuel. Particulate matter is more of a problem with diesel engines compared to gasoline even though the former always operate with excess air. This is because the fuel and air for an SI engine are premixed before entering the combustion chamber, and the fuel content is almost fully vaporized and only small portions of the mixture contain liquid fuel. Although this portion does contribute to particulate matter production in gasoline engines, the amount produced is considered to be negligible. Since the fuel in a diesel engine is injected into the combustion chamber after the air has been compressed, a brief period of time elapses where a high concentration of atomized (but not vaporized) fuel is present in the injection region creating significant amounts of soot emission. Actual formation of soot concentration is largely dependent on the design of the injection system and combustion chamber as well as engine operating speed and load [7].

The 1970 Clean Air act and subsequent legislations set the limit to the amount of pollutants that can be released into the atmosphere from the tail pipe of a vehicle. United States Environmental Protection Agency (USEPA) has categorized emission levels into tier levels based on vehicle mass. For example, light duty vehicles with a maximum curb mass of 2721 kg (6000 lb) and a maximum gross vehicle weight rating of 3855 kg (8500 lb) must meet USEPA tier 2 emissions requirements. Tier 2 emissions requirement has been broken up into 10 bins having different limitations of exhaust emissions for different compounds classified as harmful. The legislation specifies maximum limits on pollutants irrespective of the fuel type. Two of the bins were part of a phase-in policy and have since been deleted at the end of the 2006 model year. Remaining eight bins are permanent. Table 10.2 shows the allowable emission limits for tier 2 vehicles per bin for the eight permanent bins. USEPA specifies that vehicle manufacturers are required to meet a light duty vehicle fleet average to the tier 2, bin 5 specifications. While production vehicles are allowed to be sold at the bin 8 specifications, their sales must be offset by an equal sales volume of bins lower than 5.

Sulfur in diesel fuel is not of particular concern because its content can be controlled during the manufacture of the fuel, and hence not included

TABLE 10.2

USEPA Legislation for Tier 2 Classified Vehicles

Bin	USEPA Allowable Emission Limits Per Bin (g/mi)				
#	NMHC	CO	NO_x	PM	HCHO
8	0.125	4.2	0.20	0.02	0.018
7	0.090	4.2	0.15	0.02	0.018
6	0.090	4.2	0.10	0.01	0.018
5	0.090	4.2	0.07	0.01	0.018
4	0.070	2.1	0.04	0.01	0.011
3	0.055	2.1	0.03	0.01	0.011
2	0.010	2.1	0.02	0.01	0.004
1	0.000	0.0	0.00	0.00	0.000

NMHC, non-methane hydrocarbons; CO, carbon monoxide; NO_x, nitrogen oxides; PM, particulate matter; HCHO, formaldehyde.

among the pollutants in the bins. USEPA has mandated that all diesel fuel produced for highway use shall contain less than 15 ppm (0.015% weight) of sulfur after June 30, 2006 [8].

The legislation as well as installation of emission control components led to the dramatic reduction of pollutant emissions into the atmosphere. For example, CO emission levels dropped from 87 to 1.7 g/mi from pre-control days of 1966 to the emission controlled days of 2004 [6]. Similarly, HC and NO_x emission levels dropped from 8.8 and 3.6 to 0.09 and 0.07 g/mi, respectively during the same period. The 2004 numbers are corporate averages for all emissions.

10.4.2 Effect of Air-Fuel Ratio on Emissions

Improvements in exhaust emissions have been achieved through both the advancements in IC engine technology and the developments of emission control components. Advances in IC engine technology through the use of fuel injection systems, precision ignition timing, accurate fuel flow metering, and computerized engine management system have led to more complete combustion of fuels and increased efficiencies that helped reduce emissions. Despite all the advances in engine technology, the amount of pollutants in the exhaust stream of automobiles is still significant without additional treatment. Emission control components placed in the path of the exhaust stream helps bring the level of pollutants within acceptable levels.

For a given IC engine, the parameters that can be controlled by the hybrid control strategy to affect the fuel efficiency and emissions are the engine speed and load. In order to maximize fuel economy, the best region to operate the engine is where BSFC is the lowest. For hybrid vehicles operating in the series mode, the IC engine operating point should be in this lowest BSFC region when driving the electric generator for reduced fuel usage. The high fuel efficiency in an IC engine is a result of high thermal efficiency; consequently, the engine chamber temperature is also the highest in this region. However, this may not be the desired operating region as far as emissions are concerned. Emissions are strongly related to the air–fuel ratio of the mixture for combustion in the IC engine. Gasoline engines are typically operated at the stoichiometric ratio to maximize the fuel efficiency. Stoichiometric means that the air–fuel ratio of the mixture is just right for complete combustion. An equivalence ratio λ, which is the ratio of stoichiometric mixture and the actual air–fuel mixture, is often used in describing the relative air–fuel ratio in practical engines. In mathematical terms, the equivalence ratio is

$$\lambda = \frac{\text{Stoichiometric ratio}}{\text{Actual air/fuel ratio}}$$

The relative effects of the equivalence ratio in a gasoline engine on the three major pollutants CO, HC, and NO_x are shown qualitatively in Figure 10.9. The graph shows that the highest levels of NO_x are produced for $\lambda \approx 1.1$, which is also the ratio for best fuel economy [9]. NO_x generation is expected to be high in these areas of high fuel economy, since the temperatures are also higher. On the other hand, gasoline engines develop the best power around $\lambda \approx 0.9$ where the NO_x emissions are lower, but the CO and HC emissions are higher. Thus, a compromise between fuel efficiency and emissions is necessary to

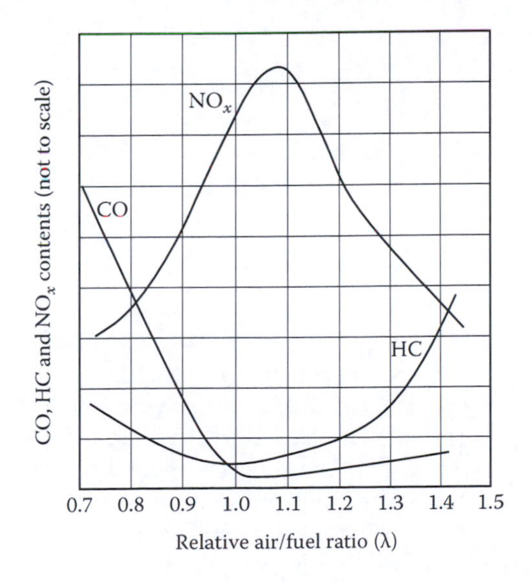

FIGURE 10.9
Effect of air–fuel ratio on emission pollutants in a gasoline engine under full load conditions.

set the engine operating point so that acceptable results may be obtained for both parameters. The control objective is often to run the engine at the stoichiometric ratio, i.e., at $\lambda \approx 1$ so that there is no oxygen in the exhaust stream. An oxygen sensor placed in the exhaust stream provides feedback to the engine control unit so that appropriate air–fuel ratio can be maintained for the engine to minimize all three emissions. In a hybrid electric vehicle, the hybrid control strategy must function in conjunction with the engine management system to achieve the desired goals.

10.4.3 NO_x Flow Rate

The total mass flow rate of exhausted NO_x is the sum of the mass flow rates of exhausted NO and NO_2.

$$\dot{m}_{\mathrm{NO}_x,exh} = \dot{m}_{\mathrm{NO},exh} + \dot{m}_{\mathrm{NO}_2,exh} \tag{10.14}$$

In terms of engine mapping, nitrogen oxide mass flow rate can be presented as simply the NO_x mass flow rate values obtained from Equation 10.14 over the relevant domain. This will facilitate computational modeling of a vehicle's NO_x output per mile over a particular drive cycle. However, expression of NO_x mass flow rate relative to power output allows the performance mapping of an engine that can be used in standardized comparison among several engines. As with fuel use analysis, brake-specific values are again calculated. *Brake-specific NO_x emission* ($BSNO_x$) is simply the NO_x mass flow rate divided by the engine's measured brake power [10].

$$\mathrm{BSNO}_x = \frac{\dot{m}_{\mathrm{NO}_x,exh}}{P_b} = \frac{\dot{m}_{\mathrm{NO}_x,exh}}{T_b 2\pi N} \tag{10.15}$$

where
- P_b is the output power
- T_b is the brake torque
- N is the engine speed (in rps)

The baseline mapping of a warmed engine under steady-state conditions in an engine dynamometer test stand is useful to gain an understanding of the engine's NO_x emission characteristics and examine the trade-off of reduced emission and fuel economy. The test stand with emissions sampling, fuel flow measurement, and engine diagnostics recorder can be configured as shown in Figure 10.10. Results from tests conducted in the laboratory with a dynamometer on a 1.9 L diesel engine over its expected range of operation is shown in Figures 10.11 and 10.12 [10,11]. In the tests designed for collecting data to develop a hybrid control strategy, diesel engine torque load and

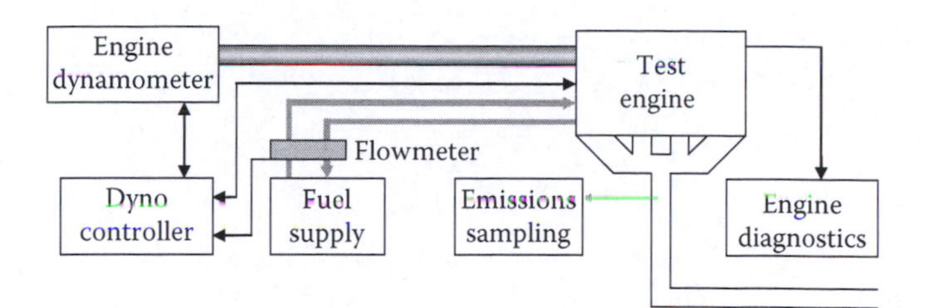

FIGURE 10.10
Dynamometer test stand for collecting emissions data.

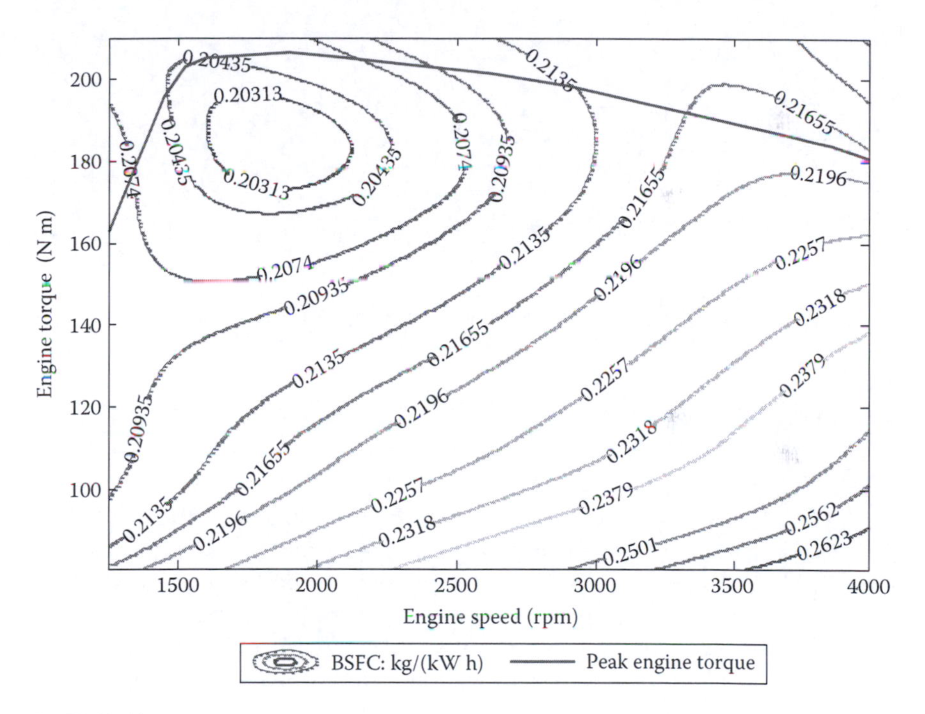

FIGURE 10.11
Brake-specific fuel consumptions for 1.9 L diesel engine. (Adapted from Hasan, S.M.N. et al., *IEEE Ind. Appl. Mag.*, 16(2), 12, March–April 2010.)

speed were modulated and read by the dynamometer controller. Fuel use was measured by the flow meters, and read off the dynamometer controller. The mass air flow rate was read from a sensor via the engine's diagnostics port. A sample of exhaust gas was taken downstream of the turbocharger. The collected data for the engine's intake air and fuel mass flow as well as volumetric exhaust content data was used to determine the levels of engine-out mass flow of NO_x over the engine's operating domain. The BSFC and $BSNO_x$ raw data, calculated using Equations 10.5 and 10.15, was regressed to

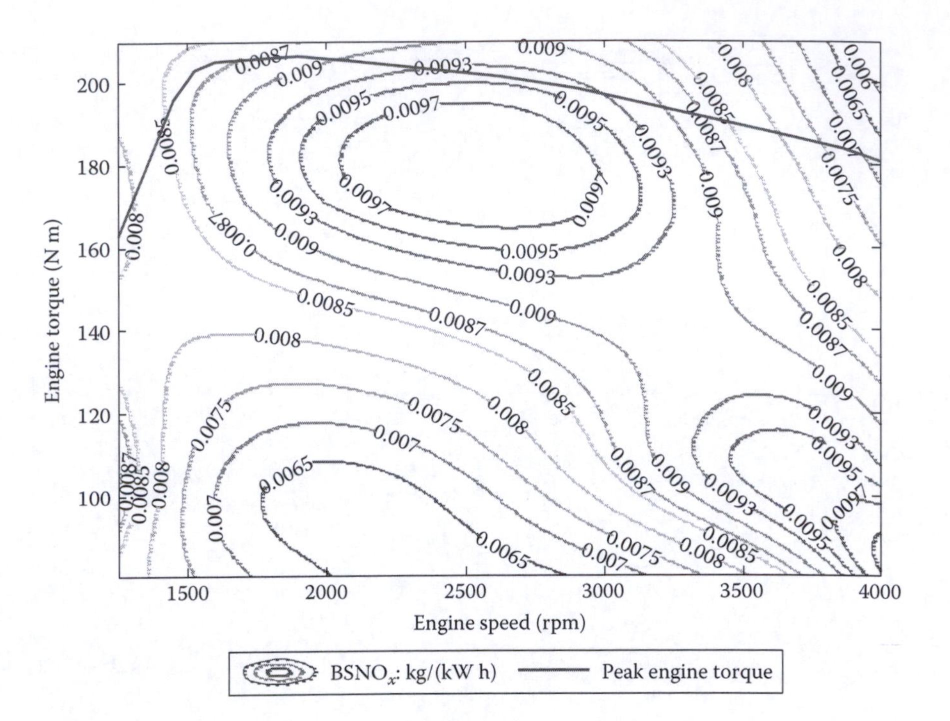

FIGURE 10.12
Brake-specific NO_x emissions for 1.9 L diesel engine. (Adapted from Hasan, S.M.N. et al., *IEEE Ind. Appl. Mag.*, 16(2), 12, March–April 2010.)

develop the baseline engine maps given in Figures 10.11 and 10.12. The BSFC is required for IC fuel usage comparison as discussed earlier in Section 10.2.

In a hybrid electric vehicle, a starter/generator can be coupled to the IC engine through a transmission for starting the engine as well as for supplying propulsion energy to the electric powertrain. The transmission is a single stage speed reducer designed to match the operating speed ranges of the electric machine and the IC engine. The combined efficiency of engine and starter/generator is useful to develop the hybrid control strategy for such hybrid systems, which is given by

$$\text{Combined efficiency} = \frac{\text{Generator output power}}{\text{Fuel power potential}}$$

$$= \frac{P_{gen,out}}{\dot{m}_{fuel} \times \text{HHV}} = \frac{V_{DC} I_{DC}}{\dot{m}_{fuel} \times \text{HHV}} \quad (10.16)$$

In the series mode of hybrid vehicle operation, the IC engine and the starter/generator are cycled on and off to maintain the energy storage system within the desired SoC band. The "on" operating point may be chosen based on fuel economy, emissions, and the characteristics of the starter/generator electric machine. If only high fuel economy is targeted, the engine should operate in a region of low BSFC. The engine maps of Figures 10.11 and 10.12 show that the regions of operation with low fuel consumption also yield high NO_x generation. Therefore, the optimal IC engine operating point has to be selected with a compromise between high fuel efficiency and reduced emissions. The selection of the "on" operating point for the hybrid control strategy is discussed in Chapter 13.

Example 10.2

The mass flow rate during series mode operation of a diesel hybrid electric vehicle is 2.3 kg/h. A permanent magnet starter/generator is belt coupled to the engine with a gear ratio of 2.8:1. The starter/generator input torque and speed are 28.93 N m and 3500 rpm, respectively. Assume the coupling losses between the engine and starter/generator shafts to be negligible. Generated output voltage and current are 300 V DC and 31.8 A, respectively. Calculate engine BSFC, engine efficiency, and combined engine–starter/generator efficiency.

Solutions

IC Engine brake torque is $T_b = 28.93 \times 2.8 = 81$ N m

IC Engine speed is $N = 3500/2.8 = 1250$ rpm

Therefore, IC brake power output, neglecting belt coupling losses is

$$P_b = 81 \times 1250 \times \frac{\pi}{30} = 10.6 \text{ kW}$$

IC Engine brake-specific fuel consumption is

$$\text{BSFC} = \frac{\dot{m}_f}{P_b} = \frac{2.3}{10.602} = 0.217 \text{ kg/kW h}$$

This is equal to 0.357 lb/hp h.

Diesel HHV is 19,733 BTU/lb, which is equivalent to

$$\text{HHV} = 19{,}733 \times 1{,}055 \times 2.205 \text{ J/kg} = 45{,}904.38 \times 10^3 \text{ J/kg}$$
$$= 12.75 \text{ kW h/kg}$$

IC engine input power calculated from the fuel potential is

$$P_{in} = \dot{m}_f \times \text{HHV} = 2.3 \times 12.75 = 29.32 \text{ kW}$$

Therefore, IC engine efficiency is $\eta_{ICE} = \dfrac{10.602}{29.32} = 36.2\%$

The combined efficiency using (13.20) is $\eta_{combined} = \dfrac{300 \times 31.8}{29.32} = 32.5\%$

10.4.4 Emission Control Components

The emission control components in a vehicle depend on the type of IC engine used, although some of the components are similar in both gasoline and diesel engines. Emission is controlled is gasoline engines through providing a chamber in the exhaust stream path for further oxidation and reduction of pollutants, and reintroducing part of the exhaust stream back into the engine compartment. The component used for the oxidation and reduction of pollutants is known as the catalytic converter, while the second method is known as exhaust gas recirculation (EGR). A catalytic converter and EGR is highly effective in the reduction of all pollutants in gasoline engine vehicle. The diesel engines require additional after-treatment components for bringing the pollutant levels within acceptable levels. We will discuss the EGR and catalytic converter in general in this section, and then address the specific needs for emission controls in diesel engine vehicles.

10.4.4.1 Exhaust Gas Recirculation

Exhaust gas recirculation is used for reducing the NO_x levels at the tail pipe of the vehicle. The method uses the EGR valve to reintroduce part of the exhaust stream into the air–fuel intake system to reduce the engine chamber temperature. The exhaust gas added dilutes the air–fuel ratio and reduces λ, which helps reduce the combustion chamber temperature. The lower the engine operating temperature, the lower will be the amount of NO_x produced. EGR action dilutes the rich air–fuel mixture (rich with fuel) to a lean mixture (lean with fuel), which has the negative effect of reducing engine performance. Therefore, EGR action is disabled when full engine power is required or when the engine is cold.

10.4.4.2 Catalytic Converter

Catalytic converters convert harmful pollutants in the exhaust stream into more acceptable compounds through oxidation and reduction. Catalytic converter designed only for two oxidation processes is known as a two-way catalytic converter, while that using two oxidation and one reduction processes is known as a three-way catalytic converter. The two oxidation processes reduce the carbon monoxide and hydrocarbons in the exhaust stream. The two-way catalytic converters first came into use for gasoline engines, but their inability to reduce NO_x in the emissions led to the development of three-way catalysts, which are now universally used. Currently, two-way catalytic converters are used in diesel engines along with additional after-treatment components to reduce NO_x and particulate matter.

Three-way catalytic converters use catalysts for the following three actions:

1. Reduction of NO_x into nitrogen and oxygen; $2NO \rightarrow O_2 + N_2$ and $2NO_2 \rightarrow 2O_2 + N_2$
2. Oxidation of carbon monoxide (CO) to carbon dioxide (CO_2); $2CO + O_2 \rightarrow 2CO_2$
3. Oxidation of hydrocarbons to carbon dioxide (CO_2) and water (H_2O); $[C_xH_y] + [O_2] \rightarrow [H_2O] + [CO_2]$

The chemical reactions are aided by catalysts that speed up the process. Platinum and rhodium are used as reduction catalysts, while platinum and palladium are used as oxidation catalysts.

Three-way catalytic converters are highly effective in reducing CO, NO_x, and HC emissions when IC engines operate within a narrow band around the stoichiometric ratio. Engine management and fuel injection systems work in conjunction with the catalytic converter to effectively reduce the amount of pollutants. The oxygen sensor in the exhaust stream provides feedback to the engine management system, which controls the ratio of the air–fuel mixture through the fuel injection system within the narrow band around the stoichiometric ratio. However, at part-load conditions of the engine or during heavy demand conditions, it may not be possible to maintain the ratio around the stoichiometric ratio. Outside the band around stoichiometric ratio when the air–fuel mixture is lean, oxidation is favored since the exhaust stream has more CO and HC. Again, when the air–fuel mixture is rich, reduction is favored since the exhaust will contain more NO_x due to the high engine chamber temperatures. Metal oxide (cerium oxide) used within the catalytic converter has the ability to store oxygen when there is excess available and release when required. This type of oxides is known as stabilizers. Metal oxide stores oxygen during lean operating condition where there is excess oxygen in the exhaust stream, and release it during rich operating condition when sufficient oxygen is not available to aid the oxidation process.

In addition to the catalytic converter, secondary air injection (SAI) method is applied to the exhaust stream in SI engines under certain conditions to burn off excess hydrocarbons. The method is predominantly used during engine start-up to control emissions when the catalytic converter does not have sufficient time to warm up.

10.4.5 Treatment of Diesel Exhaust Emissions

Diesel engines are attractive for hybrid vehicles since these can deliver a higher level of fuel economy compared to gasoline engines. Some of the latest technology diesel engines utilizing high speed direct injection for fuel atomization can achieve up to 35% lower volumetric fuel consumption than equivalent performance gasoline engines [12]. Smaller, efficient diesel engines for passenger vehicles are already available particularly from the

European auto industry. The use of a diesel engine as the primary power source for a hybrid vehicle can lead to a significant improvement in fuel economy when compared to common gasoline-fueled vehicles. However, if diesel engines are to be used in hybrid vehicles, an after-treatment system must be employed to meet EPA requirements.

While it is possible to reduce emissions output in diesel engine exhaust by varying engine operating parameters, achievements are typically at the cost of fuel efficiency. This section focuses on the after-treatment components of diesel exhaust; after-treatment refers to the methods of emission reduction in the exhaust stream, which is post combustion.

10.4.5.1 Diesel Oxidation Catalysts

The significant reduction of unburned hydrocarbons and carbon monoxide can be achieved by the use of oxidation catalysts in the exhaust stream. Similar to the gasoline exhaust catalytic converters, diesel exhaust catalytic converters oxidize hydrocarbons and carbon monoxide in the exhaust stream through the promotion of the chemical reaction using precious metal catalysts. However, the design of the oxidation catalytic converter for diesel engines differs significantly from those of gasoline engines, since performance has to be optimized for operation at much lower exhaust temperatures seen in the diesel exhaust [13]. The lower temperatures are due to the fact that diesel engines always operate with excess air.

10.4.5.2 Diesel Particulate Filters

Particulate mass can be reduced by treating with a diesel oxidation catalyst, since it contains hydrocarbons. Currently, oxidation catalysts are the only type of particulate control used in production light-duty diesel vehicles. However, to effectively control particulate materials, which mostly contain soot, a diesel particulate filter (DPF) is necessary. Several designs for DPFs exist, but the most common and effective designs are the wall-flow type. The filter provides channels for exhaust flow that force the gases through blocked ends requiring it to flow through a clay-derived material that traps the particles. Filters of this design are reported to have trapping efficiency of almost 100% [14]. The problem in DPF use lies in rapid clogging and blockage that require periodic cleaning of the filter.

The method for clearing particle mass from DPFs is known as regeneration, which simply involves combusting the soot particles. Regeneration can occur either passively or actively. Passive regeneration is accomplished by maintaining a high enough exhaust temperature so that particle oxidation can take place during normal vehicle operation. Unfortunately, the required temperature is in the range of 550°C–600°C, which is often unreasonable for the diesel exhaust stream. Heavy-duty diesel vehicles used in the commercial industry have engines sized for fuel economy; high load operation is common as most driving is done on the highway, making DPF usage ideal for such vehicles. Light duty passenger vehicles operate more frequently in urban driving conditions where high load operation creating high exhaust temperatures is not frequently encountered. Lower exhaust temperatures are not enough for regeneration, hence the lack of DPF usage in typical production diesel passenger vehicles. In active regeneration, catalyst additives are added to the fuel to reduce particulate combustion temperature. Catalyst addition can be combined with a complex engine management strategy in which fuel injection rates are altered to provide higher exhaust temperature.

An advantage of using light-duty diesel engines in hybrid vehicles is that hybrid control strategy can ensure high load on the engine even in urban driving conditions by operating in the series mode. The electric generator can load the engine sufficiently for all of city driving with excess energy being used for battery charging when possible. With high enough engine temperature at high load, passive regeneration of DPF is possible. In the development of a vehicle of this type, engine selection or development should consider the criteria for passive regeneration.

10.4.5.3 Methods of NO_x Reduction

NO_x reduction through the catalytic converter is not so effective in diesel engines as compared to gasoline engines, since the diesel exhaust temperature is lower and the exhaust stream is oxygen rich. Chemical reactions always occur in the simplest way possible, and excess oxygen in the diesel exhaust stream will bond before promoting dissociation of NO_x in the catalyst. The reduction of NO_x compounds in the diesel exhaust by catalytic promotion is further complicated when the operating scenario of a hybrid vehicle is considered. Hybrid control strategies may require high efficiency from an engine, such as in the series mode of operation. High fuel efficiency is accomplished by running the engine at high load for a given speed, a situation in which high NO_x is also generated. This operating scenario for hybrids would be typical in urban driving. Highway driving would always require engines to be operated at high loads for hybrid vehicles. This would result in the IC engine for a hybrid vehicle to be always operated with high load conditions, which would dramatically increase NO_x levels over a drive cycle compared to a conventional IC engine vehicle.

Two modern technologies applied for diesel NO_x control are lean traps that are capable of storing NO_x at low operating temperatures until higher loads are seen and a process called selective catalyst reduction (SCR) in which ammonia is used to treat engine exhaust for NO_x reduction [10].

Lean NO_x trap is a technology for NO_x adsorption under lean (oxygen rich) operating conditions and then later decomposing into nonharmful water and nitrogen in a fuel rich environment. The system operates by the use of a catalyst that promotes the adsorption and storage of NO_x in a lean environment by forming a new compound [15]. The inherent problem is that in order to create the fuel rich environment, excess fuel must be injected into the exhaust stream frequently for regeneration, which would reduce fuel economy. Lean NO_x traps are being introduced into diesel passenger vehicles only recently. Mercedes Benz was the first auto maker to introduce a model equipped with a lean NO_x trap in the 2007 model year production vehicle.

SCR involves the injection of a reducing agent into the exhaust, which is capable of bonding with NO_x compounds to form nonharmful ones. The most effective reducing agent that can be used in the process is ammonia (NH_3). The term "selective" is demonstrated in ammonia's unique ability to selectively react with NO_x compounds rather than be oxidized to form N_2, N_2O, and NO [16].

$$4NH_3 + 4NO + O_2 \rightarrow 4N_2 + 6H_2O \tag{10.17}$$

$$2NH_3 + NO + NO_2 \rightarrow 2N_2 + 3H_2O \tag{10.18}$$

$$8NH_3 + 6NO_2 \rightarrow 7N_2 + 12H_2O \tag{10.19}$$

Greater than 90% of NO_x from diesel emissions is composed of NO and thus reaction (10.17) accounts for most of the reduction as it occurs with NO and NH_3 at a 1:1 ratio in excess oxygen. Reaction (10.18) is most desirable because it occurs at a lower temperature than the others, but requires a 1:1 ratio of NO and NO_2. Reaction (10.19) takes care of the remaining NO_2 that cannot be reduced by (10.18) due to insufficient NO.

The efficient bonding of ammonia and NO_x compounds is only possible with the aid of a specially designed catalytic converter. Three types of catalysts have been developed for commercial use: noble metals, metal oxides, and zeolites. Noble (precious) metal catalysts have proven highly active in NO_x reduction, but also effectively oxidize NH_3, rendering it almost useless as a reducing agent. For this reason, slightly less effective metal oxide (compounds composed of metals and oxygen) catalysts are the most common for conventional SCR applications. Zeolite catalysts that are composed of minerals having microporous structures can also be used in SCR systems; however, their use is more suited to gas-fired cogeneration plants rather than diesel engines.

While ammonia can be an effective NO_x reduction agent, storage and transportation becomes an issue as ammonia itself is a toxic chemical regulated by the EPA. The solution is to inject an aqueous urea solution into the hot exhaust stream, which will quickly decompose into ammonia and carbon dioxide [17]. The dissociation of excess urea yielding ammonia leads to a discharge of raw ammonia out the tail pipe termed as ammonia slippage. Additionally, it is desired that only the correct amount of urea be added to the exhaust stream so that the supply be conserved, thus maximizing the time before refill of urea is necessary.

An injector unit comparable to common fuel injectors found on gasoline engines can be used for urea injection. The injection rate can be controlled by sending the correct injection frequency through a pulse width modulated (PWM) signal to the injector. The required injection pressure is usually maintained by a pressure regulator fitted to an air reservoir. Commercial systems intended to retro-fit an application typically have a compressor and motor that re-pressurizes the air tank as needed. For the control of urea injection, monitoring of exhaust temperature provides adequate feedback information to determine the injection rate. More sophisticated control systems include interfacing with the engine electronic control unit (ECU), which provides engine speed and fuel injection rate feedbacks for controlling the urea injection rate. A sensor for determining the ammonia content downstream of the catalyst is ideally suitable for developing the feedback controller for urea injection; however such sensors are currently in the prototype phase.

Problem

10.1 A hybrid electric vehicle has the following parameter values: $\rho = 1.16\,kg/m^3$, $m = 692\,kg$, $C_D = 0.2$, $A_F = 2\,m^2$, $g = 9.81\,m/s^2$, $C_0 = 0.009$, $C_1 = 1.75 \times 10^{-6}\,s^2/m^2$. The type of IC engine that will be used for the vehicle has the force (at wheel) versus velocity characteristics of $F = 2.0\sin(0.0285x)$ N for $5 < x < 100$, where x is the vehicle speed in mi/h. Determine the displacement of the ICE for a rated cruising velocity of 60 mi/h on a 2% slope.

References

1. M.J. Moran and H.N. Shapiro, *Fundamentals of Engineering Thermodynamics*, 6th edn., John Wiley & Sons, Inc., New York, 2008.
2. J.P. O'Brien, *Gas Turbines for Automotive Use*, Noyes Data Corporation, Park Ridge, NJ, 1980.
3. J.B. Heywood, *Internal Combustion Engine Fundamentals*, McGraw-Hill, Inc., New York, 1988.
4. U. Bossel, Well-to-wheel studies, heating values, and the energy conservation principle, in *Proceedings of the European Fuel Cell Forum*, 2003. [Online]. Available www.efcf.com/ reports, (E10).
5. G.C. Koltsakis and A.M. Stamatelos, Catalytic automotive exhaust aftertreatment, *Progress in Energy and Combustion Science*, 23, 1–39, 1997.
6. R. Stone and J.K. Ball, *Automotive Engineering Fundamentals*, SAE International, Warrendale, PA, 2004.
7. J.P.A. Neeft, M. Makkee, and J.A. Moulijn, Diesel particulate emission control, *Fuel Processing Technology*, 47, 1–69, 1996.

8. Technical Amendments to the Highway and Nonroad Diesel Regulation; Final Rule and Proposed Rule, Environmental Protection Agency, 40 CFR Part 80, 2006.
9. H.H. Braess and U. Seiffert, *Handbook of Automotive Engineering*, SAE International, Warrendale, PA, 2005.
10. R.N. Paciotti, An evaluation of nitrogen oxide emission from a light-duty hybrid electric vehicle to meet U.S.E.P.A. requirements using a diesel engine, MS thesis, Mechanical Engineering, The University of Akron, Akron, OH, August 2007.
11. S.M.N. Hasan, I. Husain, R.J. Veillette, and J.E. Carletta, Power Generation in Series Mode: Performance of a Starter/Generator System, *IEEE Industry Applications Magazine*, 16(2), 12–21, March–April 2010.
12. R.B. Krieger, R.M. Siewert, J.A. Pinson, N.E. Gallopoulos, D.L. Hilden, D.R. Monroe, R.B. Rask, A.S.P. Solomon, and P. Zima, Diesel engines: One option to power future personal transportation vehicles, SAE International 972683, 1997.
13. J.C. Clerc, Catalytic diesel exhaust after-treatment, *Applied Catalysis B: Environmental*, 10, 99–115, 1996.
14. P. Walker, Controlling particulate emissions from diesel vehicles, *Topics in Catalysis*, 28(1–4), 165–170, April 2004.
15. A. Hinz, L. Andersson, J. Edvardsson, P. Salomonsson, C.J. Karlsson, F. Antolini, P.G. Blakeman, M. Lauenius, B. Magnusson, A.P. Walker, and H.Y. Chen, The application of a NO_x absorber catalyst system on a heavy-duty diesel engine, SAE International 2005-01-1084, 2005.
16. P. Forzatti, Present status and perspectives in de-NO_x SCR catalysis, *Applied Catalysis A: General*, 222, 221–236, 2001.
17. M. Chen and S. Williams, Modeling and optimization of SCR-exhaust after-treatment systems, SAE International 2005-01-0969, 2005.

11

Powertrain Components and Brakes

Internal combustion (IC) engines transfer power to the wheels through a power transmission path known as powertrain. The key mechanical power transmission components of clutch, gear, transmission, and differential in the powertrain are presented in this chapter. The IC engine and the powertrain together form the mechanical power transmission path. Some of these powertrain components are also essential for the electric vehicles and for blending power from IC engine and electric motor. The electric vehicle power transmission includes the electrical path from the energy storage system to the electric motor as well as the mechanical path from the motor shaft to the wheels. The powertrain of a hybrid electric vehicle is more complex than that of the electric vehicle, because of the coupling necessary to blend the torque outputs of the electric motor and the IC engine. This chapter also discusses the hybrid powertrain after the primary powertrain components are presented.

Near the end of the vehicle power transmission path lays a critical element, which is the vehicle brake. The braking system dynamics are similar to the propulsion dynamics except that braking forces have to be evaluated in order to stop the vehicle from motion. This chapter also presents the brake force dynamics and the components of a vehicle braking system.

Modern vehicles are seeing the increasing usage of electrical components replacing hydraulic-based systems. In many cases, this electrification relates to replacing hydraulic actuators with electrical motor-driven actuators. Electric power steering and electronic valve control are examples of such components. This trend in IC engine vehicles resulted in improved performance and better fuel economy. While some of these electrical components are already in usage, others are in queue for implementation in future vehicles. Electromechanical brake system is an example of such a system. These components are the enabling technologies for the by-wire systems envisioned to integrate modern vehicle system functionalities through electronic control and communications. Vehicle control enhancements stem from the relatively easier controllability of electrical systems that are predicated on the highly dynamic behavior from the motor drives in these systems. A brief overview of the electromechanical brake system is presented in this chapter following the description of conventional brakes.

11.1 Powertrain Components

The *powertrain* of a vehicle starts with the IC engine or the electric motor that processes the stored energy and ends with the delivery of the power at the wheels. The powertrain is also often referred to as the *drivetrain* of the vehicle. The engine, transmission, electric motor, gearset, clutch, transmission, driveshafts, and the final drive are the major components of the powertrain. The energy in the vehicle is initially stored in the diesel or gasoline fuel for the IC engines or in the chemicals of a battery for the electric motors. The IC engine or the electric motor converts the stored energy into mechanical form. The mechanical power or energy still needs to be delivered to the end point that is at the wheel of the vehicles. The next component in the powertrain is the transmission with an IC engine, while it could be a simple gear-box for speed reduction with an electric motor. The transmission and the differential consists of the different types of gears and gearsets. The components of the powertrain excluding the propulsion unit (IC engine and transmission or electric motor and gearset or the combination of the two sets) is known as the *driveline*.

The final part of the powertrain is the *final drive* that reduces the output speed of the transmission/gear-box or drive shaft. It consists of a gearset made of two or more gears working together. The final drive includes the *differential* that accommodates the unequal speeds of the inside and outside wheels when the vehicle turns around a corner. The differential is necessary for the driven wheels on an axle since they are linked together so that the propulsion unit can transmit power to both the wheels. This is not a concern for the nondriven wheels that are not linked and can spin freely. The power from the differential is transferred to the final drive at the driven wheels by the *half-shafts*. The half-shafts transmit half of the power coming down the powertrain to the differential. The transmission or the gear-box out of the propulsion unit is not usually in the same plane as the differential; power needs to be transferred across the driveline at an angle. The universal joints (U-joints) and constant velocity joints (CV-joints) are used to accomplish the task.

The powertrain components used also depend on whether the vehicle is a front-wheel drive or a rear-wheel drive. The powertrain assembly after the IC engine or the electric motor in a *front-wheel drive* is compact and usually integrated into one device. Front-wheel drives, which are popular in passenger vehicles, have their transmissions located behind the engine with a change in output direction. In a *rear-wheel drive*, the power is transferred across the length of the vehicle using the *driveshaft* when the propulsion unit is located in the front of the vehicle. The driveshaft delivers power to the differential using joints in the rear-wheel drive vehicle. Rear-wheel drives are used in some passenger vehicles and more commonly in light trucks. Vehicles where power is transferred to both the front wheels and the rear wheels are known as *all-wheel drive* vehicles. All-wheel drives have a front differential and a rear differential.

In some passenger vehicles, the transmission, driveshaft, differential, and the final drive are combined into one lightweight device known as *transaxle*. It is typically used with transverse mounted engines such that its rotating axis is parallel to that of the drive axles and engine crankshaft. The arrangement doesn't require a change in direction, enabling more efficient power transmission.

In the following, we will review the powertrain of electric vehicles to present an overview of the integration of components with simpler configurations. The discussions on the components along with relevant analysis appear in the subsequent sections.

11.1.1 Electric Vehicle Powertrain

The components for an electric vehicle powertrain consist of the electric motor, gear box, driveshaft, differential, half-shafts, and wheels. The ability of electric motors to start from zero speed and operate efficiently over a wide speed range makes it possible to eliminate the clutch that is used in IC engine vehicles. A single gear ratio is sufficient to match the wheel speed with the motor speed. Electric vehicles can be designed without a gear, but the use of a speed reducer allows the electric motor to operate at much higher speeds for given vehicle speeds, which minimizes the motor size because of lower torque requirement at higher speeds.

In the case of a front-wheel drive, the electric motor drives the gear box that is mounted on the front axle as shown in Figure 11.1. This configuration is for a electric vehicle using a single propulsion motor. The single motor drives the transaxle on a common axis, delivering power to the two wheels differentially through a hollow motor shaft [1]. The powertrain is more complex in the case of a rear-wheel drive, which requires a differential to manage the cornering of the vehicle. A typical rear-wheel drive electric vehicle powertrain configuration is shown in Figure 11.2.

The use of two motors driving two front wheels simplifies the powertrain and eliminates the differential. Several configurations are possible with two

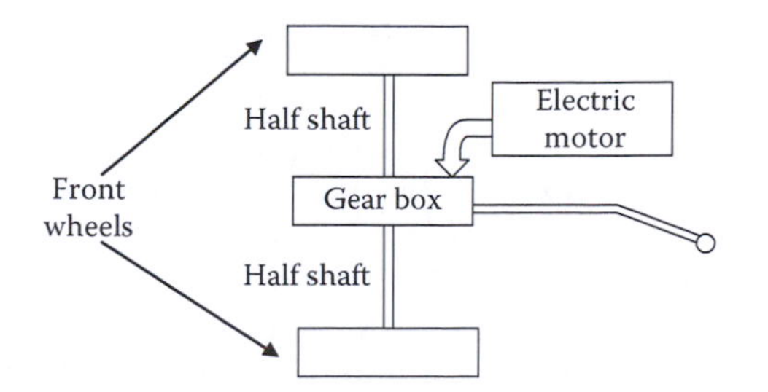

FIGURE 11.1
Typical front-wheel drive.

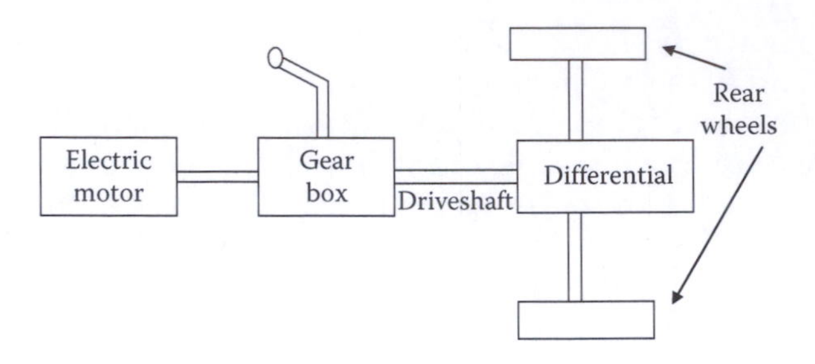

FIGURE 11.2
Typical rear-wheel drive.

propulsion motors driving two wheels. In one arrangement, the motors, mounted to the chassis, can be connected to the wheels through two short half-shafts. The suspension system of the vehicle isolates the wheels and its associated parts from the rest of the components of the vehicle for the easier handling of the vehicle depending on roadway condition. The wheels are able to move freely without the weight of the motors when they are mounted on the chassis. In an alternate arrangement, the motors are mounted on the half-shafts with the motor drive shaft being part of the half-shaft. The half-shafts connect the wheels on one side and the chassis through a pivot on the other side. In-wheel mounting of motors is another arrangement possible in electric vehicles. The difficulty in this case is that the unsprung weight of the vehicle increases due to motors inside the wheels, making traction control more complex. To minimize the unsprung weight of the vehicle and because of the limited space available, the in-wheel motors must be of high power density. As mentioned at the beginning, the use of a speed reducer is desirable, which adds to the constraint of limited space. The cost of a high power, high torque motor is the primary impediment in using in-wheel motors for electric vehicles. Another problem with in-wheel motor is the heating due to braking compounded by the limited cooling capability in the restricted space. Nevertheless, the powertrain simplicity has lead to many projects for the development of in-wheel motors for electric vehicles.

11.2 Gears

The gear is a simple machine used for mechanical power transmission with a mechanical advantage through increase in torque or reduction in speed. This mechanical device uses the law of conservation of energy maintaining the steady flow of power or energy, since torque times speed is power that remains constant in the transmission process. In an ideal gearbox, the motion is frictionless and the power and energy supplied at the input point of the gear is equal to the power and energy available at the delivery point.

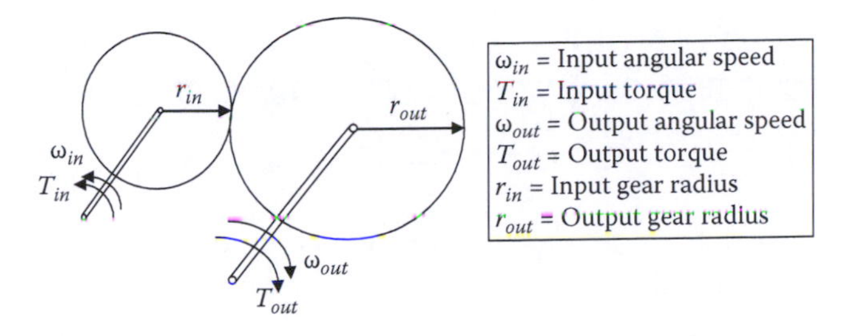

FIGURE 11.3
Gear mechanism.

The gearbox is not used to increase the shaft speed of an electric motor, since this means that a high torque motor is unnecessarily designed where the size of a motor is proportional to the torque output. Therefore, the gear can be used as a torque multiplier or speed reducer. A typical gear mechanism is shown in Figure 11.3.

A gear structurally is a round disc with teeth cut at equal intervals around the rim designed to engage with the similar teeth of another disc. The round discs, placed in combination, transmit power from one gear to another. The teeth at both the discs lock the driving and the driven shafts together to transfer the energy through contact with little, if any, loss.

The four principle types of gears are: Spur, helical, bevel, and worm gears. Almost all types of gears can be found in an automobile. Spur gears are the simplest of all type whose teeth connect in parallel to the axis of rotation. Power is transmitted to parallel shafts connected by the spur gear. Helical gears have teeth inclined to the axis of rotation, but transmit power between parallel shafts just like the spur gear. Bevel gears transmit power between shafts that intersect, but are not in parallel. The differential of an automobile uses bevel gears. Hypoid gears are a type of bevel gear whose teeth form circular arc and the shafts are nonintersecting. These gears connect shafts that are neither parallel, nor do they intersect such as in the final drive of the powertrain. The gear schematic in Figure 11.3 represents either a spur gear or a helical gear. This representation is used below to develop the gear input–output relationships.

The power transmission equations through gears will be established using an ideal gear box assumption, which are as follows:

1. $P_{\text{losses}} = 0 \Rightarrow \text{Efficiency} = 100\%$
2. Gears are perfectly rigid
3. No gear backlash (i.e., no space between teeth)

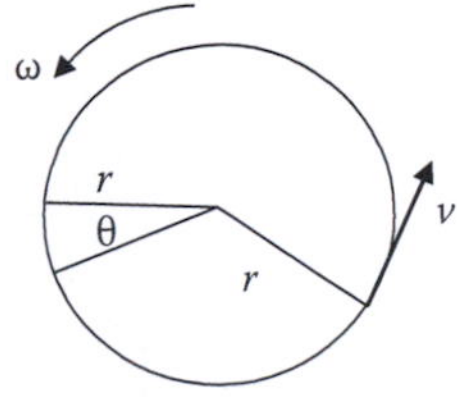

FIGURE 11.4
Variables in a gear.

The variables used in deriving the steady-state model are given in Figure 11.4. Additional variables are shown in Figure 11.4 with respect to a single disc.

11.2.1 Gear Ratio

For a disc with radius r, the tangential and the angular velocity are related by

$$\omega r = v$$

The tangential velocity at the gear teeth contact point is the same for the two gear discs shown in Figure 11.3 with different radius.

$$r_{in}\omega_{in} = v = r_{out}\omega_{out}$$

The gear ratio is defined in terms of the ratio of speed transformation between the input shaft and the output shaft.

$$\mathrm{GR} = \frac{\omega_{in}}{\omega_{out}} = \frac{r_{out}}{r_{in}} \tag{11.1}$$

Assuming 100% efficiency of the gear train

$$P_{out} = P_{in}$$

$$\Rightarrow T_{out}\omega_{out} = T_{in}\omega_{in}$$

The gear ratio in terms of the torque at the two shafts is

$$\mathrm{GR} = \frac{T_{out}}{T_{in}} = \frac{\omega_{in}}{\omega_{out}} \tag{11.2}$$

The gear ratio can be alternately derived with the help of Figure 11.5. At the point of gear mesh, the supplied and delivered forces are the same. This is an example of Newton's third law of motion, which states that every action has an equal and opposite reaction. The torque at the shaft is the force at the

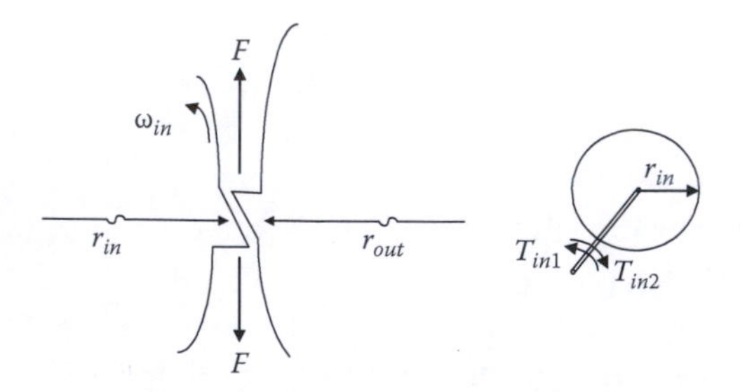

FIGURE 11.5
Force and torque working in a gear.

mesh divided by the radius of the disc. In the two-gear combination, the change in torque between the two gears is proportional to the ratio of gear discs radius.

The torque of the inner disc in terms of its radius and force at the gear mesh is

$$T_{in} = Fr_{in}$$

$$\Rightarrow F = \frac{T_{in}}{r_{in}}$$

Similarly, for the outer disc with radius r_{out}, the force at the gear mesh is

$$F = \frac{T_{out}}{r_{out}}$$

The two forces and the torques are working in opposite directions. The gear ratio can be obtained from equating the two equal but opposite forces as

$$\text{GR} = \frac{T_{out}}{T_{in}} = \frac{r_{out}}{r_{in}} \tag{11.3}$$

For any gear, the diameter is proportional to the number of teeth, which gives

$$\frac{d_{out}}{d_{in}} = \frac{r_{out}}{r_{in}} = \frac{N_{out}}{N_{in}} \tag{11.4}$$

where

N_{out} and N_{in} are the number of teeth

d_{out} and d_{in} are the diameters of the output and input gears

Using Equations 11.2 through 11.4, the gear law can be written as

$$\frac{T_{out}}{T_{in}} = -\frac{r_{out}}{r_{in}} = -\frac{N_{out}}{N_{in}} = -\frac{\omega_{in}}{\omega_{out}} \tag{11.5}$$

The negative sign signifies that the direction of torque in the two gear meshes is opposite.

The gear law can be extended to multiple gear meshes such as the four gear device shown in Figure 11.6. The arrangement of multiple gear meshes to transfer power is known as gear train. In the gear train shown, gear 2 is the driver, while gear 4 is the driven member. If N_2, N_3, and N_4 are the number of teeth in the four gears, the input–output torque ratio is given by

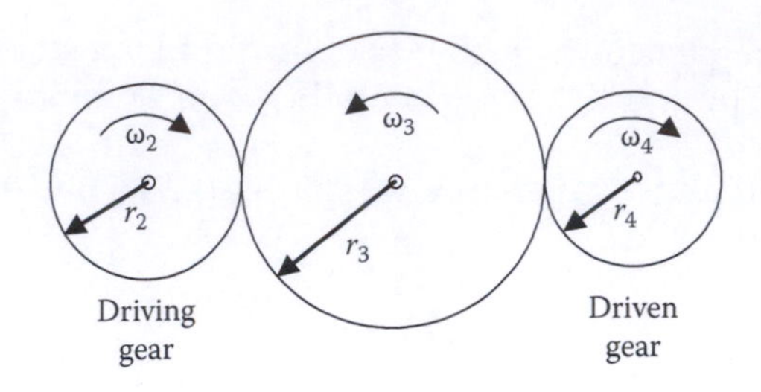

FIGURE 11.6
Two-mesh gear train.

$$\frac{T_4}{T_3}\cdot\frac{T_3}{T_2}=\left(-\frac{N_4}{N_3}\right)\left(-\frac{N_3}{N_2}\right)=\left(-\frac{\omega_3}{\omega_4}\right)\left(-\frac{\omega_2}{\omega_3}\right)$$

$$\Rightarrow\frac{T_4}{T_2}=\frac{N_4}{N_2}=\frac{\omega_2}{\omega_4} \tag{11.6}$$

Notice that the middle gear has no effect on the final torque ratio, but it made the output shaft rotate in the same direction as the input shaft. This technique can be used to get the desired direction of rotation out of a gear train.

A compound gear train made up of five gears is shown in Figure 11.7 [2]. Gear 3 serves the same function as in the middle gear of Figure 11.6, which is to affect the direction of rotation for the final gear 6 without affecting the input–output torque ratio. Gears 4 and 5 form a compound gear stage with both gears mounted on the same shaft. The input–output torque and speed relation for the gear train is

$$\frac{T_2}{T_6}=-\frac{\omega_6}{\omega_2}=-\frac{N_2}{N_3}\cdot\frac{N_3}{N_4}\cdot\frac{N_5}{N_6} \tag{11.7}$$

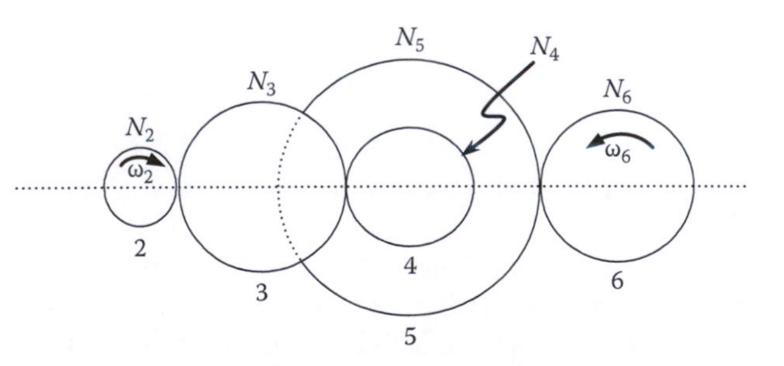

FIGURE 11.7
Compound gear train.

Gears 2, 3, and 5 are drivers, while gears 3, 4, and 6 are driven members in this gear train. The angular speed and torque directions between the input and output gears are opposite since the number of gear meshes is 3. This type of gear train is commonly used in the transmission of a vehicle. In general, the sign of the direction is given by $(-1)^n$ where n is the total number of gear meshes; the gear ratio or the train value is given by

$$\text{GR} = \frac{\text{Product of driving tooth numbers}}{\text{Product of driven tooth numbers}} \tag{11.8}$$

11.2.2 Torque–Speed Characteristics

The advantage of using a gear will be shown in this section considering a DC-motor-driven electric vehicle system. Part of the electric vehicle drivetrain including the electric motor coupled with a gear box is shown in Figure 11.8. The drivetrain will be analyzed in terms of the system torque–speed characteristics.

Electric motors are typically designed to operate at higher speeds to minimize the size of the motor. The gear box functions as a torque multiplier to deliver high torque at a reduced speed at the vehicle wheels. Let the overall gear ratio between the electric motor and the vehicle wheel be GR with ω_m and T_m representing the motor speed and torque, respectively. The torque and speed at the wheels are ω_{out} and T_{out}, respectively. For a separately excited DC motor, the speed–torque relationship at steady state is

$$\omega_m = \frac{V_t}{k\phi} - \frac{R_a}{(k\varphi)^2} T_m \tag{11.9}$$

However,

$$\frac{\omega_m}{\omega_{out}} = \text{GR} = \frac{T_{out}}{T_m}$$

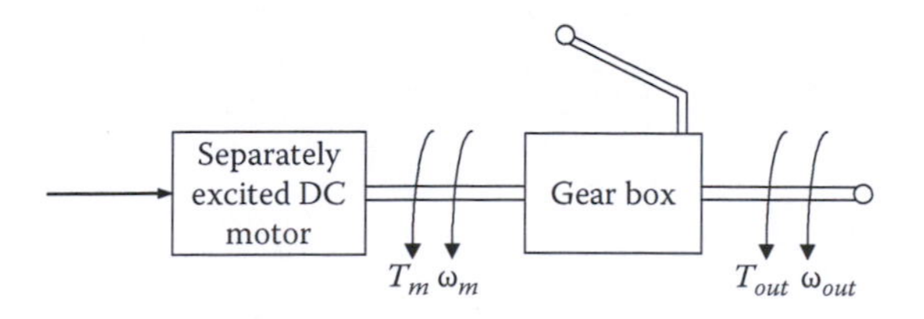

FIGURE 11.8
Connection of a gear with the motor.

Substituting in (11.9),

$$\omega_{out} = \frac{V_t}{\text{GR}(k\phi)} - \frac{R_a}{(\text{GR}k\varphi)^2} T_{out} \tag{11.10}$$

Example 11.1

The powertrain of an electric vehicle is shown in Figure 11.9.

Given, $V_t = 50\,\text{V}$, $R_a = 0.7\ \Omega$, $k\phi = 1.3$, $mg = 7848\,\text{N}$.

Gearbox gears: First $\text{GR}_1 = 2$, second $\text{GR}_2 = 1$, r_{wh} = wheel radius = 7 in. = 0.178 m

Find the vehicle maximum % gradability for each gear.

Solution

$$\text{Max. \% grade} = \frac{100F_{TR}}{\sqrt{(mg)^2 - F_{TR}{}^2}}$$

The torque–speed relationship is

$$\omega_{out} = \frac{V_t}{\text{GR}(k\phi)} - \frac{R_a}{(\text{GR}k\varphi)^2} T_{out}$$

To find F_{TR}, set $\omega_{out} = 0$.

$$0 = \frac{V_t}{\text{GR}(k\phi)} - \frac{R_a}{(\text{GR}k\varphi)^2} T_{out}$$

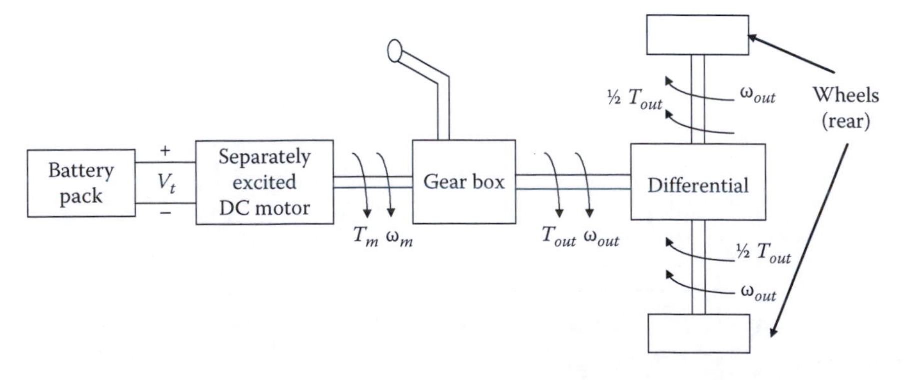

FIGURE 11.9
EV powertrain for Example 11.1.

$$\Rightarrow T_{out} = \frac{V_t}{GR(k\phi)} \times \frac{(GRk\varphi)^2}{R_a} = \frac{GR(k\phi)V_t}{R_a}$$

The relationship between F_{TR} and T_{out} is

$$F_{TR} = 2(F_{TR} \text{ per rear wheel})$$

$$F_{TR} \text{ per rear wheel} = \text{torque per wheel}/r_{wh} = (1/2\ T_{out})/r_{wh}$$

Therefore,

$$F_{TR} = 2\frac{(1/2)T_{out}}{r_{wh}} = \frac{T_{out}}{r_{wh}}$$

Substituting,

$$F_{TR} = \frac{(GRk\varphi)V_t}{r_{wh}R_a} = \frac{GR(1.3)50}{(0.178)(0.7)} = 521.7GR$$

In the first gear,

$$F_{TR} = 1043\ \text{N}$$

$$\text{Max. grade} = \frac{100(1043)}{\sqrt{(7878)^2 - (1043)^2}} = 13.4\%$$

In the second gear,

$$F_{TR} = 521.7\ \text{N},\ \ \text{max. grade} = 6.7\%$$

The steady-state torque–speed characteristics for the separately excited DC motor connected to the gear box for the first two gears are shown in Figure 11.10. The maximum percentage grades for the two gears are also shown in the figure.

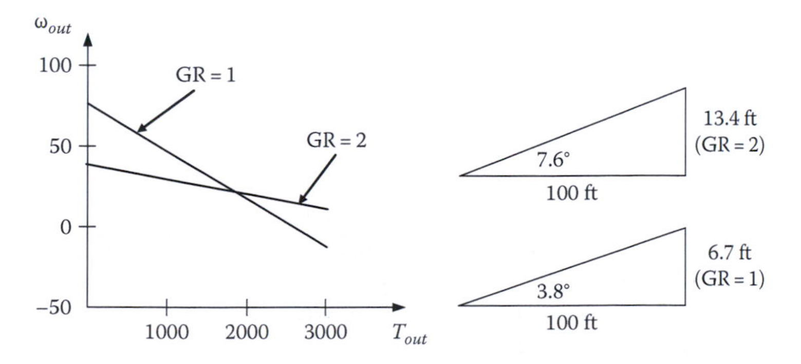

FIGURE 11.10
Plots for Example 11.1.

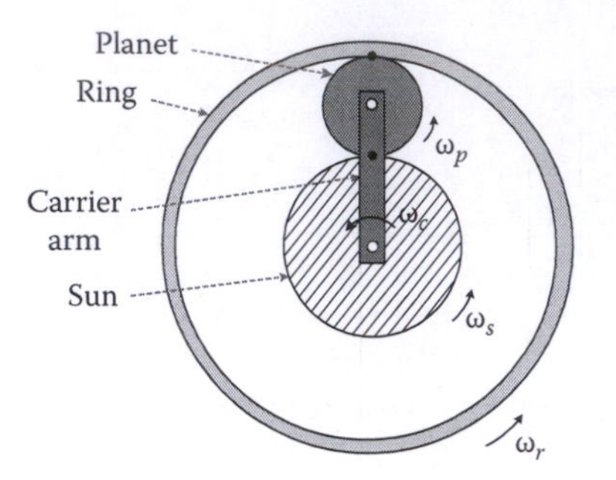

FIGURE 11.11
Planetary gear set.

11.2.3 Planetary Gear Set

The planetary gear set is a gear system consisting of three components: the outer ring gear, the planet carrier, and the central sun gear as shown in Figure 11.11. The planet gears are mounted on a movable arm or carrier that itself may rotate with respect to the ring or sun gear. In a planetary gear set, one of the basic components is typically held stationary, while power is transferred through the other two components; however, all three components can actively participate in the power transfer such as in the power split hybrid vehicles.

The planetary gear has the advantages of high power density and large gear reduction in a small volume. In hybrid vehicles, these gears enable multiple kinematic combinations and coaxial shafting of three powertrain components. The disadvantages of planetary gears are high bearing loads and design complexity.

There are several ways in which the input rotation applied through one of the components can be converted to the output rotation of the one or two of the other gears. The speed relationship of the three gears is given by

$$(r_r + r_s)\omega_p = r_r\omega_r + r_s\omega_s \quad (11.11)$$

where
r_r is the radius of the ring gear
r_p is the radius of the planet carrier
r_s is the radius of the sun gear

The torque is transferred from the input device to the output devices through the force acting at the gear teeth. If a force F is working on the planet, then equal but half of the forces act as reaction forces on the contact area of the ring and sun gears. This is shown in Figure 11.12.

The planet applies torque to the ring and sun gear in proportion to their radii. The torque relationships among the three gears are

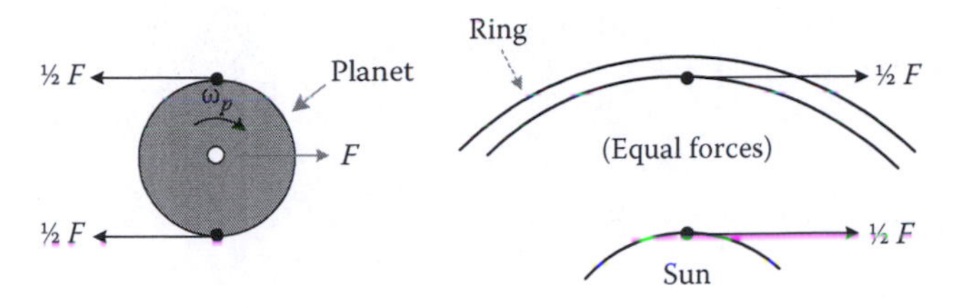

FIGURE 11.12
Reaction forces on sun and ring gears due to an applied force in planet carrier.

$$T_p r_r = T_r(r_r + r_s) \quad \text{and} \quad T_p r_s = T_s(r_r + r_s)$$

where
- T_p is the planet carrier torque
- T_r is the ring gear torque
- T_s is the sun gear torque

From the balance of torques in the three gears, we have

$$\begin{aligned} T_p &= T_r + T_s \\ &= \frac{r_r}{r_r + r_s} T_p + \frac{r_s}{r_r + r_s} T_p \end{aligned} \tag{11.12}$$

The ratios of the radii for the three gears are set by design, which results in a fixed ratio of torque distribution between the ring gear and the sun gear. If the power is supplied through the planet carrier, and delivered to the ring and sun gears, then we must have

$$\begin{aligned} P_p &= P_r + P_s \\ \Rightarrow T_p \omega_p &= \frac{r_r \omega_r}{r_r + r_s} T_p + \frac{r_s \omega_s}{r_r + r_s} T_p \end{aligned} \tag{11.13}$$

where
- P_p is the planet carrier power
- P_r is the ring gear power
- P_s is the sun gear power

In the power-split hybrid vehicles, the planet carrier is connected to the engine, the sun gear is connected to the starter/generator and the ring gear is connected to the electric motor that is also connected to the differential and hence the wheels. The torque applied to the planet carrier is transferred to the wheels and the starter/generator through the ring and sun gear. The control strategy for the power-split hybrid using the planetary gear set is discussed in Section 13.2.1.

11.3 Clutches

The clutch is a mechanical device used to smoothly engage or disengage the power transmission between a prime mover and the load. The most common use of a clutch is in the transmission system of an automobile, where it links the IC engine with the transmission system of the vehicle. The clutch allows the power source to continue running while the load is freely running due to inertia or is idle. The clutch engages and disengages the IC engine from the road load as the gear ratio of the transmission is changed to match vehicle speed with the desired IC engine speed. Clutches can be eliminated in electric vehicles, since the motor can start from zero speed and operate all the way to its maximum speed using a single gear ratio.

The type of clutch used to link the transmission with the engine in a vehicle is known as friction clutch. The clutching action brings the rotational speed of one disc to that of the other disc. The clutching action can be brought about by electrical, pneumatic, or hydraulic action in addition to mechanical means discussed here.

The friction clutch used in a manual transmission of an automobile is engaged and disengaged by the driver. The components of the clutch used in the manual transmission are: cover, pressure plate, and a friction disc. The cover is bolted to the flywheel of the engine; hence, it always rotates with the engine. The pressure plate is inside the cover, which also rotates with the engine. The disc with friction facings is positioned in between the cover and the pressure plate. The disc is connected to the transmission through a splined shaft [3]. The transmission is engaged for the power to be transmitted from the engine with the help of a series of springs attached to the pressure plate. When engaged, the pressure plate squeezes the friction disc against the flywheel to complete the mechanical linkage. The driver must depress the clutch pedal to disengage the transmission when the pressure plate moves away from the friction disc.

The mechanical brakes of both the disc and drum types are the examples of friction clutch. The brakes will be discussed in detail in Section 11.6.

11.4 Differential

The automobile differential provides a mechanism for the differential movement of the wheels on the rear axle. When a vehicle is turning a corner, the rear wheel to the outside of the curve must rotate faster than the inside

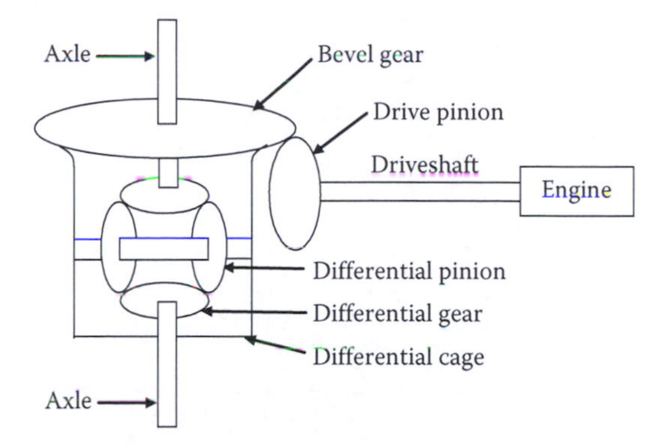

FIGURE 11.13
Simplified schematic of an automobile differential.

wheel, because the former has a longer distance to travel. The type of gear used in automobile differential is known as the planetary gear, where a set of gear train operates in a coordinated manner. A simplified schematic of an automobile differential is given in Figure 11.13. The figure omits the teeth of the gears for simplicity. The drive shaft connected to the engine drives the pinion that is connected to the bevel (perpendicular) gear. The bevel gear is connected to the differential cage, which drives the wheel axles. The cage is connected to only one of the wheel axles, connecting the other axle only by means of the differential pinion. The differential pinion connecting the differential gears does not rotate as long as the speeds of the two wheels on the axle are the same. If one of the wheels slows down due to cornering, the differential pinion starts rotating to produce a higher speed on the other wheel. The system described above is a simple differential that is not suitable for full torque transfer in low traction condition, such as on ice [4].

11.5 Transmission

The mechanical assembly that delivers the power from the IC engine to the driveshaft most effectively under the various driving conditions of speed, acceleration, roadway grades, and vehicle turns is called the *transmission*. A transmission assembly includes various rotating parts, such as gears, clutches, levels, bearings, and shafts. The transmission allows the vehicle to operate over the entire speed range of the vehicle starting from standstill to the maximum vehicle speed. The transmission is designed and controlled to operate to match the engine speed and vehicle speed. The transmission engages different gear ratios to satisfy the torque and speed requirements of the driver while satisfying the conflicting requirements of acceleration,

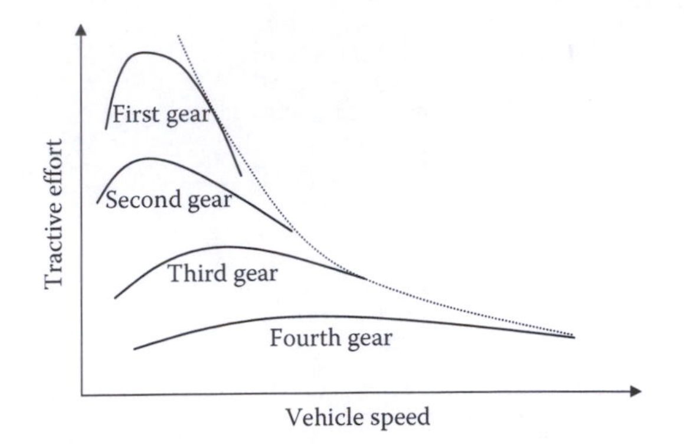

FIGURE 11.14
Tractive force versus vehicle speed for a four-speed transmission.

speed, and fuel economy. The tractive force delivered by the transmission for four gear ratios of a four-speed transmission is shown in Figure 11.14 plotted against vehicle speed.

The three primary types of transmission are: *manual, automatic,* and *continuously variable* transmissions. With the manual transmission, the driver shifts the gears manually in relation to the vehicle speed using the clutch pedal for engagement and disengagement. The driver skill plays a big role with manual transmissions for maximizing the performance of the vehicle. In automatic transmissions, the gear shifting is accomplished through the vehicle controllers without any intervention of the driver. These transmissions also allow the engine to idle when the vehicle is stopped. The continuously variable transmission also does not require any driver intervention, but provides an infinite number of gear ratios rather than a fixed set. A general overview of the three transmissions is given in the following.

11.5.1 Manual Transmission

The manual transmission is a gear train with several shafts, countershafts, spur gears, synchronizers, and safety devices to facilitate multiple gear ratios for power transmission from engine to the drive shaft. Different gear meshes lock and unlock through the use of clutch, gear lever, synchronizer, and the lock/unlock mechanisms in the transmission [3]. The simplified schematic of a four-speed manual transmission is shown in Figure 11.15. The transmission has the input shaft, output shaft, one countershaft, eight gears, an idler, and two synchronizers. The functions of the synchronizer are to ensure that the driving gear and the shaft are rotating at the same speed prior to engagement, and then to lock the gear to the shaft after engagement. The synchronizer eliminates the need for double clutches.

The forward and reverse gear ratios for the gear train and synchronizer arrangement shown in the Figure 11.15 can be calculated based on the

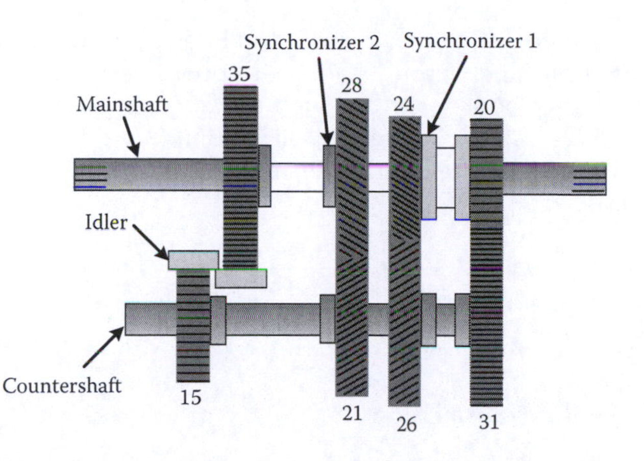

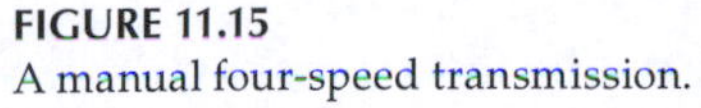
FIGURE 11.15
A manual four-speed transmission.

number of teeth in the gears. For example, the number of teeth on the gears 1 through 8 are assumed to be 35, 15, 28, 21, 24, 26, 20, and 31, respectively. In the first gear, the input shaft transmits power to the countershaft through the 20/31 gear mesh. The countershaft transmits power to the output shaft through the 15/35 gear mesh. Both the synchronizers remain disengaged in a neutral position. Therefore, the first gear ratio is

$$GR_1 = \frac{T_{out}}{T_{in}} = \frac{31}{20} \cdot \frac{35}{15} = 3.617$$

In the second gear, synchronizer 2 engages the second gear (28 teeth) to the output shaft. The input shaft transmits power to the output shaft through the right half of the countershaft and the second-gear counter gear (21 teeth). The second gear ratio is

$$GR_2 = \frac{T_{out}}{T_{in}} = \frac{31}{20} \cdot \frac{28}{21} = 2.067$$

In the third gear, the synchronizer 1 locks the third gear (24 teeth) to the output shaft. The input shaft transmits power to the countershaft and then to the output shaft through the 26/24 third gear mesh. The third gear ratio is

$$GR_3 = \frac{T_{out}}{T_{in}} = \frac{31}{20} \cdot \frac{24}{26} = 1.431$$

In the fourth gear, synchronizer 1 locks the input shaft to the output shaft. Power flows directly from the input to the output and the fourth gear ratio is 1.

In the reverse gear, the idler is used to change the direction of rotation between the input and output shafts. The power is transmitted from the input shaft to the countershaft through the 20/31 gear mesh. The idler is engaged in between the 15/35 gear mesh to change the direction of rotation. Both the synchronizers remain disengaged in this condition. The reverse gear ratio is

$$GR_{Rev} = \frac{T_{out}}{T_{in}} = -\frac{31}{20} \cdot \frac{35}{15} = -3.617$$

11.5.2 Automatic Transmission

The automatic transmission uses a fluid coupling and a planetary gear set to transmit power from the engine to the driveshaft. The fluid coupling device is known as the *torque converter*, which is central to the operation of the automatic transmission. Torque converters allow the IC engine to idle without additional driver intervention when the vehicle is stopped. This eliminates the need for the clutch pedal that exists in manual transmission vehicles.

The other significant component of the automatic transmission is the gear set. The planetary gear set used produces different gear ratios using the unique torque and speed relationships among the sun, ring, and planet carriers. One of the elements in the planetary gear set must be held fixed for torque transfer. In automatic transmission, this is achieved using clutches and bands. A set of clutches, actuated by hydraulics, engage and disengage the different gears of the planetary set. The bands are also actuated by hydraulics to wrap around a gear train and connect to the housing. The planetary gear set along with the clutches and bands establishes the various gear ratios for the transmission.

The construction and operation of the torque converter is discussed in further detail below since this is a key component in the automatic transmission.

11.5.2.1 Torque Converter

Torque converter is a type of fluid coupling device that allows power transfer from the engine to the transmission at different vehicle speeds. When the vehicle is stopped and the engine is idling, only a small amount of torque is transferred to the transmission. As the engine speed increases with increased power demand from the driver, the torque transfer amount also increases.

The torque converter consists of a turbine, an impeller or pump, the housing and transmission fluid. The parts of a torque converter are shown in Figure 11.16. The pump is the driving member of the converter, while the turbine is the driven member. The pump is of centrifugal type and is connected to the engine crankshaft. The turbine is splined to the input shaft of the transmission. The converter is completely filled with transmission fluid

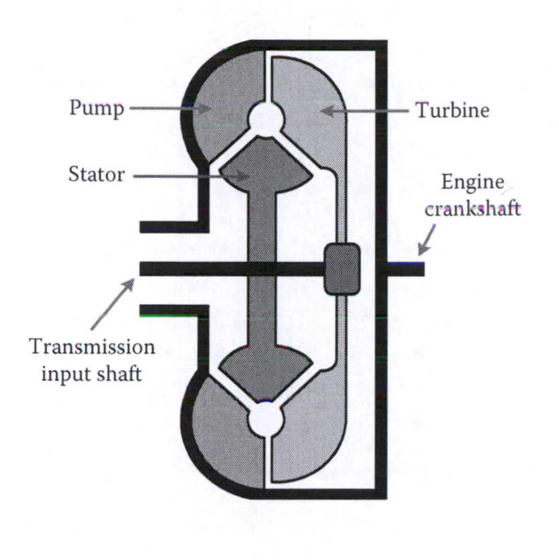

FIGURE 11.16
Torque converter.

that serves as the only connection between the pump and the turbine. It is the dynamics of the fluid that transmits power from the engine to the transmission. Engine rotation causes fluid to enter the pump through the middle and exit through the periphery. In the turbine, fluid enters through the peripheral blades and exits through the middle. As the fluid enters the turbine and pushes against the turbine blades, the turbine starts to rotate in the same direction as that of the pump. As the engine speed increases, more fluid is pumped into the turbine and the torque transmitted increases. When the pump speed is much higher than that of the turbine, the fluid enters the turbine with great force.

Torque converter has a *stator* that prevents the returning fluid from the turbine from hitting the pump with a high velocity in the opposite direction. The stator is mounted on a stationary shaft attached to the transmission and is positioned in between the pump and the turbine. The stator remains stationary and locked to the shaft in one direction when the pump velocity is high relative to the turbine. In this position, the stator redirects the fluid toward the pump in the same direction of rotation. This prevents the drag on the pump and loss of engine power. When the pump is rotating at almost the same velocity as the turbine at higher vehicle speeds, the fluid leaves the turbine with little residual velocity. The fluid strikes the back of the stator, which is then able to rotate freely in the opposite direction.

As mentioned previously, the engine, and hence, the pump speed is almost the same as turbine velocity at higher vehicle speeds, which is highway speeds. The transmission output shaft speed is the same as the turbine speed. Ideally, the transmission speed should match the engine speed exactly since the difference corresponds to power loss. However, in practice, there is always slippage resulting in power loss in the torque converter.

The power dissipation appears in the form of heat in the fluid. The heat is removed in passenger vehicles by air cooling the torque converter housing, or in trucks and utility vehicles by a adding a transmission cooling loop with a secondary radiator. The power loss in the torque converter is the primary reason why automatic transmissions are less efficient than a manual transmission.

Modern torque converters can multiply the engine torque by two to three times when the engine speed is higher compared to the output shaft speed. This allows the torque converter to deliver high torque when a vehicle accelerates from a stop. At higher vehicle speeds, the engine speed matches the output shaft speed, and there is no torque multiplication.

11.5.2.2 Automatic Transmission in Hybrids

Automatic transmissions including the torque converter were used in hybrid vehicles prior to the development of the power-split hybrid architectures. For crankshaft mounted and belt-driven starter/generator hybrid vehicles, the automatic transmission is still a preferred choice. Three main types of step-ratio automatic transmissions considered for hybrid vehicle applications are Simpson, Wilson, and Lepelletier types [5]. The three types differ mainly in the number and type of planetary gear sets, and number and type of clutches and bands required. In crank shaft mounted starter/generator systems, the electric machine is packaged around the torque converter with the rotor mounted to the torque converter at the flex plate of the impeller.

11.5.3 Continuously Variable Transmission

Continuously variable transmission (CVT) does not have a gearbox with a fixed set of gear ratios, but instead uses a sliding mechanism that allows an infinite number of gear ratios between the highest and lowest gear ratios possible with the system. The two common type of CVTs used are the belt-pulley system and the toroidal system. In either type, there are three major components: a driving shaft part connected to the engine, a driven shaft part connected to the transmission, and a coupling mechanism providing the variability. In the belt-pulley system, the input and output shafts are connected to two pulleys and a belt is used for the coupling. In the toroidal systems, two discs are connected to the shafts, and wheels or rollers slide over the discs for coupling the input to the output. In addition, the CVTs have an electronic control unit and various sensors to provide feedback to the controller. The controller adjusts the operating point of the CVT to maximize fuel economy and vehicle acceleration. The belt-pulley system, which is also known as Van Doorne system, is more suitable for low power systems, especially in front-wheel drive vehicles. The toroidal CVTs have been built in larger systems [3].

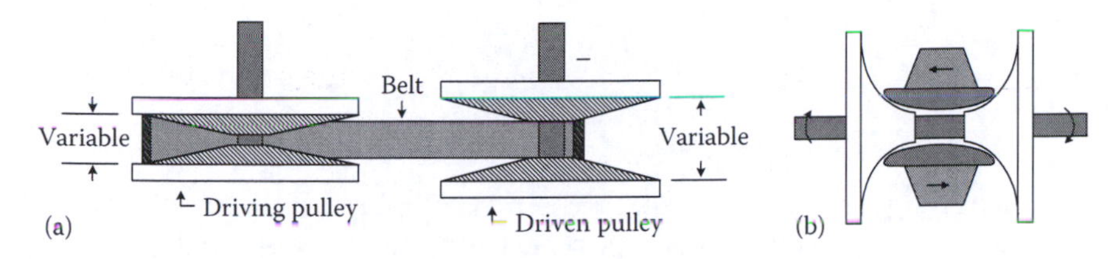

FIGURE 11.17
Continuously variable transmissions: (a) belt-pulley type and (b) toroidal type.

The schematic of a belt-pulley CVT is shown in Figure 11.17a. The three primary components shown are: a high power metal or rubber belt, an input driving pulley, and an output driven pulley. Both the pulleys have cone-shaped structures and the belt rides on top of the cones. Each pulley has two cones that can get closer or move away from each other; this movement varies the diameter of the pulleys. When the cones get closer, the diameter of the pulley increases. When the cones move further apart, the belt sits low in the groove with a smaller diameter. By adjusting the cone positions of the two pulleys, a continuous variation of the gear ratio is possible from a maximum gear ratio (low gear; when the drive pulley diameter is the smallest and the driven pulley diameter is the largest) to a minimum gear ratio (high gear; when the drive pulley diameter is the largest and the driven pulley diameter is the smallest). The drive pulley is connected to the engine crankshaft, while the driven pulley is connected to the transmission.

The toroidal CVT, shown in Figure 11.17b, has a cone-shaped disc connected to the engine (drive disc) and another one connected to the transmission (driven disc). The rollers or wheels slide on the two discs, transferring power from the drive disc to the driven disc. The rollers function as the coupling device for power transmission similar to the belt in the belt-pulley CVT. The gear ratio is 1 when the rollers are in the middle with the two discs rotating at the same velocity as shown in the figure. The gear ratio is higher when the rollers in the driving disc are in contact near the center, while those in the driven disc are near the edge. The gear ratio is lower when the contact scenario is the opposite.

11.5.4 eCVT/HEV Transmission

In the power-split hybrid vehicles, the planetary gear set along with electronic controls is used for power transmission between the engine and electric motor to the wheels. The planet carrier of the planetary gear set is connected to the engine, the sun gear is connected to the starter/generator, and the ring gear is connected to the electric motor that is also connected to the differential and hence the wheels. The torque applied to the planet carrier is transferred to the wheels and the starter/generator through the ring and sun gears. From Equation 11.12

$$T_p = T_r + T_s$$

$$\Rightarrow T_{engine} = T_{wheels} + T_{gen}$$

The use of the planetary gear set in the power-split hybrid vehicles results in a fixed ratio of torque distribution between the wheels and the starter/generator gear. A split ratio of 72%–28% between the motor/wheels and the starter/generator is used in some production hybrid vehicles.

The power relationship in the power-split hybrid vehicle is

$$T_p\omega_p = \frac{r_r\omega_r}{r_r + r_s}T_p + \frac{r_s\omega_s}{r_r + r_s}T_p$$

$$\Rightarrow P_{engine} = P_{engine\ to\ ring} + P_{engine\ to\ sun}$$

In the power-split hybrid vehicles, all of the power comes from the stored fuel through the IC engine. The control strategy algorithm varies the generator loading to split the power from the engine between the wheels and the starter/generator. The electric motor can add power to the wheels in addition to what comes directly from the IC engine through electronic control. Since the power-split is controlled electronically, these types of transmissions are known as electronic-CVT (eCVT). The details of the control strategy for the power-split hybrid vehicle are given in Section 13.2.1.

11.6 Vehicle Brakes

Vehicle braking system converts the kinetic energy of the vehicle into thermal (heat) energy to provide the driver with a mechanism to stop the vehicle or reduce its speed. The brakes in automobiles are mechanical clutches that use friction to slow down a rotational disc. The driver controls the brake action through a foot-operated linkage. The friction clutch is composed of two discs, each connected to its own shaft. As long as the discs are not engaged, one disc can spin freely without affecting the other. When the rotating and the stationary discs are engaged through the operator action, friction between the two discs reduces the speed of the rotating disc. The kinetic energy of the vehicle transfers directly between the discs and is wasted due to friction.

11.6.1 Conventional Brake System

Conventional braking systems use the hydraulic technology to actuate the friction clutches for the braking action. The entire hydraulic system is filled

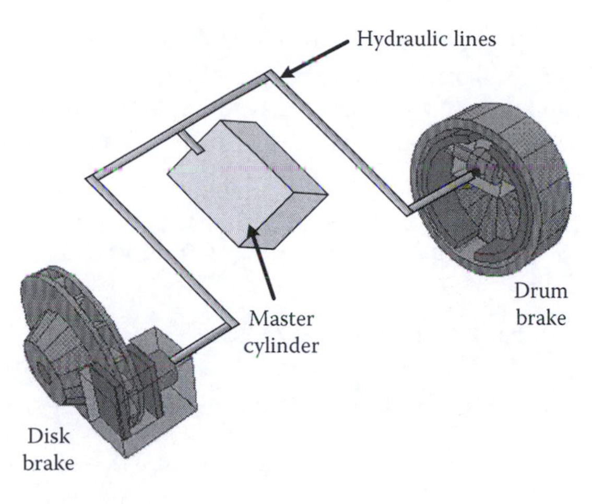

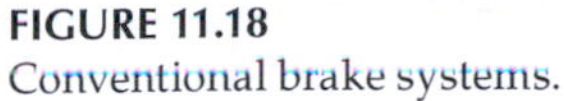
FIGURE 11.18
Conventional brake systems.

with a special brake fluid, which is pressurized and forced through the system by the movement of the master cylinder pistons. The simplified views of the two main types of conventional brakes, which are the disc-type brakes and the drum-type brakes, are also shown in Figure 11.18 [6].

The disc brakes are the preferred choice for both the front and rear wheels of passenger vehicles. The pads of the disc-type brakes are forced against the machined surfaces of a rotating disc called the rotor that is attached to the wheel. The disc brakes have friction pads controlled by a caliper arrangement, which when engaged, clamps to the rotor and the wheel. The brake pads are designed to assist cooling and resist fading. Fading causes the braking friction coefficient to decrease with temperature rise. The high force required for caliper actuation is typically supplied from a power assist device following the brake command input from the driver.

Drum brake, which has been gradually replaced by the disc brakes use an internally expanding brake shoes that are forced against the inside machined surface of a rotating drum. The drum-type brake units have cylindrical surfaces and shoes instead of pads that hold the friction material. Drum brakes are highly suitable for hand brakes; they are still used in the rear wheels of some vehicles. The shoes press against the drum cylinder on a brake command input from the driver. The shoes can be arranged to press against the outer or inner surface of the rotating drum to retard the wheel rotation. If the shoes are applied to the inner surface, then the centrifugal force of the drum due to rotation will resist disc engagement. If the shoes are pressed against the outer drum surface, the centrifugal force will assist engagement, but at the same time, may cause overheating.

When the driver applies the brakes, a command is executed to exert a brake force acting in opposition to the motion of the car. The dynamics of the braking forces are similar to the vehicle dynamics studied in Chapter 2, except that braking forces are now applied at the wheels instead of tractive forces from the propulsion unit. The braking conditions are governed by the tire-road interface friction coefficients. Let us consider F_{bf} and F_{br} as the braking forces applied at the front and rear wheels to understand the braking dynamics. From Newton's second law

$$\sum F_{xT} = m_v a_{xT} = F_{bf} + F_{br} + m_v g \sin(\theta) + F_{roll} + F_{AD} \tag{11.14}$$

where a_{xT} is the linear acceleration or deceleration in the x_T direction. The rolling resistance force F_{roll} and aerodynamic drag forces F_{AD} assist in braking action, and the function of the gravitational force depends on the roadway slope.

The time and the distance traveled during a velocity change with uniform deceleration can also be derived from Newton's second law as

$$t = \frac{m_v}{\sum F_{xT}} (V_0 - V_f)$$

$$x = \frac{m_v}{\sum F_{xT}} (V_0^2 - V_f^2)$$

where V_0 and V_f are the initial and final velocities, respectively of the vehicle during braking. The uniform deceleration approximation indicates that the time to stop is proportional to the vehicle velocity and the stopping distance is proportional to the square of the vehicle velocity.

Let us consider the maximum sustained deceleration requirement of 0.65 g, which corresponds to a panic stop from 60 mi/h [3]. Braking forces will be much larger compared to the rolling resistance and aerodynamic drag forces for harder braking actions. Also, if we assume a level roadway, then for the panic stop situation, Equation 11.14 simplifies to

$$F_{bf} + F_{br} = m_v a_{xT} \tag{11.15}$$

The average power dissipated with a uniform deceleration is

$$P_{b,avg} = (F_{bf} + F_{br}) \times \frac{V_f + V_0}{2} \tag{11.16}$$

The initial power with uniform deceleration is going to be twice the value of $P_{b,avg}$. The rate of heat dissipation must be evaluated for braking system design, since the friction brakes convert the kinetic energy of the vehicle into heat energy.

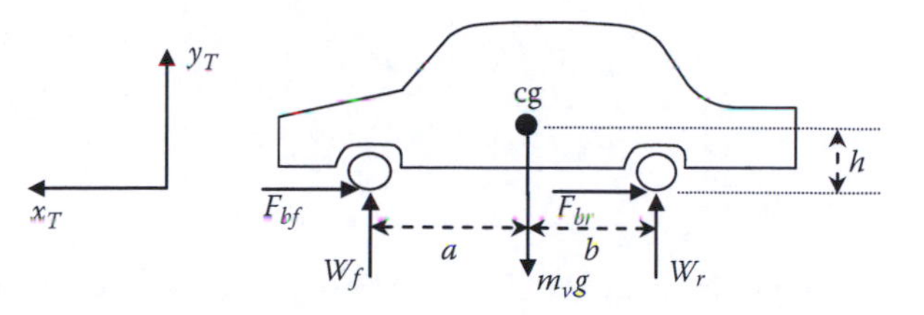

FIGURE 11.19
Free-body diagram of a vehicle under braking action.

The free body diagram of the vehicle under braking action is shown in Figure 11.19. The static weight distribution of the vehicle between the front and rear axles are

$$W_f = \frac{m_v g b}{(a+b)}$$

$$W_r = \frac{m_v g a}{(a+b)}$$

There will be a dynamic shift in vehicle weights under braking action. Let the dynamic weights of the vehicle be W_f' and W_r'. Applying Newton's second law in the y_T-direction

$$\sum F_{yT} = 0 = W_f' + W_r' - m_v g$$

$$\Rightarrow W_f' + W_r' = m_v g \tag{11.17}$$

The dynamic weights can be calculated using the moment of the forces around the vehicle center of gravity. The moment forces are

$$\sum M_{cg} = 0 = F_{bf} \cdot h - W_f' \cdot a + F_{br} \cdot h + W_r' \cdot b$$

$$\Rightarrow m_v a_{xT} h = W_f' \cdot a - W_r' \cdot b \tag{11.18}$$

Equation 11.15 has been used in deriving the above relationship. Multiplying both sides of (11.17) with b and adding with (11.18), we can obtain

$$W_f' = \frac{m_v g b}{(a+b)} + \frac{m_v a_{xT} h}{(a+b)} \tag{11.19}$$

and

$$W_r' = \frac{m_v g a}{(a+b)} - \frac{m_v a_{xT} h}{(a+b)} \tag{11.20}$$

The first terms on (11.19) and (11.20) are the same as the static weights; the second terms are the dynamic weights added or subtracted from the front and rear axles due to braking action. The equations show that during braking, the front axle bears heavier weight than the rear axles. This means that the front axles sustain higher braking forces than the rear axles.

The ratio of dynamic weights between the front and rear axles is

$$\frac{W_f'}{W_r'} = \frac{b + h a_{xT}/g}{c - h a_{xT}/g} \tag{11.21}$$

The front braking action to rear braking action can be distributed as

$$\frac{F_{bf}}{F_{br}} = \frac{W_f'}{W_r'} \tag{11.22}$$

The braking force acting on the vehicle is also the function of the tire-to-road interface, as mentioned earlier in Chapter 2. The maximum braking force possible for a wheel is equal to the coefficient of friction times the normal force.

$$F_{b,\max} = \mu W_{wh}' \tag{11.23}$$

where

μ is the tire-to-road friction coefficient

W_{wh}' is the dynamic weight on the wheel

Friction coefficients for three different road conditions are shown in Figure 11.20. The friction coefficient limits the maximum braking force that can be applied in different road conditions beyond which wheels lock-up. In addition, the friction force limit is a nonlinear function of the vehicle slip speed. Therefore, the force applied to the brake pads in each wheel of the vehicle has to be carefully optimized in order to ensure maximum effectiveness.

Brake systems are designed to apportion the applied brake force between the front and rear brakes according to the dynamic weight ratios. The front and back wheels of the vehicle will consequently have different braking behaviors and requirements due to the dynamic load transfer during braking in two-axle vehicles. Front brakes are designed to carry out higher

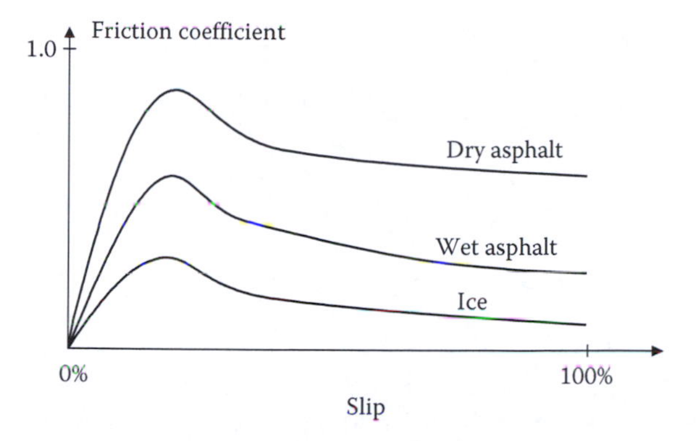

FIGURE 11.20
Friction coefficient variations as a function of tire-to-road slip.

percentage of braking due to the increase in dynamic weight during braking in the front axles. The brakes must also be prevented from locking up for good braking performance. The vehicle loses steering control when the front brakes lock up, and the vehicle tends to sway when the rear wheels lock up. Anti-lock braking systems (ABS) use control algorithms to prevent wheels from locking up.

Additional brake system analysis reveals that to obtain maximum braking power, it is necessary to consider front and back braking forces together. The front and rear brakes are highly interdependent in contrast to the simplified analysis presented here. Advanced designs also have to consider overall vehicle stability, and the optimization of the force applied to the rotor, along with critical safety considerations.

Example 11.2

A 2000 kg vehicle traveling at 85 mph has to be stopped with the maximum sustained deceleration of 0.65 g. The vehicle has a wheelbase of 2.5 m and a front/rear static weight distribution of 49%/51%. The center of gravity is at a height of 0.5 m from the ground.

(a) Find the total force and average power required to bring the vehicle to a stop.
(b) Find the average power absorbed by the brakes.
(c) Find the front and rear weight distribution during braking.

Solution

(a) The force required to stop the vehicle is

$$F_b = F_{bf} + F_{br} = m_v a_{xT} = 2000 \times 0.65 \times 9.81 = 12.753\ \text{kN}$$

(b) The average power dissipated in the brakes is

$$P_{b,avg} = \frac{1}{2}F_bV_f = \frac{12.753 \times 37.97}{2} = 242.12\ \text{kW}$$

(c) The front and rear weight distribution is

$$\frac{W_f'}{W_r'} = \frac{1.275 + 0.5 \times 0.65\ \text{g/g}}{1.225 - 0.5 \times 0.65\ \text{g/g}} = 64\%/36\%$$

In addition to the previous considerations, the controller design of the brake system must consider the nonlinear characteristic of the rotor and pad interaction. The braking pad displacement has a nonlinear relationship with rotor clamping force and is also temperature dependent. The desired braking force is achieved by applying the corresponding pad displacement according to the actuator characteristic. The use of a highly dynamic and controllable electromechanical actuator will allow vehicle manufacturers to achieve better performance and higher reliability. An electromechanical brake system is discussed next.

11.6.2 Electromechanical Brake System

Electromechanical brake (EMB) system is an alternative to conventional brake system where electromechanical components replace the hydraulic actuators and controls [7,8]. EMB avoids the use of hydraulic fluid in the brake system and facilitates improved performance with modern braking control systems such as ABS or vehicle stability control (VSC). EMB technology allows for the introduction of pure brake-by-wire systems that remove brake fluids and hydraulic lines entirely. The braking force is generated directly at each wheel by high-performance electric motors that are controlled by an electronic control unit (ECU) and actuated by signals from an electronic pedal module. This specific mechanism in conjunction with the underlying control structure translates into the most direct and individual wheel control, and therefore, increases braking performance.

The layout and communication between the control unit of an EMB system and its actuators is shown in Figure 11.21. EMB system allows the complete separate corner assembly of the components near the wheels while commands are to be executed through electronic signals only. The feel of the brake pedal can be easily tuned and adjusted to individual needs since brake pedal displacement is now completely independent from changes in brake actuator travel. An EMB system does not exhibit the often undesirable noise and vibration pedal feedback experienced in conventional hydraulic brake systems during ABS and other control modes. In addition, EMB systems are environmentally friendlier than conventional systems since no hydraulic fluid is required, while at the same time, they allow for an easy and straightforward "plug & bolt" assembly process.

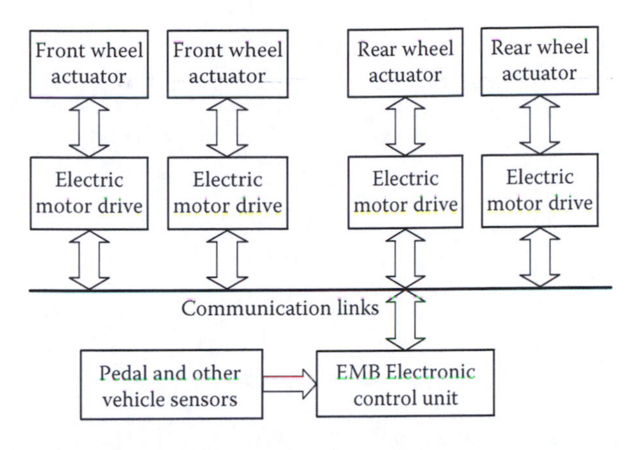

FIGURE 11.21
Electromechanical brake system in a vehicle.

Electric machines and their controllers enable electromechanical actuation to meet the specifications required in braking systems. Each wheel of the vehicle is equipped with its own brake actuator, which can be controlled independently from the other wheels. The individual control of each wheel's braking behavior can then be used to achieve optimal braking ability and improved vehicle stability. This makes it possible for the vehicle to remain stable during a braking phase even when wheels are subject to different surface conditions. Electric motor-driven actuators allow the implementation of advanced control algorithms to improve the braking behavior of the entire vehicle, which is otherwise not possible with hydraulic actuation. In addition to that, software upgrades can provide for an easy tuning of the entire system, while minimizing mechanical changes.

EMB systems have very specific requirements in terms of dynamic response and steady-state operation. When a braking operation is required, the motor must have a fast response and a sufficiently high no-load speed to bring the disengaged brake into contact with the brake rotor. On the other hand, in order to apply the necessary braking force, the motor has to generate high torques at low speeds down to a complete stall, once the pad is applied. The electric machine must have fast response in both the directions of rotation, since both engagement and disengagement functions have to be randomly available. Therefore, it is necessary to have a controller and machine suitable for four-quadrant operation, which can be appreciated from the typical operation cycle for an EMB motor as shown in Figure 11.22. The electric machine is operated in the torque-controlled mode with maximum speed limits.

EMB system shown in Figure 11.23 consists of a rotary motor coupled to a planetary-gear and ball-screw assembly that converts the motor torque and rotation into linear force and travel suitable for operation in a caliper brake. Components of one actuator to be located in a corner wheel assembly are shown in Figure 11.23. There will be four such actuators in the four corner

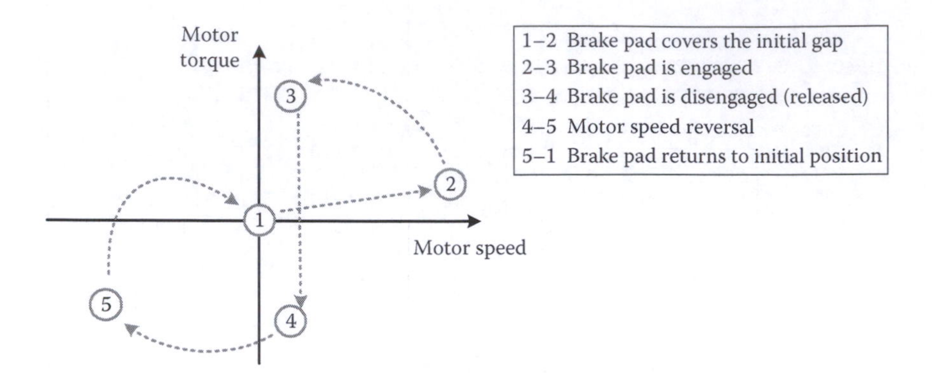

FIGURE 11.22
Four-quadrant operation of electric motor to engage and disengage electromechanical brake.

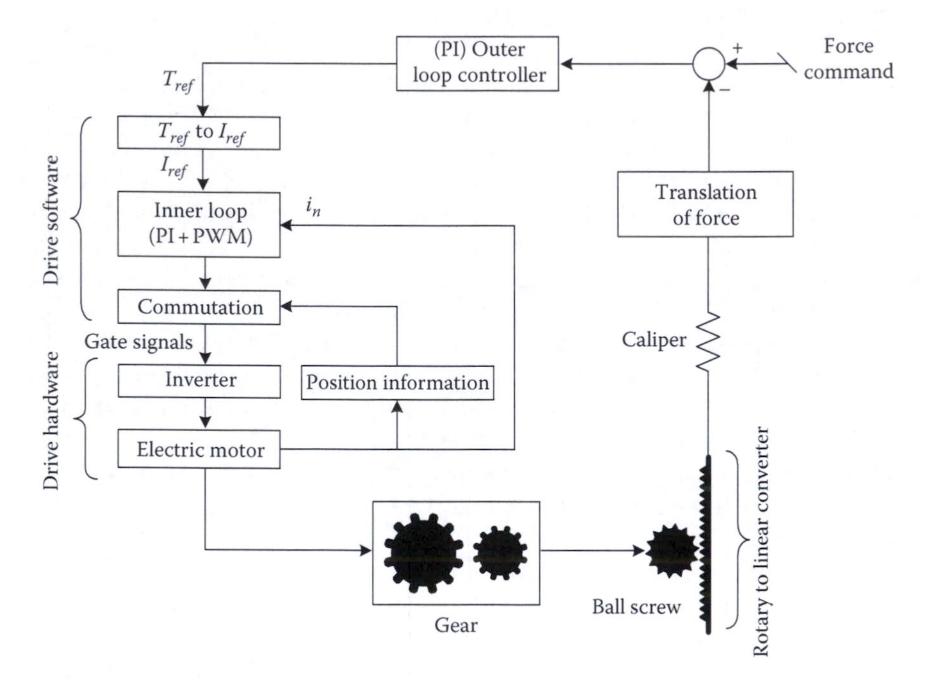

FIGURE 11.23
Electromechanical brake system with an electric actuator.

wheel assemblies in a four-wheel vehicle equipped with EMB system. A high speed/low torque electric machine along with the gears helps minimize the package size. The gears reduce the high motor speed and increase the output torque for brake actuation. The system controller can be operated in a force-controlled mode in the outer loop with force feedback from the calipers. The force commands are issued through the brake pedal. The force controller located in the electronic control unit issues signals to the individual wheel actuators to achieve desired performance. The force controller generates the torque command for each of the electric motor controllers. The motor controller processes the command signal and the feedback signals to generate the gating signals for the power converter. Pulse width modulation or hysteresis current controllers are commonly used for this purpose. The power converter drives the motor, which produces the desired torque and creates the desired force on the brake calipers.

The type of electric machines suitable for EMB systems are either the PM machine or the switched reluctance machine. PM machine does offer a higher power density, but temperature sensitivity has to be carefully evaluated since the friction heat generated due to braking could damage the magnets. On the other hand, switched reluctance machine offers simple and low rotor inertia and is also amenable to discrete position sensorless operation.

EMB is not yet widely accepted mainly because of the safety critical nature of brake systems. A first alternative to the conventional braking systems could be to combine the advantage of electronic control with hydraulic actuators. This makes it possible to implement control algorithms that will improve the performance of the braking system and user comfort. With electrohydraulic braking (EHB), the hydraulic link between the brake pedal and the wheel brake is replaced with a by-wire-transmission that offers considerable advantages. The hydraulic structure of the system is simplified, and the control of the actuator is made easier. EHB control unit receives inputs from sensors connected to the brake pedal. In normal operation, a backup valve is kept closed and the controller activates the brakes of the wheel through an electric motor-driven hydraulic pump. When the controller goes into a fail-safe mode, the backup valve is opened, which allows the brakes to be controlled through a conventional hydraulic circuit. This redundancy makes the system very reliable in the case of an electrical failure.

Problem

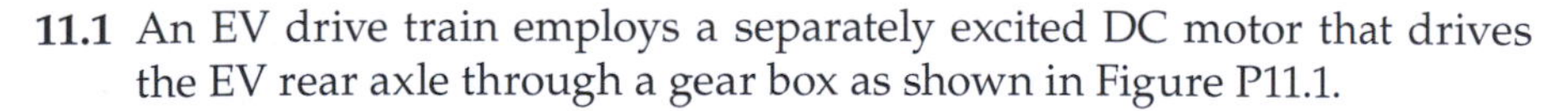

11.1 An EV drive train employs a separately excited DC motor that drives the EV rear axle through a gear box as shown in Figure P11.1.

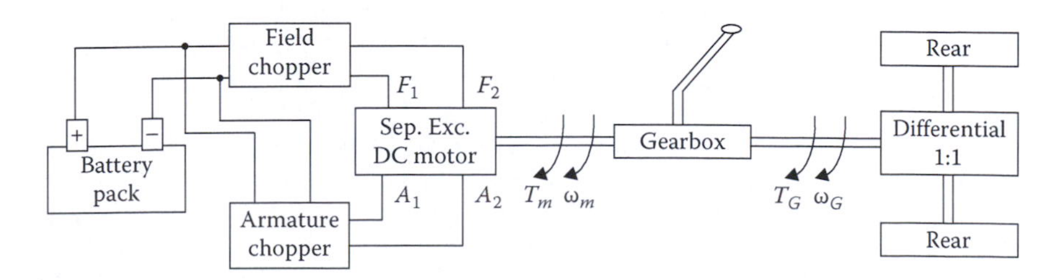

FIGURE P11.1

The vehicle is traveling in fourth gear on a level road at constant velocity of 60 mi/h. All necessary parameters are

$$mg = 6867\ \text{N},\ C_0 = 0.009,\ C_1 = 1.75\ \text{E}-6\text{s}^2/\text{m}^2$$

$$A_F = 2\text{m}^2,\ C_d = 0.2,\ \rho = 1.16\ \text{kg/m}^3$$

$$r_{wh} = \text{wheel radius} = 7.5\ \text{in., gear ratio GR} = 0.4$$

Motor parameters: $R_a = 0.2\ \Omega$, $R_F = 150\ \Omega$, $K = 0.8$ V s/Wb, $\phi = 3.75I_F$, $Bg = 1.2I_F$.

In the following calculations, assume no power loss from motor output to wheels and use motor power loss model given in class. Assume chopper outputs are pure DC.

(a) Calculate the operating speed and torque of the motor.
(b) For $0.5 \leq I_F \leq 4A$, plot I_A vs. I_F.

References

1. R.L. Willis and J. Brandes, Ford next-generation electric vehicle powertrain, *12th Electric Vehicle Symposium*, Anaheim, CA, December 1994, pp. 449–458.
2. R.G. Budynas and J.K. Nisbett, *Mechanical Engineering Design*, McGraw Hill, New York, 2008.
3. R. Stone and J.K. Ball, *Automotive Engineering Fundamentals*, SAE International, Warrendale, PA, 2004.
4. T.E. Scott, *Power Transmission Mechanical, Hydraulic, Pneumatic, and Electrical*, Prentice Hall, Upper Saddle River, NJ, 2000.
5. J.M. Miller, *Propulsions Systems for Hybrid Vehicles*, Institute of Electrical Engineers, London, U.K., 2004.
6. S. Underwood, A. Khalil, I. Husain, H. Klode, B. Lequesne, S. Gopalakrishnan, and A. Omekanda, Switched reluctance motor based electromechanical brake-by-wire system, *International Journal of Vehicle Autonomous Systems*, 2(3–4), 278–296, 2004.
7. A. Omekanda, B. Lequesne, H. Klode, S. Gopalakrishnan, and I. Husain, Switched reluctance and permanent magnet brushless motors in highly dynamic situations: A comparison in the context of electric brakes, *IEEE Industry Applications Magazine*, 15(4), 35–43, July/Aug 2009.
8. H. Klode, A. Omekanda, B. Lequesne, S. Gopalakrishnan, A. Khalil, S. Underwood, and I. Husain, The potential of switched reluctance motor technology for electro-mechanical brake applications, *Simulation and Modeling Mechatronics*, SAE Publication SP2030, SAE International, Inc., Warrendale, PA, 2006.

12

Cooling Systems

The focus of the book has thus far been on powertrain architectures and components. There are several systems within a vehicle that are required as supporting systems for powertrain components; the mission of some of these systems could be as critical as the powertrain itself. Some ancillary systems in a vehicle are designed for the safety and comfort of passengers. While a vehicle has many ancillary systems, the objective here is not to present a comprehensive coverage but to highlight the importance of these ancillary systems. The two systems covered in this chapter are related to cooling necessary in an electric or hybrid electric vehicle. The climate control system presented is also required in an IC engine vehicle.

Ancillary systems unique to electric and hybrid electric vehicles are either adapted from existing ones or developed from basics. An electric motor drive powered by the traction battery may be required to drive the compressor of the air-conditioning system in electric and hybrid vehicles. The electric machines and inverters for propulsion require a powertrain electronic cooling system to be developed. The traction batteries require adequate cooling for their operation and safety.

12.1 Climate Control System

Cabin climate in a vehicle is controlled using one of the refrigeration cycles. A working fluid circulates in a closed loop through two heat exchangers to cool the vehicle cabin. A typical air-conditioning unit in a passenger vehicle requires about 2 kW of power under steady-state conditions, but could require up to three times that during peak conditions. The work input to the system is through the component called *compressor* that compresses the working fluid.

In conventional IC engine vehicles, the compressor is belt driven by the IC engine, but in hybrid vehicles, an electric motor drive is required for electric-only operation modes when the engine is off. Hybrid electric vehicles that do not employ an electric motor drive for the compressor must turn the engine on when air-conditioning is demanded by the user. In electric vehicles, an electric motor drive has to be used to drive the compressor. The remaining components in the climate control system are the same in electric and hybrid electric vehicles as those in IC engine vehicles.

In this section, the thermodynamics of the refrigeration cycle used in cabin climate control is discussed followed by descriptions of vehicle air-conditioning system components.

12.1.1 Vapor-Compression Refrigeration Cycle

The most common refrigeration systems of today use the vapor-compression refrigeration cycle, which has been derived from the ideal Carnot vapor refrigeration cycle. Reversing the Carnot vapor power cycle gives the ideal refrigeration cycle. This cycle is completely reversible, but not a practical one. Let us first evaluate the ideal vapor cycle before discussing the practical vapor cycle. In the ideal cycle, refrigerant circulates through a series network of components while operating between a cold region and a warm region. The schematic of the ideal cycle and the associated temperature–entropy (T–s) diagram is shown in Figure 12.1 [1]. The components in the system are evaporator, compressor, condenser, and turbine. Evaporators and condensers are heat exchangers where heat is exchanged between the refrigerant and the two regions at different temperatures. The refrigerant enters the evaporator as a two-phase liquid vapor mixture at state 4. In the evaporator, the refrigerant receives heat from the cold region while some of it changes state from liquid to vapor. The temperature and pressure remain constant during this process from state 4 to state 1. The refrigerant is then compressed adiabatically (constant entropy) from state 1 to state 2 by the compressor. The compressor requires work input to increase the pressure and temperature of the refrigerant. In state 2, the refrigerant is in a saturated vapor condition. The refrigerant then enters the condenser where heat is transferred to the warm region under constant temperature and pressure

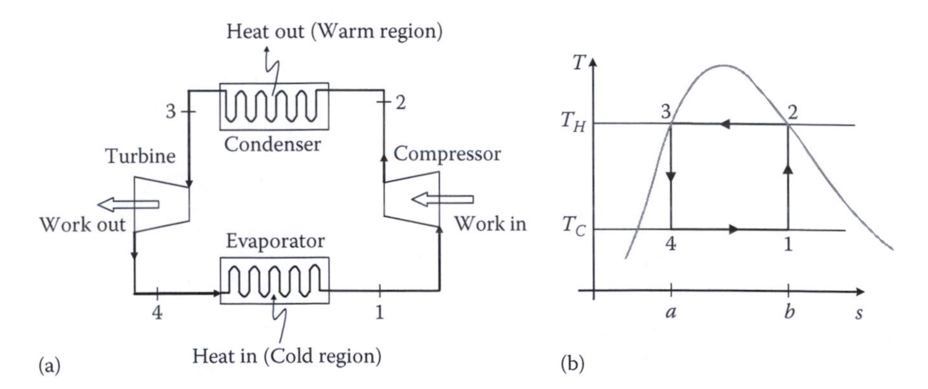

FIGURE 12.1
Carnot vapor refrigeration cycle: (a) components; (b) temperature–entropy diagram.

conditions. The refrigerant changes state from saturated vapor to saturated liquid as a result of the heat transfer in the condenser. The liquid refrigerant at state 3 enters the turbine where it expands adiabatically. Both temperature and pressure drop in the turbine as the refrigerant returns to its initial state 4. In this ideal Carnot cycle, all processes are internally reversible; heat transfer between refrigerants and the two regions takes place without any change in temperature.

In the practical vapor-compression refrigeration cycle, the most significant departure from the ideal condition is related to the heat transfers that cannot be accomplished reversibly in the heat exchangers. In particular, the temperature of the refrigerant at the evaporator needs to be maintained several degrees below the cold region temperature T_C with a practical size evaporator to maintain a sufficient heat transfer rate. Also, the temperature in the condenser needs to be maintained several degrees higher than the warm region temperature T_H. Another important point of departure from the ideal cycle is in the phase of the refrigerant during compression. As mentioned earlier, the refrigerant during compression is in a liquid–vapor mixed phase; this is known as wet compression. In practice, wet compression is avoided since liquid droplets damage the compressor. In the practical vapor cycle, refrigerant has to be completely vaporized; such a process is known as dry compression. Another impractical aspect of the ideal cycle is the expansion process in the turbine from state 3 to state 4 where saturated liquid is converted to liquid–vapor mixture. This process produces very little work output by the turbine compared to the work input at the compressor. This is because turbines working under these conditions have very low efficiencies. In the practical vapor refrigeration cycle, the turbine is replaced with a simple expansion valve. This is a cost-effective solution, although the work output of the turbine has to be sacrificed.

The vapor-compression refrigeration system and the operating (*T*–*s*) diagram are shown in Figure 12.2 [1]. Let us analyze the heat transfers and work

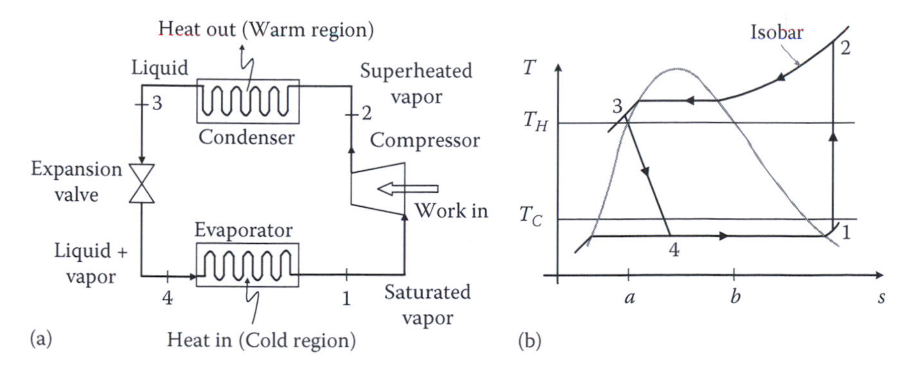

FIGURE 12.2
Vapor-compression refrigeration cycle: (a) components; (b) temperature–entropy diagram.

done in the system components starting with the evaporator. The work and energy transfers are positive in the directions indicated. Kinetic and potential energy changes in the processes will be neglected. We will also assume that there is no extraneous heat transfer and the pressure changes take place only in the expansion valve and the compressor.

In the evaporator, heat is transferred from the refrigerated space to the refrigerant which results in the vaporization of the refrigerant. The heat transfer rate per unit mass of refrigerant is given by the enthalpy change between the two states as

$$\frac{\dot{Q}_C}{\dot{m}} = H_1 - H_4 \tag{12.1}$$

where

$\dot{Q}_C$ is the heat transfer rate

$\dot{m}$ is the mass flow rate of the refrigerant

$\dot{Q}_C$ is also known as refrigeration capacity with units of kW in SI units or Btu/h in English units. Another commonly used unit for refrigeration capacity is *ton of refrigeration* which is equal to 12,000 Btu/h.

The compressor pump compresses the vapor refrigerant out of the evaporator into high-pressure vapor while simultaneously increasing the temperature. Assuming there is no heat transfer to or from the compressor, the rate of energy input per mass of refrigerant is given by the enthalpy change between the two states as

$$\frac{\dot{W}_C}{\dot{m}} = H_2 - H_1 \tag{12.2}$$

In the condenser, heat transfer takes place from the refrigerant to the warm region under constant pressure (isobar). The heat transfer rate per unit mass of refrigerant is

$$\frac{\dot{Q}_H}{\dot{m}} = H_2 - H_3 \tag{12.3}$$

The refrigerant then expands to the evaporator pressure from state 3 to state 4 in the expansion valve. This process is irreversible when the pressure decreases along with an increase in entropy. Again, assuming there is no heat transfer to or from the compressor, energy transfer rate per unit mass of refrigerant is

$$\frac{\dot{W}_{out}}{\dot{m}} = H_3 - H_4 \tag{12.4}$$

Since there is no work output in the expansion valve, $\dot{W}_{out} = 0$, and hence,

$$H_4 = H_3$$

The objective of the air-conditioning system or refrigerator is the removal of heat, called the cooling load from a low-temperature environment. The performance of a refrigeration cycle is measured in terms of the coefficient of performance, which is the ratio of the refrigeration effect to the net work input required to achieve that effect [2]. For the vapor-compression cycle, the coefficient of performance is

$$\mathrm{COP}_R = \frac{\text{Cooling effect}}{\text{Net work input}} = \frac{\dot{Q}_C}{\dot{W}_C} = \frac{H_1 - H_4}{H_2 - H_1} \tag{12.5}$$

The above equations apply to both practical vapor-compression cycles with irreversibilities and idealized compression cycles where such effects are absent since the above equations have been derived from mass and energy rate balances. Although practical cycles deviate from the ideal cycles due to the irreversibilities in evaporator, compressor, and condenser, it is insightful to evaluate systems with the idealized cycles; the idealized evaluations do establish the upper limit on the performance of a practical refrigeration system.

12.1.2 Vehicle Air-Conditioning System

The automotive air-conditioning system runs on the vapor-compression refrigeration cycle with similar components shown in Figure 12.2a. The only addition is a receiver/dryer which is placed in between the condenser and the expansion valve. The types of components and their functionalities in an automotive air-conditioning system are described below.

Compressor is responsible for compressing the low-pressure refrigerant gas to high pressure and high temperature. The compressor is essentially a pump with a suction side and a discharge side. The suction side draws the low-pressure refrigerant from the outlet of the evaporator. In some cases, the suction is drawn through an accumulator. The high-pressure compressed fluid at the discharge side is delivered to the condenser.

Compressors play a determining role in the cabin climate control of a vehicle. Either a belt coupled to the IC engine output shaft or an electric motor drives the compressor. The common compressor types used for climate control in the automotive industry are scroll compressor, rotary vane compressor, and swashplate compressor. Of these, scroll compressors deliver high efficiency at medium speeds and also have the ability to withstand high rotational speeds. Hence these are suitable for use with electric motors which can operate at higher speeds. Electric motor driven compressor also have the advantage of adjustable speeds through motor control that would improve performance of the air-conditioning system.

A unidirectional, speed-controlled electric motor drive is required for compressors that are operated without an IC engine. Either PM machines or SR machines are suitable for the application. The drive structure is similar to that presented earlier in Chapters 8 and 9 except that regeneration is not required. PM synchronous machines are in use currently for the compressor drives used in the Toyota and Lexus hybrid electric vehicles. Pressure sensors from the high-pressure and low-pressure sides provide feedback to the motor controller to regulate the motor speed and torque for maintaining the desired cabin temperature. Position sensorless vector control with space vector PWM is the common choice of motor control used for these PM electric motor drives. Indirect position sensing is essential since there is not enough space available for mounting mechanical position sensors with the compressor integrated with the electric motor. The cold side of the compressor is conveniently used for cooling the IGBTs of the electric drive inverter.

The condenser provides the area for the heat release of the refrigerant into the atmosphere. As the refrigerant fluid goes through the condenser, it cools off and condenses into high-pressure liquid which exits from the bottom. The condenser must be located in a position where there is sufficient airflow. Engine cooling fan can be utilized to radiate the heat from the condenser; alternatively, one or more electric cooling fans can be added for effective heat radiation.

The receiver-drier is located in the high-pressure side of the system right after the condenser. Its function is to store liquid refrigerant and remove moisture and dirt as the refrigerant circulates through the system. The removal of dirt is critical so as not to clog the expansion valve orifice which is next in line of the air-conditioning system.

The expansion valve receives the high-pressure liquid from the receiver-drier and allows the refrigerant pressure to drop so that it can change phase and become liquid. The valve controls the amount of refrigerant passing through its orifice so that liquid refrigerant can be delivered to the evaporator for evaporation.

The evaporator serves as the heat absorption unit and is located inside the vehicle. The liquid refrigerant evaporates by extracting heat from the warm air blown over the evaporator fins thereby cooling the cabin air. Refrigerants have very low boiling point and hence can absorb a large amount of heat from the passing air and boil to low-pressure gas as they exit the evaporator. The heat carried off by the refrigerant is eventually released to the atmosphere at the condenser. The evaporator also serves as a dehumidifier adding to the passenger comfort level inside the cabin. As warmer air travels through the cooler fins of the evaporator, the moisture in the air condenses on its surface. The evaporator also acts as an air filter attracting dust and small particles from the air to stick to its wet surfaces, which drains off to the outside of the vehicle.

Accumulators are connected at the outlet end of the evaporator to collect liquid refrigerant and also to remove moisture and dirt from the system similar to the receiver-drier. The function of the accumulator is critical to the compressor which can only compress gas and not liquid.

The climate control unit maintains the cabin temperature in a vehicle by controlling refrigerant pressure and flow into the evaporator. Pressure sensors, one in the high-pressure line and one in the low-pressure line, provide feedback to the control unit to maintain low pressure in the evaporator to prevent it from freezing.

The refrigerant used in the automobiles is R-134a, which replaced the previously used refrigerant R-12 due to environmental concerns. R-134a operates at higher pressure levels than the R-12 which requires a more powerful compressor.

12.2 Powertrain Component-Cooling System

Thermal system design and thermal management are critical components in the performance and reliability of power electronics and electric machines. Electric and hybrid vehicles require adequate thermal management for the electric machines and drives as well as for the additional power electronics component used such as the DC/DC converter for the high voltage to low voltage interface.

Heat energy is removed from the electric motor, motor drive, and power electronic components by the methods of radiation and convection. *Thermal radiation* is the process by which energy is transported by electromagnetic waves with or without any intervening medium. This means that thermal radiation can take place even in vacuum. Electronic configuration changes in the atoms or molecules within a material result in thermal radiation. The rate of energy transfer from a system through thermal radiation is complicated involving the properties of the radiating surfaces and the energy absorbing medium. A modified form of the Stefan–Boltzmann law macroscopically quantifies the rate of energy transfer as [1]

$$\dot{Q}_{rad} = \varepsilon\sigma A T_b^4 \quad (12.6)$$

where

A is the surface area

ε ($0 \leq \varepsilon \leq 1$) is the emissivity of the surface

σ is the Stefan–Boltzmann constant

T_b is the absolute temperature of the surface

Convection refers to energy transfer between a solid surface at one temperature and an adjacent moving gas or liquid (to be referred to as fluid) at another temperature. Heat energy is conducted from the solid surface to the moving fluid through the combined effects of conduction within the fluid and the bulk motion of the fluid. The rate of energy transfer through convection can be evaluated through the following empirical formula [1]:

$$\dot{Q}_{conv} = hA(T_b - T_f) \tag{12.7}$$

where

A is the surface area

T_b is the temperature on the surface

T_f is the fluid temperature

The heat transfer coefficient h is an empirical parameter relating the properties and the flow rate of the fluid, and the geometry of the system; this is not a thermodynamic property. Convection can be of two types; *forced* and *natural*. Fans or pumps used for forced convection would result in a higher heat transfer coefficient than in natural convection.

The development of the cooling system for electrical components starts with identifying the requirements which depend on the vehicle architecture and the rating of components within that architecture. The liquid-cooled components and the air-cooled components in the vehicle need to be identified first. The type of cooling to be employed depends on the cooling requirement as well as on the location of the component within the vehicle. Most of the power electronics and electric powertrain traction components use liquid cooling for compact packaging. In some cases, a liquid cooling system may already be available for other components in the vehicle, and it then becomes only an extension to add the electric drive components in the loop. The IC engine and automatic transmission are also liquid cooled. Traction batteries, especially those of NiMH types, and small electronics may be air cooled. However, advanced Li-ion batteries need to be liquid cooled. In determining the cooling requirements, the worst-case operating scenarios of the components have to be evaluated. These include steep grades, hill holding, extreme temperatures, and altitudes. The duty cycle of operation in the worst-case scenarios have to be evaluated to distinguish between transient and continuous cooling loads. The continuous and maximum temperature ratings of the powertrain and electronic components vary widely depending on the design and mission of the vehicle, and the type of component. In order to design for reliability and durability, the component cooling systems are often sized for peak operating conditions at maximum allowable temperatures. Power electronic components are highly sensitive to excessive temperatures compared to mechanical devices such as IC engines and transmission due to their much lower thermal mass.

Representative temperatures of powertrain and electronic components in a light-duty vehicle are given in Table 12.1 as a general reference. The allowable operating temperatures and liquid/airflow rates are typically specified by the component manufacturer. The performance characteristics provided by the manufacturer showing efficiencies at various loads are very useful in targeting the operating points through the vehicle supervisory controller. Efficiencies along with the power ratings give a good starting point to calculate the heat load to be removed by the cooling system.

The liquid-cooled system includes a heat exchanger to reject heat into the environment and a coolant pump for the forced circulation of the liquid coolant. Thermal analysis of the system yields the heat rejection, coolant mass flow rate, and pump size required. In the case of air-cooled systems, fans and blowers are sized based on the amount of heat removal required.

In a hybrid electric vehicle, multiple loops may be desired due to significantly different allowable operating temperatures. The number of coolant loops depends on the maximum component operating temperatures, coolant flow requirements, packaging of the component, and proximity to other components within the same loop. The coolant loop design objective is to minimize plumbing complexity, length, and restriction. The fill points and coolant reservoirs have to be incorporated within the coolant loops ensuring appropriate venting and degassing.

An example of a hybrid electric vehicle powertrain component cooling system is shown in Figure 12.3. The powertrain architecture has two electric machines and an IC engine that needs to be liquid cooled. The

TABLE 12.1

Example Temperatures in Vehicle Components

Component	Continuous	Maximum
IC engine and transmission	95°C	120°C
Traction electric motor	65°C	75°C
Traction inverter, DC/DC converter	65°C	75°C
Electronic controllers	65°C	75°C
Traction batteries	30°C	55°C

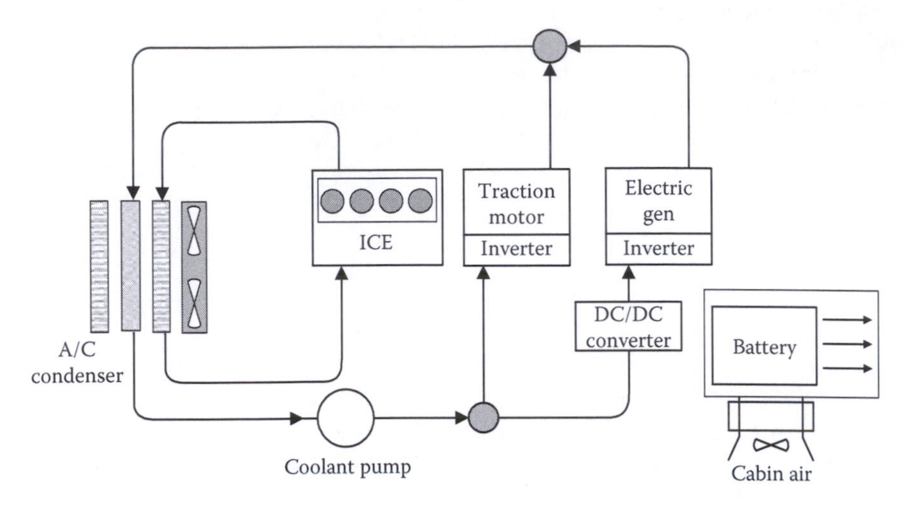

FIGURE 12.3
Cooling loops in a hybrid electric vehicle.

inverters for the electric machines and the DC/DC converter also need to be liquid cooled. The battery is air cooled using cabin air. The liquid cooling system is arranged in three loops: the first loop is for cooling the IC engine and transmission; the second loop is for the traction motor and its inverter; and the third loop is for the generator, its inverter, and the DC/DC converter. Temperatures are monitored using sensors in the coolant loop for controlling the speed of the coolant pump. Devices will have internal temperature monitoring for thermal management, but it is advisable to monitor coolant temperatures at least in inlet locations especially during the development stages.

Three heat exchangers are shown in Figure 12.3 stacked in series, which would be located in the front of the vehicle. The IC engine has a separate heat exchanger (i.e., radiator), while two electrical component cooling loops use one heat exchanger. The third heat exchanger is for the air-conditioning system of the vehicle, which is not shown in the figure. A series arrangement of the heat exchangers provides efficient packaging, but requires higher total fan power due to increased airside pressure loss, and shared fan control. In the series stacking, the lowest temperature heat exchanger must be placed first in the direction of airflow. Parallel stacking of heat exchangers maximizes the frontal area. Airflow into each heat exchanger can be optimally controlled by individual fan control. Forward facing heat exchangers can benefit greatly from dynamic air pressure created by vehicle motion, but adequate airflow is necessary at lower speeds. Lower vehicle speeds provide the more challenging cooling system design requirements due to higher loads for the devices and inadequate airflow.

References

1. M.J. Moran and H.N. Shapiro, *Fundamentals of Engineering Thermodynamics*, 6th edn., John Wiley & Sons, Inc., New York, 2008.
2. R. Stone and J.K. Ball, *Automotive Engineering Fundamentals*, SAE International, Warrendale, PA, 2004.

13

Hybrid Vehicle Control Strategy

The control of electric and hybrid electric vehicles entails translating the pedal input into power and torque commands for the powertrain propulsion components. In electric vehicles, the task is fairly straightforward since there is only one power transmission path. The control of the hybrid electric vehicle is much more complex since the powertrain is composed of multiple vehicle power sources. In either case, there should be a vehicle controller above the electric motor drive controller to oversee vehicle operation and to impose the operating limits of the components at the system level. This controller is referred to as the *vehicle master* or *supervisory controller*. The function of the vehicle supervisory controller is to receive drive demand information from the accelerator and brake pedals, and various other vehicle feedback signals to generate command signals for the various powertrain components ensuring that these are within the operating limits of the respective components. This chapter will primarily address the control strategy of hybrid electric vehicles. The control in electric vehicles is a subset of that control strategy and will not be addressed separately.

A hybrid electric vehicle powertrain comprises both electrical mechanical power transmission paths. The powertrain consists of multiple power sources and energy converters; the operation of each of these components needs to be coordinated. The powertrain components are arranged in various configurations depending on the vehicle architecture, although the most general ones are the series, parallel, and power-split architectures. The control problem is to manage the power flow through the electrical and mechanical transmission paths through the coordinated use of the multiple components with a certain objective, such as maximizing fuel efficiency. The control strategy developed for the purpose is implemented in the vehicle supervisory controller. The supervisory controller shields the driver from a complicated control problem and ensures that the driver demand is continuously and consistently satisfied.

The control strategy is essentially an algorithm that determines the operating points of the vehicle powertrain components. The primary objective of the control strategy is to ensure that the driver demand is met, while the powertrain system efficiency is optimized. The primary inputs to the supervisory controller are the vehicle speed and the pedal inputs. Secondary inputs of roadway, traffic, and GPS information may also be used in more advanced supervisory controllers. The outputs of the controller are the command signals for the powertrain components, such as electric motor or IC engine torque requests, and on/off commands for certain components.

The battery pack or energy storage system state of charge levels must also be maintained within levels that extend the life of the system. The control strategy determines when and how much of deep discharge of the energy storage system will be allowed to meet a driver demand. Moreover, vehicle kinetic energy can be recuperated and stored in the energy storage system using regenerative braking when vehicle brake commands are in effect from the driver.

Additional goals targeted by the control strategy algorithm include all or some of the following: enhancing powertrain efficiency and/or fuel economy, reducing emissions, and maintaining good drivability. Increasing fuel efficiency and reducing emissions are contending requirements; an optimization algorithm can be used to determine the best operating points of the IC engine. Drivability is also a key parameter for customer acceptance of hybrid vehicles; the controller must ensure that driver feel is almost the same as that of a conventional IC engine vehicle. The command changes for the various propulsion units must be carefully coordinated to maintain the high drivability quality of the vehicle. Drivability can be measured in terms of gear shifting and driveline vibrations.

13.1 Vehicle Supervisory Controller

Vehicle supervisory controller sits in the middle of a multilevel hierarchical structure for the control of the hybrid vehicle powertrain. At the top control level is the driver, who uses speed and position feedback to set the accelerator and brake pedal inputs. The supervisory controller accepts the driver commands and accordingly controls the various powertrain subsystems. The subsystems, such as the electric motor drive, energy storage device, IC engine, and starter/generator drive have their own controllers, but the operating set-points depend on the commands issued by the supervisory controller. The control strategy in the supervisory controller is implemented using feedback signals from all of the subsystem control modules. The supervisory controller communicates with the subsystems through a controller area network (CAN), which will be discussed in the next chapter. In addition to the CAN messages, the supervisory controller can receive data directly from sensors through A/D channels. The sensor data are not only used for control purposes, but can also be logged for performance analysis or for failure analysis if one or more subsystem components malfunction in the vehicle.

The supervisory controller executes commands based on the steady-state characteristics of the subsystems. These would include battery discharge

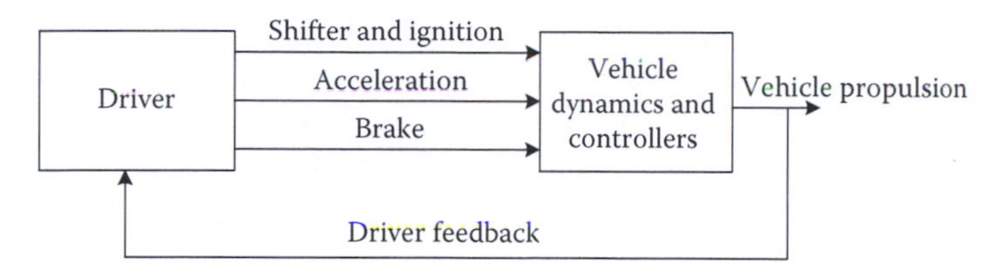

FIGURE 13.1
Driver interface with SCM and vehicle.

characteristics, motor/generator torque-speed characteristics, and engine torque/power-speed characteristics.

The driver interfaces with the supervisory controller through the inputs of ignition, transmission shifter, accelerator pedal, and brake pedal. The driver interface is shown in Figure 13.1. Once the vehicle is ready to be driven, the pedal inputs are converted from position to driver demand. The supervisory controller–driver interface makes the control system transparent by interpreting the accelerator and brake pedal depressions as propulsion/braking power and torque requests. The interpretations closely approximate what a driver would expect when operating a conventional vehicle. When both pedals are depressed, only the brake pedal request should be passed for safety.

The complexity of the control strategy depends on whether the hybrid architecture is compound, such as series–parallel or simple, such as series or parallel. The control strategy in a compound hybrid architecture can be implemented in a two-level structure: Mode selection strategy and modal control strategy. The mode selection strategy determines the powertrain mode of operation, such as series or parallel. The modal control strategy addresses the controls within a specific hybrid operating mode. In simple hybrid architectures, such as in series or parallel hybrids, a one-level algorithm is sufficient; the supervisory controller implements the modal control strategy for hybrid operation. The two levels of control strategy are discussed in separate sections of this chapter. The mode selection strategy applies to only compound hybrid architectures, while the modal selection strategy applies to both compound and simple hybrid architectures.

13.2 Mode Selection Strategy

The mode selection strategy discussed in the context of a compound hybrid architecture assumes that there is one IC engine and at least two electric machines in the powertrain of the vehicle. Up to four fundamental operating modes are possible in the hybrid vehicle, which are electric-only, series, parallel, and power-split. The mode selection strategies will be discussed in general since the constraints on the series mode of operation will be similar for a series hybrid electric vehicle to that of a vehicle that allows additional modes. Furthermore, the control strategies discussed are examples of a possible method of implementation. There can be various other mode selection strategies depending on the powertrain components available in the compound hybrid architecture, the relative sizes of the components, and the goal of the control strategy.

The first decision to be made by the supervisory controller is the mode of operation of the powertrain. The decision is made primarily based on the driver demand and the vehicle speed. The torque request issued by the driver through the acceleration pedal input is converted into a driver power demand through the multiplication of the torque request with the vehicle speed. A propulsion mode is selected based on the driver power demand and the vehicle speed.

In the electric-only mode, the stored energy in the battery or other energy storage system is used to drive one or more of the electric drive motors for propulsion. In the series mode, the energy stored in the fossil fuel is converted by the IC engine to mechanical form that runs an electric generator to produce electrical energy. The generated electricity is used by the electric drive motor to deliver traction power to the wheels. In the power-split mode, the IC engine provides traction power to the wheels through both the series electrical path and the direct mechanical path. The power from the IC engine is split through the mechanical and electrical transmission paths. For either series or split operation, the energy storage system can supply or absorb power depending on the power demand, the storage system state of charge (SoC), and IC engine efficiency considerations. In the parallel mode, the IC engine is operated up to its maximum capacity and the electric machines are operated in the motoring mode to provide propulsion power using stored energy from both the energy storage system and fossil fuel. The power-split mode differs from the parallel mode in that one of the electric machines is operated as a generator.

The mode selection for hybrid operation is governed by rule-based control laws that depend on the current and past states of the vehicle operating point, driver power demand, the vehicle speed, and the energy storage system SoC. The rules are generated based on the operating limits of the subsystems, models or performance maps of subsystems, designer experience, and heuristics. The concept of "load-leveling" for energy management is often used in rule-based strategies. In this type of controls, one of the energy converters, such as the IC engine, is primarily used to supply propulsion power, while a subsidiary subsystem component, such as an electric motor, is called in to assist the IC engine to shift the operating point of the latter to a different location of increased fuel economy and reduced emissions. The rule-based strategies can be further divided into two categories as follows [1]:

Deterministic rule-based methods: Deterministic rules are developed to determine the mode of hybrid operation and the conditions for switching from one

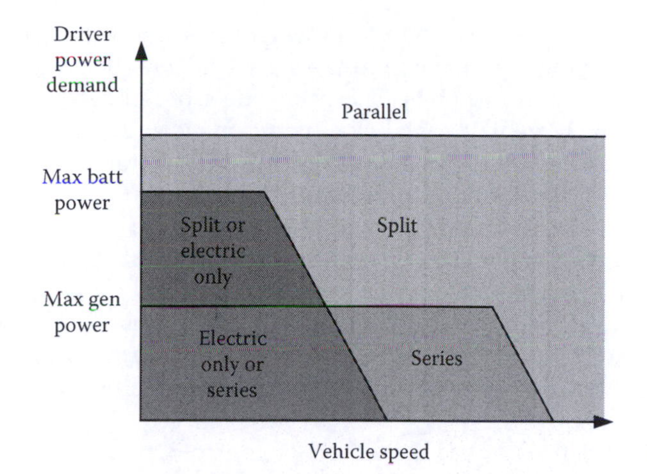

FIGURE 13.2
Hybrid vehicle modes of operation.

mode to another [2,3]. The most popular deterministic-rule-based strategy that is used in production hybrids is the power follower approach. The driver input is interpreted as a power request to decide the operating mode. Figure 13.2 shows an example qualitative dependence of the operating mode on the driver power demand (accelerator input) and the vehicle speed. In a parallel or series–parallel hybrid vehicle, at low power-demand levels only the electric motor is used below a minimum vehicle speed. If the power demand exceeds the minimum level, the engine turns on. In a parallel hybrid, the electric motor would assist the engine only if the power demand is greater than the maximum engine power. In a series–parallel or power-split architecture, the powertrain would be in a power-split mode after the electric-only mode. A series mode may be used between the electric-only and the split mode for low power demands. Only at very high power-demand levels will the parallel mode be used. The battery SoC is also maintained within a preset band by using the drive electric motor in the generator mode in a pure parallel hybrid or using the generator of a series–parallel hybrid vehicle. The main disadvantage of the method is that the overall powertrain efficiency is not optimized.

The supervisory controller may also use a state machine to implement the mode selection strategy. The state machine executes the transitions between the various hybrid modes based on driver demand, vehicle speed, and subsystem operating limits and faults, if any. The implementation of supervisory controller through state machines facilitates fault management. Although powertrain efficiency optimization is typically not addressed at the state transition levels, such optimization can be addressed at the modal control algorithms.

Fuzzy-rule-based methods: The hybrid powertrain is a nonlinear, multidimensional, time-varying plant, where fuzzy-logic-based mode selection strategy can be particularly useful [4–6]. The fuzzy-logic-based control system uses the driver demand, vehicle speed, SoC, component torque outputs, and estimated road load as inputs and a fuzzy-rule base to determine the operating mode of the vehicle. The main advantage of the fuzzy-rule base compared to a deterministic-rule base is its robustness toward imprecise measurements and parameter variations. A well-designed fuzzy-rule base can give the suboptimal performance of the powertrain in real time. The fuzzy-rule base includes weighted parameters that can be adapted in real time depending on the driving condition. A fuzzy-rule-based control algorithm can also be effectively utilized to develop a modal control strategy within a hybrid operating mode.

The deterministic rule-based method of control strategy is further elaborated with the help of series–parallel hybrid electric vehicle architectures. Series–parallel hybrid architectures with two electric machines and one IC engine can be implemented either using the mechanical device of planetary gear set or using electronic controls. The mode selection strategies for the two types of series–parallel hybrid vehicles appear below.

13.2.1 Mechanical Power-Split Hybrid Modes

The mechanically enabled power-split architecture uses a planetary or epicyclic gear set without any driveline clutches. The concept of power-split using planetary gear set was first introduced by TRW [7,8]. The non-shifting transmission in a planetary gear set offers continuously variable transmission (CVT)-like performance while enabling the mounting of three powertrain components in a single driveline. The mechanical power-split architecture has rapidly become the design choice for hybrid sedans, light trucks, and SUVs. The mechanical power-split design offers compactness and ease of assembly with the motor, generator, and IC engine mounted in a single driveline.

A series–parallel power-split powertrain architecture using a planetary gear set similar to that used in the Toyota Prius is shown in Figure 13.3. As was presented earlier in Section 11.2.1, the planetary gear set has three gears, the ring, the sun, and the planet carrier. The output shaft of the IC engine is connected to the planet carrier of the planetary gear set. The mechanical power is transferred to the final drive and the wheels from the IC engine indirectly through the ring gear. There is no direct coupling between the IC engine and the final drive. The generator is connected to the sun gear; the motor along with the wheels is connected to the ring gear. The IC engine power can be split and part of it be transferred to the final drive, and the wheels through the sun gear and the electrical power transmission path. The IC engine drives the generator which either charges the battery or supplies electrical power to the motor or does both. The generator operation

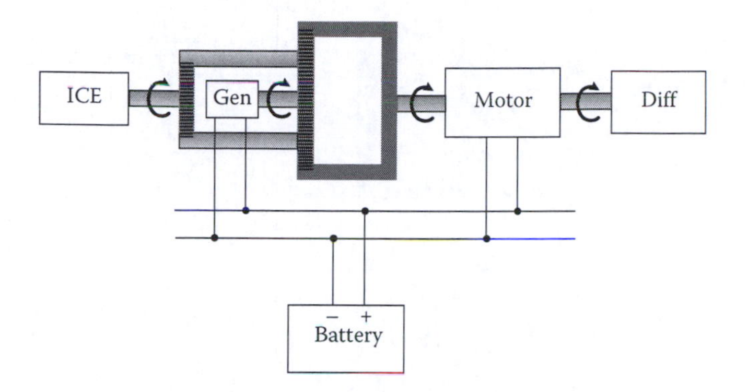

FIGURE 13.3
Series–parallel HEV powertrain component arrangements with a planetary gear set.

develops the reaction torque against the planet gear to which the shaft of the IC engine is coupled. The motor can add torque to the ring gear to provide additional power to the wheels. The generator can also be operated in the motoring mode during peak acceleration demand to add torque to the planet gear which then gets transmitted to the wheels through the ring gear. Both the electric machines can act either in the motoring mode or in the generating mode. During vehicle braking, the motor acts as a generator to capture regenerative braking energy, which is converted to electrical energy and stored in the battery. The several modes of operation possible through the planetary gear coupling and dual modes of the electric machines are described in further detail [9].

The speed–torque relationships among the ring, sun, and planet carrier gears discussed earlier in Section 11.2.1 apply to the hybrid powertrain architecture presented here, and will be used in this section. The power relationship in the power-split hybrid vehicle from Equation 11.13 is

$$P_{engine} = P_{engine\ to\ ring} + P_{engine\ to\ sun}$$

$$\Rightarrow T_p\omega_p = \frac{r_r\omega_r}{r_r + r_s}T_p + \frac{r_s\omega_s}{r_r + r_s}T_p \tag{13.1}$$

The rotational directions shown in Figure 13.3 for the three gears connected to the three powertrain components will be assumed to be the positive direction of rotation.

13.2.1.1 Electric Only (Low Speeds, Reverse, Battery Charging)

In this mode, the IC engine is off and the planet carrier is stationary, i.e., $\omega_p = 0$. The starter/generator rotates freely under a no-load condition in the

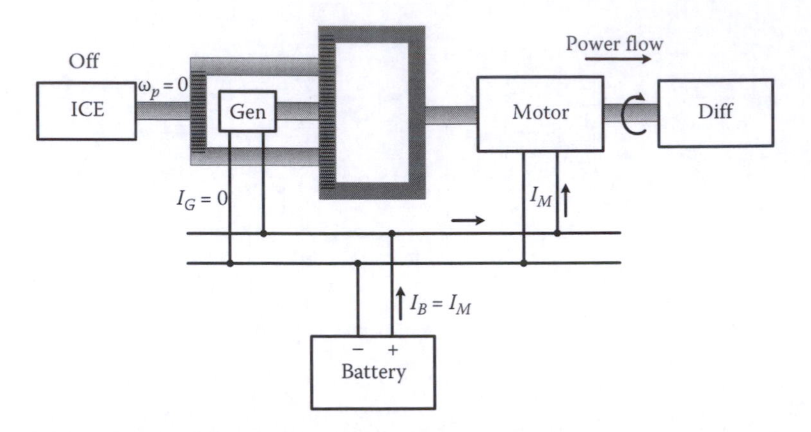

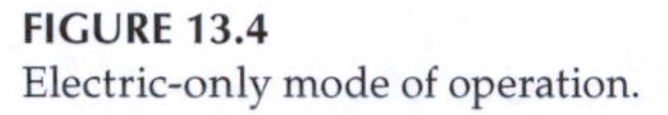

FIGURE 13.4
Electric-only mode of operation.

reverse direction. The speed relation between the starter/generator and the electric motor is given by (from Equation 11.11)

$$r_r \omega_r = -r_s \omega_s$$

The electric motor supplies the traction power to the wheels being driven by the battery energy as shown in Figure 13.4. No power is transmitted through the planetary gear set. The accelerator position (power demand) and vehicle speed determine the motor current set point.

$$I_{motor} \propto T_{motor} = \frac{P_{motor}}{\omega_r} \tag{13.2}$$

The batteries can be recharged using the regeneration control strategy with the electric machine operated in the generator mode when the vehicle is coasting or braking and the engine turned off.

13.2.1.2 Engine Starting (Low Speeds)

The starter/generator operates in the motoring mode to crank the engine. The motoring mode is enabled with a positive torque command while it is rotating in the reverse direction, causing the power to flow from the battery to the starter/generator to the IC engine. The vehicle may be moving during engine starting with the traction power provided by the electric motor. The source of energy during engine starting is the battery.

13.2.1.3 Parallel Mode (Heavy Acceleration)

The powertrain operates in the parallel mode for heavy acceleration which can be demanded during vehicle launch. The starter/generator and motor

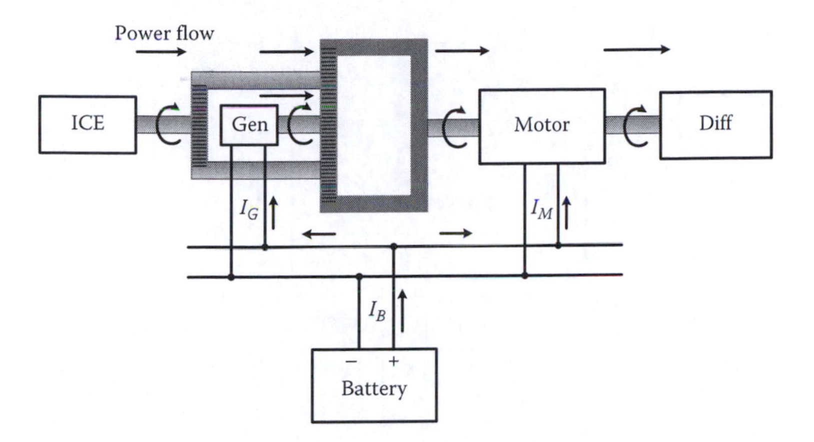

FIGURE 13.5
Parallel operation of electrical and mechanical powertrain components.

both assist the IC engine using battery power. The power flow in the parallel mode is shown in Figure 13.5.

The starter/generator power using Equation 13.1 is given by

$$P_{sun\ to\ ring} = \frac{r_s \omega_s}{r_r + r_s} T_p, \quad \omega_s < 0 \tag{13.3}$$

The sun gear rotates in the reverse direction, i.e., counterclockwise looking into the generator shaft from the gear, which makes $P_{sun\ to\ ring} < 0$. The reverse rotation of the sun gear increases IC engine torque, enabling the starter/generator to deliver motoring torque to the final drive. The starter/generator input is electrical power from the battery.

13.2.1.4 Power-Split Mode (Cruise, Light Acceleration)

In this mode, the IC engine power is split between the starter/generator and the motor shaft as shown in Figure 13.6. The motor assists the engine by adding more torque to the differential input shaft of the final drive. At the same time, some engine power may be diverted to the sun gear to run the starter/generator in the generating mode to charge the battery. The starter/generator is rotating in the forward direction. The IC engine power flow to the starter/generator, and hence to the battery, is regulated by controlling the IC engine speed, since the percentage of engine torque transferred to the starter/generator is mechanically fixed. The battery power required determines the starter/generator current, which in turn sets the starter/generator torque. The IC engine speed control will be discussed further later.

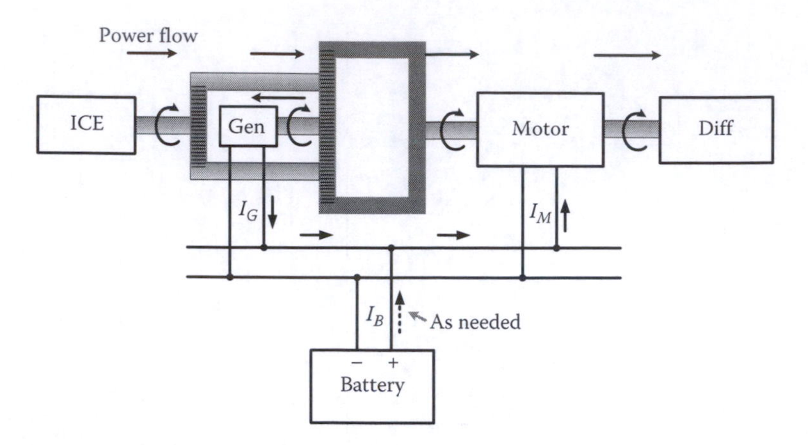

FIGURE 13.6
Power-split mode of operation.

$$I_{gen} \rightarrow T_s = \frac{r_s}{r_r + r_s} T_p \tag{13.4}$$

This sun gear torque in turn fixes the engine torque T_p; therefore, the IC engine must be speed controlled.

13.2.1.5 Engine-Brake Mode (Driver Selectable Mode)

In this mode, the IC engine is idling and absorbs some power. The motor is rotating freely ($I_{motor} = 0$). The starter/generator is rotating in the reverse direction and generating. The power flow in this mode is shown in Figure 13.7.

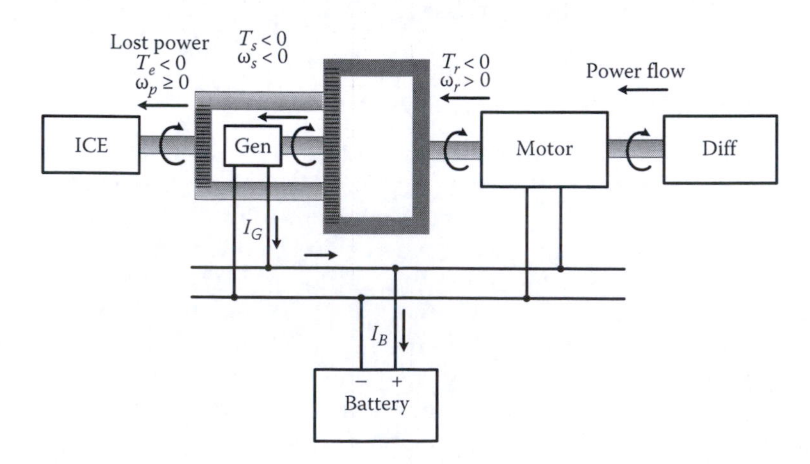

FIGURE 13.7
Engine-brake mode of operation.

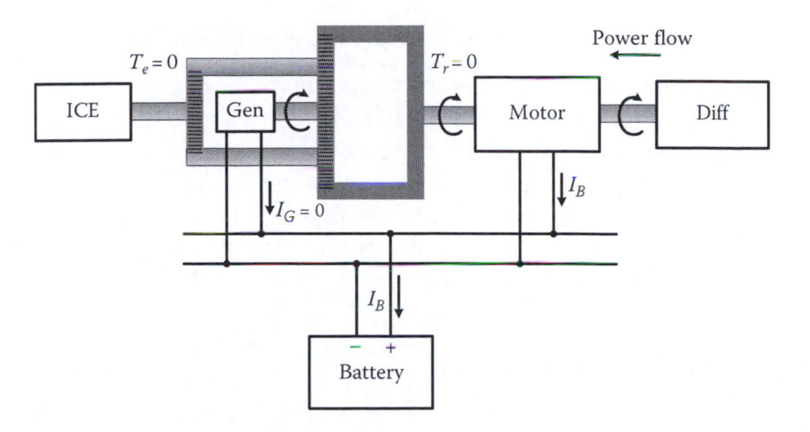

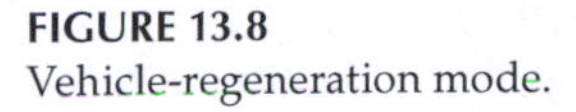
FIGURE 13.8
Vehicle-regeneration mode.

13.2.1.6 Regeneration Mode (Vehicle Braking)

In this mode, the generator is rotating freely ($I_g = 0$). The motor is being driven by the kinetic energy of the vehicle and is regenerating delivering energy to the battery. The power flow in this mode is shown in Figure 13.8.

13.2.2 Series–Parallel 2 × 2 Hybrid Modes

The series–parallel 2 × 2 architecture is an electronically controlled power-split architecture (Figure 13.9). The IC engine is coupled to the front wheels through a clutch and transmission while the starter/generator is coupled to an output shaft of the IC engine. The electric motor powers the rear wheels providing a four-wheel-drive capability. The architecture allows for great flexibility in the modes of operation of the hybrid electric vehicle.

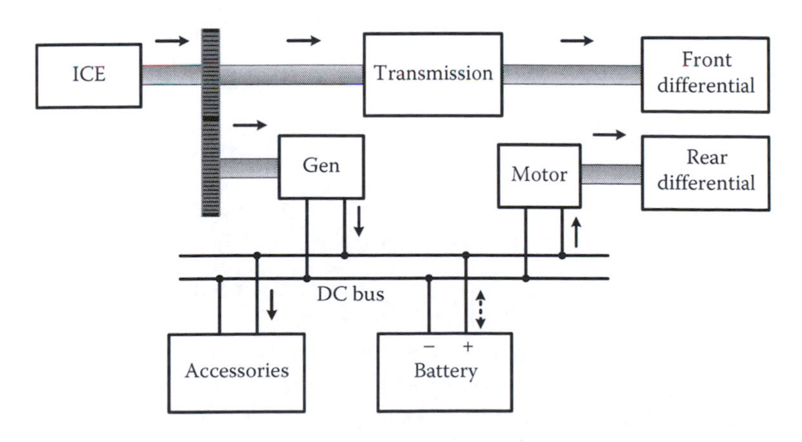

FIGURE 13.9
Series–parallel 2 × 2 architecture.

13.2.2.1 Electric-Only (Low Speeds, Reverse, Battery Charging)

Keeping the IC engine off and leaving the clutch disengaged, the vehicle can be operated as an electric vehicle by using the rear traction motor. The stored energy in the battery or any other type of storage device is used to power the electric motor which supplies traction power to the rear wheels through the rear differential. The batteries can be recharged using the regeneration control strategy with the electric motor operated in the generator mode when the vehicle is coasting or braking and the engine turned off.

13.2.2.2 Series Mode (Low Speeds)

The vehicle can also be operated as a series hybrid within the limitations of the electric generator especially in urban driving conditions when the driver power demand is relatively low. The total IC engine torque is delivered to the generator which supplies the electric motor with the storage system acting as the buffer. The electric motor delivers traction power to the rear wheels. The transmission is held in neutral in the series mode.

13.2.2.3 Power-Split Mode (Cruise, Light Acceleration)

When the speed reaches a threshold or the power requested is more than that can be provided in the series mode, the clutch is engaged and the IC engine provides propulsion power through the transmission. In this mode of operation, the IC engine speed is set by the power-split control mode algorithm which will be discussed later. Power from the IC engine is delivered to the front wheels and the rear wheels through the mechanical transmission path and the electrical transmission path, respectively. Using both the electric machines this power-split operation is similar to the mechanical power split that has a planetary gear set. The power-split ratio between the mechanical and electrical transmissions can be arbitrarily chosen by the generator loading in contrast to the mechanical power-split hybrid. The power-split proportions may be chosen to optimize energy storage system usage and to maintain a desirable front-to-rear traction force ratio, subject to transient smoothing considerations.

13.2.2.4 Parallel Mode (Heavy Acceleration)

In the parallel mode which is used during high driver demands, the IC engine and the two electric machines all work in parallel to deliver traction power to the wheels. The electric generator is operated in the motoring mode, and the IC engine and electric generator torque blended at the IC engine-generator coupling is delivered to the front wheels. The electric motor simultaneously delivers traction power to the rear wheels. The energy for both the electric machines comes from the energy storage system; this is a charge-depleting operation, and hence, is not sustainable for a long duration. This can be used as a boost mode of the vehicle utilized during heavy acceleration demands.

13.3 Modal Control Strategies

The modal control algorithms use the driver demand and system feedback inputs to satisfy the demand while optimizing the driveline system efficiency and minimizing the emissions. The energy storage SoC is also maintained within a preset band to meet the performance requirements without damaging the system. Control laws within the modal control strategy determine the best reference torques for the energy converters and the best gear ratio for the transmission. The control laws can be generated based on an optimization algorithm where a cost function representing fuel economy, emissions, motor efficiency, or battery life is minimized. A global optimization to achieve the best fuel efficiency depends on an a priori knowledge of power demand and driving conditions, which is only possible in simulation. A global optimization based on fixed driving cycles can be used to determine the rules for real-time power-flow management. In real driving scenarios, a real-time optimization-based control strategy can be developed [10,11]. An instantaneous cost function defined in terms of real-time variables is used for such an optimization strategy.

The powertrain subsystems place constraints on the maximum continuous and peak power that can be provided in the series, parallel, and power-split modes. However, the modal control algorithms share common calculations to determine the maximum power that the IC engine, motor, and generator can provide. The calculations are based on the device ratings, temperature, DC-link bus voltage, vehicle speed, and other parameters that determine the limits on the operating points for each subsystem. For example, the motor power limit can be reduced as a function of bus voltage. The bus voltage depends on the energy storage SoC. The objective of motor power reduction for reduced voltages is to ensure acceptable drivability by eliminating sudden reduction in acceleration if the electric motor enters a shutdown mode when the battery runs out of energy. The power limit for the electric motor also depends on the temperature. The motor ratings are derated for the elevated temperatures of operation.

Series, parallel, and power-split modes require different control algorithms. The optimization techniques, if used, for the different modal control strategies are also different. The following discusses modal control strategies along with examples of optimization techniques. It must be mentioned that the modal control strategies discussed in this section are only examples; there can be various other ways of implementing a modal control strategy depending on the size of the powertrain components available, and the desired optimization strategy.

13.3.1 Series Control

In a series hybrid or in the series mode of a multimode hybrid, the electric motor provides all of the traction power demand measured off of the accelerator pedal. The IC engine, the generator, and the energy storage system must ensure that sufficient electrical power is available for the electric motor at all times to meet the driver demand. In urban driving, the power demand varies greatly due to variations in vehicle speed and frequent start/stop operation; the hybrid electric vehicle undergoes a highly variable battery discharge/charge profile. The average power required by the motor is relatively low, compared to the peak transient power requirements during the peak acceleration demands.

The IC engine is buffered from the vehicle load variations by a passive control algorithm of the energy storage system charge which is independent of the active vehicle powertrain control. The energy storage system control is based on an on/off control method, which is also known as the thermostat control of the IC engine. The criterion for on/off control can be set by the state of charge (SoC) limits of the battery. The maximum and minimum SoC limits are set such that the driver demand can be met by the electric motor at all times. The thermostat control of the engine to maintain the SoC within specified limits is shown in Figure 13.10.

The objective in the series mode is to maintain the IC engine speed and torque at the most efficient operating point. An example selection of the IC

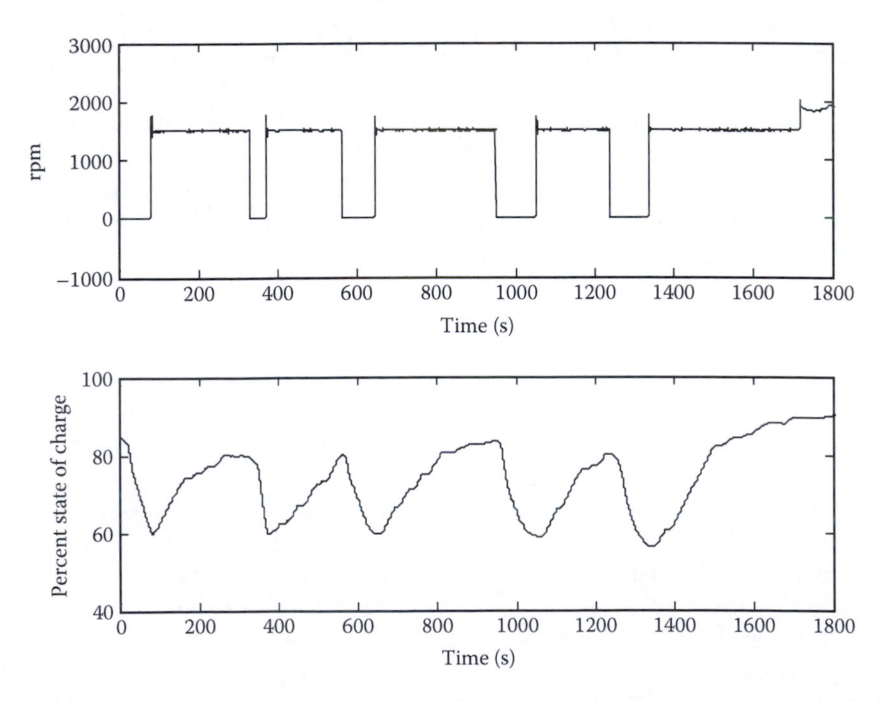

FIGURE 13.10
IC engine usage and battery SoC.

engine "on" operating point in series mode is explained with the help of the torque-speed characteristics of the engine and the generator. The "on" operating point of the IC engine may be chosen based on fuel economy, emissions, and the characteristics of the electric generator. If only high fuel economy is targeted, the engine should operate in a region of low brake-specific fuel consumption (BSFC). The baseline engine maps including the BSFC and BSNOx data of a 1.9 L diesel engine are shown in Figures 13.11 and 13.12 [12]. The continuous and peak torque curves of a permanent magnet electric generator reflected through engine-generator coupling gear is superimposed on both plots. The engine maps show that the regions of operation with low fuel consumption also yield high NO_x generation. Considering the generator characteristics, the target operation should be at the point where the generator goes from torque to power limited along the continuous generation threshold where its efficiency is also relatively high. The generator is torque limited below the corner speed and not capable of its maximum possible continuous power generation; operation in this region will increase the time that the IC engine must run to recharge the energy storage system in series mode before it is shut down.

Selecting the "on" operating point near the corner point of the PM generator will cause a slight increase in the BSFC from the minimum of 0.20313 kg/(kW h), but it works out quite favorably in terms of emissions. The

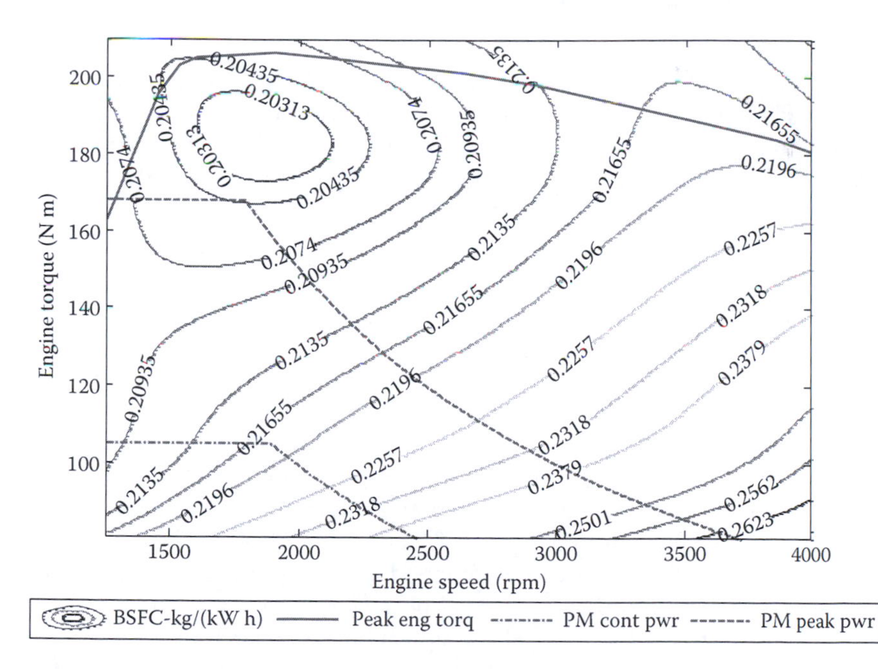

FIGURE 13.11
Brake-specific fuel consumptions for 1.9 L diesel engine. (From Hasan, S.M.N. et al., *IEEE Ind. Appl. Mag.*, 16(2), March–April 2010. With permission.)

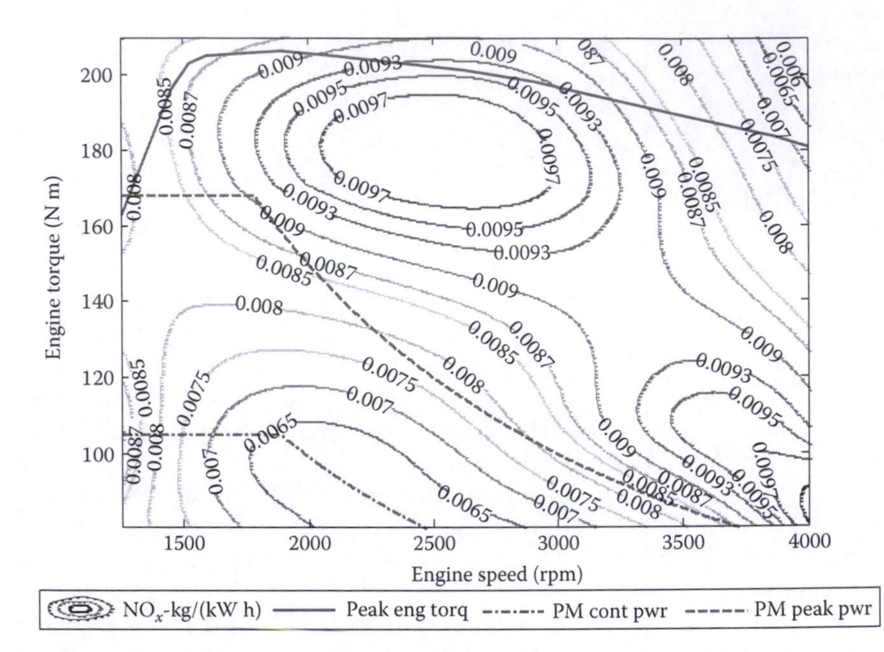

FIGURE 13.12
Brake-specific NO_x emissions for 1.9 L diesel engine. (From Hasan, S.M.N. et al., *IEEE Ind. Appl. Mag.*, 16(2), March–April 2010. With permission.)

NO_x emissions around the corner point are significantly below the maximum of 0.0097 kg/(kW h). The optimum "on" operating point is in the corner region where it produces relatively low emissions, and exhibits reasonable efficiency while allowing the generator to run close to its maximum continuous power rating (37.5 Nm, 4750 rpm) [12].

13.3.2 Parallel Control

The parallel mode control algorithm delivers power to wheels through both the electric machine and the engine to meet the driver demand. The torque from the two devices may be blended before being delivered to a set of wheels or could be separately delivered to the front and rear wheels. In either case, the control algorithm must additionally ensure that the energy storage SoC is maintained within a band and a compromise operating point between fuel efficiency and tailpipe emissions is selected for the IC engine.

The IC engine operating speed is matched with the vehicle speed through the transmission gear ratio. The transmission gear-shifting strategy is an essential component in the parallel control strategy. Following the gear selection, the control strategy determines the torque commands for the engine and the electric motor to meet the driver demand and manage the energy storage. The goal of the shifting strategy is to match the vehicle speed with

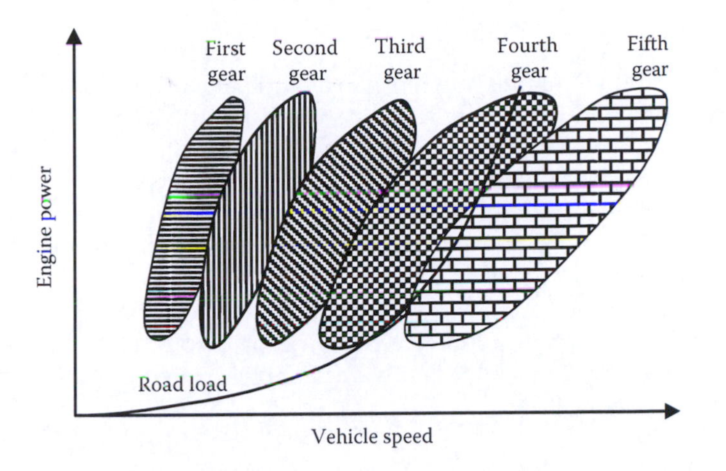

FIGURE 13.13
Regions of higher IC engine efficiency reflected through each gear ratio.

the IC engine speed at its best operating region. There must also be sufficient hysteresis built into the strategy to prevent frequent shifting for an acceptable drivability.

In an IC engine powertrain, the regions of higher IC engine efficiency in the engine power–vehicle speed plane are smaller for lower gears than it is for higher gears. A qualitative plot showing regions of higher engine efficiency operation reflected through each gear ratio as a function of vehicle speed is shown in Figure 13.13 [13]. The plot shows that the overlap regions decrease significantly for lower gears, which makes it more difficult to prevent frequent shifting at lower vehicle speeds. A sufficiently large hysteresis band even at the cost of sacrificing efficiency is needed at lower speeds to prevent frequent shifting for acceptable drivability. At higher vehicle speeds, the larger hysteresis band poses much less of a problem since there is a lot of overlap. The road-load characteristic for a 0% grade roadway is shown in the figure to emphasize that engine efficiency has to be sacrificed at lower speeds.

The IC engine and the electric motor have their own high-efficiency operating regions. A parallel hybrid vehicle will operate in electric-only mode for very low speeds and low power. The IC engine alone is used for propulsion above the minimum power-demand level in most parallel control strategies. When the power demand exceeds the capacity of the IC engine, the electric motor is called in to assist the IC engine in the pure parallel mode. In the parallel mode there is not much of an opportunity to optimize the algorithm for fuel-efficient operation, since the operating region of the IC engine is set by the vehicle speed and transmission gear, while the electric motor speed is directly related to vehicle speed. In a series–parallel hybrid vehicle, the parallel mode of operation can be used as the boost mode for the vehicle where meeting the high power demand becomes critical.

13.3.3 Series–Parallel Control

The series–parallel architecture has two electric machines and an IC engine; the power flow can be through the series path, the parallel path, or the combination of the two, which is known as the power-split path. The various power flow path options available in this architecture make the control strategy design complicated, but at the same time they provide the flexibility to choose one or more of the available power flow paths to maximize the vehicle efficiency for the given driving condition.

The series–parallel architecture with a power-split path can be implemented using a mechanical device coupling or through an electronic coupling of the powertrain components. The control strategies of the two types are discussed below [9].

13.3.3.1 Mechanical Power-Split IC Engine Control

The fuel flow rate is determined by the power requirement, which corresponds to the accelerator pedal input. The load torque can be controlled by the generator through its torque control. The control equation presented earlier in Equation 13.4 is repeated here for convenience

$$I_{gen} \rightarrow T_{gen} = T_s = \frac{r_s}{r_r + r_s} T_p \quad (T_p = \text{engine torque})$$

where generator torque T_{gen} is the same as the sun gear torque. It has been assumed that the engine power requirement is constant; also that all power comes from the IC engine, and that there are no transmission losses. The primary input-output variables for the IC engine operation in shown in Figure 13.14. The constant engine power curve is shown in Figure 13.15. The starter/generator command I_{gen} can be manipulated to operate the engine at the best operating point along the curve. When I_{gen} is set at a higher level, the IC engine torque T_{ICE} (which is the same as the planet gear torque T_p) is also high; consequently, more power is delivered to the wheels directly, and less power to the starter/generator. The lower T_{ICE} operating point for a smaller I_{gen} command represents more power to the starter/generator and less power to the wheels from the IC engine; the electric motor torque command can be increased to maintain the same power to be delivered to the wheels, if required, but with increased power flow through the electric power transmission path. Therefore, the power-split ratio between the sun gear (starter/generator) and the ring gear (wheels) can be adjusted by varying the I_{gen} command.

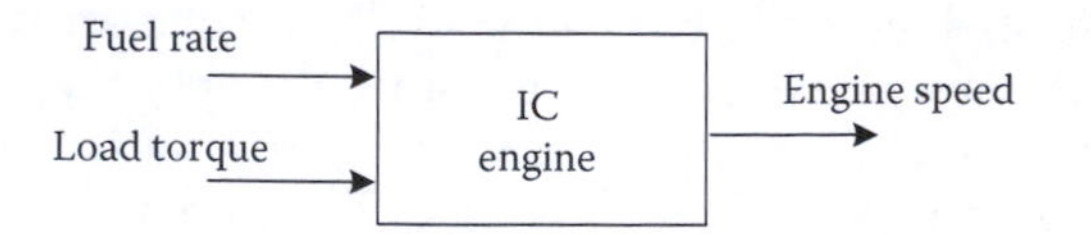

FIGURE 13.14
IC engine operation.

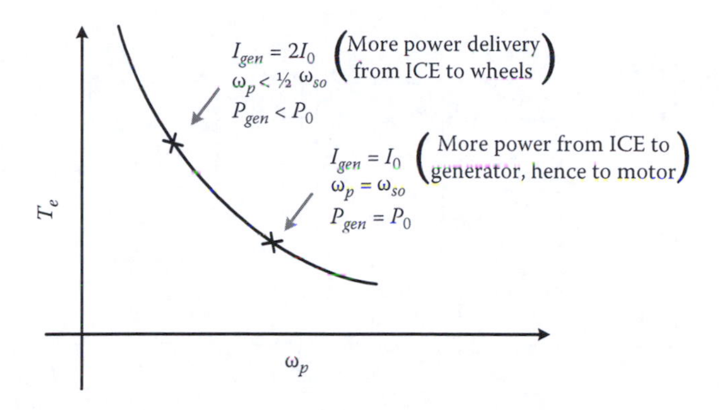

FIGURE 13.15
Constant IC engine power curve.

A possible criterion to choose the operating point is the maximization of engine fuel efficiency, which is given by

$$\eta_{fuel} = \frac{\text{Engine power}}{\text{Fuel flow rate} \times \text{HHV}}$$

The constant efficiency curves on the engine performance map are shown in Figure 13.16. The best operating point is along a given curve of constant power where that curve is a tangent to a constant efficiency curve. The

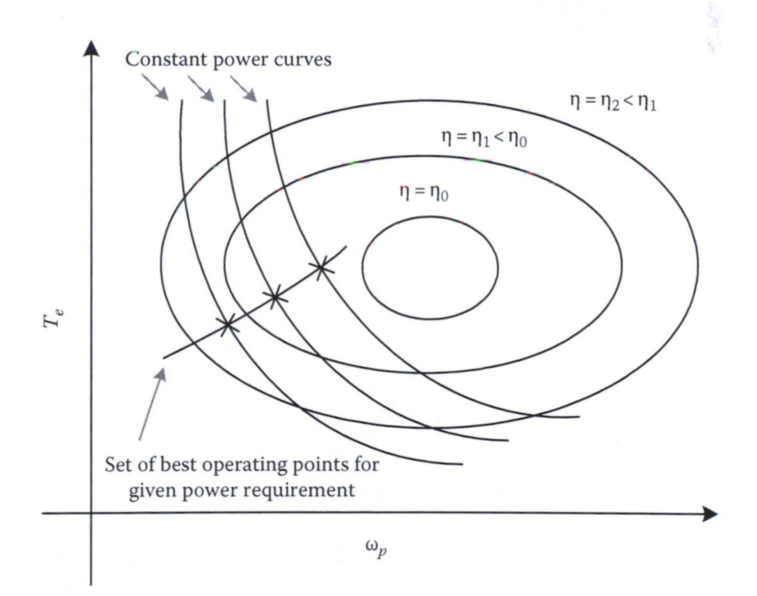

FIGURE 13.16
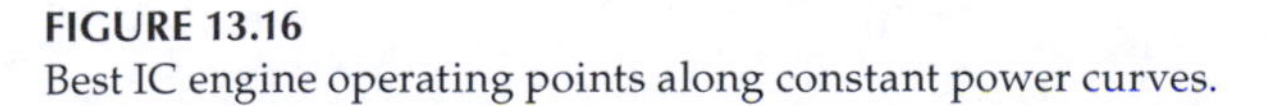
Best IC engine operating points along constant power curves.

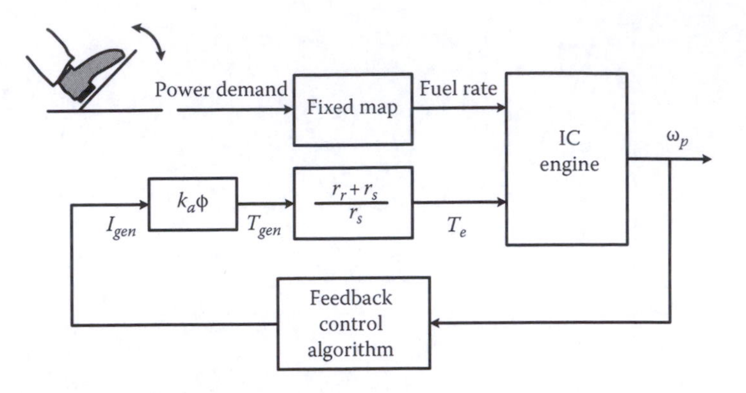

FIGURE 13.17
IC engine and starter/generator control system.

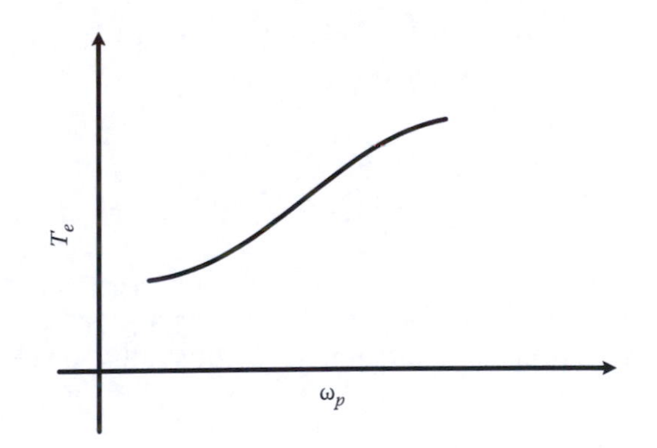

FIGURE 13.18
Best IC engine operating points.

feedback control system that can be used to keep the IC engine operating along the set of optimal (i.e., the most engine fuel-efficient) points is shown in Figure 13.17. The IC engine speed and torque is maintained at the desired best operating point by controlling the starter/generator current. The power demand translated from the accelerator pedal input is used to control the fuel flow rate. The desired operating point trajectory is shown in Figure 13.18.

13.3.3.2 Series–Parallel 2×2 Control

In the hybrid architecture shown in Figure 13.9, the electric generator is coupled to the IC engine through a fixed-gear ratio while the electric motor is coupled to the wheels through another fixed-gear ratio. The torque relationships between the powertrain components are derived based on the electric generator and motor speeds reflected to the IC engine side through these fixed-gear ratios. The speed relationships when the transmission gear is engaged are

$$\omega_{gen} = \omega_{ICE}$$

$$\omega_{motor} = \omega_{vehicle} = n_{TX} \cdot \omega_{ICE}$$

where

ω_{gen}, ω_{ICE}, ω_{motor}, and $\omega_{vehicle}$ are the electric generator, IC engine, electric motor, and final drive speeds, respectively
n_{TX} is the transmission gear ratio

Since the final drive and electric motor speeds are the same when the transmission is engaged, it follows that

$$\omega_{motor} = n_{TX} \cdot \omega_{gen}$$

Also, assume that all of the generator power is transmitted to the wheels through the electric motor using the electric power transmission path. Neglecting all the drivetrain losses, we have

$$T_{motor} = \frac{T_{gen}}{n_{TX}}$$

The traction torque from the hybrid powertrain is

$$T_{TR} = \frac{T_{TX}}{n_{TX}} + T_{motor} = \frac{T_{TX}}{n_{TX}} + \frac{T_{gen}}{n_{TX}}$$

$$\Rightarrow T_{TR} = \frac{1}{n_{TX}} \cdot (T_{TX} + T_{gen}) \tag{13.5}$$

However, the torque produced by the IC engine is delivered to the generator and the transmission giving

$$T_{ICE} = T_{TX} + T_{gen} = n_{TX} \cdot T_{TR} \tag{13.6}$$

Using Equation 13.5, the IC engine torque given by Equation 13.6 is independent of the generator torque. It can be concluded from the analysis that the IC engine setpoint is determined by the vehicle speed $\omega_{vehicle}$ and the transmission gear ratio n_{TX}. The control strategy is shown graphically in Figure 13.19. The proportions of torque delivered to the wheels in the power-split mode through the mechanical and electrical power transmission path may be arbitrarily determined by the generator loading. The proportions for power split and torque split are equal. In the extreme cases of power-split mode, the vehicle operation reverts to using only the mechanical power transmission path or the electrical power transmission path. When the generator torque command

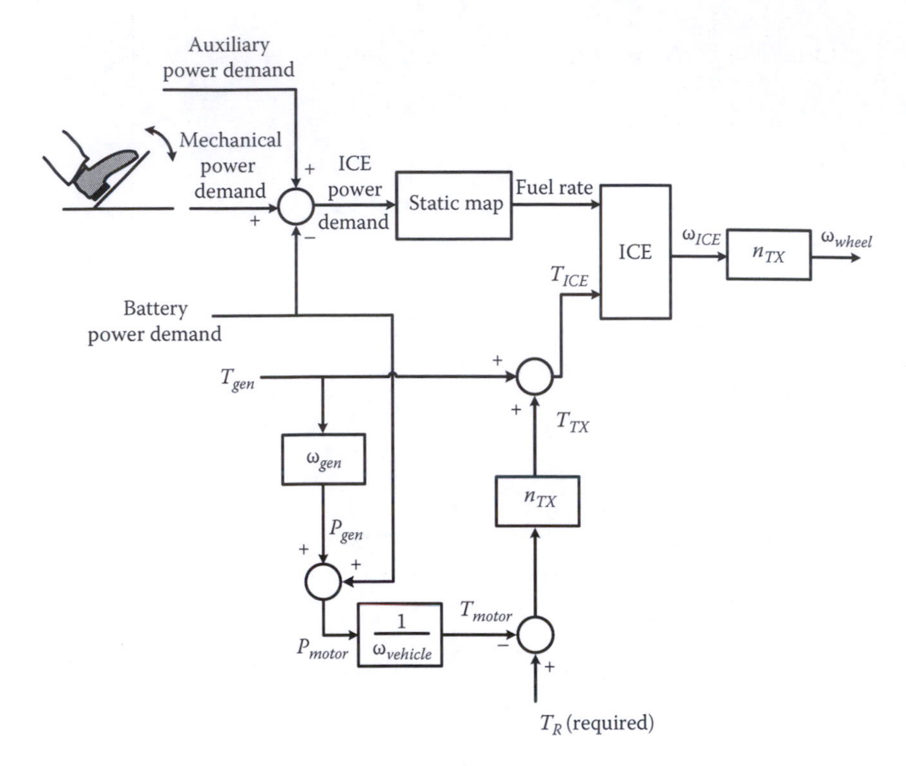

FIGURE 13.19
IC Engine control strategy for series–parallel 2 × 2 hybrid vehicle.

is zero, i.e., $T_{gen} = 0$, the IC engine alone delivers traction power to the wheels; when the IC engine is loading only the generator, i.e., $T_{gen} = T_{ICE}$, the vehicle is in the series operation mode, and the transmission can be disengaged.

13.3.4 Energy Storage System Control

The energy storage SoC is maintained by recharging the batteries through either a generator coupled to the engine or running the electric motor in the generating mode if the vehicle has only one electric machine. For a series or series–parallel hybrid architecture, the IC engine power demand has to be increased by the amount of electrical power needed to recharge the energy storage. These vehicles have at least two electric machines in the powertrain; the energy storage devices can be recharged while delivering propulsion power through both the IC engine and the electric motor simultaneously. For a parallel vehicle, the parallel mode of operation must be ceased before the energy storage device can be recharged if there is no separate generator present in the hybrid powertrain.

The above discussion leads to the fact that the power demand from the IC engine in a hybrid vehicle depends not only on the driver demand for vehicle propulsion, but also on the energy storage system charge control algorithm. The thermostat control of the IC engine for a series vehicle, discussed earlier in Section 13.3.1, is a method of energy storage system charge control where the vehicle powertrain alternates between the series mode and the electric-only mode for the "on" operating point and "off" condition of the IC engine, respectively depending on the SoC limits set in the control algorithm. With wider SoC limits, the battery pack reaches a lower SoC level before the IC engine restarts for the battery recharging using the starter/generator. A narrower band results in a lighter use of the battery pack and a more frequent on/off operation of the IC engine. The cycling period of the IC engine also depends on the driver power demand. A variable set of SoC limits depending on the driver power demand can be used to attain a compromise between the cycling of IC engine and the use of the battery pack.

Let us now discuss the possible variations in power demand that can be used in a series–parallel vehicle for better IC engine efficiency by energy buffering the energy storage system while delivering the same traction power at the wheels. We will consider the series–parallel 2 × 2 hybrid architecture, shown in Figure 13.9, for this discussion. The thermostat control of the IC engine can be applied for mode selection between series and electric-only for lower driver demands. For driver demands below the maximum generator power, the SoC limits essentially determine the vehicle mode. This is the region shown in the lower-left section of Figure 13.2. For higher driver demand levels, the power-split strategy in the series–parallel 2 × 2 hybrid can be designed to operate the IC engine at an optimum operating point given the speed of the IC engine and the SoC of the energy storage system. In the power-split mode, both the electric machines are available for operation to assist in satisfying the driver demand and maintaining the energy storage SoC within the acceptable limits. The best operating power level for the IC engine is first selected based on the vehicle speed, the transmission gear ratio, and the driver demand. If the driver power demand matches the best IC engine operating power level, then the IC engine alone can be used to meet the driver demand and zero-torque commands are dispatched to the electric generator and the electric motor. If the driver power demand is larger than the best power level for the IC engine, then the electric motor can be called in to provide additional power to the wheels. If the drive demand is less than the best IC engine power level, then the electric generator can be used to recharge the energy storage thereby providing additional load to the IC engine. This example of power-split control strategy for following the best IC engine operating power level is shown graphically in Figure 13.20.

13.3.5 Regeneration Control

The regeneration control algorithm allows the recovery of a part of the propulsion energy converted into kinetic energy through the electric machine

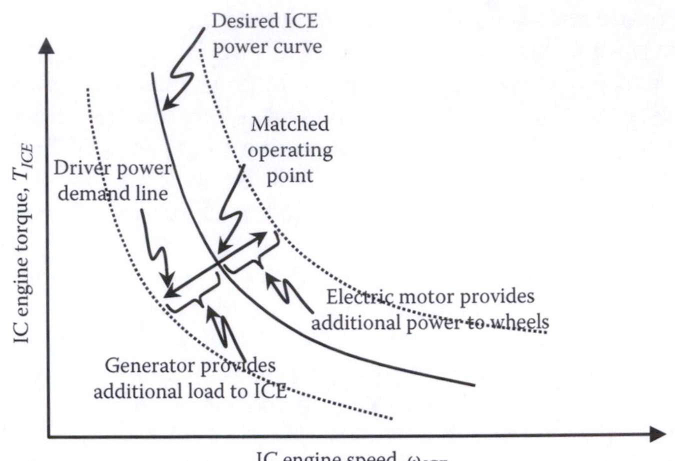

FIGURE 13.20
Power-split strategy for a given speed when energy storage system is within limits.

instead of being wasted in the friction brakes of the vehicle. The electric machine runs as a generator in this mode following a negative torque request command from the supervisory controller converting mechanical energy into electrical energy and storing it in the energy storage device.

The regenerative control algorithm can be designed based on three components: the maximum amount of regeneration torque, the rate of change of torque command, and the relationship of the pedal position to the torque command [13]. The maximum amount of regenerative torque which is typically interpreted as negative torque is limited by the capacity of the electric machine and its power electronic drive. The maximum allowed regenerative torque can be made a function of the vehicle speed to meet the drivability requirements. A rate limit on the torque command change can be imposed for driver comfort and acceptable drivability. A limit on the jerk is necessary for driver comfort during changes in negative torque command. The pedal position input has to be mapped to a driver request curve based on the maximum torque and the maximum jerk.

In a hybrid vehicle, the regenerative algorithm could implement full regenerative braking where driver inputs from both the acceleration pedal as well as the brake pedal are utilized in determining the negative torque request for the electric machine. For safety concerns, the vehicle must be fitted with friction brakes even though the electric machines are capable of providing braking action. The regenerative braking causes drivability concerns at very low vehicle speeds, and is not capable of braking action at zero vehicle speed. The travel of the brake pedal is mapped into a braking command which is distributed between the mechanical friction brakes and the regenerative brakes. If the electric motor is connected to the front wheels of the hybrid, then applying regenerative braking to the front wheels would assist the friction brakes and still maintain an acceptable bias between front and rear-wheel braking. If the electric machine is connected to the rear wheels, then most of the rear-wheel braking can be accomplished through regenerative braking, since only about a quarter of the total braking action is executed in the rear wheels. The full regenerative braking scheme would contribute to improved fuel economy at the expense of complicating the vehicle braking system. In hybrids converted from conventional vehicles, this entails modifying hydraulic pressures on the brakes to reduce the friction brake command by the amount compensated by the regenerative braking. The control algorithm must ensure that vehicle-braking safety is not compromised under any circumstances.

The driver comfort may be compromised by the use of excessive regeneration at low vehicle speeds; therefore, the regeneration command should be disabled for those speeds. As the vehicle speed increases, the percentage of maximum allowable regeneration power or torque can be gradually increased. A piecewise linear increase of allowed regeneration torque between 5 and 15 mph is shown in Figure 13.21 for the Akron hybrid vehicle [12]. The absolute value of vehicle speed has been used to account for vehicle driving in reverse. Regeneration strategy following a profile of constant power, constant torque, or a combination of the two is possible, but test drivers felt most comfortable with a constant torque regeneration profile as shown in the figure.

The acceleration pedal input must also be mapped into a regeneration request; a piecewise linear mapping used in the Akron hybrid vehicle is shown in Figure 13.22 [13]. The regeneration strategy affects the drivability of the vehicle during both acceleration and deceleration. The drivers of

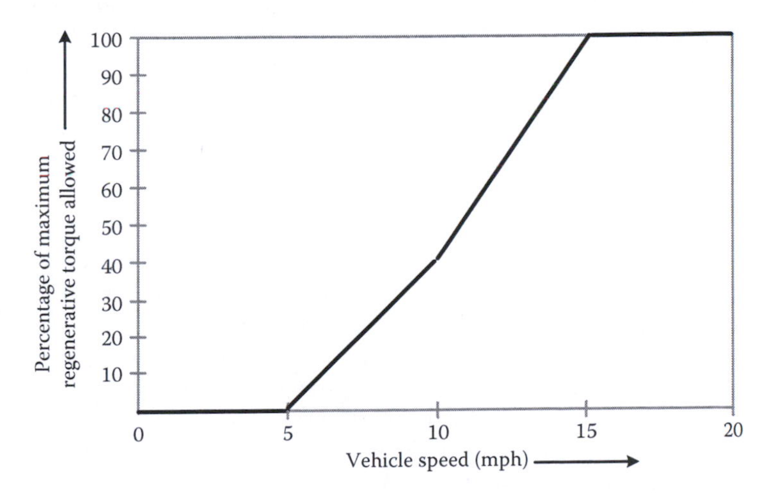

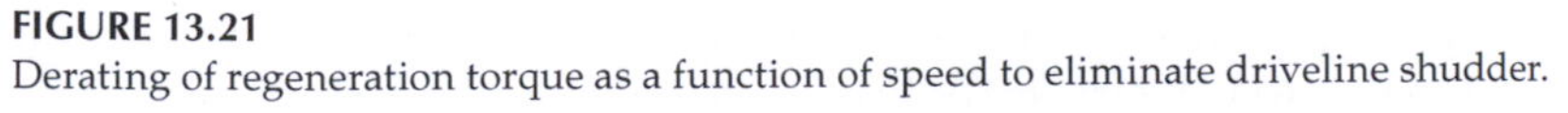

FIGURE 13.21
Derating of regeneration torque as a function of speed to eliminate driveline shudder.

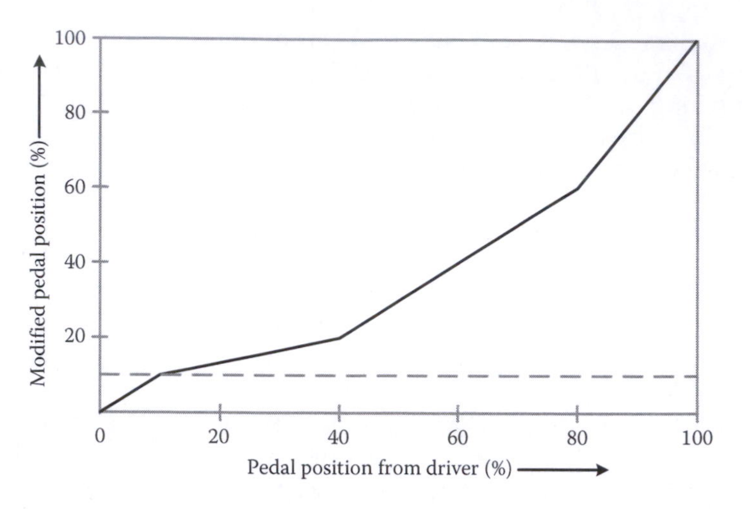

FIGURE 13.22
Piecewise linear mapping of pedal position from driver input.

hybrid vehicles feel comfortable with a constant negative torque when letting up the accelerator pedal. The dotted line in Figure 13.22 indicates the division between requests for propulsion and regeneration. Pedal positions less than 10% of the total pedal travel correspond to regeneration request or negative torque command for the electric machine. The linear mapping inversely translates 0% pedal position to the maximum regeneration torque request $T_{regen,max}$ and 10% pedal position to zero regeneration torque request. $T_{regen,max}$ request also has to be a function of the energy storage SoC. If the SoC is at its upper limit at SoC_{max}, then $T_{regen,max}$ has to be reduced to zero; whereas, if the SoC is at its lower limit at SoC_{min}, then $T_{regen,max}$ can be set to the maximum regeneration torque capability of the electric machine. The segment of the curve in Figure 13.22 representing the pedal position for propulsion has been shifted and scaled using piecewise linear mapping from the range [10%, 100% of modified pedal position] to [0%, 100% of available power]. The driver power demand can be converted to torque demand through dividing the power demand by the vehicle speed.

The propulsion torque request from the electric machine should also be limited by the energy storage SoC condition. The torque request corresponds to the driver demand related to the accelerator pedal position and vehicle hybrid operating mode less a bias that depends on the SoC. The regeneration or negative torque request from the electric machine can be directly linked to this bias. The less the energy storage SoC, the less the electric motor should assist propulsion, and the higher should be the magnitude for $T_{regen,max}$. The energy storage SoC would have to be elevated from SoC_{min} with regeneration even when the accelerator pedal is not depressed at all, especially in the case of parallel hybrid vehicles. The bias to be subtracted from the propulsion torque request for the electric machine can be configured as

$$B = \frac{SoC_{max} - SoC}{SoC_{max} - SoC_{min}} \times 100\%$$

According to the above relation, if $SoC \geq SoC_{max}$, then no bias is subtracted and full propulsion or positive torque request is passed through to the electric machine. If $SoC < SoC_{min}$, then 100% bias is subtracted, and transmitted torque request for the electric machine is always negative to recharge the energy storage.

Problems

13.1 The supervisory control module (SCM) of a series–parallel 2×2 hybrid vehicle has the inputs and outputs as shown in Figure P13.1. The vehicle is cruising at constant speed on a level road in power-split mode. The stored energy in the energy storage system (ESS) is constant. Answer the following questions qualitatively:

a. The accessory power demand increases suddenly because the cabin air-conditioner (powered by an electric motor) turns on. How should the SCM adjust its outputs to accommodate this increase while maintaining the vehicle propulsive power level and the ESS conditions?

b. The overall power demand remains constant, but the SCM decides that the ESS stored energy needs to be increased. How should the SCM adjust its outputs to cause the ESS to be charged while maintaining the vehicle propulsive and accessory power levels?

c. The driver increases the propulsive power demand slightly while ascending a grade. How should the SCM adjust its outputs to

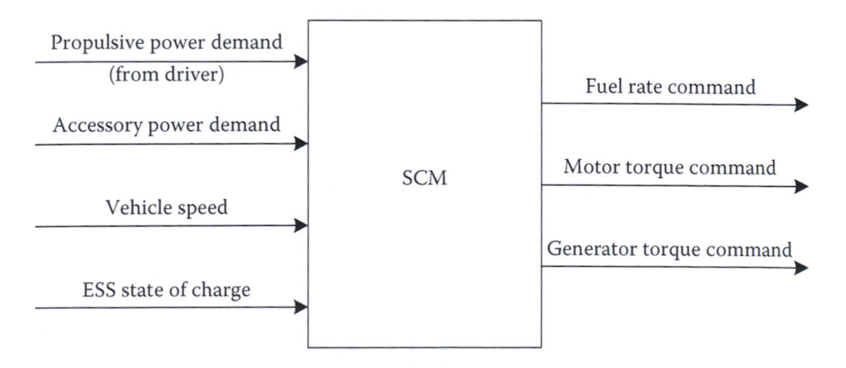

FIGURE P13.1

accommodate this increase while maintaining the accessory power level and the ESS conditions?

d. The driver demands maximum propulsive power to accelerate and pass another vehicle. How should the SCM adjust its outputs to draw maximum power from the ESS and accommodate this demand?

13.2 A planetary gear set is used in the powertrain of mechanical power-split series–parallel hybrid electric vehicle. The ring of the gear set is connected to the motor/differential, the planet carrier is connected to the IC engine, and the sun is connected to the starter/generator. In this hybrid vehicle, the ring-to-sun gear radii ratio is given as $r_r/r_s = 2.8$.

In one vehicle operating condition, the ring power demand is $P_{ring} = 32\,\text{kW}$ corresponding to the ring speed of $\omega_r = 180\,\text{rad/s}$. The ring speed is directly related to the vehicle speed, since it is connected to the electric motor and wheels. The IC engine speed is controlled at 2100 rpm.

Calculate the (a) ring torque, (b) ICE torque, and (c) generator speed.

References

1. F.R. Salmasi, Control strategies for hybrid electric vehicles: Evolution, classification, comparison and future trends, *IEEE Transactions on Vehicular Technology*, 56(5), 2393–2404, September 2007.
2. V.H. Johnson, K.B. Wipke, and D.J. Rausen, HEV control strategy for realtime optimization off fuel economy and emissions, SAE paper 2000-01-1543, 2000.
3. A.M. Phillips, M. Jankovic, and K. Bailey, Vehicle system controller design for a hybrid electric vehicle, in *Proceedings of the IEEE International Conference on Control Applications*, Anchorage, AK, September 25–27, 2000, pp. 297–302.
4. H.-D. Lee and S.-K. Sul, Fuzzy-logic-based torque control strategy for parallel-type hybrid electric vehicle, *IEEE Transactions on Industrial Electronics*, 45(4), 625–632, August 1998.
5. J.-S. Won and R. Langari, Fuzzy torque distribution control for a parallel hybrid vehicle, *Expert Systems: International Journal Knowledge Engineering and Neural Networks*, 19(1), 4–10, February 2002.
6. B.M. Baumann, G. Washington, B.C. Glenn, and G. Rizzoni, Mechatronic design and control of hybrid electric vehicles, *IEEE/ASME Transactions on Mechatronics*, 5(1), 58–71, March 2000.
7. J.M. Miller, *Propulsions Systems for Hybrid Vehicles*, Institute of Electrical Engineers, London, U.K., 2004.
8. G.H. Gelb, N.A. Richardson, T.C. Wang, and B. Berman, An electromechanical transmission for hybrid vehicle powertrains—Design, and dynamometer testing, SAE Congress Paper No. 710235, Detroit, MI, January 1971.

9. R.J. Veillette, Course Lecture Notes on *Design of Hybrid Electric Vechicles*, Department of Electrical and Computer Engineering, University of Akron, Akron, Ohio, 2005.
10. G. Paganelli, G. Ercole, A. Brahma, Y. Guezennec, and G. Rizzoni, General supervisory control policy for the energy optimization of charge sustaining hybrid electric vehicles, *Journal of Society of Automotive Engineers of Japan*, 22(4), 511–518, October 2001.
11. A. Sciarretta, M. Back, and L. Guzzella, Optimal control of parallel hybrid electric vehicles, *IEEE Transactions on Control Systems Technology*, 12(3), 352–363, May 2004.
12. S.M.N. Hasan, I. Husain, R.J. Veillette, and J.E. Carletta, Power Generation in Series Mode: Performance of a Starter/Generator System, *IEEE Industry Applications Magazine*, 16(2), 12–21, March–April 2010.
13. N. Picot, A strategy to blend series and parallel modes of operation in a series–parallel 2×2 hybrid diesel/electric vehicle, MS thesis, Electrical Engineering, The University of Akron, Akron, OH, December 2007.

14

Vehicle Communications

The components and subsystems in a vehicle are linked together through a system of networked communication. The command signals, the sensor feedback signals, and the system output variables interconnect the vehicle driver, the supervisory controller, and the subsystems and components through this automotive communication network. The type of networked communication used in automobiles is known as controller area network (CAN). CAN is a special type of local area network (LAN) that uses a serial multi-master communication protocol. CAN supports distributed real-time control with a very high level of security, and a communication rate of up to 1 Mbps. The CAN bus is ideally suitable for applications operating in noisy and harsh environments, such as in the automotive and other industrial fields that require reliable communication or multiplexed wiring. This chapter presents the basics of network communications and in-vehicle communications, leading to a more detailed treatment of CAN communications. The CAN protocol is covered in sufficient detail so that one can practice implementation in a vehicle or in other industrial systems.

To learn the CAN protocol, we first need to develop an understanding of network communications. The fundamental of networked communications is the open systems interconnection (OSI) seven-layer network model, which describes the protocol for communication among computer nodes over a common network. The model was developed by the International Standards Organization (ISO) in 1983 to provide an open communication protocol for the industry to use and build systems with defined interfaces at the physical link.

14.1 OSI Seven-Layer Model

The seven-layer OSI model is structured in a top-to-bottom arrangement starting with the application layer at the top [1]. Figure 14.1 shows the communication of two computer nodes, a transmit node and a receive node, via the seven layers of the network for each system. The application layer serves as the human–machine interface (HMI) for the computer node of a system. The bottom layer of OSI provides the physical link to the communication channel through which data are transferred to other systems built with a similar communication model. The seven layers of OSI are described below:

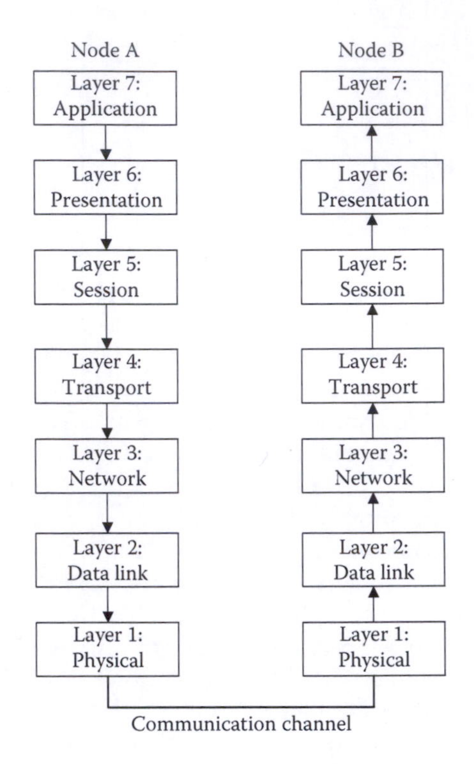

FIGURE 14.1
Communication of two nodes via seven-layer OSI network model.

Layer 7 (*Application layer*): The end user interacts with the machine through an executable application in this layer. The layer supports a collection of miscellaneous protocols for high-level applications. Examples are electronic mail, command execution, file transfer, connecting remote terminal, etc.

Layer 6 (*Presentation*): The presentation layer is concerned with the semantics of the bits. This layer isolates the application layer of sending node A from the particulars of the environment and can tell the receiving node B of the format. For example, this layer describes how floating point numbers can be exchanged between nodes with different math formats. The layer is responsible for protocol conversion, and data cryptography, if implemented.

Layer 5 (*Session*): This layer ensures that the context of all packets sent by the other layer is preserved. This is an enhanced version of the transport layer and very few applications use it.

Layer 4 (*Transport*): The transport layer acknowledges message transmission across the network and validates that the transmission has occurred without any loss of data. The layer divides the message into smaller packets, assigns sequence numbers, and transmits them. The receiving computer reassembles the packets. If a data packet is lost and arrives out of order, the layer resends the data packet placing it back in the correct sequence.

Layer 3 (*Network*): The network layer is responsible for the transmission of packets routing them either through the shortest distance route or through the fastest route whichever is the best path. The layer is responsible for managing network problems such as packet switching, data congestion, and routines. As an example, the Internet protocol (IP) resides in this layer.

Layer 2 (*Data link*): The data link layer is responsible for controlling the error between adjacent nodes and transferring the frames to other computers via the physical layer. Data link layer is used by hubs and switches for their operation. The layer groups the bits into bit stream and ensures their correct delivery. It adds few bits at the beginning and end of each frame and also adds the checksum. The receiving node verifies the checksum; if the checksum is incorrect, it sends a control message requesting retransmission.

Layer 1 (*Physical*): The physical layer is responsible for transmitting the bit stream over the physical cable. The physical layer defines the hardware items such as modems, cables, cards, voltages, etc. The bit rate and the voltage levels for "0" and "1" are defined in this layer. The transmission may be one-way or bidirectional.

14.2 In-Vehicle Communications

In-vehicle communication protocols have been categorized into classes A through D according to the data-handling speed, with class A being the slowest. Within a class there are several different types of communication protocols. The network classes, the data-handling speed ranges for the four classes, and the types of communication protocols in a class are listed in Table 14.1 [1]. The data-handling speed requirements of the various systems in a vehicle determine the class of protocol that would be used for that system; not all devices need to communicate at the same speed. For example, the powertrain components need to be networked through a high-speed protocol since these are drive-critical components of the vehicle. On the other

TABLE 14.1

Vehicle Communication Networks

Network Class	Speed	Communication Protocols
Class A	<10 kbps	CAN (basic), LIN, SAE J2602, TTP/A
Class B	10–125 kbps	CAN (low-speed), SAE J1850, ISO 9141-2
Class C	125 kbps–1 Mbps	CAN (high-speed), SAE J 1039
Class D	>1 Mbps	Flexray, MOST

hand, passenger seat operation is not drive critical, and hence, can be managed through a slower communication protocol.

Class A is the slow-speed protocol used for vehicle body electrical functions, such as power seat, power windows, power mirrors, etc. Protocols under class A are local interconnect network (LIN) and time-triggered protocol/A (TTP/A).

Class B network is used for instrumentation and data display to the user; for example, dashboard display, cabin climate control, etc. The most common class B protocol is the SAE J 1850.

Class C network is used for real-time control applications, such as those in vehicle dynamics control, powertrain controls, engine controls, and hybrid component controls.

Class D networks include the fastest communication protocol in the automobiles. This class targets future protocols for in-vehicle communications as well. Media-oriented systems transport (MOST) and Flexray are example protocols for class D, which are making their way into modern automobiles. MOST is the vehicle communication bus standard intended for interconnecting multimedia components. A reliable, high-speed communication network is also essential for safety-critical components for today's automobiles, particularly as drive-by-wire systems get introduced in a vehicle. Drive-by-wire technologies include steer-by-wire, brake-by-wire, throttle-by-wire, etc.; these technologies will replace all the existing hydraulic and mechanical systems currently in place. In addition, the control command for all of the electric powertrain in hybrid and electric vehicles is by-wire technologies. The safety-critical and propulsion components require reliable and deterministic time-triggered protocols instead of an event-triggered protocol such as that in the CAN. The Flexray protocol, which is being developed by the Flexray consortium, provides high-speed, deterministic, fault-tolerant message transmission essential in drive-by-wire and electric powertrain components.

14.3 Controller Area Network

The CAN was originally developed by Bosch in 1985 for in-vehicle communications among the devices, sensors, and actuators. Prior to CAN, the automotive industry used point-to-point wiring which required digital and analog inputs and outputs to every electronic device in the vehicle. This required a heavy wiring harness that increased the weight, cost, and complexity of the in-vehicle wiring. The problem compounded as the number of electronic devices used in vehicles gradually increased. CAN provides an inexpensive, reliable network that enables all the electric control units or the electronic interface of vehicle components to communicate with each other. The automotive industry quickly adopted CAN since its original development because of its flexibility and advantages. Realizing the advantages of CAN, other industries have also adopted CAN for networked communications. CAN is widely used in aerospace applications, such as aircraft electronic control units of components, flight-state sensors, and navigation systems. Rail transportation systems extensively use CAN for communications at various levels ranging from electronic control units of components to keeping track of the number of passengers. CAN communications is widely used in human–machine interface (HMI) units of industrial equipments. CAN protocols have applications in medical equipments and in hospital operating rooms.

The focus on CAN in this book is on in-vehicle communications where systems or devices require distributed real-time control including fault management and diagnostics. Operating points of powertrain and ancillary components are transmitted from the vehicle supervisory controller to the CAN nodes of the respective components. Similarly, feedback information from the components is transmitted into the CAN bus. An example CAN layout with three classes of networks in a hybrid electric vehicle is shown in Figure 14.2. The three classes serve three types of vehicle systems or devices categorized based on their data-handling speed requirement. The faster class C type CAN, which is also known as full CAN, is used for powertrain devices. The class A type basic CAN, which is of lower speed, is used for vehicle body communications as shown in Figure 14.2. The class B type CAN is used for devices requiring medium speed communications such as the cabin climate control unit.

Each of the vehicle components that need to receive and transfer a command or a message has a CAN controller to connect to the CAN bus. The

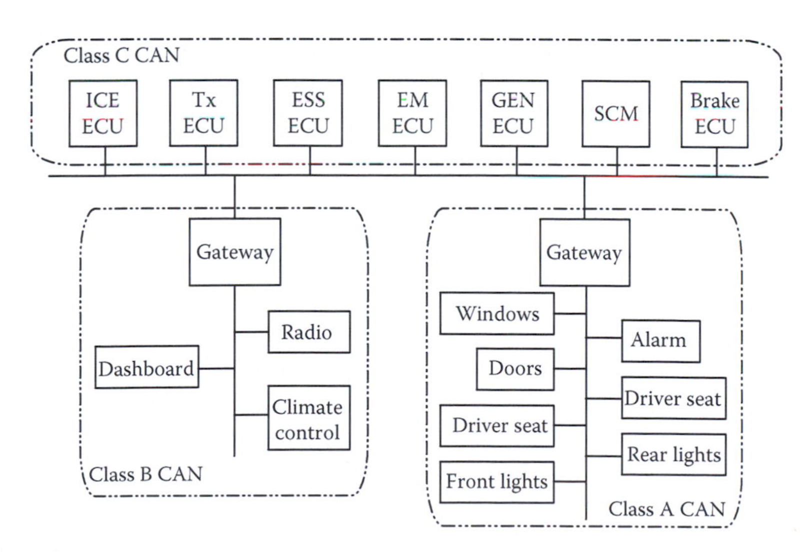

FIGURE 14.2
Typical CAN layout in a hybrid electric vehicle.

controller is often referred to as a *CAN node* for the system or the device. The requirements of a CAN controller are simple interface to the device, message filtering and buffering, protocol handling, and a physical layer interface. The electronic control units of the powertrain devices have dedicated hardware for CAN communications to reduce its primary control processor's overhead. The devices on the slow network may not have a microprocessor-based control unit, but do require a CAN controller to interface the device with the CAN bus. A device could be something as simple as a sensor providing information for the electronic control units of one of the powertrain components or a much more complex unit such as the vehicle supervisory controller.

In the CAN communication protocol, commands, feedback signals, or any information is converted into a message format for transmission into the CAN. On the other hand, the messages can be picked up by any of the devices connected to the network and decoded. Each component sees all the messages and can decide on the appropriate message to pick up. The messages have a priority structure; if two CAN nodes attempt to transmit two messages at the same time then the one with higher priority gets transmitted while the other one has to wait. On the other hand, multiple nodes may receive data and act upon the message simultaneously. The CAN is highly flexible; adding a node or removing a defective node does not require any hardware or software change of the remaining CAN.

The details of these CAN communication protocols are described below in terms of the physical layer, transfer layer, and programming.

14.3.1 CAN Transfer Protocol

The CAN protocol defines only the two lower layers of the OSI seven-layer communication model: Data link layer and physical layer. The data link layer has the transfer layer dealing with message frame coding, message arbitration, overload notification, bus access control, identifying message reception, error detection, and error signaling. The data link layer also includes an interface with the application layer of the device including deciding which messages are to be transmitted and received. The physical layer performs the actual transfer of bits between the CAN node of a device and the CAN bus. The layer has the transreceivers, wires, and connectors.

A CAN node can be either a transmitter or a receiver. A node that originates a message transmission is called the *transmitter*. The CAN communication is based on carrier sense multiple access (CSMA) protocol. Any node can start transmitting a message when the network bus is idle, which is the multi-master feature of the CAN protocol. Each node must monitor the bus for a period of no-activity before transmitting a message. If multiple nodes start transmitting at the same time then the node with the highest priority will be successful in transmitting a message. A node remains a transmitter until it finishes the transmission and the bus becomes idle or when it loses a message transmission arbitration with another node. A node becomes a *receiver* when it is not sending a message, but another node is transmitting keeping the bus busy.

The information is transmitted to the CAN bus from a node using a defined format called *messages*. The messages are framed containing the data and other relevant information pertaining to the data. For example, each message has a unique identifier that describes the meaning of the transmitted data. The identifier also defines the priority of the message during bus access. Each node can decide whether or not to act on the data upon receiving it. The application layer of the device utilizes the data that is useful. It is also the application layer that decides on the information to be transmitted.

Messages are transmitted at a fixed baud rate. This is also known as *bit rate*. The transmission speed may be different for different networks, but the *bit rate* is fixed and uniform for a given network.

The magnitude of the CAN bit signal is known as *bus value*. The CAN bus values are defined as *dominant* and *recessive* using the reverse logic.

14.3.2 CAN Transfer Layer

The CAN transfer layer protocol is described in this section in terms of bit timing, message frames, message arbitration, error detection, and error signaling.

14.3.2.1 Bit Timing

The CAN transfer layer handles the bit timing, transfer protocol, and message frames. The bit timing depends on the CAN class used. Several CAN buses exist within a vehicle for communication at different rates. For each class of CAN protocol, the message is transmitted within the time period known as the *nominal bit time*. The *nominal bit rate* is the reciprocal of the nominal bit time given by

$$\text{Nominal bit rate} = \frac{1}{\text{Nominal bit time}}$$

The nominal bit rate for the CAN bus can be anything less than 1 Mbps.

Nominal bit time in a CAN protocol is partitioned into four time segments as shown in Figure 14.3. Message transmission starts with synchronization segment *Sync_seg* which is used to synchronize the various nodes in the bus. The rising edge of a signal is expected to lie within this segment. This is the smallest time segment. The segment *Prop_seg* accounts for the physical propagation delays in the network. The remaining two segments *Phase_seg1* and *Phase_seg2* are used to compensate for the edge phase errors. *Phase_seg1* is used for positive phase edge errors and can be

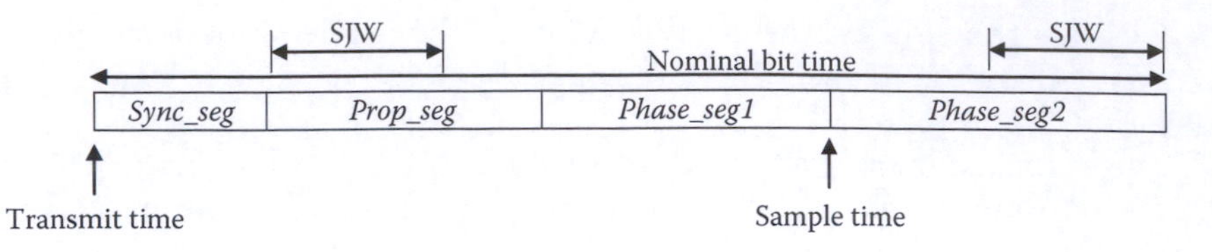

FIGURE 14.3
Four segments within the nominal bit time of a CAN message.

lengthened by resynchronization. *Phase_seg2* is used for negative phase edge errors and can be shortened by resynchronization. The data sample point is in between *Phase_seg1* and *Phase_seg2*. *Phase_seg2* length should be large enough for information processing. The lengths of the phase segments can be adjusted for edge phase errors. The lengths of the various segments are defined in terms of a fixed time quantum T_Q, which is related to the oscillation period as

$$T_Q = n \times T_{osc}$$

Here n is a prescaler integer number ranging between 1 and 32. The lengths of the segments are

$$Sync\_seg = 1 \times T_Q$$

$$Prop\_seg = n_p \times T_Q, \quad n_p = 1, 2, \ldots, 8$$

$$Phase\_seg1 = n_1 \times T_Q, \quad n_1 = 1, 2, \ldots, 8$$

$$Phase\_seg2 = n_2 \times T_Q, \quad n_2 \geq n_1$$

For the purpose of synchronization, a *synchronization jump width* (SJW) is used, which is defined as the largest allowable adjustment of *Phase_seg1* and *Phase_seg2*. Synchronization jump width is given as

$$\text{SJW} = n_{SJW} \times T_Q$$

All the CAN controllers in a common CAN bus must have the same bit rate and bit length. If the clock frequencies of the individual controllers are different, then the bit rate and bit length has to be adjusted using the T_Q.

14.3.2.2 CAN Message Frames

There are four types of message frames in a CAN protocol: data frame, remote frame, error frame, and overload frame. The *data frame* transmits data

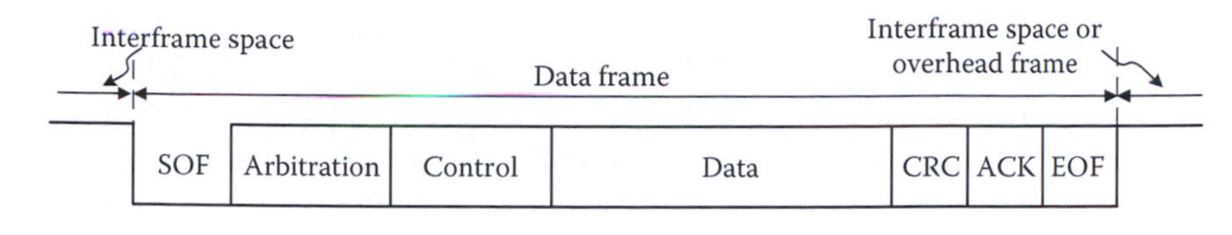

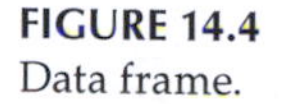

FIGURE 14.4
Data frame.

from a transmitting node to the receiving nodes. The *remote frame* is used by a node to request the transmission of a data frame by another node with the same message identifier. An *error frame* is transmitted by a node after detecting a bus error. An *overload frame* is used to provide additional delay between two data and remote frames. The error frame and overload frame are not necessarily used in all CAN implementations.

The CAN *data frame* consists of seven fields each with a varying bit number to serve a specific purpose. The data frame structure is shown in Figure 14.4. The fields are described below.

Start-of-frame (*SOF*): The SOF marks the beginning of a data frame. It has only a single dominant bit. All nodes have to synchronize with the leading edge of the SOF bit.

Arbitration field: The field has the 11-bit message identifier and a remote transmission request (RTR) bit in a standard frame format. The RTR bit is dominant for a data frame and recessive for a remote frame. In an extended message format, the message identifier is of 29 bits. There is also the RTR bit, SRR (substitute remote request) bit, and IDE (identifier extension) bit.

Control field: The control field has six bits. The first two bits are reserved for future expansion and must be transmitted as dominant. The remaining four bits indicate the length of the data.

Data field: The data field contains the data to be transmitted. The number of bits can be 0 to 8 bits with the most significant bit being sent first.

CRC field: The CRC field has 15-bit sequence followed by a CRC delimiter. The CRC delimiter is a single recessive bit.

Acknowledge field: The acknowledge field consists of two bits, one for the acknowledge slot and the other for the acknowledge delimiter. The transmitting node sends a recessive bit for the acknowledge slot, while all the receiving nodes send a dominant bit through this slot after receiving a valid message. The acknowledge delimiter is always a recessive bit.

End-of-frame (*EOF*): The end of a frame is designated by seven recessive bits.

The *remote frame* is transmitted by a CAN node to request certain data that it needs from a remote node. The remote frame has six fields: SOF,

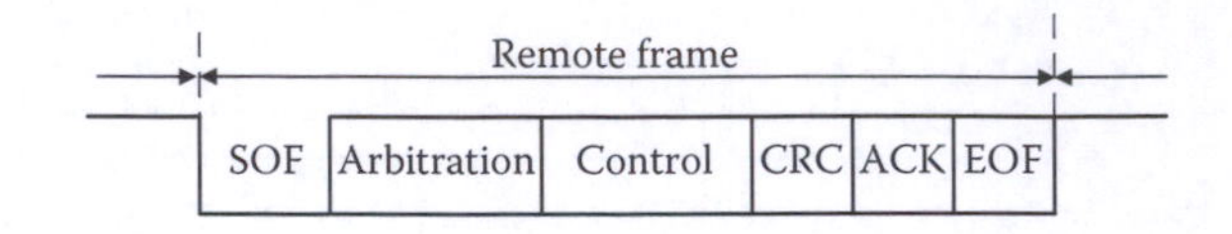

FIGURE 14.5
Remote frame.

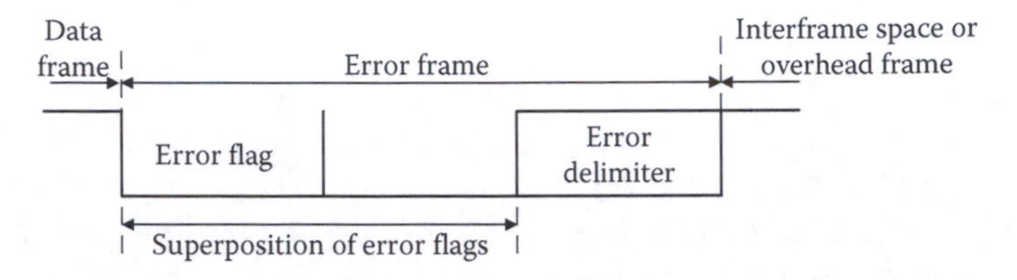

FIGURE 14.6
Error frame.

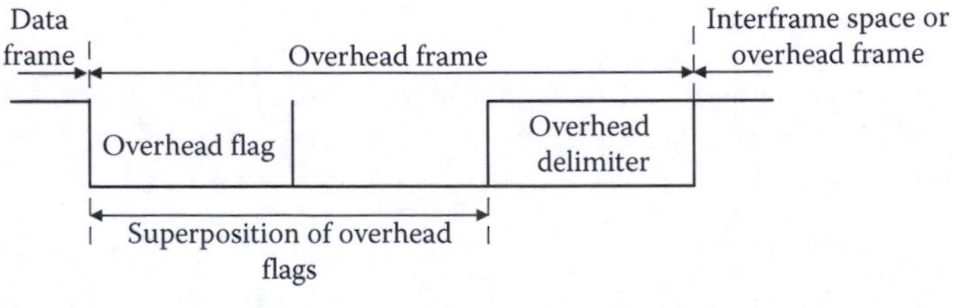

FIGURE 14.7
Overload frame.

arbitration, control, CRC, ACK, and EOF. There is no data field, regardless of the value in the data length code (DLC). The remote frame structure is shown in Figure 14.5.

The *error frame* is transmitted by a CAN node when it detects an error in the bus. The error frame structure is shown in Figure 14.6. The frame consists of two fields: Error flag field and error delimiter. The error flag field contains the superposition of the error flags generated by the different nodes. The error flags can be an active flag or a passive flag. The active error flag consists of six consecutive dominant bits. The passive error flag consists of six consecutive recessive bits, unless it is overwritten by dominant bits from other nodes. The error delimiter consists of eight recessive bits. A node sends a recessive bit after the transmission of an error flag and monitors the bus until it detects a recessive bit. After detecting the recessive bit, the node starts sending seven more recessive bits.

The *overload frame*, similar to the error frame, consists of two fields: Overload flag field and overload delimiter. The overload frame structure is shown in Figure 14.7. A node transmits an overload frame when it detects one of two conditions: (1) When receiving node requires a delay for the next

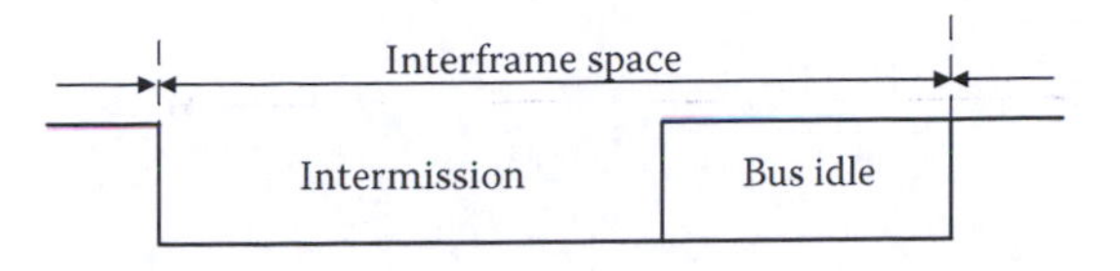

FIGURE 14.8
Interframe space.

data frame or the next remote frame due to internal conditions and (2) when a node detects a dominant bit during bus intermission. The overload flag field consists of six dominant bits, and the overload delimiter consists of eight recessive bits.

The *interframe* space, shown in Figure 14.8, separates the data frames and remote frames with the help of three recessive bits. No other node is allowed to start transmission of a data frame or remote frame during the interframe space. The only action permitted during this intermission is the signaling of an overload condition. The bus may remain idle for any arbitrary length of time after the interframe space or accept a new frame. Any node having to transmit a message can start a transmission when the bus is idle with a dominant bit which indicates the SOF. A node with a pending message due to the bus being busy starts transmitting with the first bit following the intermission.

14.3.2.3 Message Arbitration

If two nodes start to transmit at the same time a collision occurs in the bus. The resolution of the bus collision through bitwise evaluation is known as *message arbitration*. The process includes collision detection as well as arbitration so that all the messages can be transmitted eventually without having to lose any message. The transmitting node monitors the bus to see if the logic state that it is trying to send already exists in the bus or not. If the level of the bit transmitted is the same as the level monitored on the bus, then the node will continue message frame transmission. However, if a recessive level is sent by a node and a dominant level is monitored on the bus then that node has lost arbitration and must stop transmitting any further bits. The message with the dominant bit gets through, while the other node waits for the bus to become idle without destroying its message. A dominant bit state (logic "0") will always win arbitration over a recessive bit state (logic "1"), which means that the message with a lower identifier value has the higher priority. The collision is essentially resolved with the message having the lower identifier going through first.

Example 14.1

Let two nodes start transmitting messages at the same time with the following message identifiers:

Node-1—0010010...
Node-2—0010110...

Determine the message that will get through in the bus.

Solution

For the fifth identifier bit, Node-2 will notice that the bus has a dominant bit when it is trying to transmit a recessive bit. Node-2 loses arbitration at this step, and immediately stops transmitting. The node also turns itself into a message receiver from a message transmitter. Node-1 will continue to finish the transmission of its message, while Node-2 will wait for next bus idle period and try to transmit its message again.

14.3.2.4 Error Detection and Error Signaling

A high level of data integrity is maintained in a CAN through measures of self-evaluation, error detection, and error signaling at each node. Each transmitting node compares the level of bit transmitted with the level monitored on the bus. Any node that detects an invalid message transmits the error frame. These corrupted messages are aborted and retransmitted automatically.

Error detection in the CAN protocol takes place at both the message level and the bit level. At the message level the three error detection methods are: Cyclic redundancy check (CRC), frame check, and acknowledgment error check. At the bit level, there are two methods of error management: Bit error and bit stuffing.

The cyclic redundancy check (CRC) sequence included in the data frame of a transmitted message consists of the CRC calculation results. The CRC sequence is calculated by the transmitter using a predefined algorithm. A receiving node calculates the CRC using the same algorithm as the transmitter. A CRC error is detected if the calculated result does not match the transmitted CRC sequence.

The receiving node does a *format check* on a message by evaluating the bits in CRC delimiter, ACK delimiter, EOF bit-field, and interframe space. If an invalid bit is detected in one of the above positions by the receiver, a format or frame error is flagged.

All nodes on the bus acknowledge each valid message. The receiving nodes acknowledge a consistent message, but would flag any inconsistency in the message. A consistent message is acknowledged by sending a dominant bit into the ACK slot of the frame by each of the receiving nodes. If a transmitter determines that a message has not been acknowledged, then an ACK error is flagged. The ACK error may occur because of transmission error or if there is no active receiver.

A transmitting node monitors the bus simultaneously as it transmits a bit on the bus. When the bit transmitted is different from the bit monitored then the node interprets the situation as a *bit error.* The exceptions are when sending a recessive bit during message arbitration or seeing a dominant bit in the bus in the ACK slot due to the acknowledgment from the receiving nodes. A node detecting an error bit starts the transmission of an error flag immediately at the next bit transmission time.

Error is confined in CAN in terms of transmit and receive error counts. The transmit error count increases when a node transmits error flags and the receive error count increases when a receiving node detects an error. The transmit error count is decremented when the node successfully transmits a message. Similarly, when a receiving node successfully receives a message, it reduces its receive error count. A node is an Error-Active node when the error count is between 1 and 127, and it is Error-Passive when the error count is between 128 and 255. The node will be in a bus-off status if the error count is greater than 255. An Error-Active node transmits an active error flag, while an Error-Passive node transmits a passive error flag. A bus-off node is not allowed to have any influence on the CAN bus.

The CAN controllers are designed for the receiver to synchronize on recessive-to-dominant transitions. The method of *bit stuffing* is used to guarantee that there are enough transitions in the NRZ (non-return-to-zero) bit stream to maintain synchronization. If five identical and consecutive bit levels are transmitted within a message from SOF to CRC, the transmitting node automatically inserts or stuffs one bit of the opposite polarity into the bit stream. Receiving nodes will automatically delete the stuffed bits from the message. A stuff error is flagged if any node detects six consecutive bits of the same level. For example, a message with ID 0×340 will be transmitted as

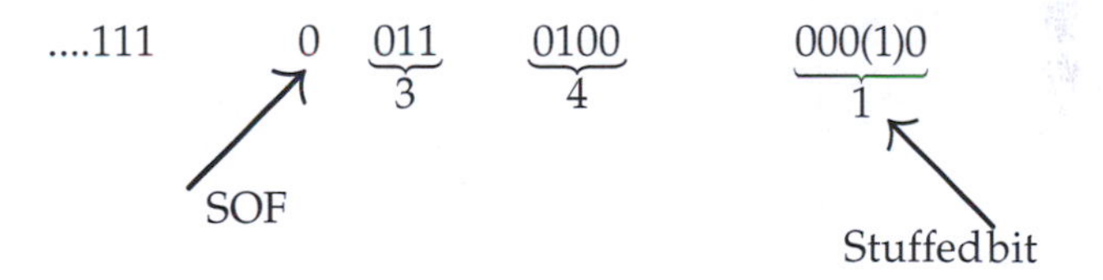

14.3.3 CAN Physical Layer

Several different physical layers can be established for the CAN. The physical layer defines the bus topology, signal methods, bus termination, maximum baud rates, and several other features. The most common physical layer is the high-speed CAN. The bus topology consists of a two-wire differential bus and allows communication transfer rates up to 1 Mbps. Class B type low-speed CAN is also implemented with two wires, which allows transmission rates up to 125 kbps. In these two-wire CAN buses, one wire carries the CANH (CAN high) signal and the other wire carries the CANL (CAN low) signal without any clock signal. Class A type basic CAN uses a single-wire plus ground CAN interface with data transfer rates up to 33.3 kbps. Basic CAN is used for devices not depending on critical communications. The CAN signals are NRZ serial data, which are inherently not self-synchronizing.

There can be long periods of inactivity in a device connected to the CAN bus. During that period the CAN node can be put to sleep to conserve resources. The activities of that node can be restored through a wake-up signal by any bus or by changes in the internal conditions of the device. CAN nodes can distinguish between short disturbances and permanent failures. Temporary disturbances may be caused by voltage spikes or other electrical noise. Permanent failures may be caused by CAN controller failures, faulty cables, bad connections, transreceiver failures, or extended external disturbances. When a CAN node becomes defective, it is switched off and will have no further effect on the CAN bus.

The two-wire CAN bus network with a straight-line topology where nodes can be connected anywhere in that line is shown in Figure 14.9. The nodes from the vehicle devices are all connected in parallel. The two ends of the bus are the terminating nodes, which in its simplest form can be a resistive termination. A terminating impedance with a bypass capacitor and two 60 Ω resistances can also be used to reduce the common-mode noise. The resistive and capacitive terminations of the CAN bus with pure resistance and with resistance-capacitance combination are shown in Figure 14.10a and b, respectively.

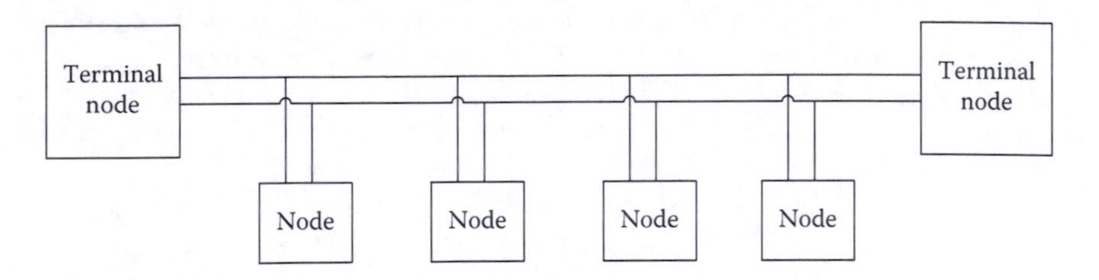

FIGURE 14.9
CAN bus network in a straight-line topology.

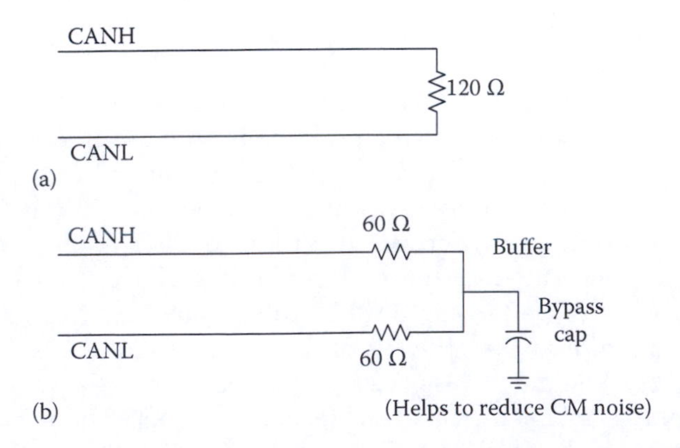

FIGURE 14.10
CAN bus termination: (a) resistive termination; (b) resistance-capacitance termination.

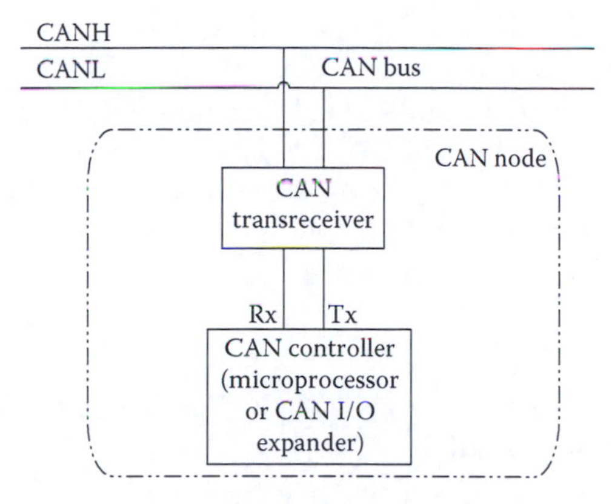

FIGURE 14.11
CAN node hardware.

The *dominant* and *recessive* CAN bus values in the network have magnitudes as follows:

Logic "0" = "dominant"; CANH = 5 V and CANL = 0 V.

Logic "1" = "recessive"; CANH = CANL= "floating" (a weak pull to 2.5 V).

Note that the CAN bus values use the reverse logic.

A typical CAN node hardware associated with the electronic control unit of a powertrain component or any other vehicle device consists of a CAN transreceiver, such as MCP 2551 and a CAN controller. CAN controllers are available either in a microprocessor or as a CAN I/O expander. A CAN node hardware is shown in Figure 14.11. The CAN controller hardware is available in many microcontrollers and digital signal processors, such as in MicroChip PIC 18Fxxx series, TI DSP 28xxx series, and Freescale MPC 55xxx series [2].

The CAN bus is in a "0" state when it is busy. Although only one node is allowed to transmit at a given time, two nodes can start transmitting at the same time when both see the bus to be idle. During the simultaneous transmission of dominant and recessive bits, the resulting bus value will be dominant. The arbitration process resolves the issue of which node gets to transmit its message first as was described earlier.

14.3.4 CAN Programming

The basic concepts of the CAN layers and protocols are essential in developing a CAN interface for the electronic units of a vehicle component. The specifics of a CAN controller in a certain microprocessor must also be evaluated for developing the CAN software program. The general structure of a CAN program is essentially the same for the various CAN controllers that are available in various microcontrollers and digital signal processors; they include register configuration, timer configuration, and interrupt service routine (ISR) programming.

The task of register configuration includes defining the bit timing message frame constants for the type of CAN protocol to be used. The transfer and receive buffer registers for managing also need to be configured. The transfer layer configuration includes the following:

- Define T_Q.
- Define SJW.
- Define *Prop_seg*, *Ph_seg1*, and *Ph_seg2* (together with $Sync\_seg = 1 \times T_Q$; this sets the bit period).
- Define other processor specific parameters.

The transmit and receive buffer register configurations include the following:

- Transmit identifiers, data length codes (DLCs).
- Receive identifiers (filters and masks).
- Receive configuration bits.

Messages are transmitted based on the times calculated in the timers. The timer registers and timer interrupts are set accordingly to transmit messages periodically. Messages are also received on an interrupt basis. Interrupts are defined and enabled on the registers associated with receiving a message.

The ISR manages the transmit and receive messages. The ISR programming includes the following:

- Load transmit buffers with data to be sent and request transmission.
- Read receive buffers, copy data to local variables, and clear the receive buffer flag.
- Reset interrupt flags and re-enable interrupts.

The number of nodes is only theoretically limited by the number of identifiers available. The drive capabilities of the devices impose the practical limitation. The bit rate depends on the total length of the bus and the delays associated with the transreceivers.

The very first step in developing a CAN is to define the messages for the various systems that need to be communicated. This is accomplished through developing a message list. An example message list segment is shown in Figure 14.12. The list includes a subset of messages from the generator control module (GCM), system health monitor (SHM), and supervisory control module (SCM) of a hybrid electric vehicle. The node that will transmit the message and the nodes that will receive and use the message are identified in the message list. Each message is given a unique identifier according to

Tx	Rx	Signal frame	CAN ID	Timing	Signal
GCM	SCM, SHM	Generator status	0 × 440	10 ms	Mode of operation
					Generator current actual
					Generator speed actual
					HV bus voltage
					Inverter temperature
GCM	SHM	GCM fault mode status	0 × 468	100 ms	System critical fault
					Local critical fault
					Local non-critical fault
					Local diagnostic fault
					Rolling count
					Critical communications with UALAN lost
					Critical inverter overheat
					Critical error on inverter phase A
					Critical error on inverter phase B
					Critical error on inverter phase C
					Critical error in computed speed
					Critical instantaneous overcurrent
					Critical continuous overcurrent
					Critical undervoltage
					Critical overvoltage
					Critical persistent invalid position sequence
					Non-critical speed high
					Non-critical inverter overheat
					Invalid position sequence
					Invalid position sensed
SCM	GCM	Generator command	0 × 340	10 ms	Connect generator
					Clear faults
					Rolling count
					Generator current commanded

CAN ID	Bytes	Bits before stuffing	Worst case bits after stuffing	Worst case transmit time	Periodic interval	Worst case %bus utilization	Best case transmit time	Best case %bus utilization per signal
0 × 440	6	92	112	0.000225	0.01	0.02245	0.00018	0.0184
0 × 468	3	68	82	0.000165	0.1	0.001645	0.00014	0.00136
0 × 340	3	68	82	0.000165	0.01	0.01645	0.00014	0.0136

FIGURE 14.12
An example CAN message list.

the priority of the message. The signal information contained within the message is detailed in the message list. Once the hardware is configured according to the message list and the CAN nodes are physically connected to the network, the messages can be evaluated through a CAN analyzer. The CAN analysis and development tools are extremely useful during the development process of a device as well as for system integration, testing, and debugging.

References

1. J.M. Miller, *Propulsions Systems for Hybrid Vehicles*, Institute of Electrical Engineers, London, U.K., 2004.
2. D. Ibrahim, *Advanced PIC Microcontroller Projects in C: From USB to RTOS with the PIC 18F Series*, Elsevier, Oxford, U.K., 2008.

Index

Q

R

T

U

V

W